NEW DIRECTIONS in ORGANIC and BIOLOGICAL CHEMISTRY

Series Editor

C.W. Rees, FRS
Imperial College of Science,
Technology and Medicine
London, UK

Activated Metals in Organic Synthesis
Pedro Cintas

Capillary Electrophoresis: Theory and Practice
Patrick Camilleri

Cyclization Reactions
C. Thebtaranonth and Y. Thebtaranonth

Mannich Bases: Chemistry and Uses
Maurilio Tramontini and Luigi Angiolini

Vicarious Nucleophilic Substitution and Related Processes in Organic Synthesis
Mieczyslaw Makosza

Radical Cations and Anions
M. Chanon, S. Fukuzumi, and F. Chanon

Chlorosulfonic Acid: A Versatile Reagent
R. J. Cremlyn and J. P. Bassin

Aromatic Fluorination
James H. Clark and Tony W. Bastock

Selectivity in Lewis Acid Promoted Reactions
M. Santelli and J.-M. Pons

Dianion Chemistry
Charles M. Thompson

Asymmetric Methodology in Organic Synthesis
David J. Ager and Michael B. East

The Anomeric Effect
Eusebio Juaristi

Chiral Sulfur Reagents
M. Mikołajczyk, J. Drabowicz, and P. Kiełbasiński

Synthesis Using Vilsmeier Reagents
C. M. Marson and P. R. Giles

C-Glycoside Synthesis
Maarten Postema

Organozinc Reagents in Organic Synthesis
Ender Erdik

Chirality and the Biological Activity of Drugs
Roger Crossley

CYCLIZATION REACTIONS

C. Thebtaranonth
Professor
Department of Chemistry
Mahidol University
Bangkok, Thailand

Y. Thebtaranonth
Professor
Department of Chemistry
Mahidol University
Bangkok, Thailand

CRC Press
Boca Raton Ann Arbor London Tokyo

Library of Congress Cataloging-in-Publication Data

Thebtaranonth, C.
Cyclization reactions / authors, C. Thebtaranonth, Y. Thebtaranonth.
p. cm. - - (New directions in organic and biological chemistry)
Includes bibliographical references and index.
ISBN 0-8493-7861-3
1. Ring formation (Chemistry) I. Thebtaranonth, Y. II. Title. III. Series.
QD281.R2T48 1994
547′.504593- -dc20 92-38244
CIP

Direct all inquiries to CRC Press, Inc., 2000 Corporate Blvd., N.W., Boca Raton, Florida 33431.

International Standard Book Number 0-8493-7861-3
Library of Congress Card Number 92-38244
Printed in the United States of America 1 2 3 4 5 6 7 8 9 0
Printed on acid-free paper

Preface

This book is conceived in response to the request that followed the overview article on cyclization reactions published as a Tetrahedron Report (Report No. 270, *Tetrahedron*, **46**, 1385, 1990). Judging from its enthusiastic reception, the Report seemed to have been a timely survey of the area. However, with the rapid developments taking place, there now appears to be a need for a more comprehensive review, and this book aspires to provide insight into the subject and deliver a more complete picture of the major trends in the chemistry of cyclization.

The contents herein are aimed at readers wishing to know about, and to use, the different types of cyclization reactions, in particular the research scientist who would want a quick update of the latest advances in the specific areas. A mechanistic format has been adopted for the book, and the chapters are organized according to the different cyclization intermediates, *i.e.* cationic, radical, anionic, and metal complex intermediates, with the exception of the last chapter which contains macrolactonization, vicinal tricarbonyl, and Bergman (enediyne) cyclization reactions, since these are in distinctive focus today.

The advantage (or perhaps disadvantage) of writing a book on cyclization is that a massive amount of information is available, reflecting, of course, the extensive interest in the construction of cyclic molecules. Relevant publications in leading scientific journals between 1989 and 1992 alone number almost 2000! Sorting, organizing, and, most difficult of all, selecting materials to be included in the book were very time-consuming. Interesting work, at times, had to be sacrificed for the sake of clarity and continuity of the writing as a whole. Through all this we have tried diligently to use the latest results for illustrations wherever possible.

We would like to thank Professor W. D. Ollis for his suggestions and critical comments on the preceding overview Report in 1990. We would also like to thank Dr. F. M. Dean for his guidance and advice, since many years back, on the art of writing and scientific presentation. Their contributions have been valuable in the preparation of this book. We must also thank our colleagues and students, Miss Wacharada Panich, Mr. Sirichai Kittivarapong, Miss Chutima Kongkittingam and Mrs. Chantana Yuvaniyama for their phenomenal input during the entire course of this writing, and especially Miss Bongkoch Tarnchompoo, without whose incredible effort this book would not be as we see it today.

C. Thebtaranonth
Y. Thebtaranonth

THE AUTHORS

Yodhathai Thebtaranonth is Professor of Chemistry in the Department of Chemistry, Faculty of Science, at Mahidol University, Bangkok. Professor Thebtaranonth received his B.Sc. and M.Sc. degrees in 1966 and 1968 respectively, from the University of Medical Sciences (which later changed its name to Mahidol University), and his Ph.D. degree in Organic Chemistry in 1972 from the University of Sheffield under the Colombo Plan Scholarship. He spent one year at Cornell University, in 1975, doing postdoctoral research.

Professor Thebtaranonth has devoted his entire career to building up organic chemistry research in the Department of Chemistry at Mahidol University, which he joined in1968. Research publications, including papers, reviews and books, testify to his fight to carry out competitive research under severely limiting conditions. His works have been recognized and presented at major international chemistry conferences.

In 1987 Professor Thebtaranonth was awarded the "Outstanding Scientist Award", the most prestigious science award of Thailand. His present research interests are in the areas of bioactive natural products, synthesis, and the study of reaction mechanisms.

Chachanat Thebtaranonth was Professor of Chemistry at the Department of Chemistry, Mahidol University during the writing of this book . Professor Thebtaranonth received her primary and secondary school education in India and England, respectively. She returned to Thailand and obtained her B.Sc. degree in 1967 from the University of Medical Sciences, and upon graduation joined the Chemistry Department of the same university. In 1968 she left for England on a Colombo Plan Scholarship and received her Ph.D. degree from the University of Liverpool in 1972. In 1974 she married Yodhathai Thebtaranonth and in 1975 accompanied him to Cornell University to do postdoctoral research.

Professor Thebtaranonth held the office of assistant Dean of the Faculty of Science from 1977 to 1980. She was associate editor of the *Journal of the Science Society of Thailand* during 1988, 1989 (with Yodhathai Thebtaranonth as editor). She shares the same research interests as her husband, and since their marriage, has carried out much of her research in collaboration with him. She became full Professor in 1989.

Since completing the writing of this book Professor Thebtaranonth has taken up her new position as Science Park Project Director at the National Science and Technology Development Agency (NSTDA).

ABBREVIATIONS

ACAC — acetonylacetone
AIBN — α,α'-azobis(isobutyronitrile)
BINAP — 2,2´-bis(diphenylphosphino)-1,1´-binaphthyl
Bn — benzyl
BOC — *t*-butyloxycarbonyl
BPO — benzoylperoxide
BSA — bis(trimethylsilyl)acetamide
Bz — benzoyl
COD — 1,5-cyclooctadiene
Cp — η^5-C_5H_5, cyclopentadienyl
DBA — dibenzylideneacetone
DBM — dibenzoylmethido
DBU — 1,8-diazabicyclo[5.4.0]undec-7-ene
DCC — 1,3-dicyclohexylcarbodiimide
DCN — 1,4-dicyanonaphthalene
DEAD — diethyl azodicarboxylate
DEIPS — diethylisopropylsilyl
DIBAL-H — diisobutylaluminium hydride (DIBAH)
Diphos — 1,2-bis(diphenylphosphino)-ethane (see DPPE)
DMAP — 4-dimethylaminopyridine
DME — 1,2-dimethoxyethane
DMF — *N,N*-dimethylformamide
DMPU — 1,3-dimethyl-3,4,5,6-tetrahydro-2(1*H*)-pyrimidinone
DMSO — dimethylsulfoxide
DPPB — 1,4-bis(diphenylphosphino)butane
DPPE — 1,2-bis(diphenylphosphino)ethane (see Diphos)
DPPF — 1,1´-bis(diphenylphosphino)-ferrocene
HMPA — hexamethylphosphoric triamide
HMPT — hexamethylphosphorous triamide
IBDA — iodosobenzene diacetate
LDA — lithium diisopropylamide
MEM — methoxyethoxymethyl
MOM — methoxymethyl
Ms — methanesulfonyl
MsCl — methanesulfonyl chloride
MSH — *O*-mesitylenesulfonyl-hydroxyllamine
NMO — *N*-methylmorpholine-*N*-oxide
OTf — trifluoromethanesulfonate
PDC — pyridinium dichromate
PMB — *p*-methoxybenzyl
PPTS — pyridinium *p*-toluenesulfonate
Py — pyridine
TASF — tris(dimethylamino)sulfur (trimethylsilyl)difluoride
TBAF — tetrabutylammonium fluoride
TBDMS — *t*-butyldimethylsilyl
TBDMSOTf — *t*-butyldimethylsilyl trifluoromethanesulfonate
TBHP — *t*-butylhydroperoxide
TBS — tri-*n*-butylsilyl
TES — triethylsilyl
Tf — trifluoromethanesulfonyl
TFA — trifluoroacetic acid
THF — tetrahydrofuran
THP — tetrahydropyranyl
TIPS — triisopropylsilyl
TMEDA — *N,N,N´,N´*-tetramethyl-ethylenediamine
TMSE — 2-(trimethylsilyl)ethyl
TMSOTf — trimethylsilyl trifluoro-methanesulfonate
Ts — *p*-toluenesulfonyl
TsOH — *p*-toluenesulfonic acid
TTMSS — tris(trimethylsilyl)silane

TABLE OF CONTENTS

Chapter 4
Anionic Cyclization

Chapter 5
Metal-Catalyzed Cyclization

Chapter 6
Miscellaneous Cyclizations

Chapter 1

CYCLIZATION REACTIONS

The challenge to create and study the properties of both natural and synthetic molecules for the benefit of mankind lies at the very heart of organic chemistry. The challenge to imitate (or even compete with) living organisms (including the simple cells) in their ability to build complex and beautiful architectural molecular structures in near-quantitative yields and with complete stereocontrol, forms an important driving force in organic synthesis. The curiosity to know how and why living organisms build these molecules, the latter's functions, their modes of action with respect to the parent organism or to other organisms, *etc.* is yet another compelling incentive for the organic chemist to synthesize these natural compounds and their synthetic analogs for a multidisciplinary investigation.

The fact that the molecules of nature mostly possess cyclic components (either as part of the molecule or as the molecular skeleton) has made the cyclization reaction one of the most important arts of organic synthesis. Continuous and all-round efforts in this area are manifested in the improvement of existing methodologies and in the discovery of new methods and new reagents. The last 10 years, in particular, have witnessed a quantum leap (which has not shown any sign of tailing off) in the art of ring construction.

In general, cyclization reactions may be grouped into four main types, *i.e.*, those involving cationic, anionic, radical, and metal complex intermediates, which might be stable species or simply transient. Scheme 1 outlines the four groups in their simplest forms.

Scheme 1

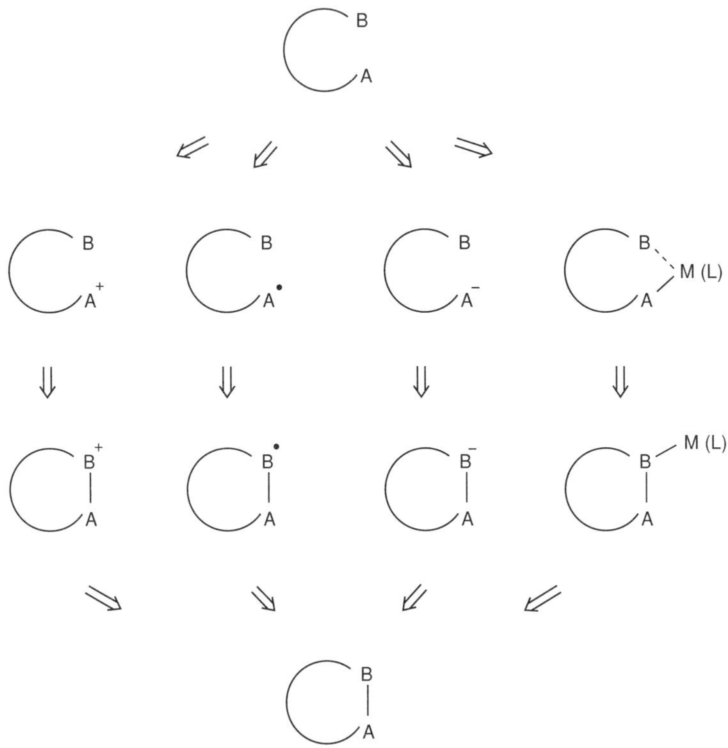

It can be seen, as depicted in the scheme, that the reaction constitutes the connection of a bond between A and B, which effectively transforms an open chain structure into a cyclic form. Extending along this basis, later studies have worked on connecting several bonds at a time to build a polycyclic structure in one operation. Such polycyclization or tandem cyclization, as the reaction is more popularly called, may lead to linear (Scheme 2) or nonlinear (Scheme 3) polycyclic compounds, depending upon the design of the cyclizing substrates.

Scheme 2

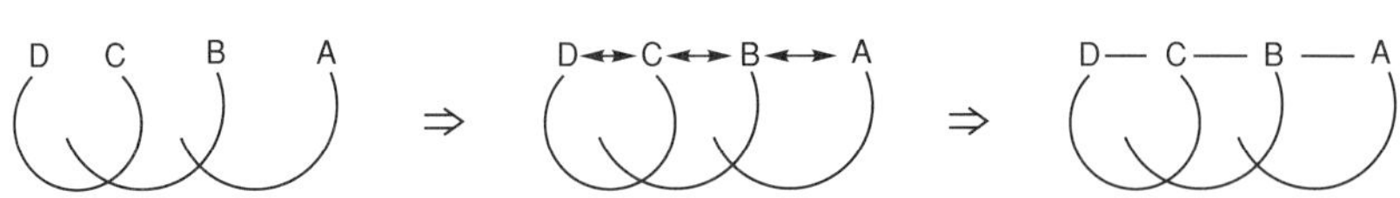

Scheme 3

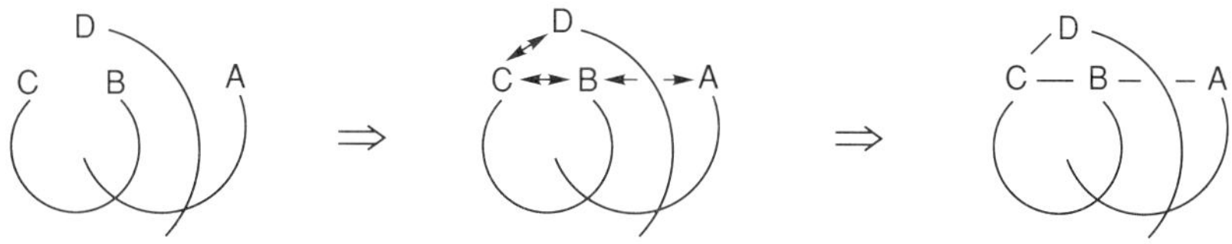

The art of cyclization will be shown in the following chapters through reactions that involve the different intermediates outlined above, by which process it is hoped that the beauty of organic synthesis will present itself in a way that can be fully appreciated.

Chapter 2

CATIONIC CYCLIZATION

I. INTRODUCTION

Cyclization reactions *via* cationic intermediates have continued to engage the attention of the chemist over the years. In this area carbocation cyclization continues to offer innovative contributions, with highlights that are regarded as methods of choice for the synthesis of cyclic molecules. Nevertheless, the study of reactions that involve "true" carbocations has lately dropped in popularity due to difficulty in the control of regio- and stereochemistry, not to mention troublesome side products. In contrast, cyclizations involving stabilized cations such as the iminium, oxonium, and thionium ions have gained increasing popularity and have been extensively investigated and successfully applied to the synthesis of complex natural products, particularly in the field of alkaloids. At the same time, vinylsilanes have been developed to such an extent that cyclizations involving silicon-stabilized carbocations now play a central role in organic synthesis.

For the sake of continuity, cation–olefin cyclizations involving vinyl-, allyl-, propargyl-, and allenylsilanes will all be discussed in this chapter despite the obvious differences between the "true" cationic character of the vinylsilane cyclization (Equations 1 and 2) and the dubious cat-/anionic character of its allyl counterpart (*e.g.*, Equation 3). True anionic allylsilane cyclizations, however, will be discussed in Chapter 4.

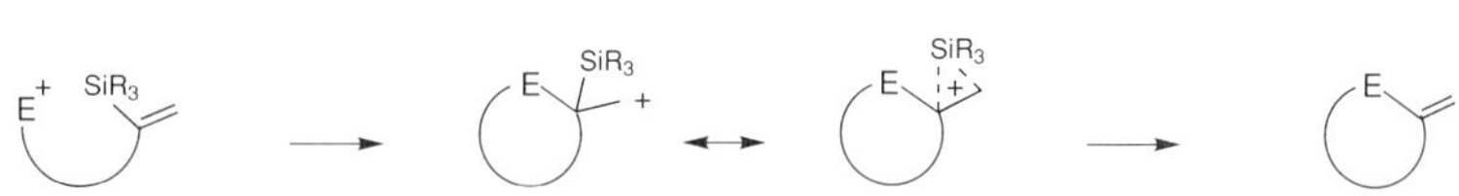

Equation 1

Equation 2

Equation 3

II. CYCLIZATION OF CARBOCATIONS

In 1984 Gassman and Singleton[1] reported that treatment of alkene **1a** with tris(*p*-bromophenyl)aminium hexachloroantimonate, or protic acids such as trifluoromethanesulfonic or *p*-toluenesulfonic acids, at fairly low temperatures, yields a single cyclized product **6a**. Seemingly a [4+2]-cycloaddition, the reaction in fact proceeds stepwise *via* carbocation intermediates,[1,2] as discovered when a similar treatment of **1b** yielded a mixture of **6b** (major product), **7b**, and **8**. The isolation of **8** implicates the carbocation precursor **5**, itself evidently the result of methyl migration in intermediate **4b**. The overall reaction is shown in Scheme 1.

Scheme 1

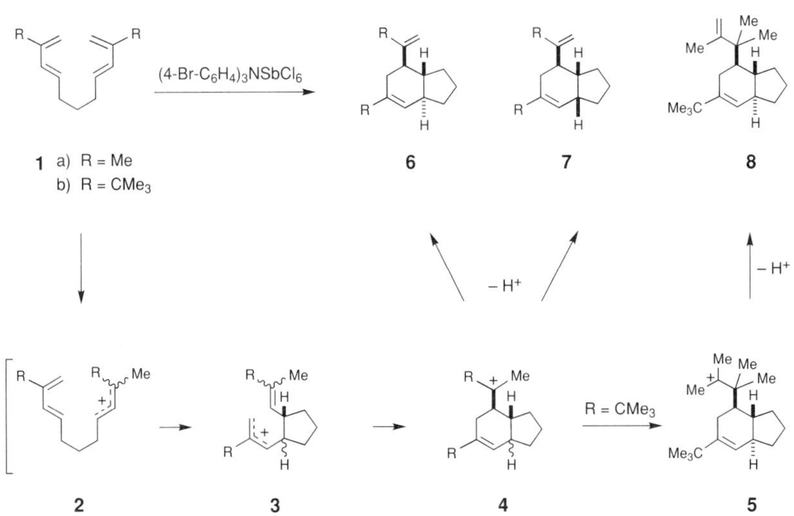

Further investigation has found that substituents such as hydroxy, ether, or allyl groups can control the regiospecificity of reaction, and thereby ring size of products, while the strength of acid catalysts can influence product distribution. For example, the allylic alcohol moiety in alkene **9** directs the site of dienophile allyl cation **10** formation and thereby induces regiospecific cyclization to yield solely product **11** (Scheme 2).[3] Likewise, methyl substituents in alkene **12** direct specific protonation upon treatment with 10 mol% triflic acid at –23°C to generate a single tetrasubstituted allyl cation **13**, which cyclizes as shown in Scheme 3 to give a mixture of 6,7-*cis*-fused carbocycles **14** (72%)

and **15** (4%) and 6,6-*trans*-fused carbocycle **16** (24%).[4] Controlled experiments reveal that under the above conditions **15** rapidly and cleanly isomerizes to the exocyclic olefin **14** but not *vice versa*. Further, the use of a weaker acid such as *p*-toluenesulfonic acid in place of triflic acid dramatically alters the product ratio, increasing the yield of 6,6-fused carbocycle **16** to a major 41%, while reducing the combined mixture of **14** and **15** to a minor 30% yield.

Scheme 2

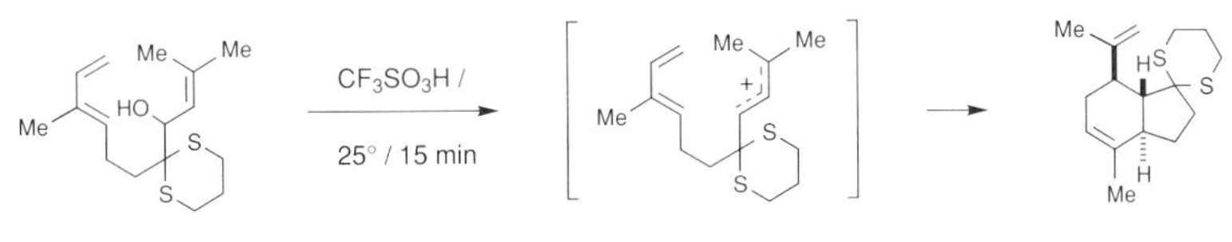

9 **10** **11** (56%)

Scheme 3

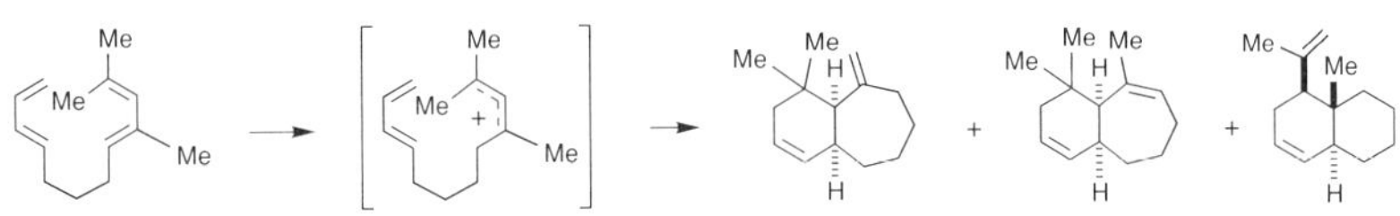

12 **13** **14** **15** **16**

A new addition to carbocation cyclization is the rather unusual but extremely useful rearrangement of tricyclo[6.3.0.0^{1,4}]undecan-5-one **17**, under acid catalysis, to provide a high yield of triquinane **20**.[5] The mechanism is proposed as involving cyclobutane ring opening to the intermediate carbocation **18** followed by hydride migration and subsequent ring closure to the observed product **20**. The recent (1992) trapping of carbocation intermediates **18** and **19** in the form of enol acetates **21** and **22** corroborates this mechanism (Scheme 4).[6] This rearrangement has been successfully applied to the synthesis of fused triquinane natural products,[7] *e.g.*, 5-oxosilphiperfol-6-ene **23** and silphiperfol-6-ene **24**. Besides protic acids, Lewis acids such as $AlCl_3$ and $SnCl_4$ also serve equally well in inducing the reaction.

Scheme 4

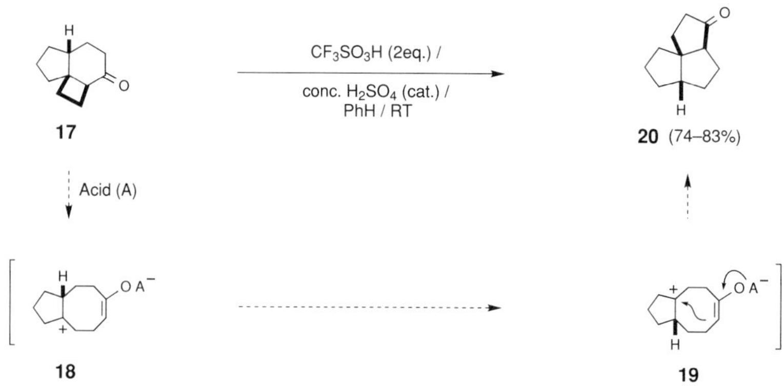

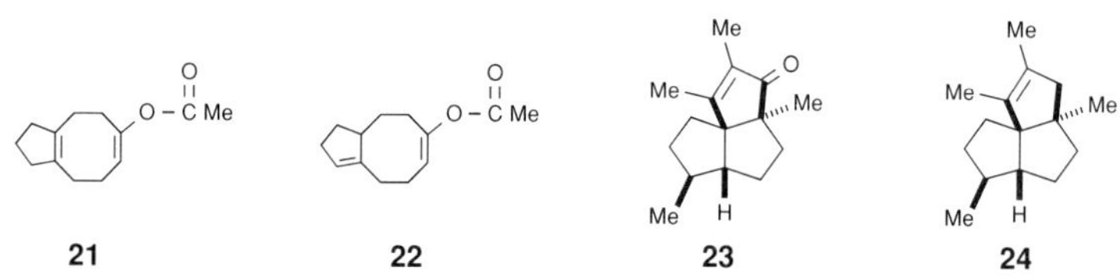

Other carbocation rearrangements are involved in the acid-induced carbocyclizations shown in Schemes 5 and 6.[8] Although sensitive to structural changes in the precursor the reaction has nevertheless been studied in detail, particularly in connection with the preparation of methyl *O*-methylpodocarpate **38** (Scheme 7), starting either from **34**[9] or **35**.[10] Formation of the desired *trans*-A/B ring junction in the product may be explained by invoking neighboring group participation of the methoxycarbonyl group, as depicted in **37**.

Scheme 5

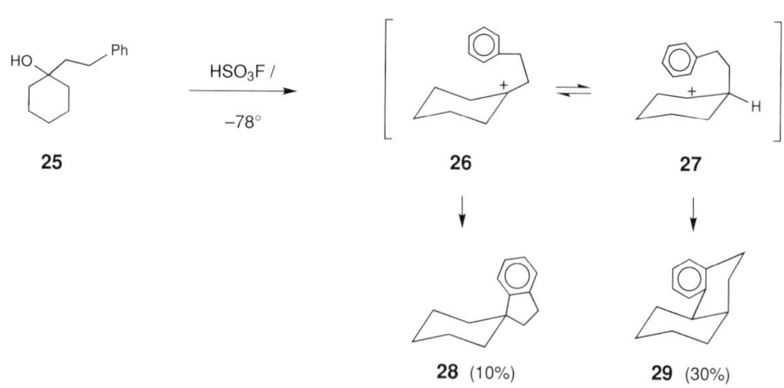

Scheme 6

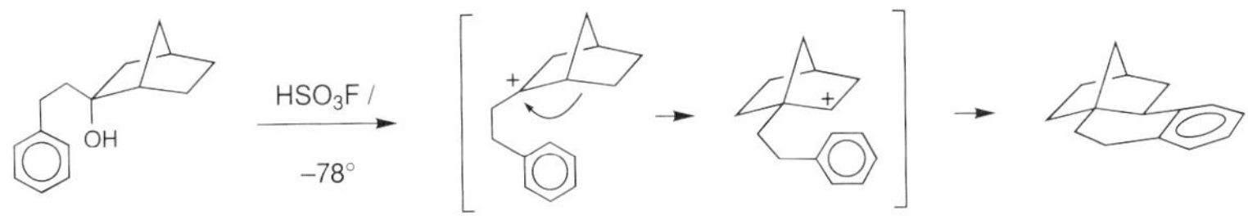

30 **31** **32** **33** (82%)

Scheme 7

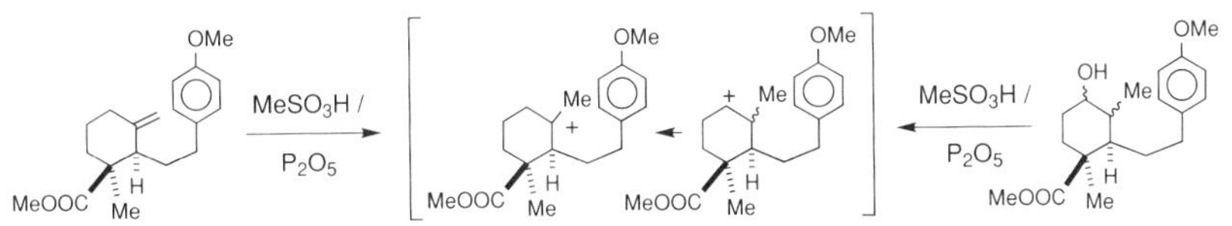

34 **36** **35**

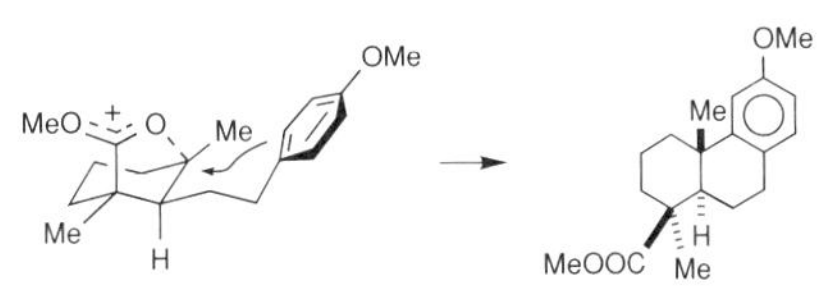

37 **38**

(92% from **34**)
(42% from **35**)

In the course of his work on the synthesis of spirotetrahydrofurans **42** and **44,** Paquette and Negri[11] reported, in 1991, a very interesting study of acid-catalyzed dehydration of diastereomeric diols **39** and **40,** as illustrated in Scheme 8 and the accompanying table. It was observed that, when R = OMe (entries 1 and 2), instead of protonation at the tertiary hydroxy group to generate the thermodynamically more favorable carbocation **45** (which would lead to a mixture of products **42** and **44**), protonation occurred selectively at the primary hydroxyl, resulting in products **42** from diol **39** and **44** from diol **40**, with over 90% stereochemical retention. This striking reactivity[11] is attributed to the large rate accelerations that accompany the intramolecular displacement of protonated primary hydroxyl by the tertiary hydroxylic oxygen as depicted in intermediates **41** and **43**. The mechanism has been confirmed by ^{18}O labeling experiments.

Scheme 8

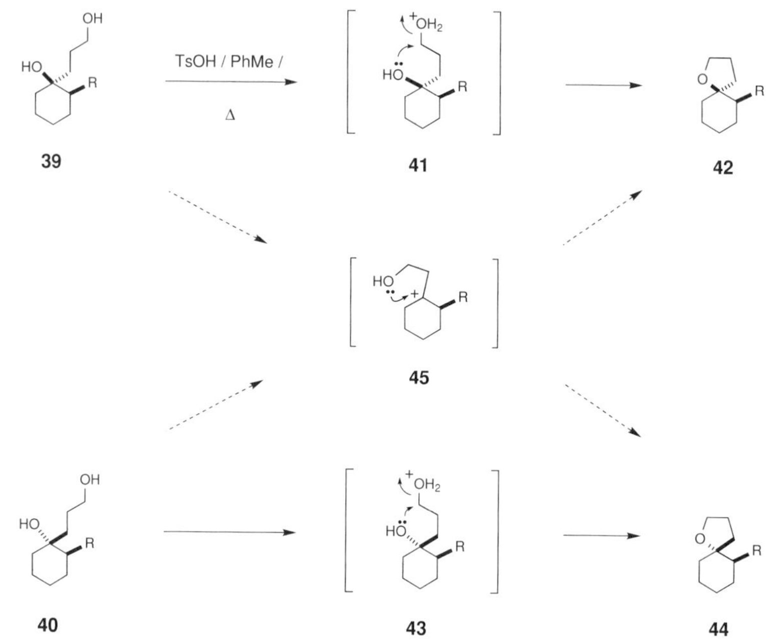

Entry	Diol	R	42 (%)	44 (%)
1	**39**	OMe	96	4
2	**40**	OMe	8	92
3	**39**	Ph	67	33
4	**40**	Ph	55	45
5	**39**	Me	40	60
6	**40**	Me	55	45

The commonly known carbocation–olefin cyclizations using the epoxide function as the initiating unit[12] can sometimes be wrought with problems such as pinacol rearrangement to carbonyl compounds, or 1,2-diol formation, precluding the cyclization. To alleviate the problems associated with activating the epoxide Corey and Sodeoka[13] have suggested the use of methylaluminum dichloride ($MeAlCl_2$) as catalyst, which probably functions as shown in Equation 4.

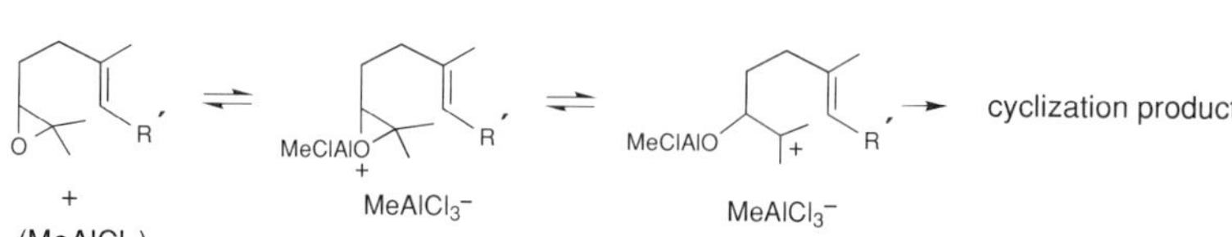

Equation 4

In an experiment with olefin **46** using the methylaluminum dichloride catalyst (0.3 to 0.5 equivalent) in methylene chloride solution at –78°C for 1 h, a mixture of bis-cyclization products **47** and **48** (total yield 75 to 79%) was obtained, together with a small amount of monocyclic ketone **49** (10 to 15%). Apart from $MeAlCl_2$ other related aluminum reagents such as Me_2AlCl and $Me_2AlOSO_2CF_3$ also afford comparable yields of the above cyclized products, although at slower reaction rates. However, incorporation of an extra double bond into the starting olefin (**50**) seems to inhibit the reaction under the above conditions, a very small amount of a mixture of products being produced (Scheme 9).

Scheme 9

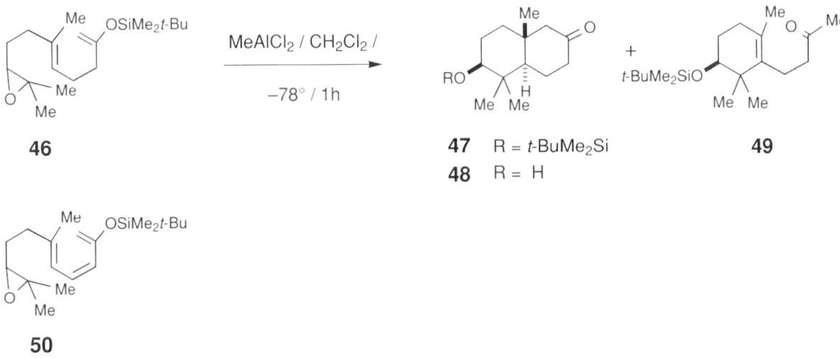

46 **47** R = *t*-BuMe$_2$Si **48** R = H **49**

50

The most recently reported catalyst is the complex $BF_3 \bullet MeNO_2$ formed between gaseous boron trifluoride and nitromethane.[14] This reagent is reported to be exceedingly effective for promoting the proton-initiated type of cascade cyclizations. Excellent results have been obtained from applications to 9-arylnona-2,6-dienes and related systems. For example, treatment of aromatic diene **51** with the $BF_3 \bullet MeNO_2$ complex yielded a single isomeric product **52**, a valuable intermediate for the synthesis of antineoplastic agent taxodione **53** (Scheme 10).[15] This novel complex is recommended as an excellent substitute for strong Lewis acids which are frequently incompatible with their intended substrates.

Scheme 10

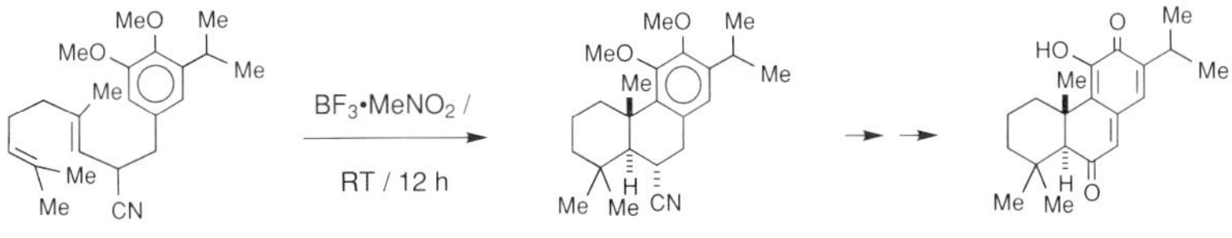

51 **52** (83%) **53**

Introduction of the $BF_3 \bullet MeNO_2$ complex to other carbocyclizations serves to further illustrate its effectiveness. As shown in Scheme 11,[16] isomeric silyl enol ethers **56** and **55** can be selectively prepared from *β*-keto ester **54** by employing the appropriate protic (*t*-BuMe$_2$SiCl / imidazole) or aprotic (*t*-BuMe$_2$SiOTf / NaH) silylating conditions. Upon subjecting the *Z*-isomer **55** to $BF_3 \cdot MeNO_2$-catalyzed reaction, bicyclic keto ester **57** is obtained as the sole product (70%). Under similar reaction conditions the *E*-isomer **56** stereospecifically gives the axial *β*-keto ester **58** in very high yield (90%). However, extension of the method to *Z*-enol acetate **59** registers a marked drop in yield and stereocontrol, the mixture of isomeric products being obtained in only 40% total yield, though still with a 5:1 predominance of **57** over **58**.

Scheme 11

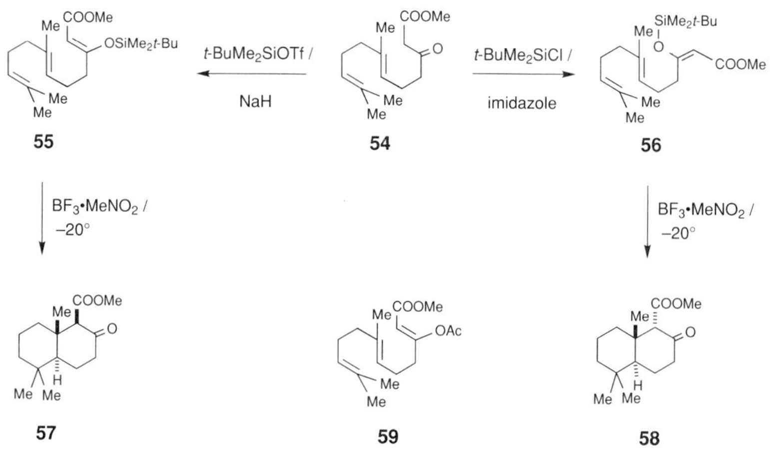

$SnCl_4$-catalyzed mono-, bis-, and tris-cyclization of optically active olefinic *β*-keto sulfoxides **60** to **62** offer further interesting examples of cascade cyclization.[17] Briefly, as shown in Scheme 12, treatment of **60** with $SnCl_4$ in methylene chloride affords optically active **63** (85%) while diene **61** and triene **62** each yields a 1:1 mixture of isomeric products: **64** + **65** (92%) from bis-cyclization of the former, and **66** + **67** (60%) from tris-cyclization of the latter.

Scheme 12

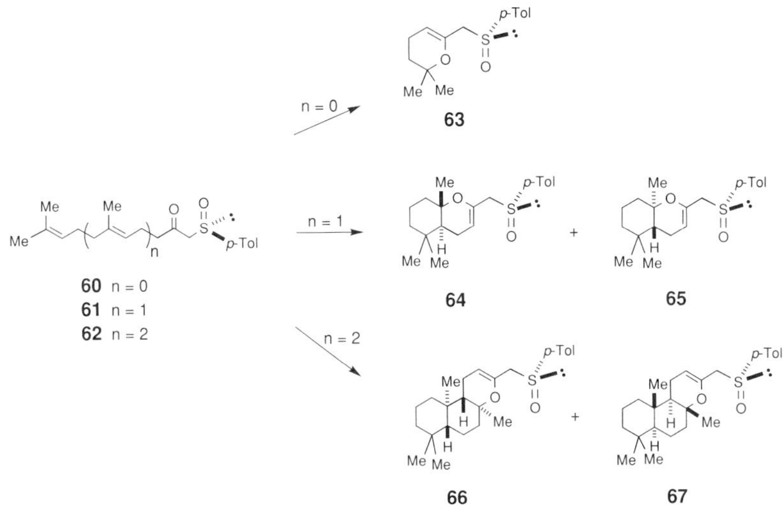

E,E-Suspensolide **69**, anastrephin **70**, and epianastrephin **71** are all key components of sex pheromones released by the *Anastrepha suspensa* and *Anastrepha ludens* species.[18] Synthesis of the 11-membered ring lactone **69** *via* macrolactonization of the hydroxy acid **68** was the first to be accomplished, but since then a Japanese group[19] has achieved the direct conversion of acid **68** to a mixture of anastrephin **70** and epianastrephin **71** (1:1 ratio, 67% yield) by treating with boron trifluoride etherate (3 equivalents) in methylene chloride. More recently another group[20] has succeeded in cyclizing lactone **69** with one molar equivalent of $BF_3 \cdot Et_2O$ at room temperature for 30 min to yield, almost exclusively, pheromones **70** and **71** in a 1:2.3 ratio, respectively. Although no yields are given, complete disappearance of starting material is claimed, with only two other minor components (2 and 3%) being detected by gas chromatography GC analysis.

The above rearrangement of macrocyclic lactone **69** to bicyclic lactone **71** can be explained as proceeding *via* carbocation **72** as a consequence of the (more favorable) *syn* conformation of **69**. Ring closure by equatorial attack of the carbonyl group on the intermediate cyclohexyl cation **72** would obviously lead to bicyclic lactone **71**. The presence of the accompanying minor isomer **70**, however, indicts concomitant participation of the *anti* conformer of the starting material. The sequence is illustrated in Scheme 13.

Scheme 13

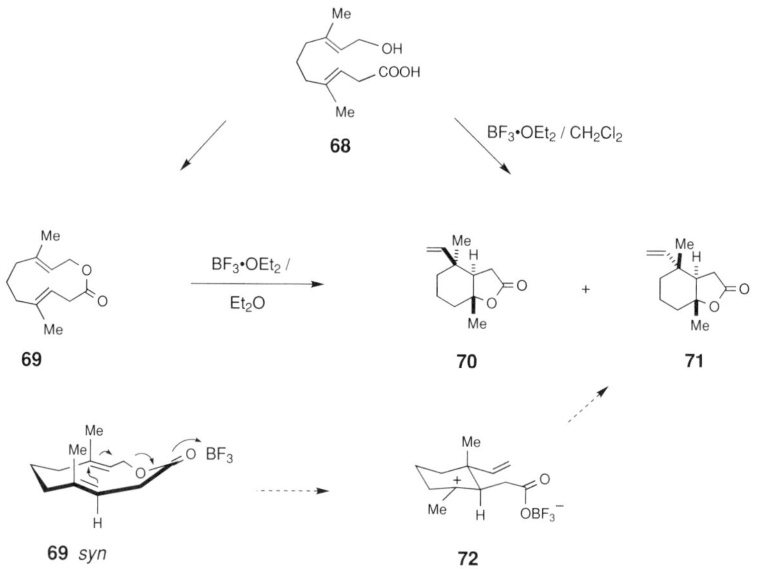

The catalytic potential of commercially available, strongly acidic Amberlyst® 15 (wet) ion-exchange resin[21] has also been investigated. The resin's unique property in promoting intramolecular cyclizations is well illustrated in its reaction with 3-vinylcycloalkenones **73**. The unactivated alkene function in **73** is known to react with Lewis acids such as ethylaluminum dichloride, to initiate successive ring closures which terminate in tricyclic structure **74** *via* carbocation intermediate **75**.[22] In contrast, the use of Amberlyst® resin deprotonates the initially formed cation **75** to yield exclusively bicyclic enone **76**[23] as summarized in the table following Scheme 14.

Scheme 14

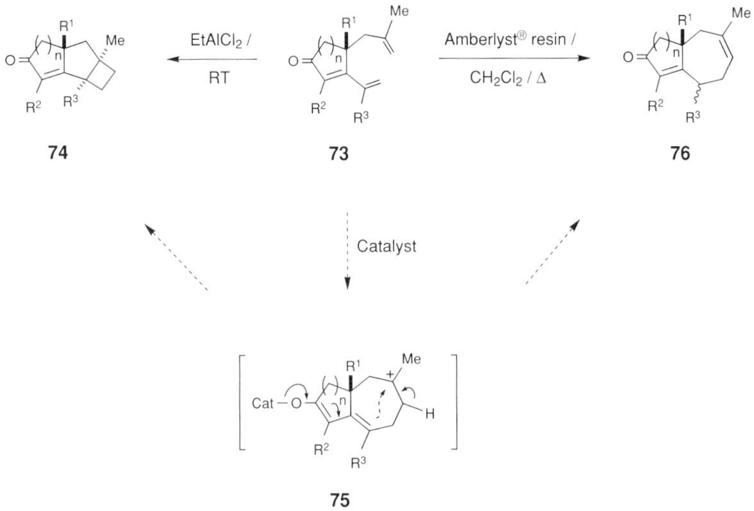

Starting material	n	R^1	R^2	R^3	Catalyst	Product (%) 74	76
73a	1	Me	Me	H	a	58	–
					b	–	60
73b	2	H	H	Me	a	85	–
					b	–	50
73c	1	H	H	Me	a	50	–
					b	–	25

a) $EtAlCl_2$ b) Amberlyst® 15 (wet) ion-exchange resin

The cyclization of pentadienyl cation intermediates to cyclopentenyl cation as depicted in Equation 5 has been thoroughly studied as, for example, in the Nazarov cyclization[24] (*e.g.*, Equation 6), yet certain related interesting work prompts further mention.

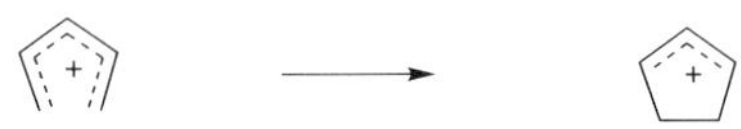

Equation 5

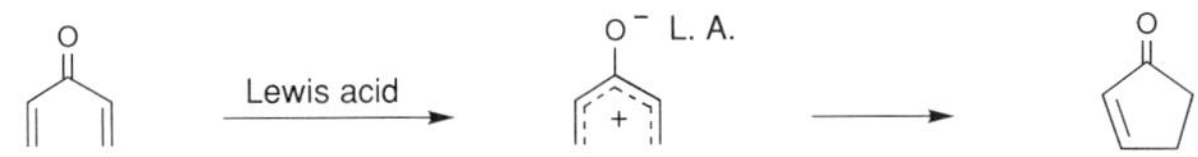

Equation 6

Over a decade of study has been carried out on the Lewis acid-catalyzed Nazarov cyclization of substrate **77** to cyclopentenone **78**, and it is established that the ester group in **77** controls the position of the double bond in product **78**.[25] Unfortunately, the success of the reaction shown in Scheme 15 is not always guaranteed especially when the ester function in **77** is derived from a bulky alcohol. In the latter situation the reaction has been found to proceed, in low yields, only upon substituting trimethylsilyl triflate (TMSOTf) for the Lewis acid. For example, when R^1 = (–)-menthyl, the triflate-catalyzed reaction yields 39% of the expected product, while the use of other Lewis acids results in complex mixtures.[26] In fact trimethylsilyl triflate is generally as effective as other catalysts for Nazarov cyclizations, stereospecifically affording the *trans*-cyclopentenone product **78** as shown in Scheme 15.

Scheme 15

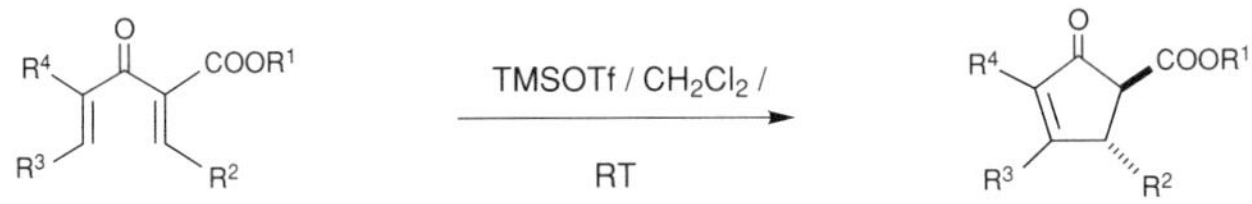

77 **78**

Cyclizations related to that shown in Equation 5 are extremely useful, particularly in the synthesis of perfluoroalkylated indenes. For example, the preparation of 1-(trifluoromethyl)indene **82** from indanone **79** is faced with the problem of sluggish dehydration of alcohol **81** to product **82**, which is also a poor-yield step in the synthetic sequence. However, by starting from trifluoroacetophenone **83** and adding vinylmagnesium bromide to obtain allyl alcohol **84** (84%), then effecting carbocyclization by treatment with methanesulfonic acid at 45°C (see Equation 5), the required product **82** can be obtained in 53% yield as outlined in Scheme 16.[27]

Scheme 16

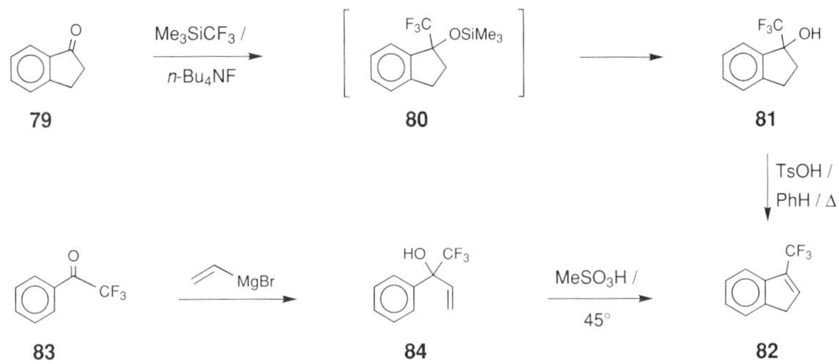

The above reaction is also very convenient for the synthesis of bis- and tris(trifluoromethyl)-substituted indenes such as **87a** and **87b** (Scheme 17). Especially interesting here is the fact that the electronically unfavorable carbonium ion **86** is capable of forming and undergoing cyclization to the product, even though, of course, more vigorous conditions are needed than in the case of alcohol **84**.

Scheme 17

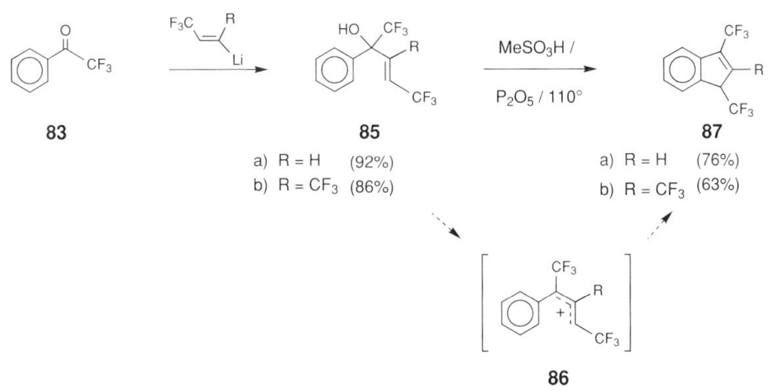

III. CYCLIZATION OF IMINIUM IONS

A. CYCLIZATION OF IMINIUM IONS BY INTERNAL NUCLEOPHILES

Intramolecular cyclization of iminium ions can be considered among the most useful reactions for the synthesis of nitrogen heterocycles. The reaction has received widespread interest and several methods have been developed for the generation of the iminium ion intermediate. Applications cover a whole range of cyclic products, from simple molecules to complex alkaloids of diverse structures containing several stereogenic centers.

The most straightforward cyclization in this category is the cyclization of iminium ions, generated *in situ*, with internal oxygen nucleophiles which can be appropriately positioned to vary the size and nature of the cyclic ether products. Taken together, the examples in Equations 7 to 11 provide a good overall picture of the reaction. Briefly, they all involve attack of internal oxygen nucleophile on the iminium ion intermediate, but differ in the manner of generation of the iminium ion, which are elimination of cyanide ion[28] (Equation 7); irradiation of cyclic amines in the presence of 1,4-dicyanonaphthalene (DCN) *via* a single electron transfer process[29,30] (Equation 8); oxidation with chlorine dioxide (ClO_2) in a mild basic medium[31] (Equation 9), which occurs preferentially at the less hindered carbon,[31] in contrast to, and therefore complementing, earlier mercuric acetate oxidations of tertiary amines to iminium salts[32] (Equation 10); and lastly, reduction of hydroxylactam to aminal with LiAlH(i-Bu)$_2$(n-Bu) complex, itself obtained from the reaction of DIBAL-H [(i-Bu)$_2$AlH] with n-BuLi and reported only recently[33] (Equation 11).

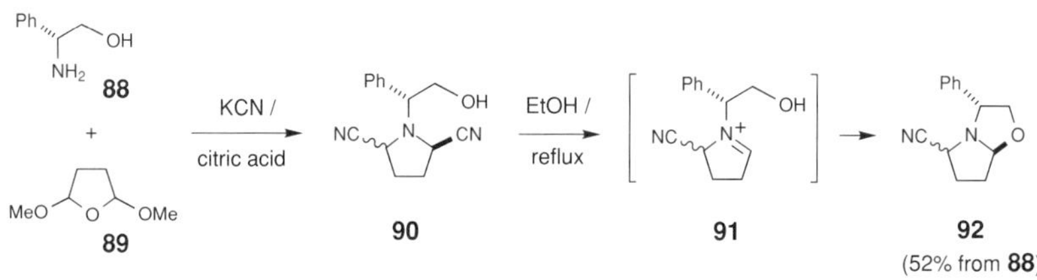

Equation 7

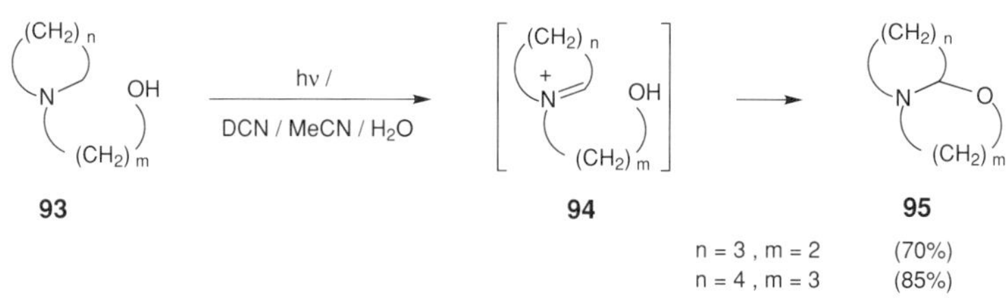

Equation 8

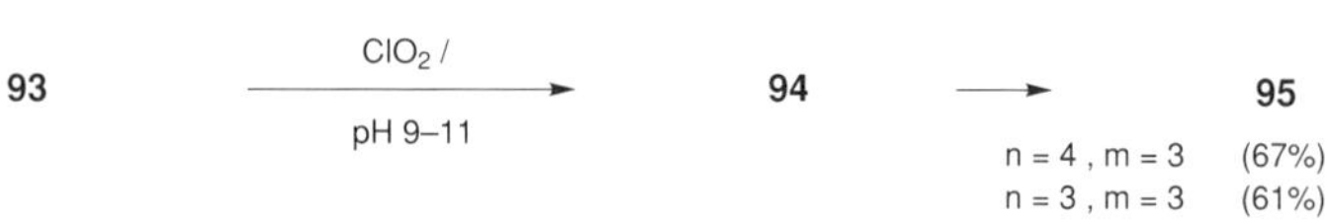

Equation 9

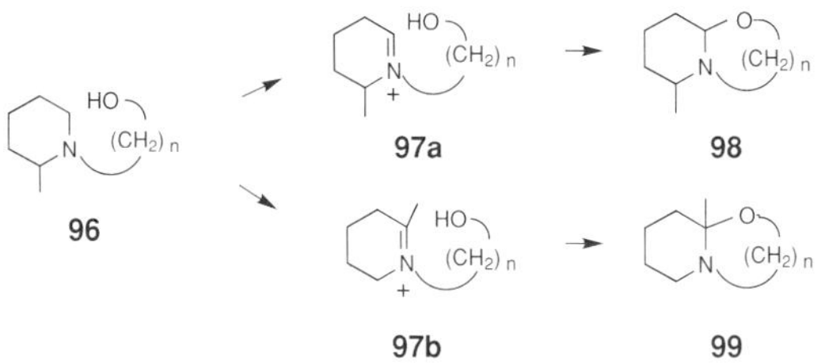

ClO₂ (%)		Hg(OAc)₂ (%)	
n = 2	n = 3	n = 2	n = 3
37	41	3	6
15	7	55	51

Equation 10

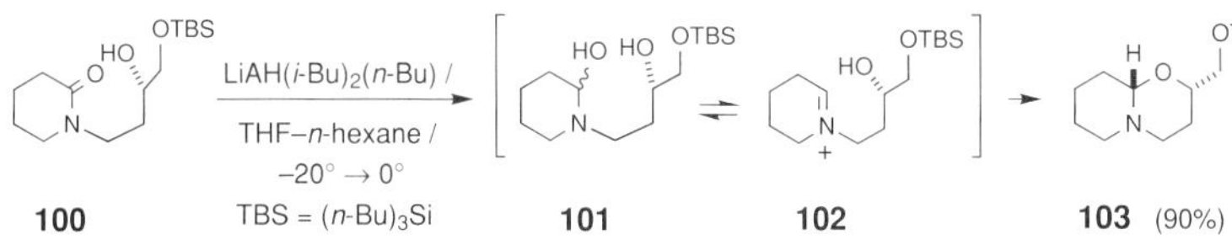

Equation 11

Related to the reaction in Equation 7 is the generation of iminium ion **105** from **88** and **104**, leading to the formation of stereospecific product **106** as depicted in Equation 12.[34] This cyclization was reported quite some time ago and has proved effective in the synthesis of optically active alkaloids.[35]

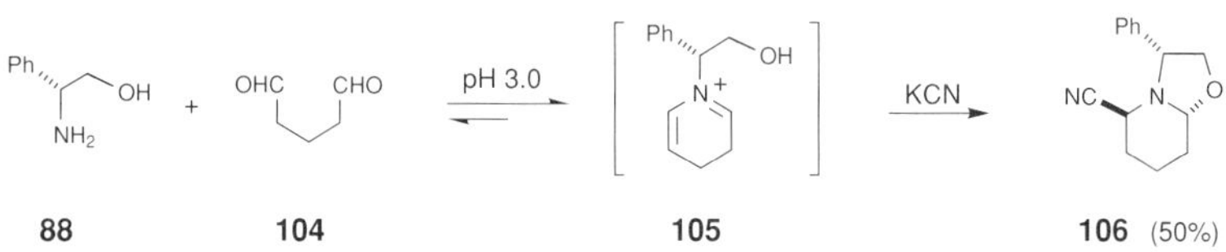

Equation 12

The first synthesis[36] of natural (+)-tetraponerine-8 **112** is one example of application of the above reaction. Here the starting material, optically active **107**, obtained from **106** (Equation 12), is first hydrolyzed with HCl then treated at pH 2 to 3 with 4-aminobutyraldehyde diethyl acetal **108** and KCN at room temperature to directly yield the tricyclic product **111**. The mechanism of reaction can be explained *via* the formation of intermediate **109** from the condensation between **107** and **108**. Under the controlled acidic conditions (pH 2 to 3) the iminium ion **110** is generated and suffers axial attack by the cyanide ion, in compliance with the stereoelectronic requirements proposed by Deslongchamps,[37] to yield **111** which can be converted easily to **112**.

Scheme 18

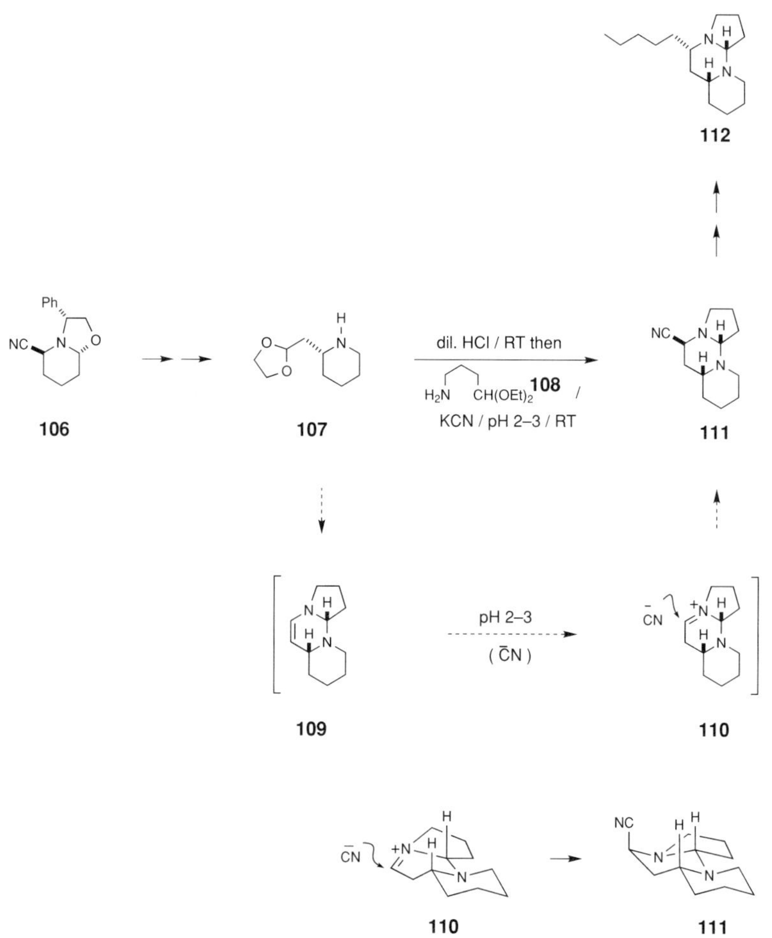

While on this subject, mention must also be made of the stereospecific synthesis of the ladybug defensive alkaloid coccinelline **120**. Although reported since 1979 by the late Professor Stevens,[38] the reaction is still regarded as one of the most classical examples of tandem cyclization of iminium ion. As briefly presented in Scheme 19, the protected amino aldehyde **113** is first hydrolyzed with hydrochloric acid at pH 1, then reacted with acetone dicarboxylic ester **115** at room temperature under controlled citrate–phosphate-buffered conditions (pH 5.5), whereupon a single isomeric product **119** simply crystallizes out from the reaction mixture in 75% yield, with relative stereochemistry of stereogenic centers as expected from a stereoelectronically controlled process.[37] Simple chemical manipulations of **119** are then all that is required to obtain the target coccinelline **120**.

Scheme 19

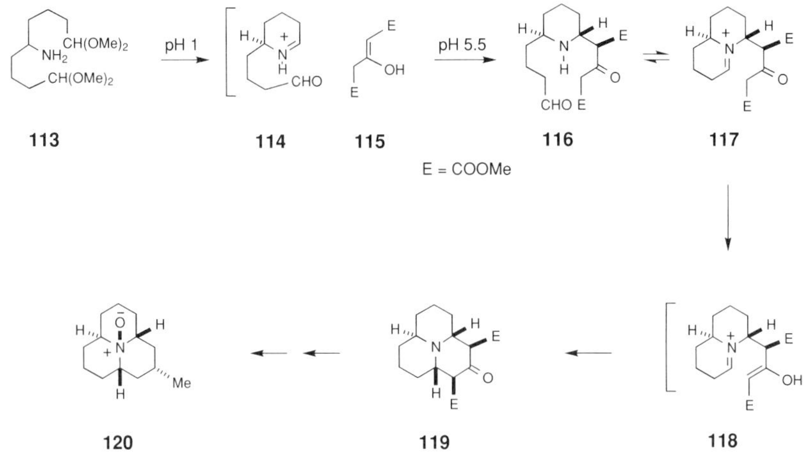

Numerous examples of iminium ion cyclizations belonging to the classical Pictet–Spengler condensation[39] class have appeared lately. These include the synthesis of alkaloid elaeocarpidine **125** by the previously mentioned ClO_2 oxidation[31] which converts cyclic amine **121** to iminium ion **123** for subsequent reaction with tryptamine **124** (Equation 13); the preparation of fluoro compound **128**, an intermediate in the synthesis of sorbitol, a well-known aldose reductase inhibitor (Equation 14);[40] the synthesis of tetracyclic amide **131** (Equation 15);[41] and the asymmetric syntheses of tetrahydroisoquinoline (–)-laudanosine **136** (Scheme 20)[42] and indole alkaloid (+)-antirhine **142** (Scheme 21)[43] briefly discussed below.

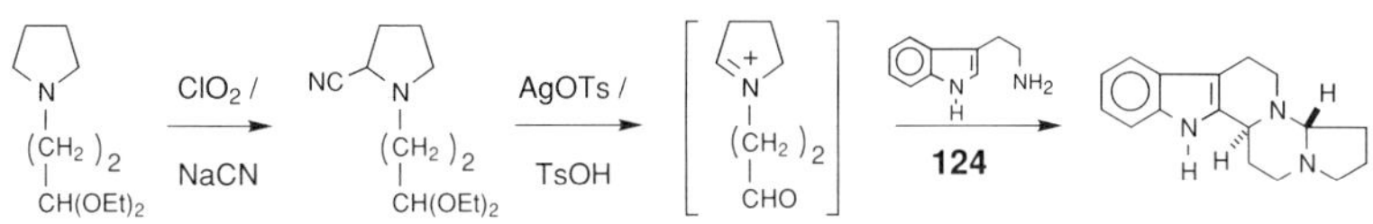

121 **122** **123** **125** (38% from **122**)

Equation 13

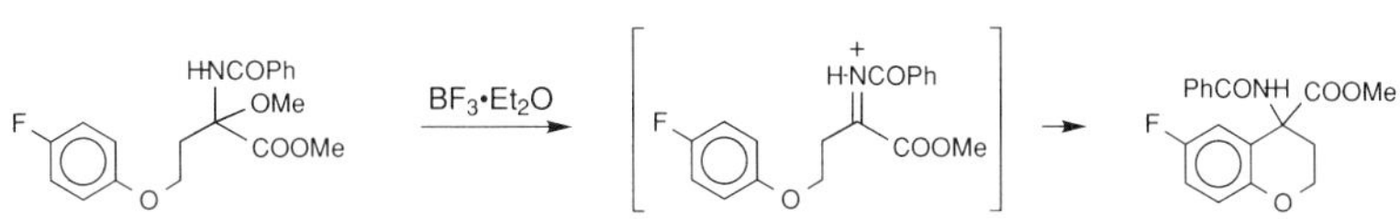

126 **127** **128** (85%)

Equation 14

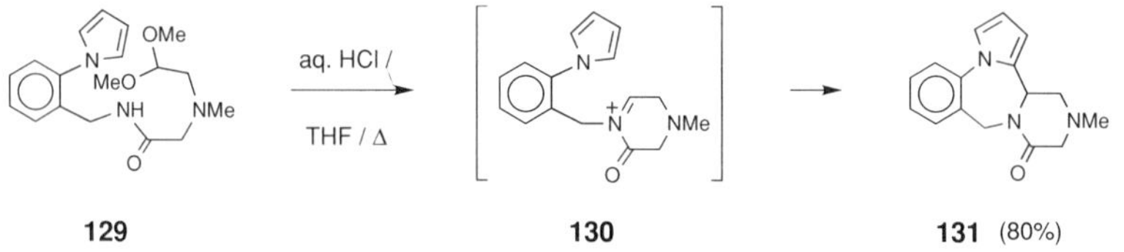

Equation 15

The asymmetric Pictet–Spengler synthesis of (–)-laudanosine **136** is accomplished by reacting optically active carbamate **132b** with protected aldehyde **133** to yield, *via* transitional iminium ion **135**, the intermediate tetrahydroisoquinoline **134b** (ee = 66%), which is then reduced with lithium aluminum hydride to afford (–)-laudanosine **136** with an ee value of 63% based on $[\alpha]_D$ comparison with an authentic sample (Scheme 20).[42]

Scheme 20

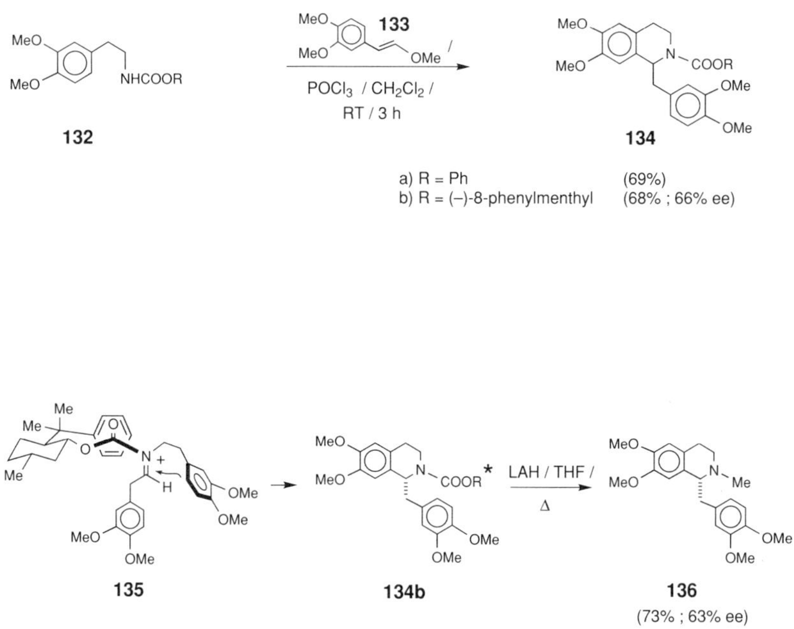

In the case of (+)-antirhine **142,** two key steps are involved in its synthesis starting from secologanin ethylene acetal **137**: first, sequential hydrolysis–selective reduction of **137** with baker's yeast at pH 6.4 to give **139** (which is converted to **140** by successive saponification, deformylation, and treatment with tryptamine) and second, stereoelectronically controlled cyclization of iminium ion **141** to yield the required product **142** (Scheme 21).[43]

Scheme 21

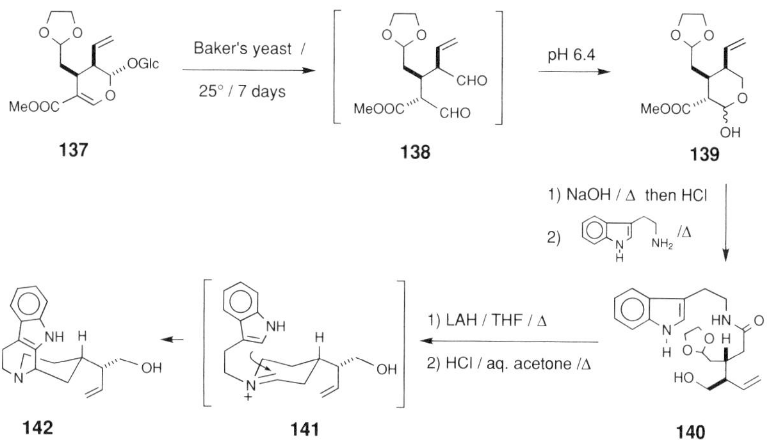

B. CYCLIZATION OF IMINIUM IONS INITIATED BY EXTERNAL NUCLEOPHILES

External nucleophiles such as the halide (particularly iodide) and hydroxide, and other nucleophilic groups such as the sulfide (RS^-), azide (N_3^-), acetate (AcO^-) or organozinc reagents, are all capable of inducing iminium ion cyclizations in both the *endo* and *exo* modes as illustrated in Scheme 22. The reactions can be carried out by generating the iminium ion *in situ* in the presence of the required nucleophile.

Scheme 22

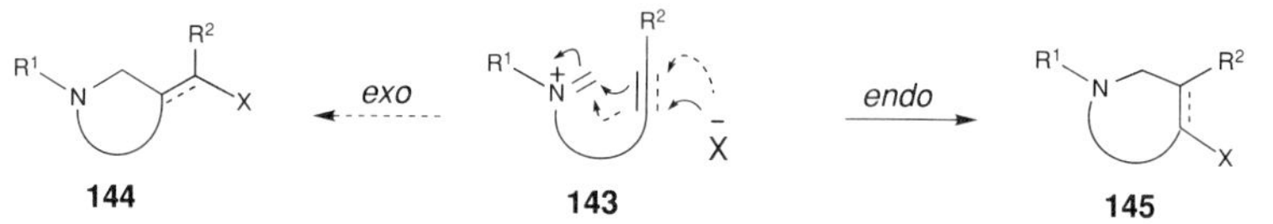

For example, iminium ion **147**, generated from the Mannich reaction of **146** with formaldehyde, undergoes exclusively *exo* cyclization initiated by the nucleophile X^- ($X^- = I^-$ or N_3^-) to afford the six-membered ring product **148** (Scheme 23). On the other hand, iminium ion **149**, with a shorter acetylenic chain, can form either a five- or six-membered ring from *exo* or *endo* cyclizations, respectively, depending upon the nature of substituent "R" (Scheme 24). Thus, when R = methyl, cyclization occurs in an *exo* fashion to give solely product **150** (in variable yields from 45–90% contingent upon X^-, *e.g.*, Br^-, I^-, N_3^-), whereas when R = hydrogen or trimethylsilyl, *endo* cyclization prevails to give the six-membered ring heterocycle **151** in reasonable yields. Although debatable in the case of R = H, the *endo* mode of cyclization when R = $SiMe_3$ is undisputably the result of the directive influence of silicon, stabilizing cation formation at the β-position.[44]

Scheme 23

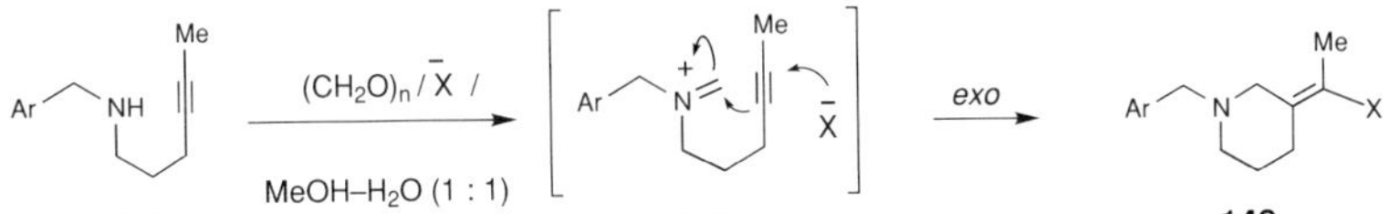

146 147 148

Scheme 24

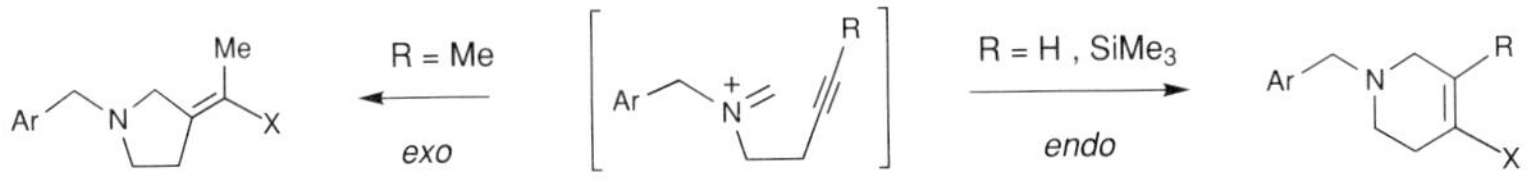

150 149 151

It should be noted that the cyclizations referred to in Schemes 23 and 24 have long been employed in the synthesis of alkaloids.[45,46] Examples are the cyclization of iminium ion intermediate **153**, generated from **152**, to yield the stereoisomeric mixture **154** (49%, Scheme 25), and the rearrangement–cyclization of **156** to **159** (Scheme 26), whose mechanism *via* [3,3]-sigmatropic rearrangement of **156** to **157** (making the latter, not the former, the cyclizing entity) has been thoroughly studied.[46,47]

Scheme 25

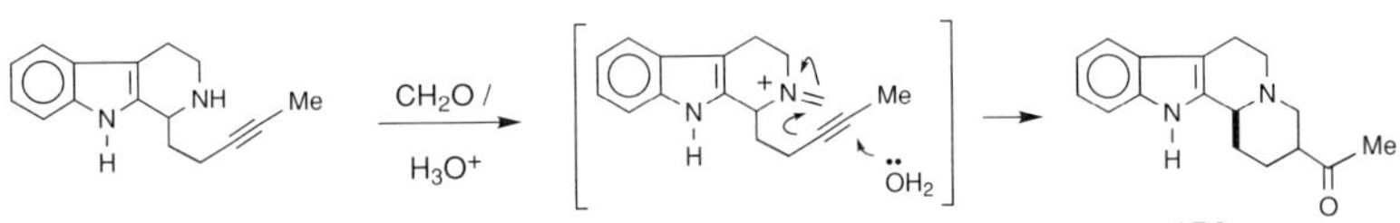

152 153 154

Scheme 26

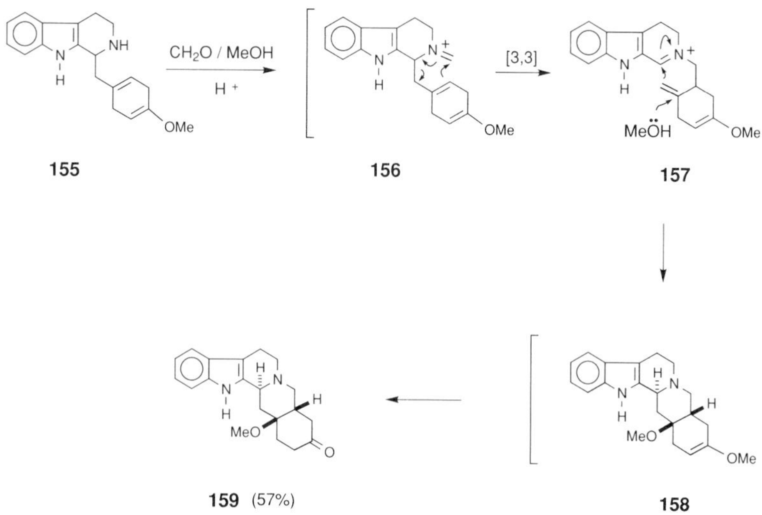

A related cyclization, shown in Scheme 27,[48] makes use of the heterogeneous reaction between allylsilane, trifluoroacetate salt of a primary amine, and aqueous formaldehyde, to produce iminium ion **164**, which, under the prevailing aqueous conditions, undergoes cyclization with water acting as external nucleophile to yield hydroxypiperidine **165**. The reaction works well with various types of allylsilanes giving good to quantitative yields.

Scheme 27

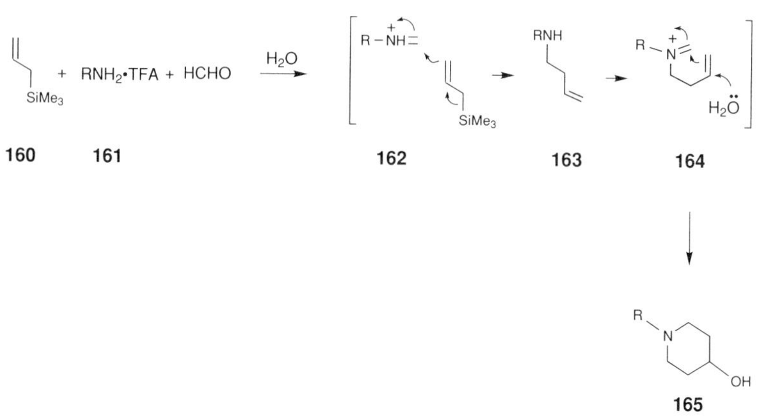

Similar reactions are shown in Scheme 28.[49] The cyclizing iminium intermediate **168** is formed from the reaction of acid on **167**, which, in turn, is prepared from the ruthenium-catalyzed oxidation of tertiary amine **166** with alkyl hydroperoxide. The same applies to the preparation of **172** from **170**.

Scheme 28

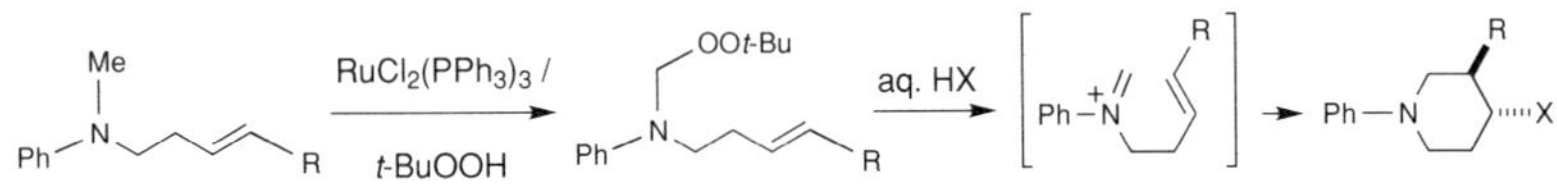

166 **167** **168** **169**

R = H (87%)
R = C_3H_7 (76%)

R = H , X = Cl (77%)
R = C_3H_7 , X = Cl (55%)

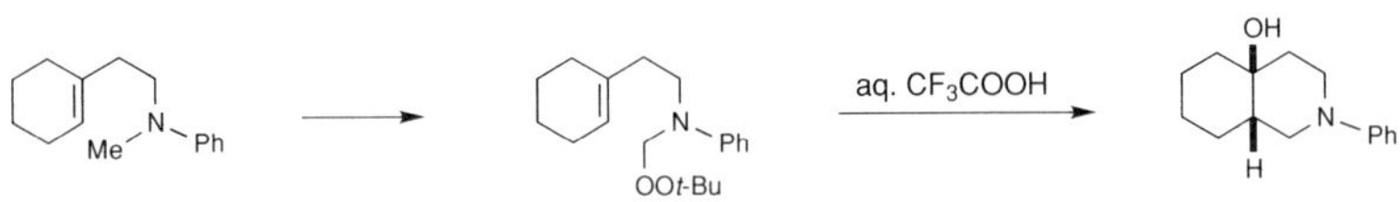

170 **171** (85%) **172** (44%)

Interesting reports of the most recent synthetic applications of iminium ion cyclizations with external nucleophiles include the asymmetric synthesis of pipecolic acid derivatives, as presented in Scheme 29.[50] Optically active amino alcohol **173** undergoes a one-pot condensation–cyclization reaction with glyoxal and thiophenol in aqueous solution to yield a 4:1 diastereomeric mixture of amino thioethers **174** and **175** (70%) *via* iminium ion intermediate **179a**. The mixture is then converted to the corresponding trimethylsilyl derivatives (**176** and **177**) and subjected to a second round of cyclization *via* the intermediacy of iminium ion **179b** (formed by elimination of SPh)[51] to furnish epimeric lactol **178**, whose stereochemistry can be rationalized in terms of transition state **180**. Although in principle it should be feasible to generate intermediate **179** directly from alcohol **173**, nevertheless this has not been possible in practice, hence the stepwise double cyclization procedure. The transformation of **181** into enantiomerically pure (2*R*,4*R*)-(–)-4-hydroxypipecolic acid **183** can be completed by simple chemical manipulations.

Scheme 29

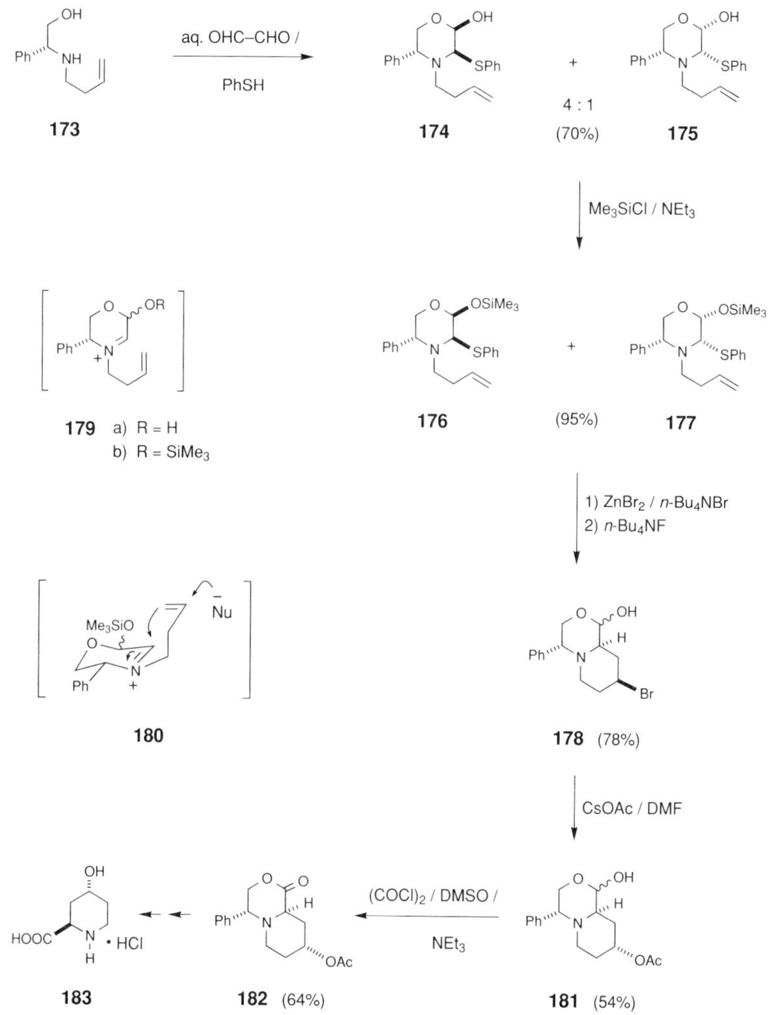

Double iminium ion cyclization is also employed in the recently reported first total synthesis of (+)-allopumiliotoxin 339A **191** (Scheme 30).[52] The reaction involves two key steps, *i.e.* iminium ion cyclization (**187** → **188**) by internal attack of oxygen nucleophile, and cyclization of intermediate **189** initiated by external iodide ion (**188** → **190**).

Scheme 30

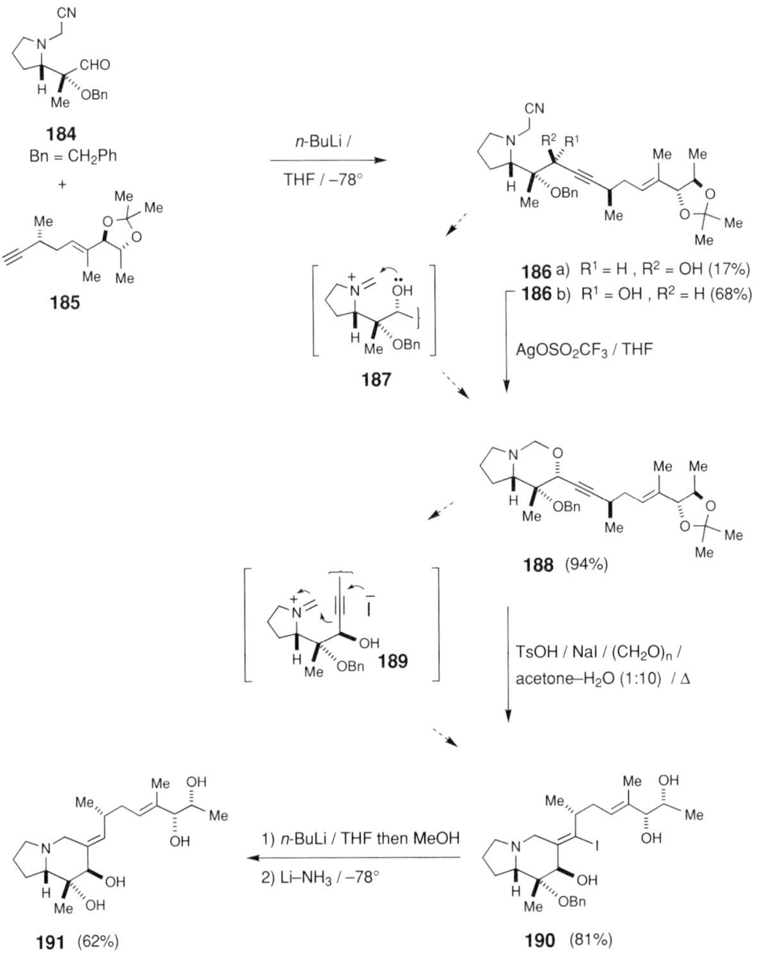

The above synthetic sequence starts with an aldol reaction between amino aldehyde **184** and the lithium acetylide derived from **185**, to provide alcohols **186a** (17%) and **186b** (68%). Alcohol **186b** is then cyclized to yield, *via* iminium ion **187**, oxygen ring compound **188**. Reopening of the latter's oxygen ring in a subsequent step triggers iodide-initiated *exo* cyclization as depicted in intermediate **189**, to give bicyclic iodide **190** (81%). Deiodination followed by benzyl ether cleavage finally affords the target (+)-allopumiliotoxin 339A **191**.

C. REARRANGEMENT–CYCLIZATION OF IMINIUM IONS

In 1979 Overman and Kakimoto[53] reported an acid-catalyzed reaction between aldehydes and homoallylic amines carrying an allylic hydroxy function, which yields a rearranged, cyclic product **193** (Equation 16).

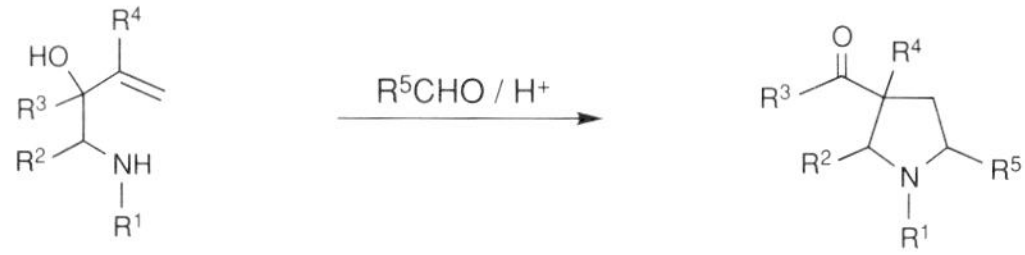

Equation 16

The wide scope of this reaction was demonstrated in a series of systematic studies,[54] and the reaction has been used as the key step in the synthesis of several complex alkaloid molecules.[55–57] The mechanism of this highly stereoselective process is now known[58–60] to involve the formation of an iminium ion intermediate **194** which undergoes a [3,3]-sigmatropic rearrangement followed by intramolecular cyclization to the observed product **193** (Scheme 31). The strategically placed –OH group serves to trigger cyclization in the rearranged iminium ion **195**, rendering irreversible the aza-Cope rearrangement process (**194** → **195**) and driving the reaction to yield only cyclized product **193**. It should be remembered here that iminium ion cyclization following on [3,3]-sigmatropic rearrangement has been mentioned once before (see Scheme 26).

Scheme 31

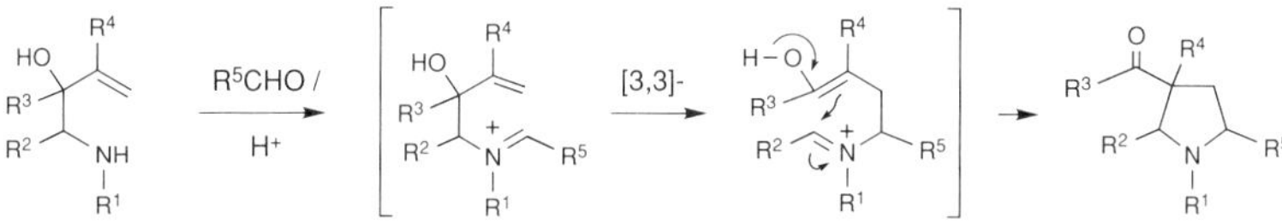

Recent applications of the above reaction include such notable examples as the syntheses of *Strychnos* alkaloid,[61] *Amaryllidaceae* alkaloid[62] and *Melodinus* and *Aspidosperma* alkaloids.[63] Inventive application of tandem cationic aza-Cope rearrangement–Mannich reaction is seen in the synthesis of dehydrotubifoline **201** (Scheme 32), in which Fevig and co-workers[61] react bicyclic amino alcohol **197**, obtained from bis-aminolysis of allylic chloride **196**, with formaldehyde in the presence of camphorsulfonic acid to obtain a single crystalline product **200** in 88% yield. The reaction proceeds *via* iminium ion **198** and successively progresses through [3,3]-sigmatropic rearrangement to **199** and cyclization to **200**. Subsequent hydrolysis with potassium hydroxide removes the amide protecting group to liberate the corresponding amine which simultaneously condenses to furnish the pentacyclic *Strychnos* skeleton **201** (70%).

Scheme 32

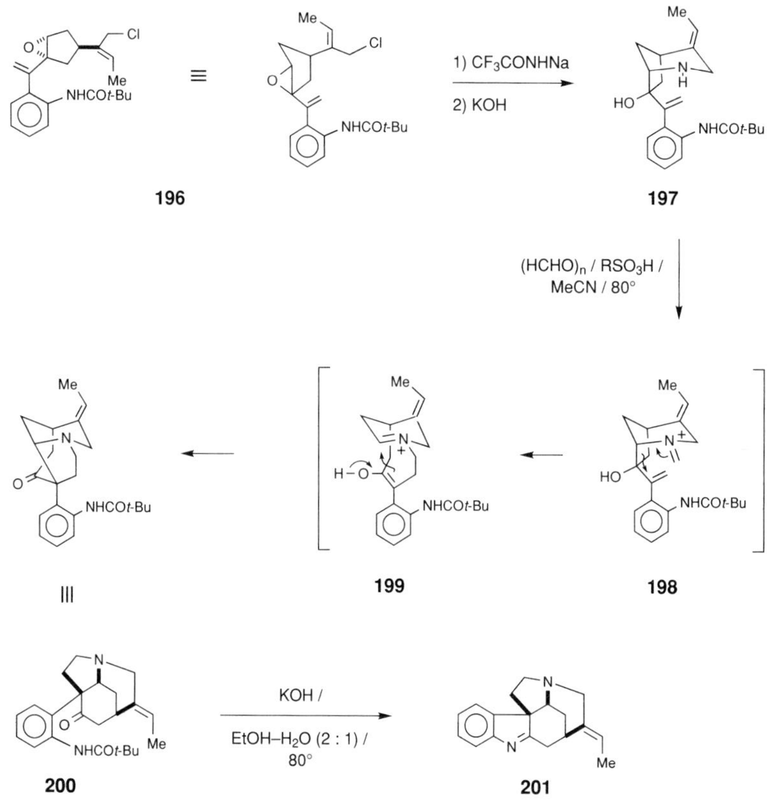

Tandem cationic aza-Cope rearrangement–Mannich reaction also constitutes the key step in the synthesis of *Amaryllidaceae* alkaloids, for example, (±)-pancracine (Scheme 33).[62] Indeed, this highly efficient and stereocontrolled synthesis takes maximum advantage of iminium ion reactions. Starting from alcohol **202**, the first in a series of iminium ion cyclizations is initiated by internal oxygen nucleophile to give **203** (81%). This, in turn, is treated with boron trifluoride etherate to effect ring opening to iminium ion intermediate **204** which rearranges to **205** before cyclizing to **206** in a superb 97% yield. Deprotection of **206** (to **207**) prepares for a final Pictet–Spengler cyclization which formally achieves the construction of tetracyclic skeleton **208**, the precursor of pancracine **209**. Standard chemical reactions then complete the remaining conversion of **208** to **209**.

Scheme 33

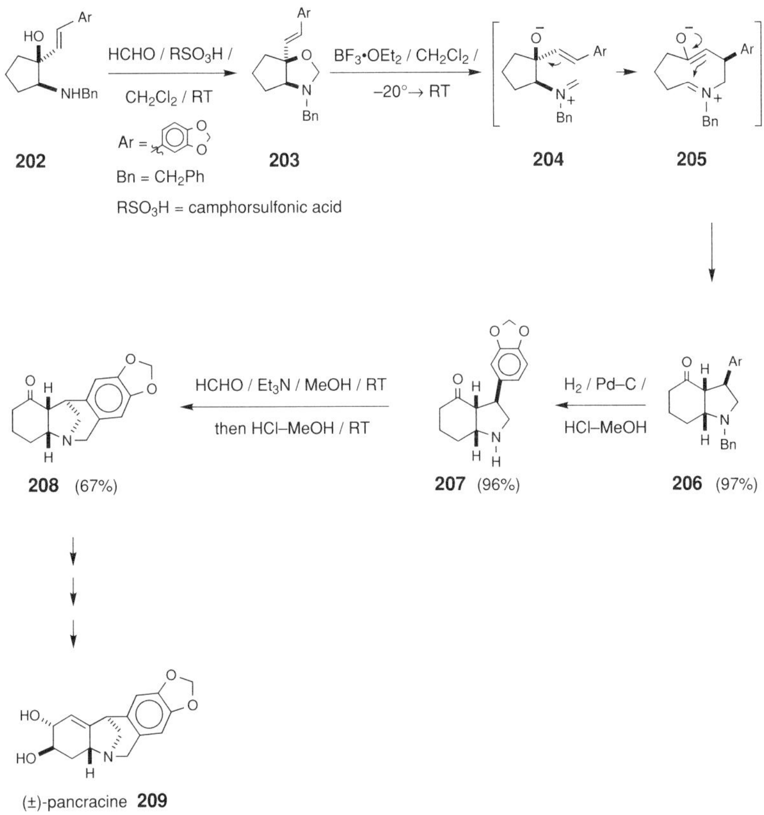

In similar manner, the syntheses of *Melodinus* alkaloids, meloscine **210** and epimeloscine **211**, and *Aspidosperma* alkaloid, deoxoapodine **212** (details of which have only recently been released)[63] all successfully employ the strategy discussed in Schemes 31 to 33 for ring construction. The retrosynthetic steps are summarized in Scheme 34. As shown, meloscine **210** and epimeloscine **211** are obtained from **213** (R = H) by straightforward amine–ester condensation to form the requisite nitrogen ring, plus appropriate conversion of the side chain to the vinyl substituent. Meanwhile, deoxoapodine **212** is derived from **214** through the introduction of the ester group and side chain cyclization to form the tetrahydrofuran ring. Retracing further back, however, both intermediates **213** and **214** converge upon the common precursor **215**, itself derived from **216**. This latter is the key intermediate constructed *via* the tandem cationic aza-Cope rearrangement–Mannich reaction of substrate **217** or **218** (see Schemes 32 and 33).

Scheme 34

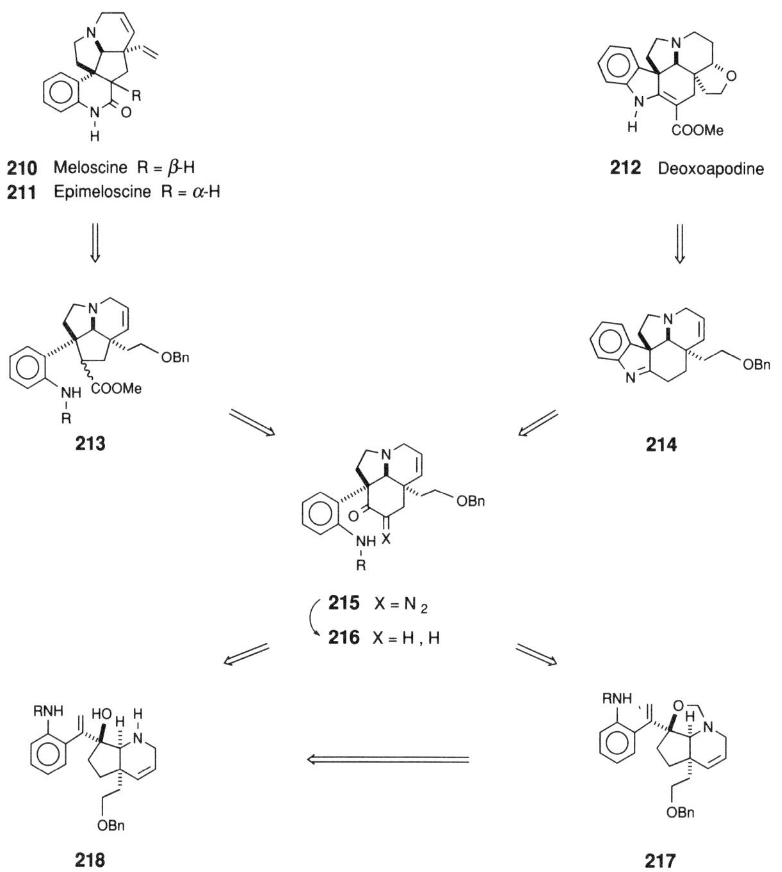

In connection with the rearrangement–cyclization of iminium ions mention should also be made of a synthetically useful process involving tandem photocycloaddition—retro-Mannich reaction of enaminone **219** to yield imine **221**, which can be further cyclized in Mannich fashion to deliver bicyclic amino ketone **222** in a 50% overall yield from **219**. Scheme 35 also shows an application of the process to the synthesis of mesembrine **227**, which was accomplished in 33% overall yield from veratrole **223**.[64]

Scheme 35

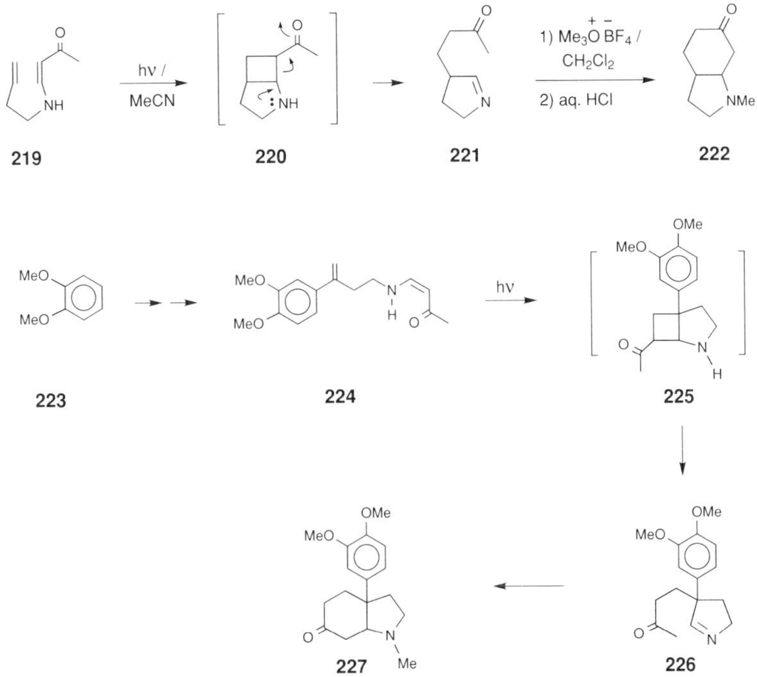

A rather interesting case is that of iminium salt **229**, generated from the corresponding bicyclic amine **228**, which undergoes simple Mannich cyclization to **231**[65] instead of the expected aza-Cope rearrangement. However, activating the transition state through the use of potassium salt **233** finally did induce rearrangement similar to that observed in the carbon analog,[66] to give **234**. Subsequent conversion to **236** followed by heating in trifluoroacetic acid accomplished cyclization *via* boat-shaped transition state **237** to yield tricyclic product **238**,[65,67] an advanced intermediate in the synthetic route to gelsemine **239** (Scheme 36). Coincidentally, another synthesis of the gelsemine skeleton, employing a similar approach, also appeared at around the same time.[68] As depicted in Scheme 37, it involves cyclization of the *N*-acyliminium ion **241** (7:3 mixture of *E*:*Z* isomers), derived from **240**, with boron trifluoride etherate in methylene chloride at room temperature, which produces a corresponding 7:3 isomeric mixture of aldehydes **242** and **243** in high yield. The products have been characterized as their respective alcohol derivatives.

Scheme 36

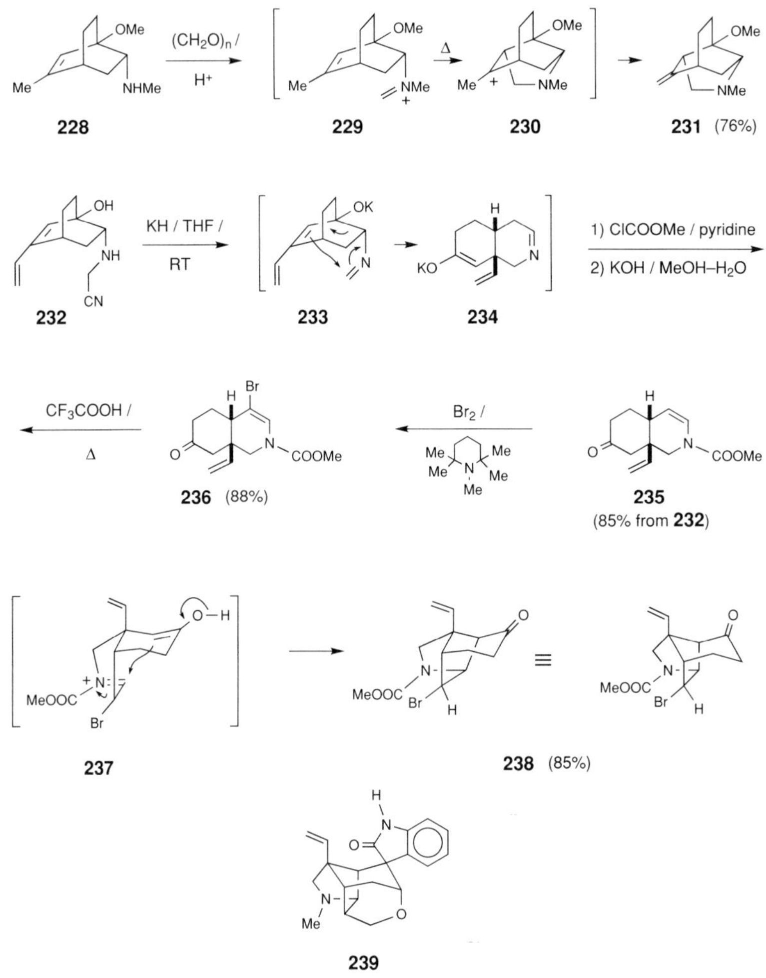
228
$(CH_2O)_n$ /
H^+
229
Δ
230
231 (76%)
232
KH / THF /
RT
233
234
1) ClCOOMe / pyridine
2) KOH / MeOH–H_2O
235
(85% from 232)
Br_2 /
236 (88%)
CF_3COOH /
Δ
237
238 (85%)
239

Scheme 37

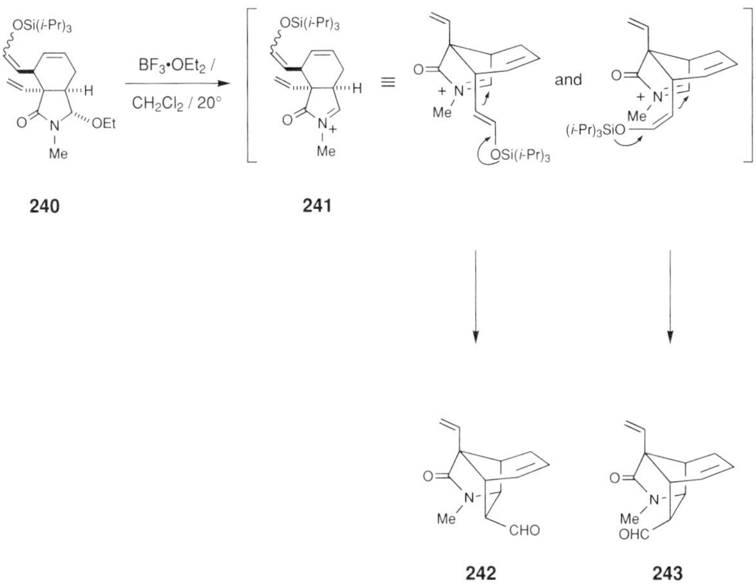

D. IMINIUM ION-INITIATED MULTIPLE CYCLIZATION OF UNACTIVATED POLYENES

Not unlike the cascade carbocyclizations of unactivated polyenes that lead to polycyclic carbon skeletons (*e.g.*, steroids, see preceding Schemes 9 to 11) multiple cyclization is possible also with the nitrogen analog, using the iminium ion, generated *in situ*, as the initiating electrophile. This high potential process, if carefully designed, can result in spectacular cyclizations that could assemble, in one operation, multiple rings of complex alkaloid structures otherwise extremely difficult to construct.

A very good example of the application of the reaction is seen in the synthesis of the *Daphniphyllum* alkaloids, *i.e.*, methyl homosecodaphniphyllate **244**,[69] (–)-secodaphniphylline **245**,[70] and daphnilactone A **246**.[71] Important steps in the synthesis of **244** and **245** are the Swern oxidation of diol **247** to dialdehyde **248** and subsequent condensation of the latter with gaseous ammonia in methylene chloride solution, followed by treatment with acetic acid to directly deliver pentacyclic compound **249** in 77% yield, as shown in Scheme 38.

Scheme 38

244 R = OMe

246

245 R = 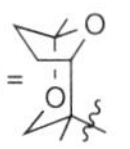

The mechanism of multiple cyclization from monocyclic **248** to pentacyclic **249** can be conceived along two pathways, diverging from the common bicyclic azadiene intermediate **250** but finally converging on the iminium ion **253**. Accordingly, following the protonation of **250** to **251** either stepwise cyclization through a series of cations **252**, **253,** and **254**, or intramolecular [4+2]-cycloaddition between the diene and olefin moieties in **251**, directly giving **253**, may occur. In either case the intermediate **253** is formed and undergoes π-cyclization of the prenyl double bond onto the iminium ion to give **254**, which then deprotonates to the observed product **249**.

Heathcock *et al.*[69] are strongly in favor of the cycloaddition pathway. First of all, the reaction is highly stereoselective. Second, both compounds **250** and **255** have been isolated from the reaction and characterized. Third, chromatographic analysis shows that **255** is formed immediately upon the addition of acetic acid to **250**, while the reaction must be heated for product **249** to be obtained. Finally, it is also found that **250** can be converted to **255** without the aid of acid, by heating in toluene alone at 110°C.

Whatever the mechanism of reaction, the above cyclization is undeniably very elegant, and can be diagrammatically summarized as shown in **256** which depicts connection of the three new bonds (dotted lines), whether they be all cationic a, b, c, in stepwise order, or concerted a, b, followed by carbocation cyclization c.

The final transformation of **249** to the desired **244** is easily accomplished by hydrogenation of the double bond with concomitant cleavage of the benzyl group, followed by Jones oxidation, then esterification. Treatment of the lithium enolate of optically active (–)-**244** with acid chloride **257** and subsequent removal of the β-keto ester group gives (–)-secodaphniphylline (Scheme 39).[70]

Scheme 38

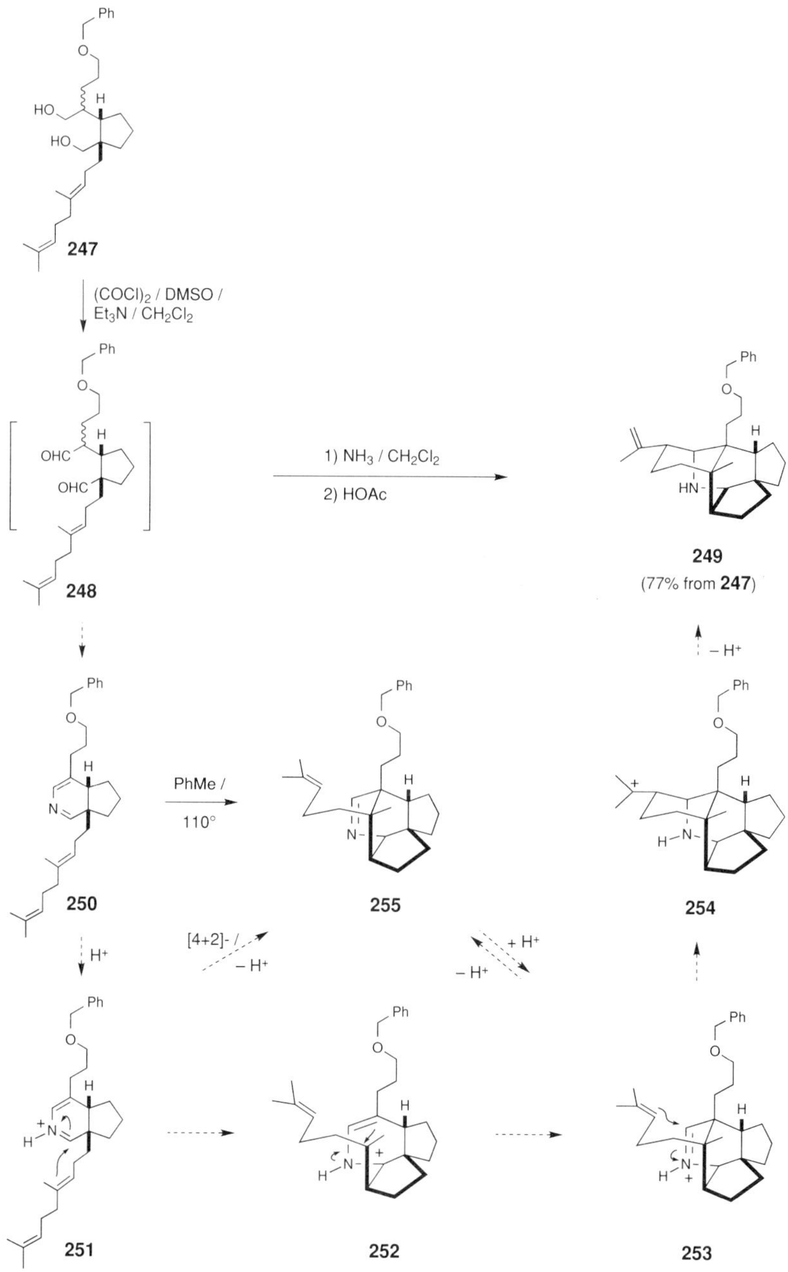
Ph
HO
H
HO
247
(COCl)2 / DMSO /
Et3N / CH2Cl2
OHC
OHC
248
1) NH3 / CH2Cl2
2) HOAc
HN
249
(77% from 247)
– H+
N
250
PhMe /
110°
255
254
H+
[4+2]- /
– H+
+ H+
– H+
251
252
253

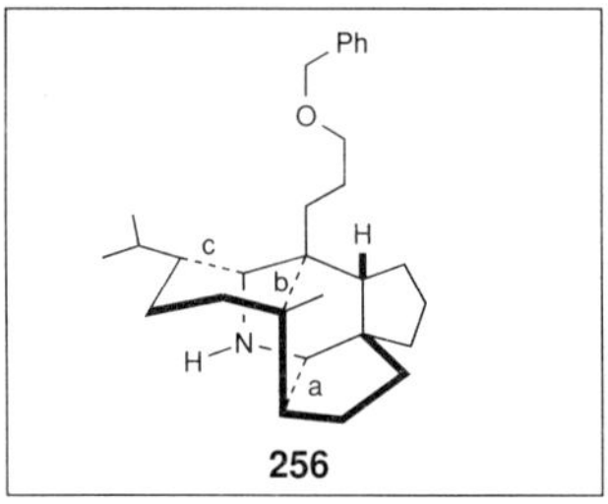

Scheme 39

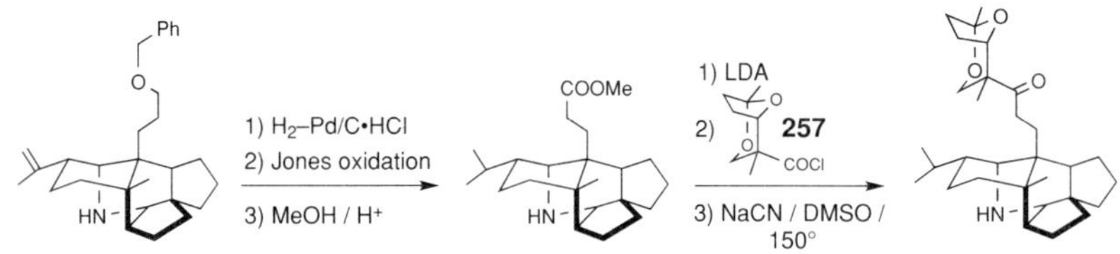

249 **244** **245**

Employing the above multiple cyclization process daphnilactone A **246**[71] has been synthesized in likewise manner, starting from **258**, as outlined in Scheme 40.

Scheme 40

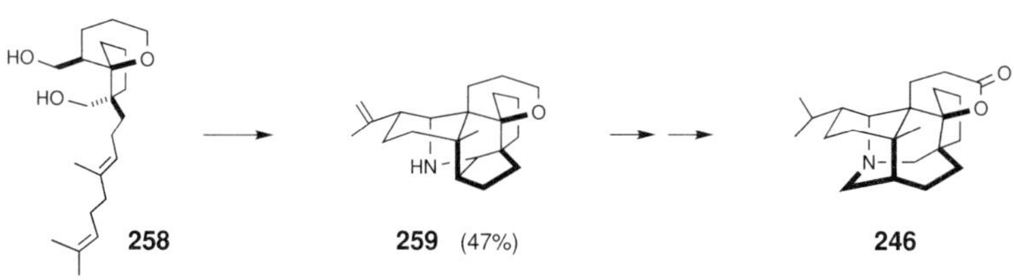

258 **259** (47%) **246**

The impact of iminium cyclization is manifested in the recent surge of interest in the total synthesis of complex alkaloid molecules, which all seem to center around the chemistry of iminium ion cyclization, regardless of whether they are purely chemical or biomimetic syntheses.[72]

IV. CYCLIZATION OF OXONIUM AND THIONIUM IONS

A. CYCLIZATION OF OXONIUM IONS

The cyclization of oxonium and thionium ions somehow has not attracted the same kind of attention as the cyclization of iminium ions. This might be interpreted in various ways, such as that there is higher interest in complex alkaloid molecules than in cyclic ethers and their sulfur analogs, or that the art of organic synthesis has reached a point where the construction of nitrogen heterocycles *via* the iminium ion has become the method of choice, while for oxygen and sulfur heterocycles many alternatives are still equally effective.

Oxonium ion cyclization can be efficiently used to construct four to eight-membered ring cyclic ethers, as shown in Equation 17. The process entails the elimination of the leaving group "X" to form oxonium ion **261** which, once generated, is attacked by the appropriately placed internal nucleophile "Y" to result in cyclic product **262**.

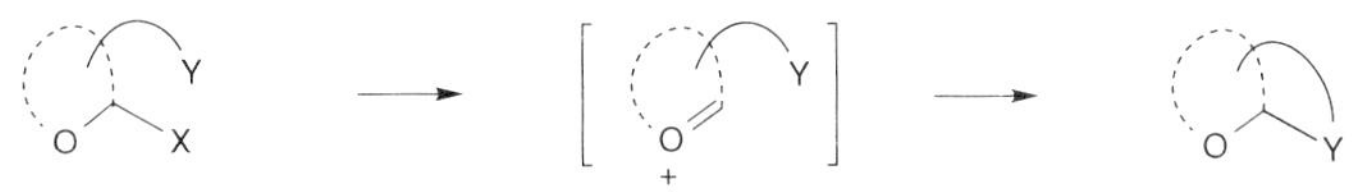

260 **261** **262**

Equation 17

The Lewis acid-catalyzed intramolecular condensation of enol silanes with acetals (Mukaiyama reaction)[73] has been known for over a decade. Exploiting the reaction, Alexakis *et at.*[74] employed titanium tetrachloride-catalyzed condensation of acetal **263** to prepare hydroazulenone **266**, thereby setting a fine example of oxonium ion cyclization in synthesis (Scheme 41).

Scheme 41

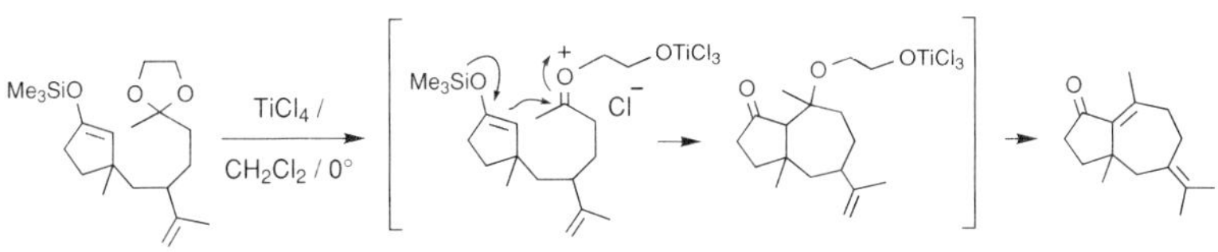

263 **264** **265** **266** (60%)

Since then the reaction has been further developed and more widely utilized in organic synthesis. All the possible modes of ring closure associated with this type of cyclization, summarized in Scheme 42, have been investigated.[75,76] *Endo* cyclization on the parts of the nucleophile (*n*) and the electrophile (*e*) are both possible, provided that "m" is sufficiently large.[77] (It should be pointed out that the cyclizations of iminium ion **143** to products **144** and **145** as diagrammed in Scheme 22 can, in principle, be classified as exo_n-$endo_e$ and $endo_n$-$endo_e$ cyclization modes, respectively.) The latest details on *endo* cyclization of oxonium ion to produce eight-membered ring oxocenes, an important unit in the polyether natural products, have just been published.[78]

Scheme 42

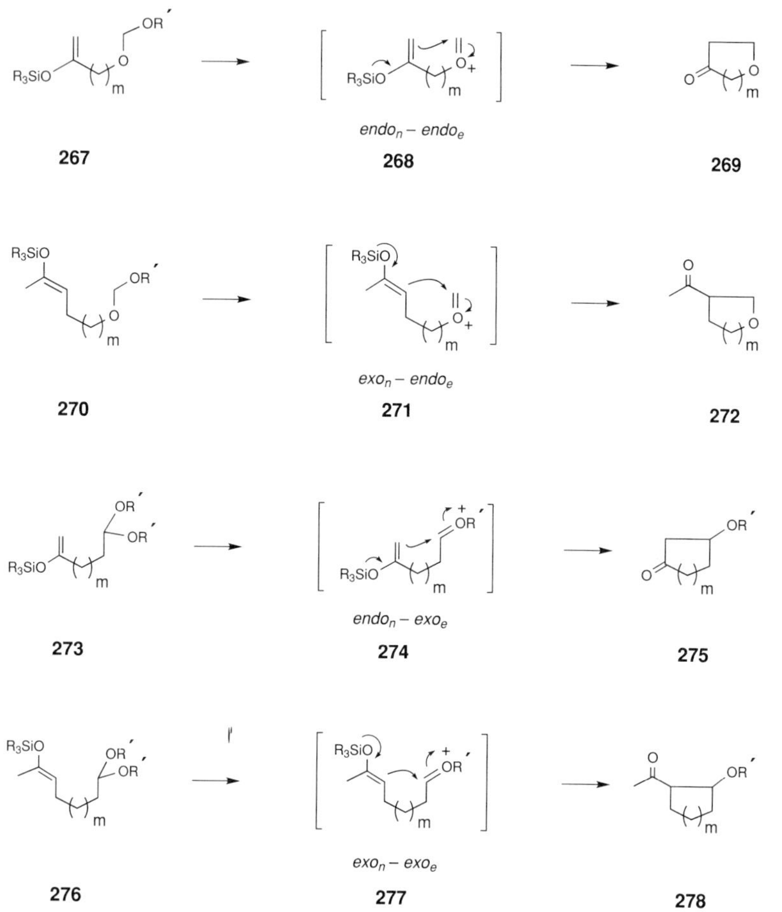

In the cyclization of (2-pyridylthio)glycoside **279**, catalyzed by silver trifluoromethanesulfonate, both the *cis*- and the *trans*-isomers were found to afford the same stereospecific product **281** in high yield,[79] indicating the intermediacy of oxonium ion **280** (Scheme 43). The reaction can also be used conveniently to construct the four-membered ring ether **283** from either the *cis*- or the *trans*-glycoside **282** in comparable yields.

Scheme 43

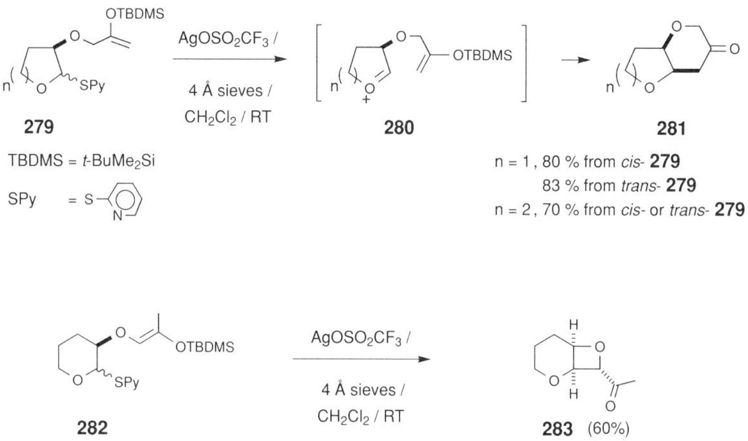

Another aspect of oxonium ion study is that which regards its disposition in rearrangement–cyclization reactions. Analogous to the iminium ion reaction depicted in Scheme 31, the generation of oxonium ion **285** from **284** results in the expected product **288** (Scheme 44). However, in contrast to the [3,3]-sigmatropic rearrangement–cyclization mechanism of the nitrogen analog,[60] the more reactive oxonium ion **285** operates *via* a different mechanism, being trapped immediately upon generation by the internal alkene to form carbonium ion **286**, which then undergoes pinacolic rearrangement to the observed tetrahydrofuran **288**.

Scheme 44

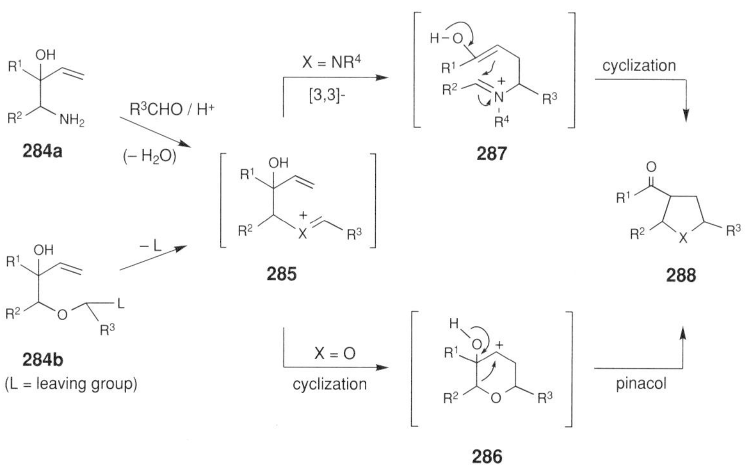

The diverging mechanisms shown in Scheme 44 are readily distinguishable upon consideration of the operating stereochemical constraints. According to the cyclization–pinacol rearrangement mechanism (**285** → **286** → **288**), reaction should occur with total retention of configuration at the carbon bearing the R^2 substituent. On the contrary, the aza-Cope–Mannich mechanism (**285** → **287** → **288**), because it involves intermediate **287**, would result in loss of stereochemical integrity of the chiral carbon. These mechanisms have been confirmed by subjecting optically active oxazolidine **289** to acid-promoted rearrangement to obtain racemic 3-acylpyrrolidine **293**,[60] and treating optically active **294** with Lewis acid (*i.e.*, $SnCl_4$) to yield, *via* **295** and **296**, optically active acyltetrahydrofuran **297** as the sole product (Scheme 45).[80,81]

Scheme 45

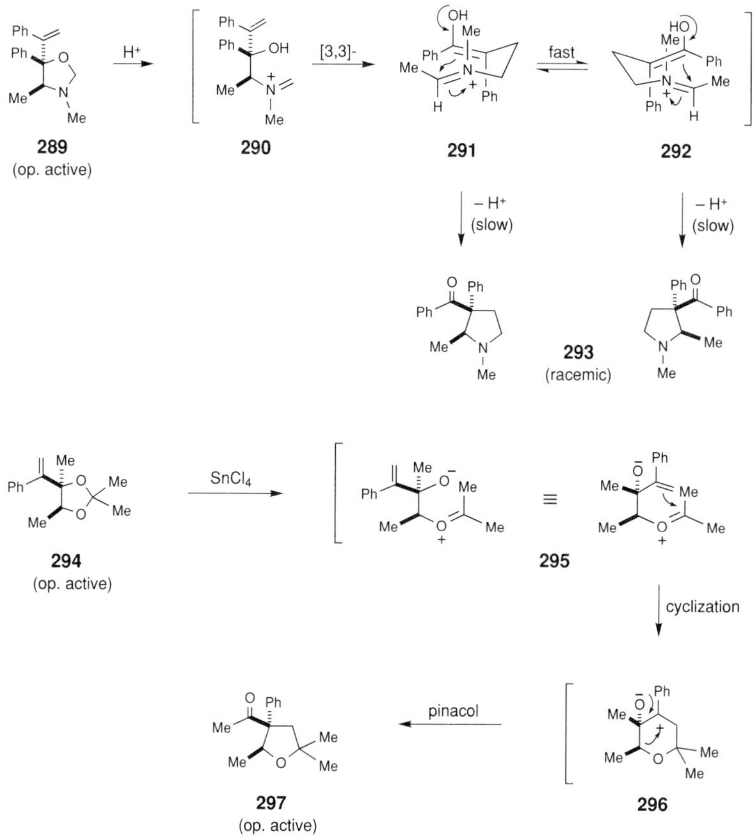

Apart from the mentioned $SnCl_4$, other Lewis acids can also induce the above cyclization–rearrangement process. The synthetic potential of the reaction is exemplified in Equations 18 and 19 which demonstrate utilization of the sequence in ring-enlarging furan annulation reactions[82,83] and in the asymmetric synthesis of (+)-muscarine tosylate **303**,[81] respectively.

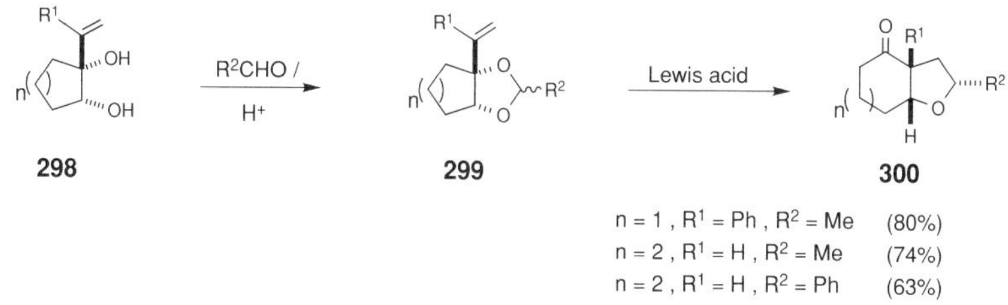

Equation 18

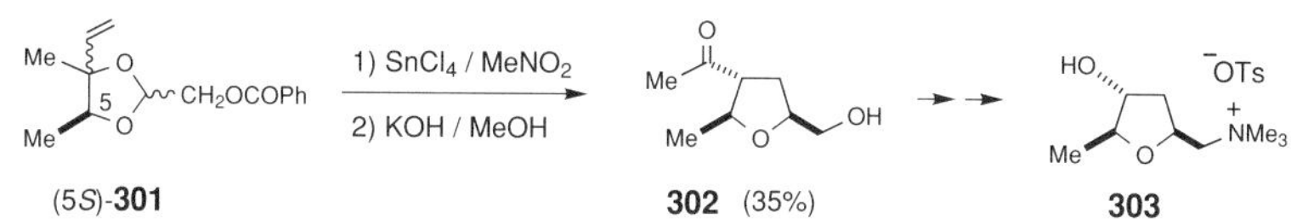

Equation 19

Recently reported related reactions are the Lewis acid-promoted ring-enlarging cyclopentane and cyclopentene annulations[84,85] shown in Scheme 46. Treatment of alkenyl acetal **305**, obtained from ketone **304**, with $SnCl_4$ effects cyclization–rearrangement to **308**,[84] while alkynyl acetals **309** and **310** yield the corresponding fused cyclopentene **311**.[85]

Scheme 46

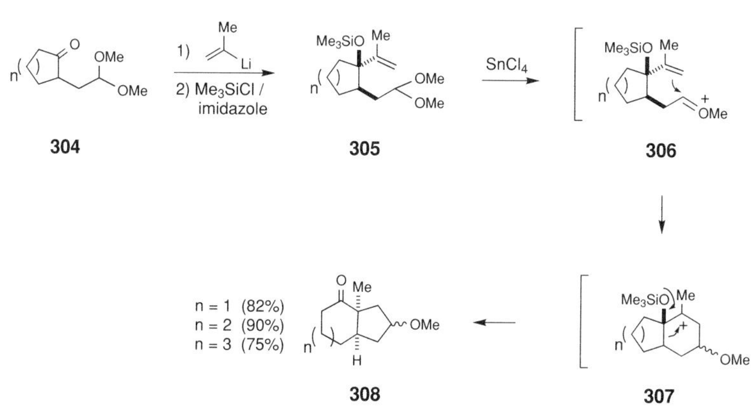

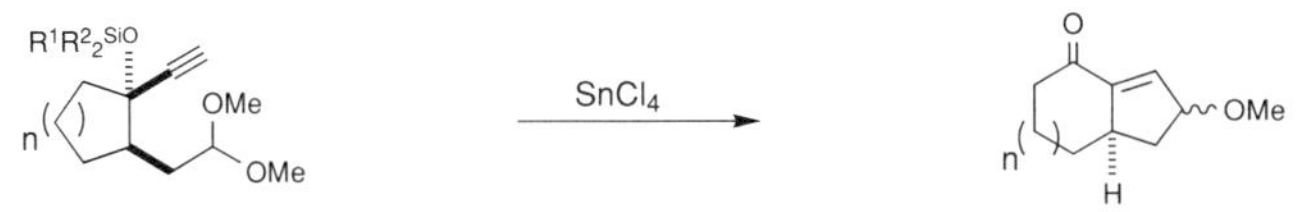

309 **311**

n = 1 , R^1 = *t*-Bu , R^2 = Me (73% ; *cis* : *trans* = 3 : 1)
n = 2 , R^1 = R^2 = Me (68% ; *cis* : *trans* = 6 : 1)

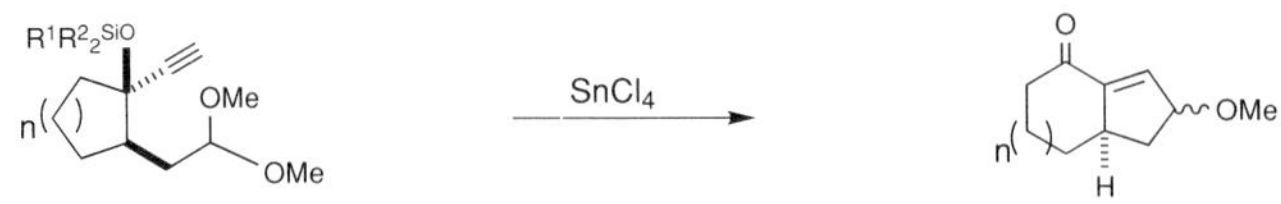

310 **311**

n = 2 , R^1 = R^2 = Me (48% ; *cis* : *trans* = 1 : 1)
n = 3 , R^1 = *t*-Bu , R^2 = Me (80% ; *cis* : *trans* = 1 : 1)

Innovative combinations of the reaction in Scheme 46 with earlier observations by Trost and Lee[86] and Sworin and Neumann[87] have proved very useful in synthesis, as illustrated in Equations 20 and 21, respectively.

312 **313**

n = 1 (93%)
n = 2 (93%)
n = 3 (64%)

Equation 20

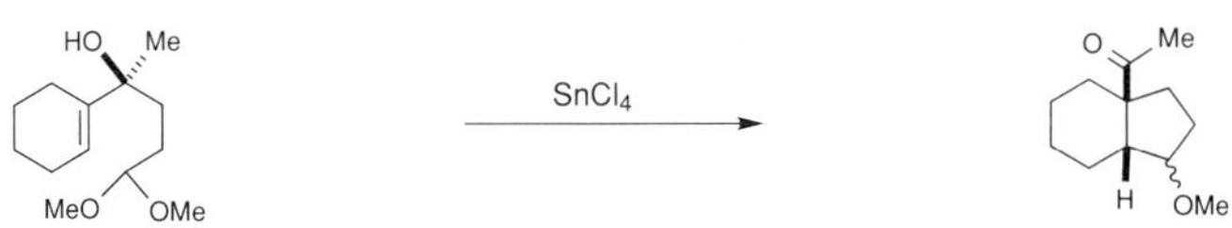

314 **315** (90%)

Equation 21

In treating the [4+2]-cycloaddition reaction and certain pericyclic reactions as reactions of ring construction (even though they may not be cyclization reactions in the true sense), it should be brought to attention that oxonium ions can very strongly promote the Diels–Alder reaction. For example, attempts to react the ethylene glycol acetal of cyclic keto allene, **316**, with 1,3-dienes result in very low yields of the cycloadducts even after prolonged heating in toluene, but, in contrast, if **316** is treated with a Lewis acid catalyst ($BF_3 \cdot OEt_2$ being the most suitable), the oxonium ion formed behaves as a forceful dienophile to drive the required cycloaddition with 1,3-diene to yield the cycloadduct **319**. Scheme 47 and the accompanying table illustrate this powerful influence of oxygen-stabilized carbocation on the adjacent allene moiety.[88]

Scheme 47

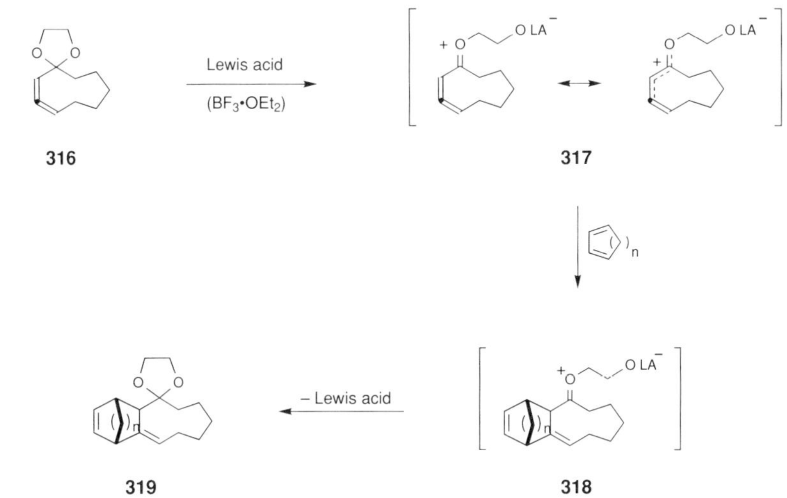

Diene	Cycloadduct	Temperature (°C)	Time (h)	% Yield
	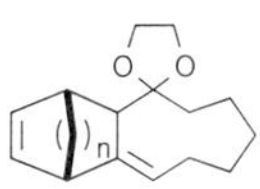 **319**			
n = 1		–78	0.5	65
n = 2		–23	2.5	62
n = 3		–10	2.5	50

A study of the related oxo-stabilized allyl cation **321** has also been reported recently (Scheme 48).[89] Here the precursor acrolein acetal **320**, upon treatment with $BF_3 \cdot OEt_2$, is found to undergo cycloaddition with di-, tri-, and tetrasubstituted olefins to afford cyclobutanes in moderate yields. The mechanism has been proposed as stepwise, proceeding through intermediate cations **321**, **323**, and **324** to finally yield cyclobutanes of various structures, **325a–d**, according to the olefin being employed.

Scheme 48

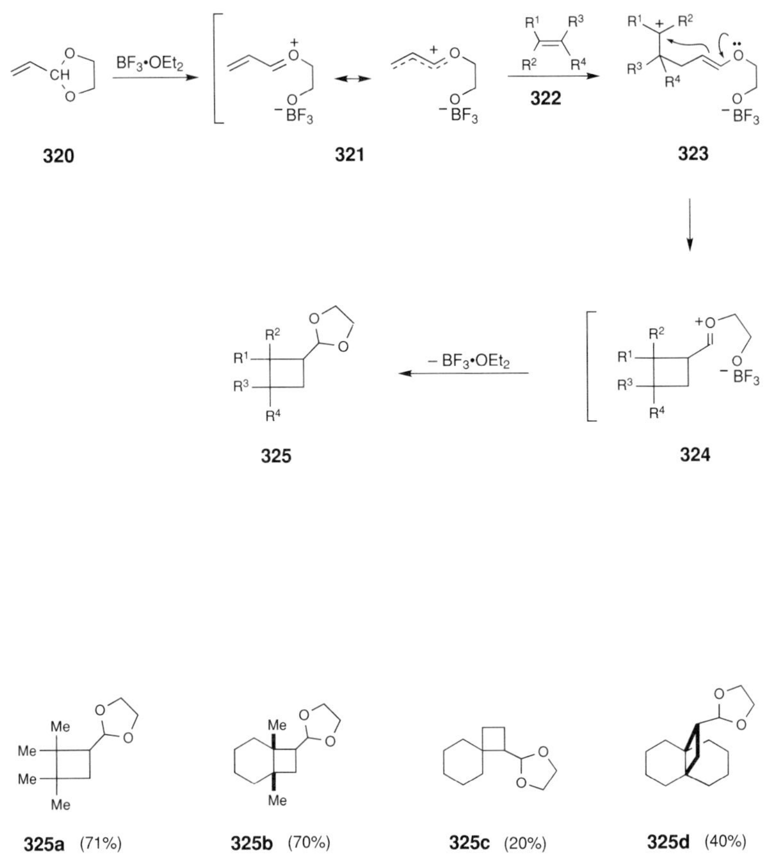

B. CYCLIZATION OF THIONIUM IONS

Although the last few years have seen few applications of thionium ion cyclization in organic synthesis, nevertheless the chemistry of thionium ion intermediates, particularly in the Pummerer reaction, has been explored by several research groups. Significant examples include the olefin cyclization to produce bicyclic ketone **329** by treating keto sulfoxide **326** with trifluoroacetic anhydride[90] (Scheme 49) and the reaction of sulfoxide **331** with a stoichiometric amount of trifluoroacetic anhydride in methylene chloride at room temperature to afford **334** in 87% yield *via* sulfonium and carbonium intermediates **332** and **333**. This last reaction constitutes the key step in the synthesis of optically active trachelanthamidine **335** from prolinol **330** (Scheme 50).[91]

Scheme 49

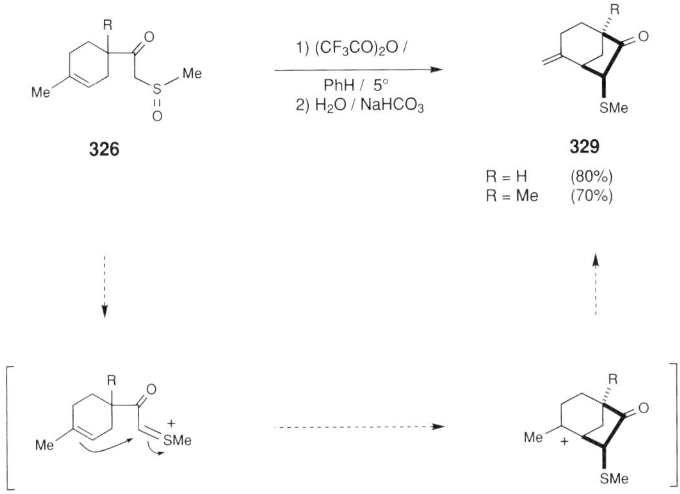

326 **327** **328** **329**

Scheme 50

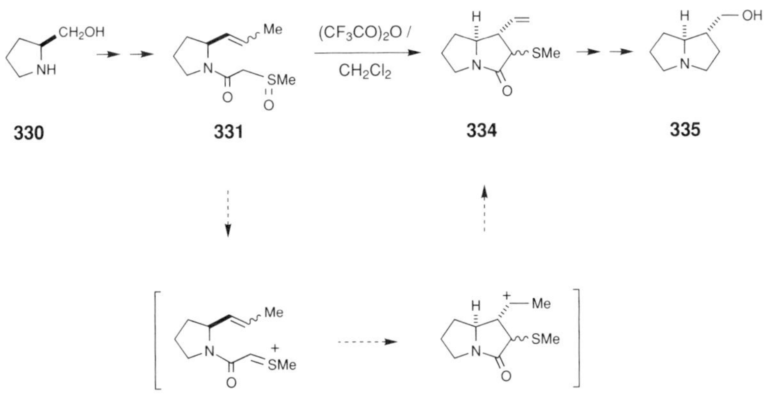

330 **331** **334** **335** **332** **333**

Magnus and his group have employed intramolecular Pummerer reactions in the total synthesis of *Aspidosperma* alkaloids,[92–94] as shown in Scheme 51. Cyclization takes place *via* the Pummerer intermediates of indoles **336** and **339** to yield the cyclized products **338**[92] and **340**,[93,94] respectively.

Scheme 51

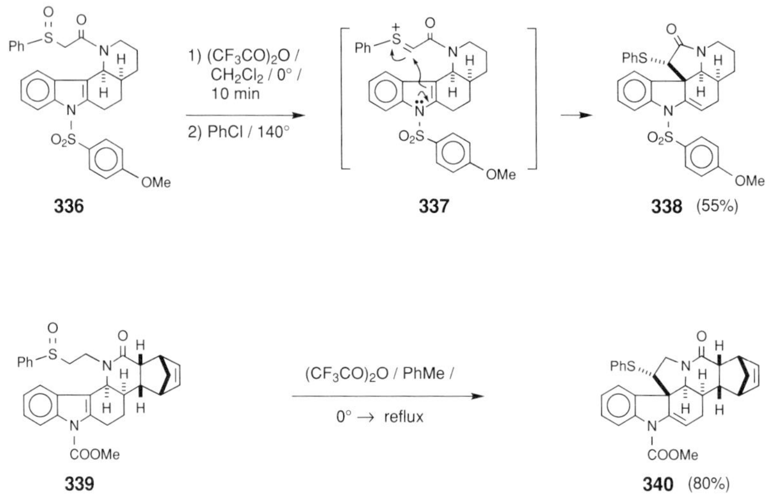

The use of trifluoroacetic anhydride to generate the Pummerer intermediate is, however, not without limitations. For instance, when 3-(phenylsulfinyl) propionamide **341** is subjected to the above-described Pummerer rearrangement conditions,[92] the reaction yields **344** instead of the cyclized product, indicating failure to form the Pummerer intermediate. However, upon substituting trimethylsilyl trifluoromethanesulfonate (TMSOTf), coupled with triethylamine, for trifluoroacetic anhydride, the required β-lactam **347** is successfully obtained.[95] It could well be that, in the presence of a hindered base, TMSOTf converts sulfoxide **345** to the Pummerer intermediate with simultaneous silylation of the amide to form the silyl imidate **346**, thus facilitating cyclization to **347** (Scheme 52).

Scheme 52

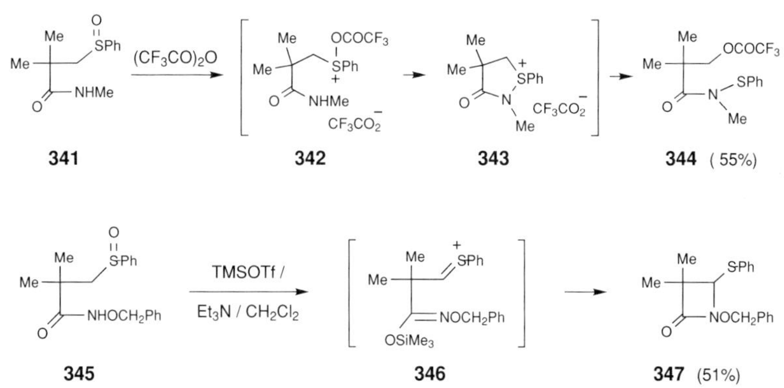

Besides the above methods, other modes of thionium ion generation are also available. Scheme 53 shows the reaction of sulfoxides **348** with silylketene acetal **349** in the presence of a catalytic amount of zinc iodide[96] to afford cyclized products **352** in yields ranging from moderate to excellent.[97] The reaction has been applied to the synthesis of heterocycles **354** and **357** (Equations 22 and 23, respectively).

Scheme 53

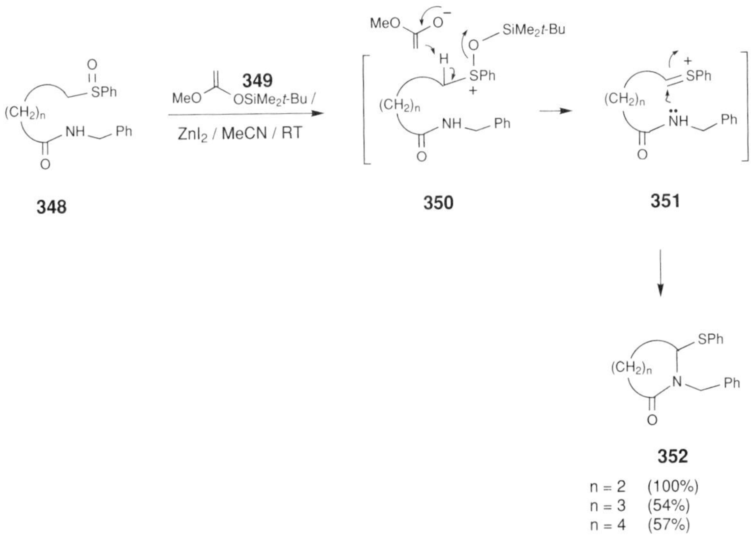

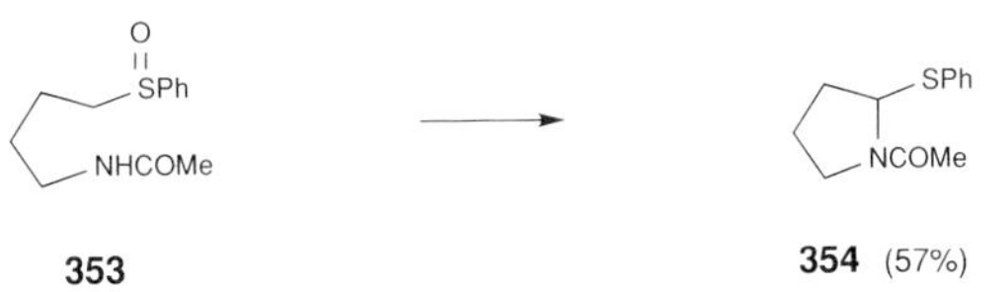

Equation 22

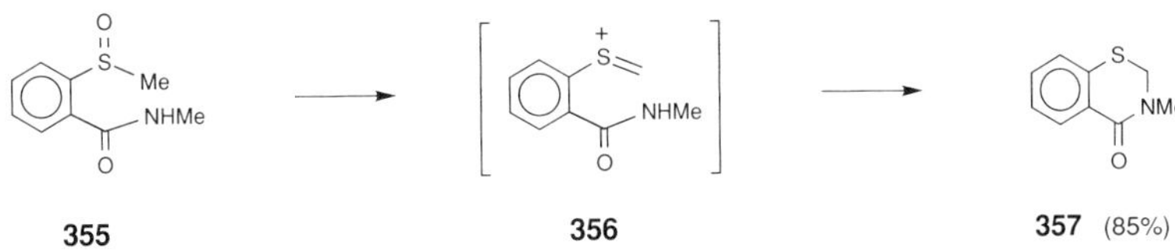

Equation 23

V. CYCLIZATION OF VINYL-, ALLYL-, PROPARGYL- AND ALLENYLSILANES

One of the currently most popular and widely used reactions in organic synthesis is the cyclization of silanes. As stated earlier in the Introduction, silane cyclizations with, or with a tendency towards, cationic character will be treated in this section, whereas reactions displaying anionic behavior will be discussed in Chapter 4. Thus cyclizations of silanes, particularly those involving allylsilanes, will also be found in the chapter entitled "Anionic Cyclization."

Due to their immense popularity and synthetic potential the silane cyclizations have been reviewed frequently,[98–101] but post-review reports are fast appearing and applications of the reaction seem to be quite endless.

A. CYCLIZATION OF VINYLSILANES

In general, the reaction of vinylsilanes follows the general pathways shown in Equations 1 and 2. The commonly observed regiospecificity associated with the reaction results from the well-known "β-effect" of silicon.[102] In actual fact silicon interacts with the positively charged carbon inclusively at the α-, β-, ,γ- and δ-positions, but the β-effect is by far the most influential.[103] Nevertheless, a few reactions have been found to exhibit unusual regiospecificity, as will be shown in later examples. Electrophiles employed have included iminium ions, oxonium ions, activated carbonyl functions, *etc.*[99,101]

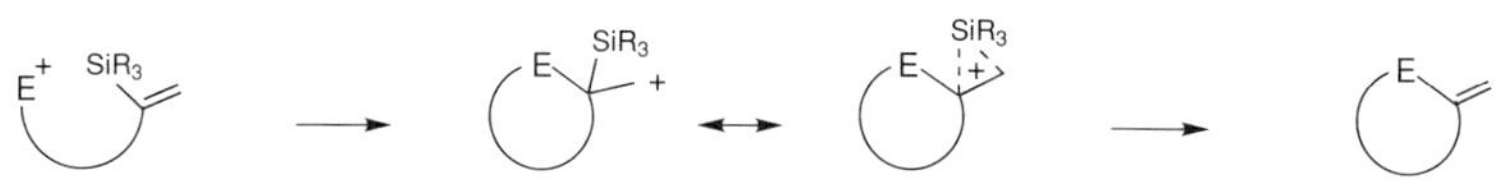

Equation 1

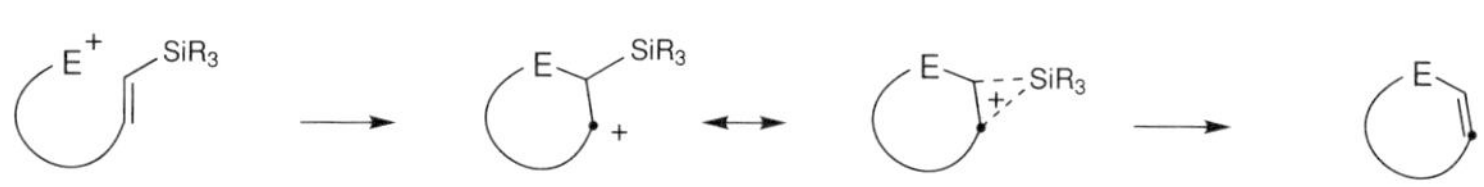

Equation 2

Scheme 54[104] and Equations 24 to 26[105] show examples of iminium ion-induced intramolecular cyclizations of vinylsilanes. Two important features of these reactions are apparent: first, all are regiospecific and proceed *via* the carbocation β- to the silane function, *i.e. via* intermediate **360** in Scheme 54, and second, the cyclization can occur either in a stereospecific manner, as in the syntheses of antibiotic streptazolin **363** (from tartrate-derived iminium ion **359**, Scheme 54),[104] geissoschizine **367** (Equation 24),[105] and isositsirikine **370**

(Equation 25),[105] or render no asymmetric induction at all (Equation 26),[104] conceivably due to the geometry of the iminium ion **372** involved.

Scheme 54

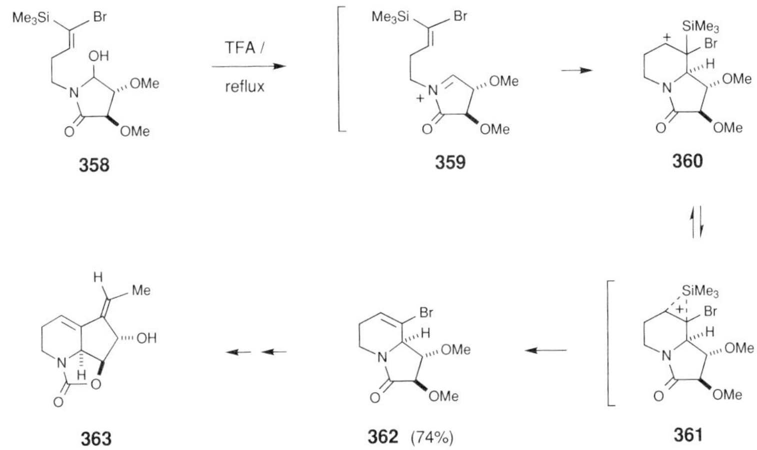

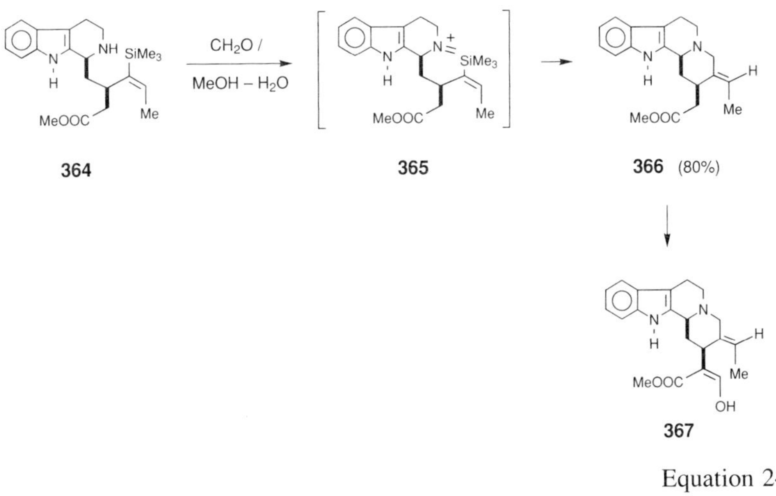

Equation 24

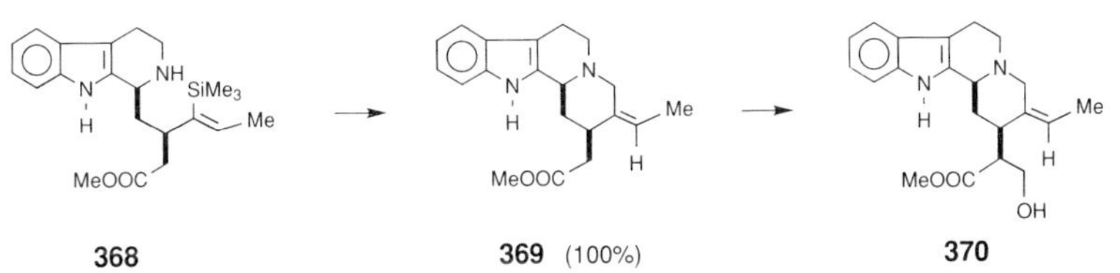

Equation 25

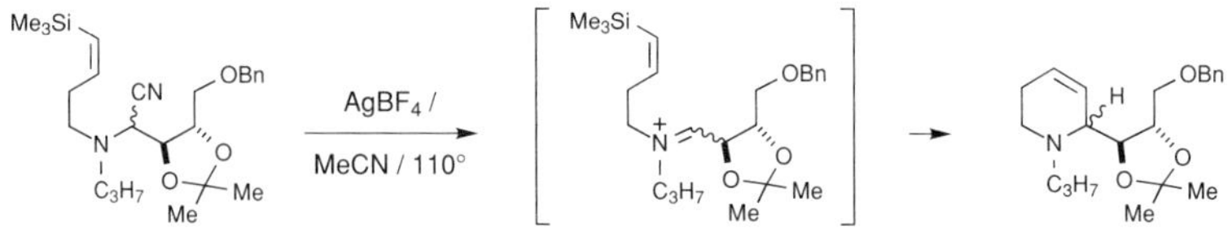

371 **372** **373** (63%)

Equation 26

Reported vinylsilane–oxonium ion cyclizations, on the other hand, seem to display an opposite regiochemistry to that of their nitrogen analogs. Strangely, here the α-silyl cation appears to triumph over its β-counterpart. This is evident in several examples, *e.g.*, the Lewis acid-catalyzed cyclization of mixed acetals **374**, **375,** and α-chloro ether **376**, which all lead to eight-membered ring oxocenes **379** in low yields.[78] Rather than proceeding *via* the β-silyl cation **380** to a seven-membered ring product, the reaction seems to prefer the route *via* α-silyl cation **378** (Scheme 55). Similarly, the reaction of **381** with $SnCl_4$ in methylene chloride followed by *O*-desilylation with fluoride ion produces **382** in 37% purified yield (Equation 27).[106]

Scheme 55

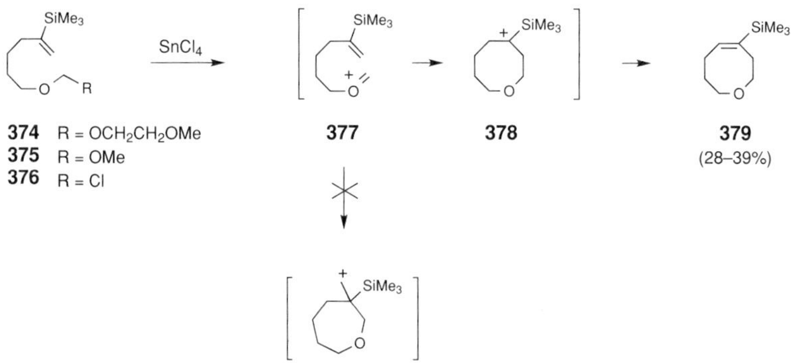

374 R = OCH_2CH_2OMe
375 R = OMe
376 R = Cl

377 **378** **379** (28–39%)

380

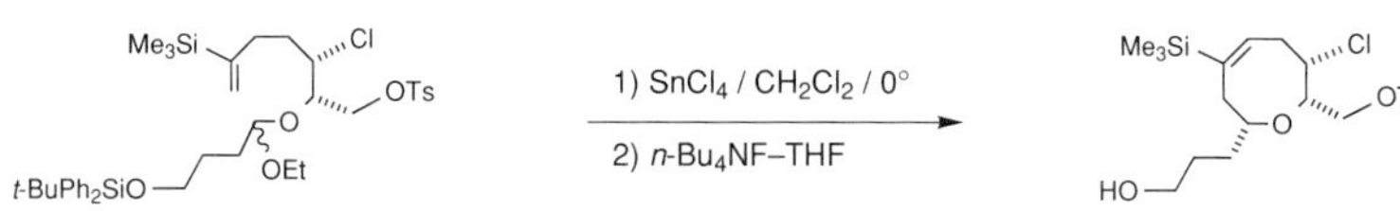

381 **382** (37%)

Equation 27

Also synthetically useful is the intramolecular acylation reaction of vinylsilanes. Examples of these have been reported since 1983 and are shown in Scheme 56.[107] The reaction constitutes a remarkably short and convenient synthesis of cyclopentenones.

Scheme 56

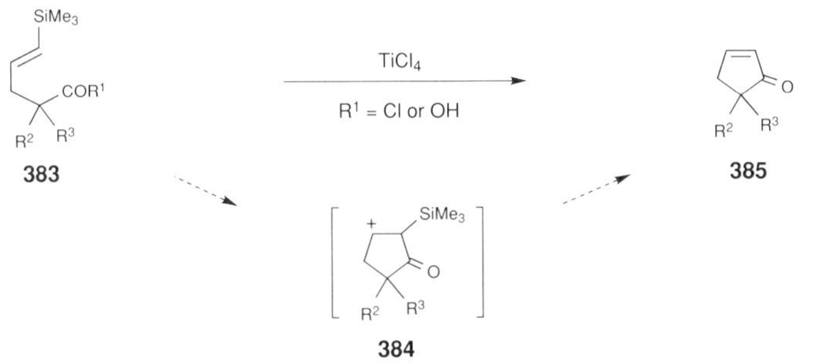

$R^2 - R^3 = n\text{-}C_5H_{10}$ (65–70%)
$R^2 = H, R^3 = n\text{-}C_6H_{13}$ (79%)
$R^2 = H, R^3 = CH_2 = CH\text{–}(CH_2)_6\text{–}CH_2\text{–}$ (60%)
$R^2 = CH_2 = CH\text{–}CH_2\text{–}, R^3 = n\text{-}C_6H_{13}$ (69%)

The above acylations follow the endocyclic cyclization pathway of vinylsilanes depicted in Equation 2. More recently, acyl cyclizations proceeding through exocyclic carbocations, equivalent to the pathway shown in Equation 1, have also been reported (Scheme 57).[108] From the latter study it became evident that regiospecificity of the vinylsilane cyclization is in fact influenced by a number of factors, and is not simply the result of the β-effect alone. Thus, contingent upon variables such as chain length and nature of substituent R^1, the cyclization of vinylsilane **386** can proceed either along route a *via* the β-silyl cation **387** (α-cyclization), or along route b *via* the α-silyl cation **389** (β-cyclization). When R^1 = alkyl the secondary β-silyl carbocation prevails (route a, entry b), whereas when R^1 = H the would-be primary β-silyl carbocation is rejected in favor of the tertiary α-silyl carbocation (route b) which may experience migration of group R^2 to give intermediate **390** before undergoing elimination and eventual double bond migration to yield the final product **392** (entries c, d). With a shorter acyl chain, however, the strain of cyclobutenone ring formation can compete with electronic inducements and, as in entry a, steer the cyclization to form the more stable cyclopentenone major product *via* the less favorable α-silyl cation. In retrospect, the above reasonings can equally well be applied to the foregoing results from iminium and oxonium ion-induced cyclizations of vinylsilanes. However, more information will be needed (such as the effect of the nature and position of substituents) before the reaction can be fully understood.

Scheme 57

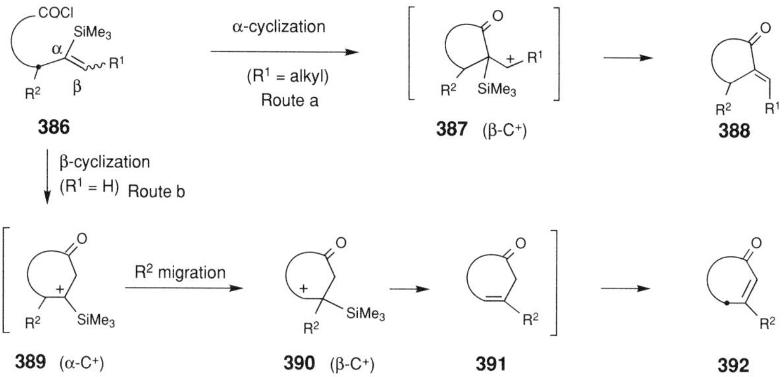

Entry

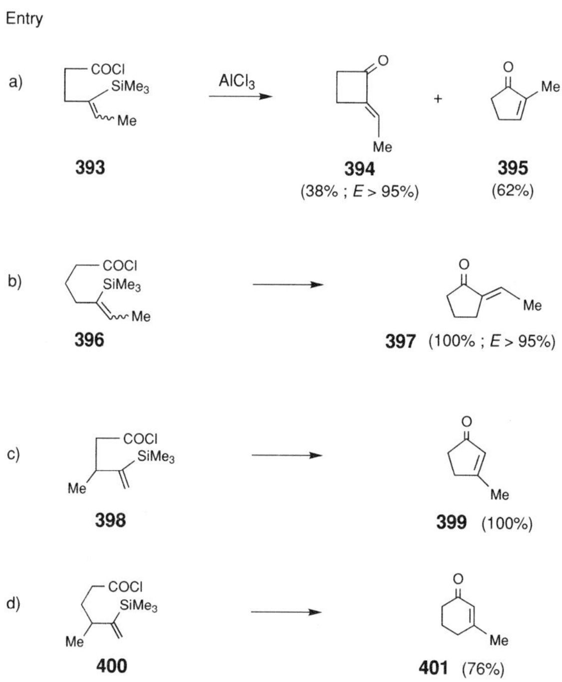

A very interesting reaction has been observed between selenovinylsilane **402** and enone **403** in the presence of $SnCl_4$ catalyst. The reaction mixture, upon quenching with triethylamine, quite unexpectedly furnished cyclopropane **405** instead of the anticipated cyclobutane **406** (Scheme 58).[109] The reaction has been rationalized in terms of the addition of vinylsilane **402** to the Michael acceptor **403** to give, initially, the zwitterionic intermediate **404** which is stabilized by the adjacent C–Si bond, as shown. Simultaneous cyclization–trimethylsilyl migration directly leads to product **405**.[109] The driving force for this unprecedented 1,2-silicon shift–cyclopropanation process probably comes from the ease of formation of cyclopropane as compared to cyclobutane. Advantage has already been taken of the reaction; product **405a** (R = Me) has been utilized as the starting material in the synthesis of the non-head-to-tail monoterpene rothrockene **407**, a constituent of the essential oil of *Artemesia tridentata rothrockii*.[110]

Scheme 58

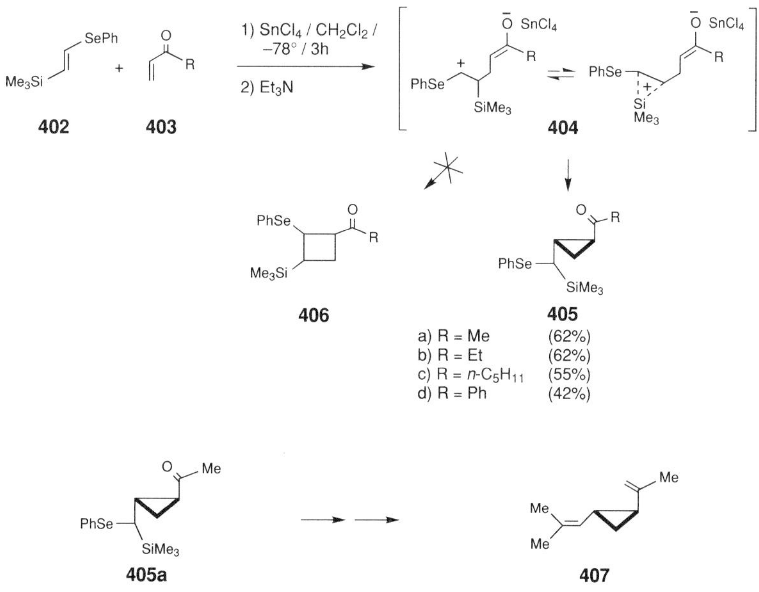

B. CYCLIZATION OF ALLYLSILANES

Allylsilanes constitute a class of nucleophiles which exhibits excellent reactivity towards electrophiles. These nucleophiles enjoy wide use in organic synthesis and their chemistry has received constant reviews.[98,100,101] The general reactions of allylsilanes can be summarized as shown in Equations 28 and 29.

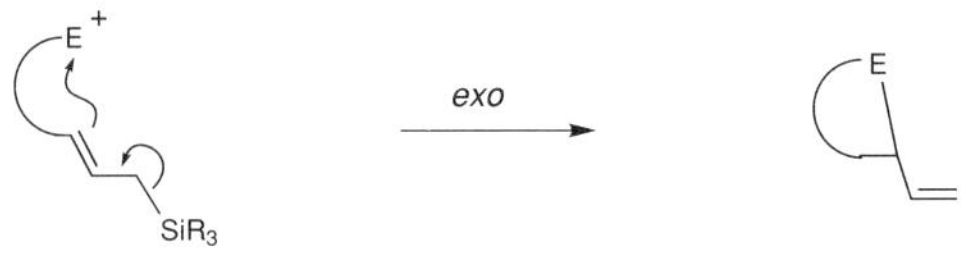

Equation 28

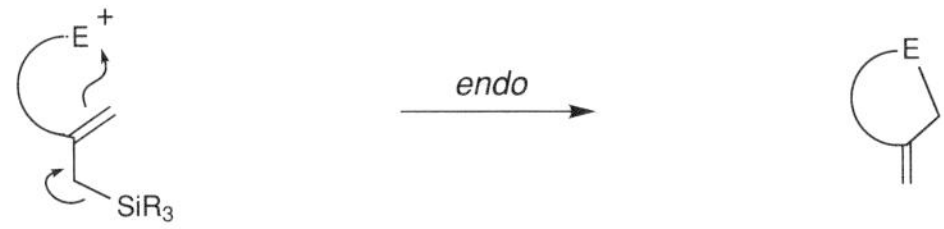

Equation 29

A popular approach to the construction of nitrogen heterocycles makes use of the intramolecular condensation of iminium ions, generated *in situ*, with allylsilanes. Examples illustrating the exo_n–$endo_e$ and $endo_n$–$endo_e$ modes of cyclization are presented in Schemes 59 and 60 respectively (see Scheme 42). The reaction proceeds well in all cases to give the respective products **410** and **413** in good yields, except where n = 1 in **412**, when strain in the $endo_n$–$endo_e$ transition state precludes the formation of a five-membered ring.[111]

Scheme 59

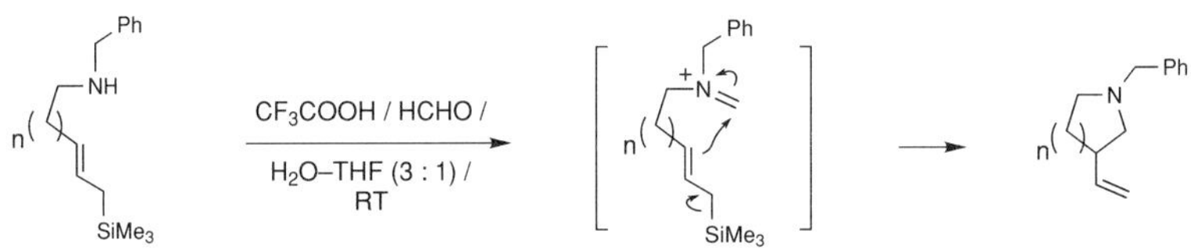

408 **409** **410**

n = 1 (81%)
n = 2 (94%)

Scheme 60

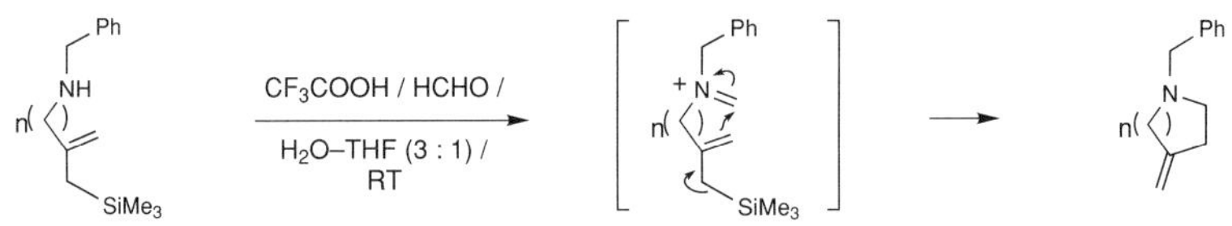

411 **412** **413**

n = 1 (0%)
n = 2 (73%)
n = 3 (96%)

Application of the intramolecular cyclization between iminium ion and allylsilane is seen in the synthesis of (±)-myrtine **419**, a quinolizidine alkaloid from *Vaccinium myrtillus*,[112] and its stereoisomer (±)-epimyrtine **420**. Here a diastereomeric mixture **415**, obtained from borohydride reduction of imide **414**, provides the substrate for cyclization. Treatment of **415** with trifluoroacetic acid in methylene chloride at 0°C generates the acyliminium ion **416** which cyclizes, probably instantaneously, to yield two separable isomeric products **417** and **418** (96%, ratio 7:3). Simple chemical manipulations performed on **417** and **418** lead to (±)-myrtine and (±)-epimyrtine, respectively. The noted superior yield of the *cis*-(**417**) over *trans*-(**418**) cyclization product has been ascribed to the less hindered axial methyl arrangement in transition state A as compared to the cumbersome crowding between the equatorial methyl and the carbonyl function in transition state B (Scheme 61).[113]

Scheme 61

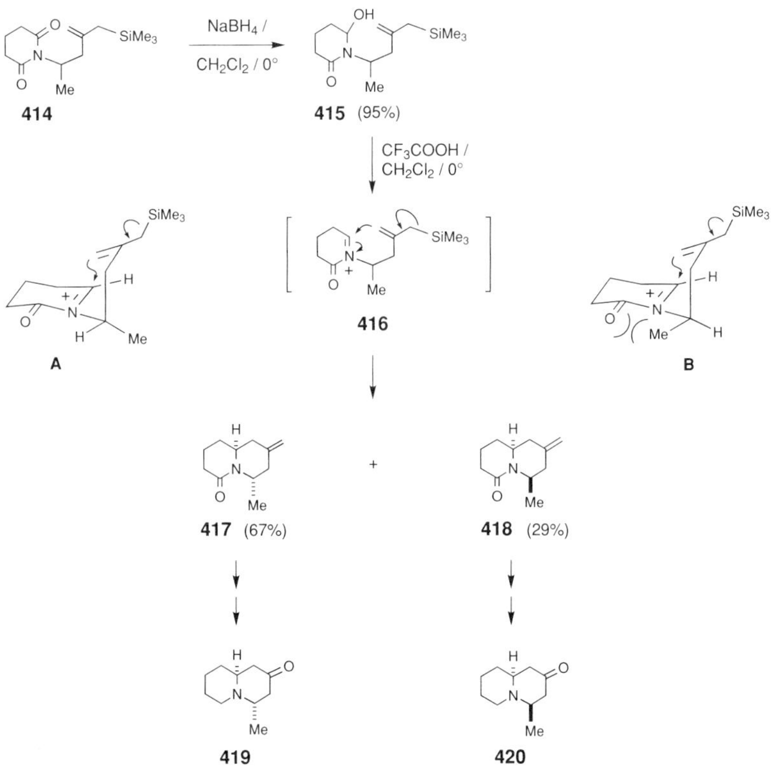

In the total synthesis of (±)-α-allokainic acid **430**,[114] shown in Scheme 62, the nucleophilicity of allylsilane is twice demonstrated, first in the intermolecular substitution reaction between **421** (X = OMe or Cl) and silane **422** to yield **423**, and again in the cyclization of substrate **426** to **429** *via* intermediates **427** and **428**. Briefly, product **423** from the first step of synthesis is obtained as a mixture of isomers and, without further separation, is subjected to lactonization and subsequent *N*-methoxymethylation to yield *trans*-amino lactone **424** as the sole product. This is then converted to the precursor for cyclization **426**. As stated above, the cyclization of **426** under $BF_3 \cdot OEt_2$ catalysis furnishes **429**, from which α-allokainic acid **430** is then easily obtained.

Scheme 62

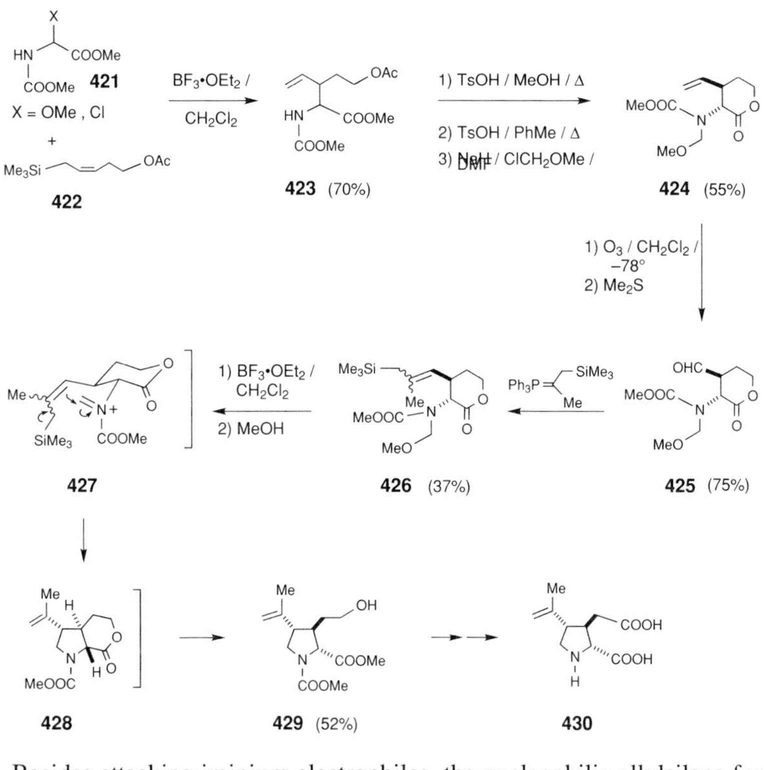

Besides attacking iminium electrophiles, the nucleophilic allylsilane function is also able to intercept the carbocationic intermediate formed in the Beckmann rearrangement process, trapping it in an intramolecular cyclization reaction of a novel kind. For example, the *Z*-oxime mesylates **431** and **435**, upon treatment with diisobutylaluminum hydride (DIBAL-H) in methylene chloride, at low temperature, yield bicyclic nitrogen compounds **434** and **438,** respectively (Scheme 63). The mechanism[115] has been proposed to involve stereoselective Beckmann rearrangement, brought on by the mild Lewis acid DIBAL-H, which produces intermediates **432** and **436**. These are trapped in the respective *exo* and *endo* manner by the allylsilane nucleophile (see Equations 28 and 29). The resulting bicyclic imines **433** and **437** then suffer stereoselective reduction by DIBAL-H to the observed products.

Scheme 63

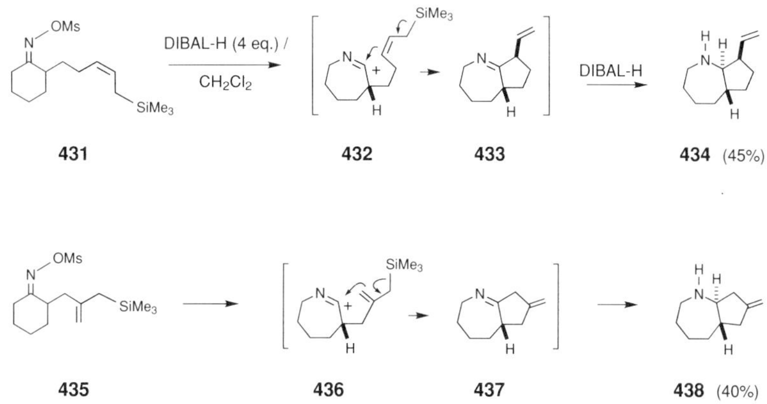

Analogous treatment of the isomeric *E*-oxime mesylates **439a** and **443a**, on the other hand, produces only decomposition of the starting materials. Nonetheless, under the same conditions the parent *E*-oximes **439b** and **443b** yield 6,6-fused quinoline derivatives **442** and **446,** respectively (Scheme 64).[115] In the same manner as the Beckmann rearrangement, this reaction is thought to proceed *via* an initial polarization of the N–O bond, induced by DIBAL-H, to first form the polarized species **440** and **444**. Thereafter, however, the similarity ends as the strategically placed and more powerful allylsilane nucleophile competes successfully with the potential Beckmann nucleophile (alkyl ring) in displacing the leaving group on the electrophilic nitrogen. Thus, ring-enlarging rearrangement is suppressed in favor of cyclization to bicyclic nitrogen compounds **441** and **445**, which are further reduced to the observed products. It should be noted that, although the stereochemistry of the ethenyl groups in **434** and **442** has not yet been confirmed, each product is the sole isomer obtained in each case.

Scheme 64

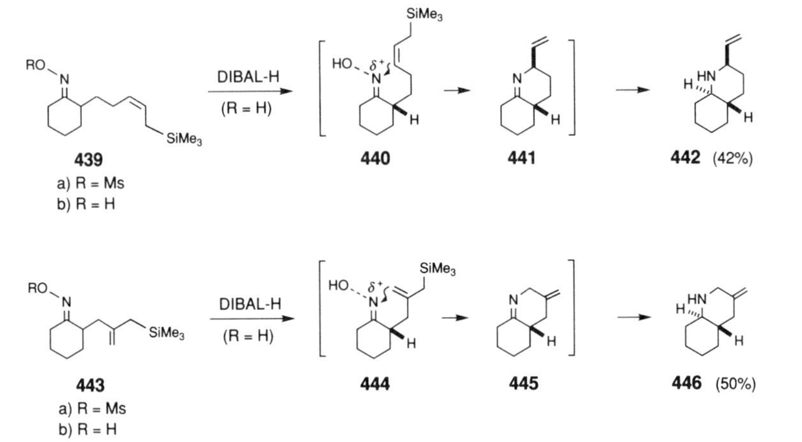

Allylsilane cyclization is also useful for the synthesis of spirocompounds, including spiroethers and spiroketals. The example in Equation 30 demonstrates a facile preparation of 5,6-spirohydrocarbon **449**[116] by heating hydroxyallylsilane **447** in benzene containing a stoichiometric amount of *p*-toluenesulfonic acid, which accomplishes construction of the desired spiroskeleton. Subsequent removal of the sulfonyl group furnishes product **449** as a mixture of two diastereoisomers.

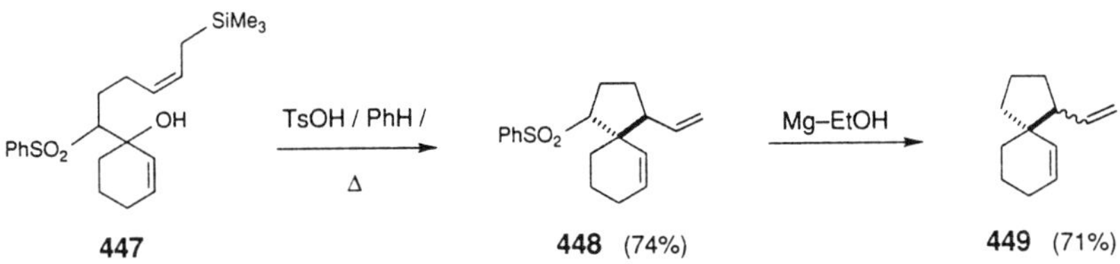

Equation 30

The enterprising application of allylsilane cyclization to the synthesis of spiro-oxygen ring compounds owes its development to the efficient and versatile three-component, one-pot reaction of carbonyl compounds with trimethylsilyl ethers and allyltrimethylsilane, which yields homoallylic ethers **454** as shown in Scheme 65.[117] The intramolecular version of this reaction has led to the convenient synthesis of tetrahydropyrans **459**, spiroether **461,** and spiroketals **463,**[118,119] as depicted in Scheme 66. The ingenuity of this approach lies in the use of **455**, readily prepared from the corresponding commercially available alcohol. Compound **455** is, in fact, equivalent to a preformed adduct between silyl ether **451** and allylsilane **452**. The reaction of **455** with carbonyl compounds to yield tetrahydrofuran **457** is catalyzed by trimethylsilyl triflate and proceeds *via* oxonium ion intermediate **456**. In certain cases the product **457a** may also be formed under these reaction conditions due to isomerization of the exocyclic double bond in **457** to the endocyclic position.[118]

Scheme 65

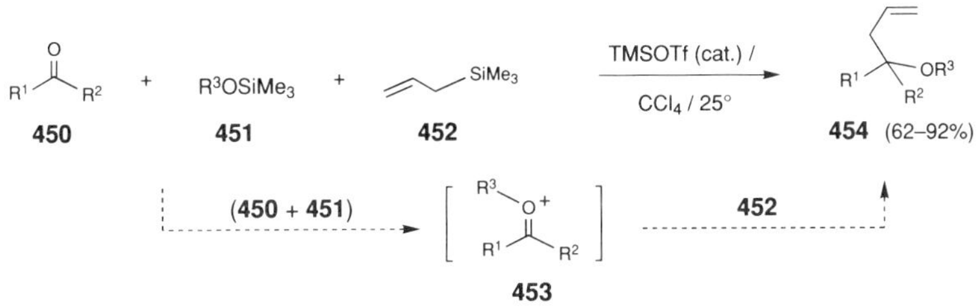

Scheme 66

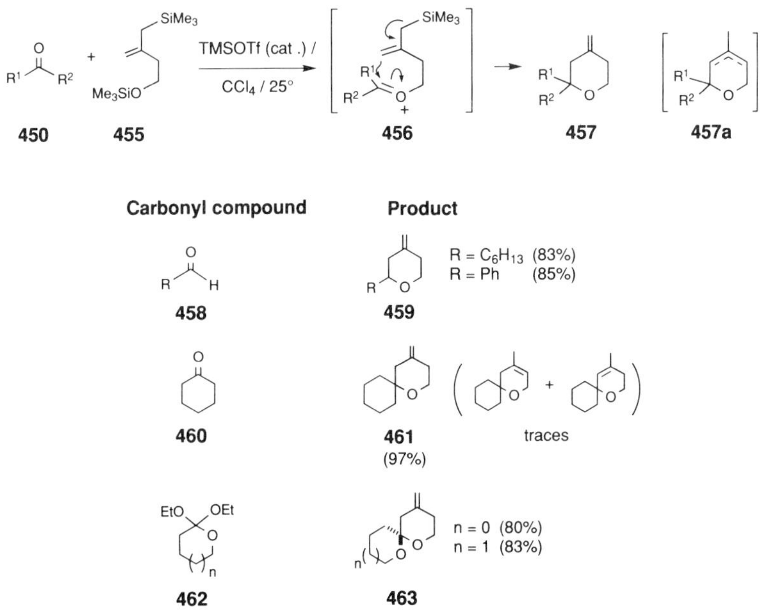

Upon irradiating a mixture of enamine carbaldehyde **464** and acrylate **465**, a fast photochemical reaction of unknown mechanism ensues to yield cyclic product **466**. When treated with acid compound **466** dehydrates, presumably *via* iminium ion **467**, to give **468**.[120] Taking advantage of this observation an allylsilane side chain was incorporated in an attempt to utilize the allylsilyl iminium intermediate in intramolecular cyclization. Indeed, subjecting allylsilane **469** to the photochemical reaction gave **470**, which, without further purification, directly yielded the bicyclic product **472** upon treatment with Lewis acid (Scheme 67). Quite noteworthy is the high diastereoselectivity of the reaction; from several possible diastereoisomers, only one, **472**, was isolated. Intermediate **471** has been proposed[120] to account for the stereochemistry of **472**, although it is still not clear whether the stereochemistry of **471** itself is determined in the photolysis reaction between **464** and **469**, or results from the iminium–enamine isomerization after the formation of **470**. Other significant features of the process include the persistent presence of by-products **473** and **474** in every case, and the superior performance of CF_3COOH and $SnCl_4$ over $TiCl_4$ with regard to the yield of products.

Scheme 67

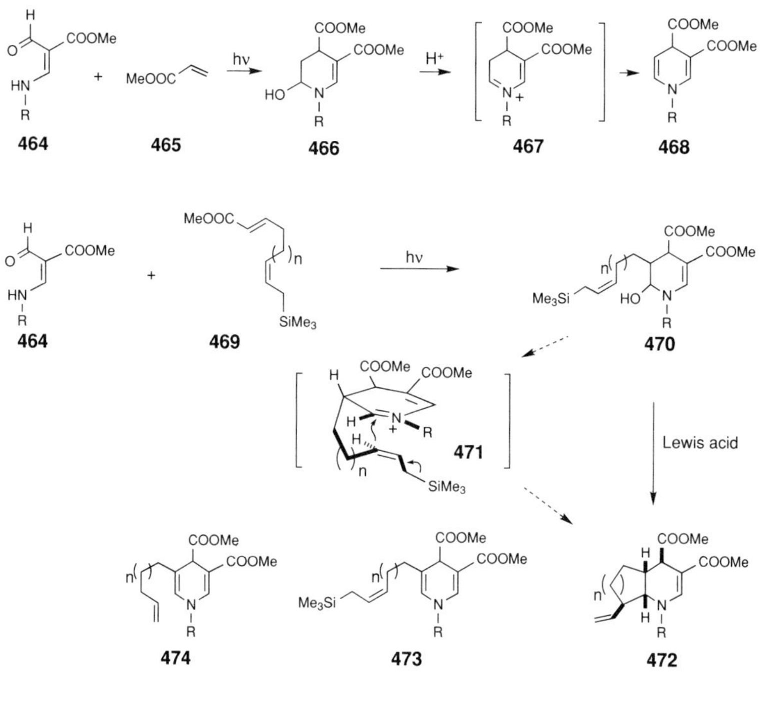

Lewis acid = CF_3COOH, TMSOTf, Me_2AlCl, $TiCl_4$ or $SnCl_4$
R = H, *t*-Bu or 4-OMe-C_6H_4
n = 1 or 2

Yield of **472** from **469** = 9–51%
Yield of **473** + **474** = 10–19%

In connection with allylsilyl iminium salts, mention should also be made of the photocyclization of iminium salt **475** to the spiroproduct **476** shown in Equation 31.[121,122] The mechanism of this reaction,[123] elucidated with the aid of deuterium-labeled starting material **477**,[124] is now proved to be dual radical and radical cation according to paths a and b, respectively, in Scheme 68.

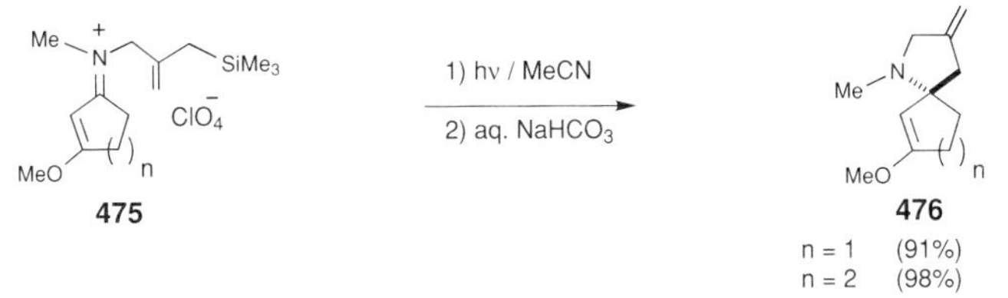

Equation 31

Scheme 68

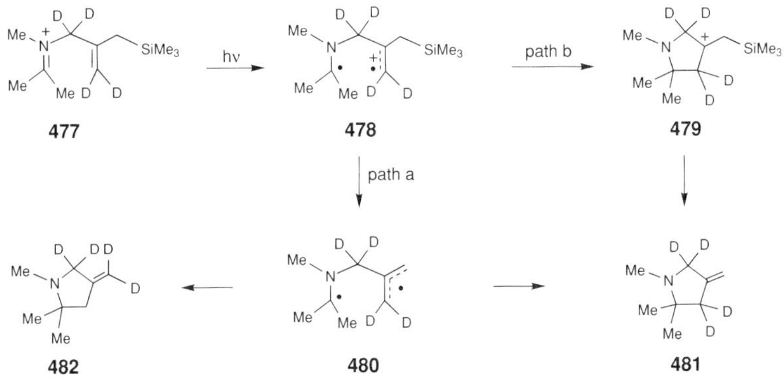

C. CYCLIZATION OF PROPARGYL- AND ALLENYLSILANES

In parallel with the vinyl- and allylsilanes, the chemistry of propargyl- and allenylsilanes has been studied and employed in various syntheses. Several reviews[99,100,125] have been written on the important reactions of these compounds. In summary, the cyclization of propargyl- and allenylsilanes can be represented as shown in Equations 32 and 33, respectively.

Although related to the foregoing cyclization of allylsilane, the intramolecular cyclization of propargylsilane (Equation 32) nevertheless differs in its production of a cyclic product **484** possessing a reactive exocyclic allene group, which functional component conveniently allows for further chemical transformations.

In the case of allenylsilanes reaction takes place as shown in Equation 33. Here silicon has the dual role of first dictating the regiospecific addition of **486** to electron acceptor **485** (β-effect of silicon) and thereafter facilitating the transfer of electrophilic character from the central to the terminal carbon (**487** $\rightarrow$ **488**), hence the cyclization to product **489**.

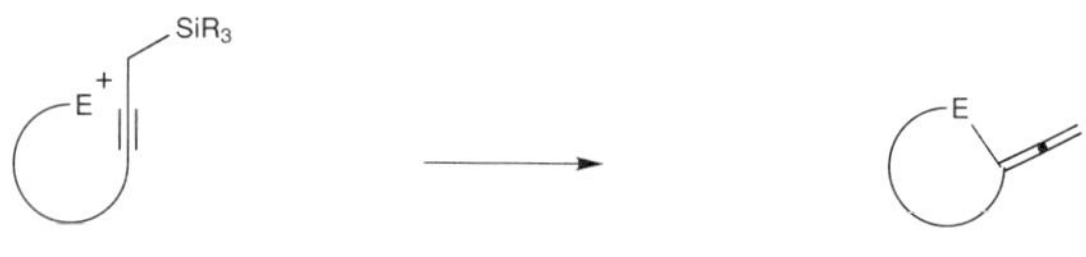

483 484

Equation 32

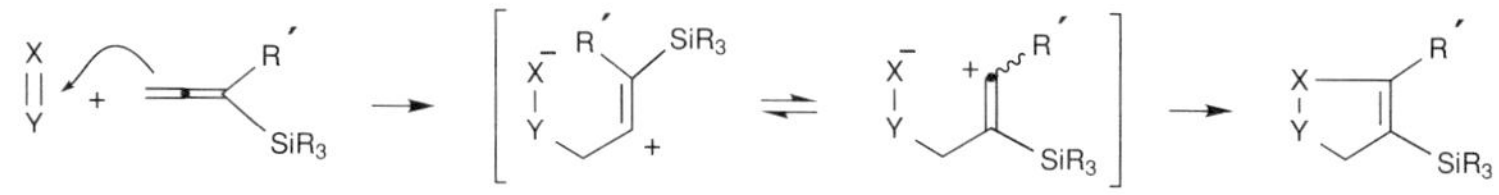

485 486 487 488 489

Equation 33

A number of interesting examples of intramolecular acyliminium ion cyclization of propargylsilane are depicted in Equations 34[126] and 35.[127] Attention should be drawn to the final production of **493** by ozonolysis of **492**, revealing the true function of propargylsilane here to be that of a masked internal acyl anion. The synthesis of bicyclic lactams **496** in Equation 35 further illustrates the application of this type of cyclization. In addition, propargylsilanes also readily undergo intramolecular cyclization with Michael acceptors under Lewis acid catalysis (Equation 36).[128] However, this latter reaction is not categorized as cationic cyclization and will be discussed later under anionic cyclizations (Chapter 4).

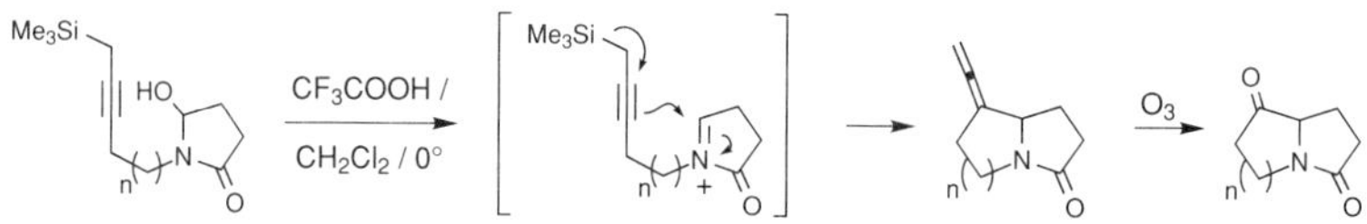

490 491 492 493

n = 1, 3 (70–90%)

Equation 34

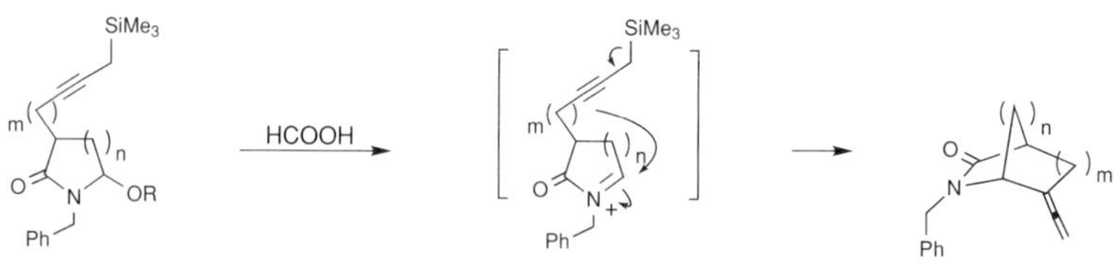

494
(R = Me , Et)

495

496
n = 1 , m = 2 (97%)
n = 1 , m = 3 (89%)
n = 2 , m = 2 (93%)

Equation 35

497 **498** (62%)

Equation 36

In another study of the intermolecular reaction between **499** and propargylsilane **500** in the presence of Lewis acid, cyclized product **504** was obtained.[129] The reaction is explained as proceeding progressively *via* cations **501**, **502** and **503**, as shown in Scheme 69. That is, initial formation of iminium ion **501** provides the electrophile for propargylsilane addition, which results in intermediate **502**, triggering cyclization by the carbonyl oxygen to give the stabilized cation **503** before losing R´ to yield the final product **504**.

Scheme 69

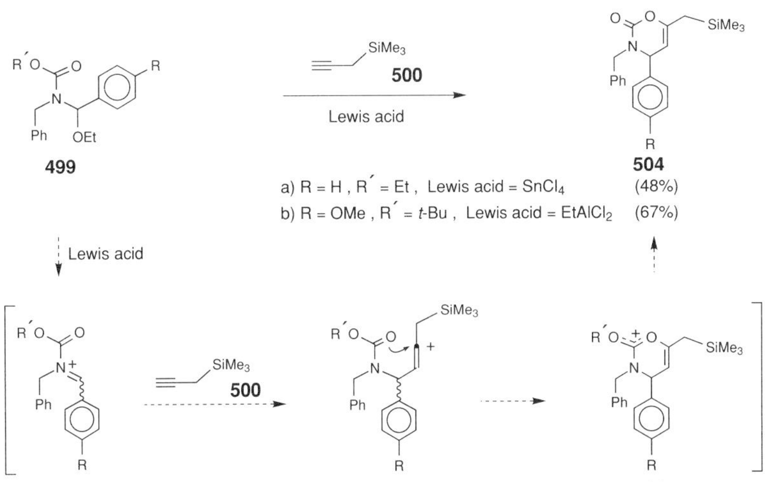

A related reaction to that shown in Scheme 69 has been reported, employing monocyclic starting material **505** with $SnCl_4$, and producing a mixture of tricyclic products **509** and **510**. Thus the assembly of two rings is accomplished in one operation, by way of tandem cyclizations, as depicted in Scheme 70.[130]

Scheme 70

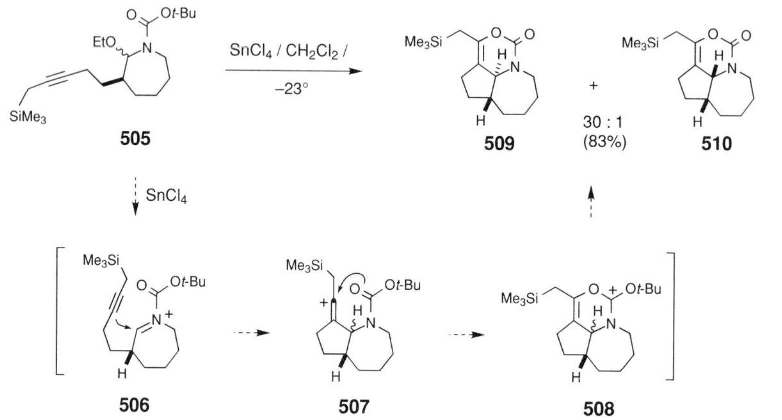

As regards the allenylsilane cyclization shown in Equation 33, this type of reaction has also recorded successful applications, for example, in the synthesis of isoxazoles **517** and furans **523**, employing the nitrosonium (N≡O⁺) and acylium (RC≡O⁺) ions as allenophiles, as illustrated in Schemes 71[131] and 72,[132] respectively.

Scheme 71

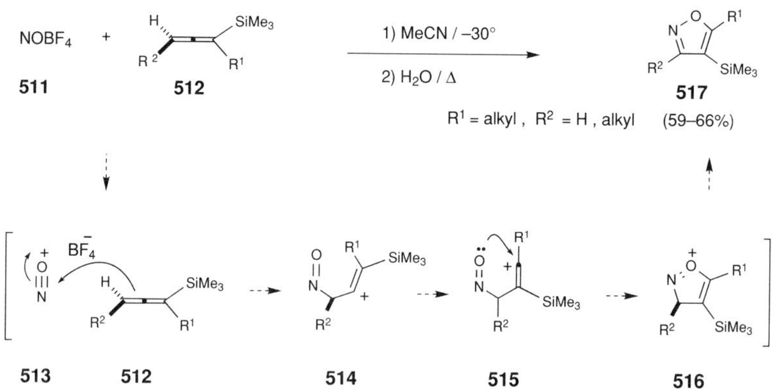

Scheme 72

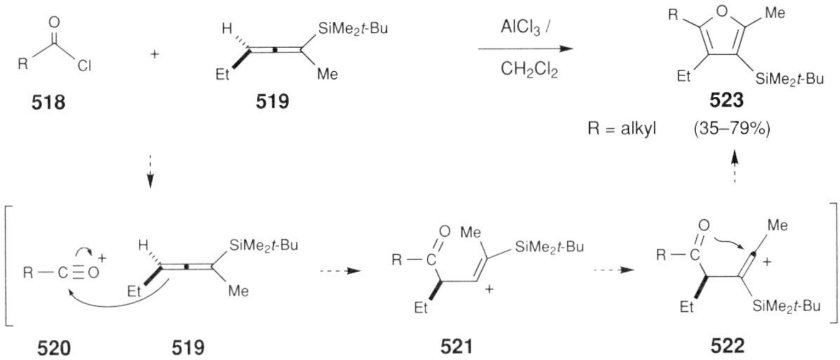

518 **519** **523** (35–79%) **520** **519** **521** **522**

The behavior of allenylsilane **526**, homolog of the foregoing **512** and **519**, has also been studied. Rather than following the general allenylsilane reaction pathway, the homolog behaves in a manner similar to propargylsilanes in intramolecular cyclization, with the product dependent upon the Lewis acid and reaction conditions. To elaborate: treatment of keto–allenylsilane **526** with a 1.1 equivalents of ethylaluminum dichloride in methylene chloride at –78° to 0°C yields the desilylated cyclization product **527**, while exposing **526** to a 1.1 equivalents of trimethylaluminum in methylene chloride at room temperature results in the methylated cyclization product **528**[133] (Scheme 73).

Scheme 73

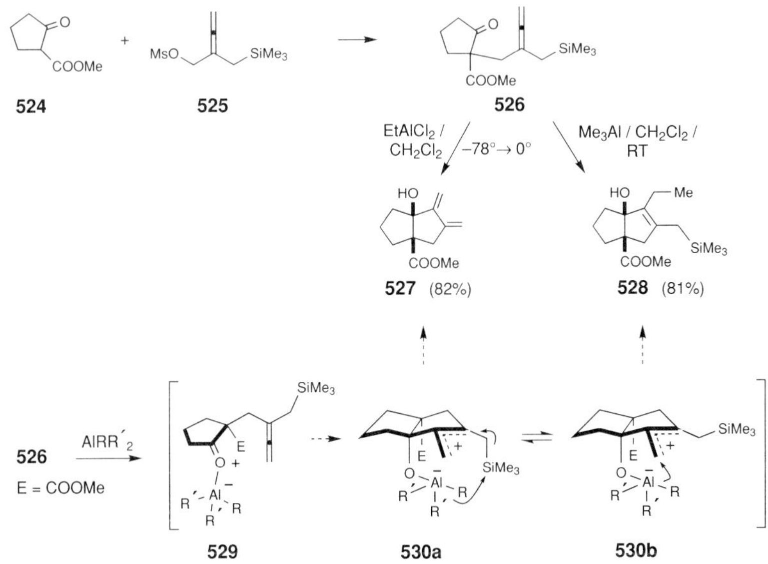

524 **525** **526** **527** (82%) **528** (81%) **526** **529** **530a** **530b**

The mechanism of the reaction has been proposed in terms of the silyl-stabilized allyl cation **530**, which key intermediate results from the preferential attack of electrophile on the central, rather than terminal, allene carbon. Desilylation of **530a** and (when R = Me) transfer of group R in **530b** would lead to products **527** and **528,** respectively. The reaction is very useful in providing a high-yield synthesis of compounds such as **527**, a diene of high reactivity in the Diels–Alder reaction and thus an important intermediate in the synthesis of polycyclic compounds.[133]

Finally, mention should be made of the cyclization reaction of the "propargyl ↔ allenyl" silane carbocation such as **532** (Scheme 74). Generated *in situ* by reacting acetylenic silane **531** with titanium tetrachloride in methylene chloride, cation **532** reacts with substituted alkene **533** to afford cyclopentene **536** in moderate yield. The reaction is thought to proceed as shown in Scheme 74. Quite notable is the absence of any product from silyl migration in the intermediate carbocation **535**. That this is due to the fast attack of chloride ion on **535** is supported by the isolation of intermediate acyclic chloride **537** when boron trichloride catalyst is substituted for $TiCl_4$ in the reaction. As expected, **537** could be converted easily by further treatment to the product **536**.[134]

Scheme 74

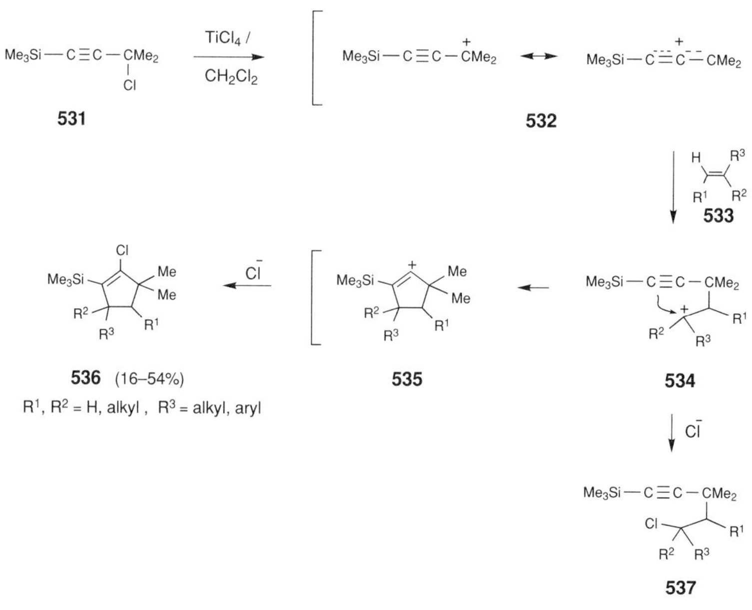

VI. CONCLUSION

Despite the myriad known variations of cationic cyclizations the field is far from exhausted. New and improved reactions regularly appear; for example, the Lewis acid-catalyzed rearrangement of acylcyclobutane to cyclopentanone derivatives[135] (which resembles the reaction shown in Scheme 4), the Nazarov-type cyclizations[136,137] (see Equation 6, Scheme 15); the squalene polycyclization[138,139] (see Schemes 11 and 12), and cyclizations *via* iminium ion intermediates,[140–142] to name just a few. A common observation with most of these later developments is the very high degree of stereocontrol that they manage to achieve in spite of the architecturally ever more complex substrates. The future of cationic cyclizations is thus expected to grow increasingly more sophisticated, in keeping with the demands of current synthetic practice.

REFERENCES

1. Gassman, P. G., and Singleton, D. A., Acid-catalyzed intramolecular "Diels–Alder" reactions. The cycloaddition of allyl cations to 1,3-dienes, *J. Am. Chem. Soc.*, 106, 6085, 1984.
2. Gassman, P. G., and Gorman, D. B., Stepwise mechanism for the formation of 2π+4π cycloadducts in the ionic Diels–Alder reaction, *J. Am. Chem. Soc.*, 112, 8624, 1990.
3. Gassman, P. G., and Singleton, D. A., Control of regiospecificity in ionic Diels–Alder reactions. The use of allylic alcohols and allylic ethers as precursors of dienophilic allyl cations, *J. Org. Chem.*, 51, 3075, 1986.
4. Gassman, P. G., and Gorman, D. B., Use of alkyl substitution to control ring size in the intramolecular ionic Diels–Alder reaction, *J. Am. Chem. Soc.*, 112, 8623, 1990.
5. Ue, M., Tsukahara, H., Kobiro, K., Kakiuchi, K., Tobe, Y., and Odaira, Y., Unusual rearrangement of tricyclo[6.3.0.$0^{1,4}$]undecan-5-one, *Tetrahedron Lett.*, 28, 3979, 1987.
6. Kakiuchi, K., Fukunaga, K., Jimbo, M., Yamaguchi, B., and Tobe, Y., Construction of the bicyclo[6.3.0]undecane skeleton *via* ring opening of tricyclo[6.3.0.$0^{1,4}$]undecan-5-ones, *J. Org. Chem.*, 57, 1021, 1992.
7. Kakiuchi, K., Ue, M., Tsukahara, H., Shimizu, T., Miyao, T., Tobe, Y., Odaira, Y., Yasuda, M., and Shima, K., Novel rearrangement of 5,6-disubstituted bicyclo[4.2.0]octan-2-ones with $AlCl_3$. Application to total synthesis of (±)-5-oxosilphiperfol-6-ene and (±)-silphiperfol-6-ene, *J. Am. Chem. Soc.*, 111, 3707, 1989.
8. Bright, S. T., Coxon, J. M., and Steel, P. J., Reactions of 2-phenylethyl and 3-phenylpropyl carbinols with fluorosulfuric acid, *J. Org. Chem.*, 55, 1338, 1990.
9. Node, M., Hao, X., Nagasawa, H., and Fuji, K., General entry to the synthesis of optically active diterpenoids of C-20β series, *Tetrahedron Lett.*, 30, 4141, 1989.
10. Banerjee, A. K., Nasipuri, D., and Pakrashi, S. C., A simple and highly stereoselective route to (±)-podocarpic acid, *J. Org. Chem.*, 55, 3952, 1990.
11. Paquette, L. A., and Negri, J. T., Acid-catalyzed dehydration of diols. Kinetic and stereochemical ramifications of spirotetrahydrofuran synthesis, *J. Am. Chem. Soc.*, 113, 5072, 1991.
12. van Tamelen, E. E., Bioorganic chemistry: sterols and acyclic terpene terminal epoxides, *Acc. Chem. Res.*, 1, 111, 1968.

13. Corey, E. J., and Sodeoka, M., An effective system for epoxide-initiated cation–olefin cyclization, *Tetrahedron Lett.*, 32, 7005, 1991.
14. Harring, S. R., and Livinghouse, T., A concise biomimetic total synthesis of (±)-taxodione *via* a $BF_3 \bullet MeNO_2$ promoted cationic cascade annulation, *J. Chem. Soc., Chem. Commun.*, 502, 1992.
15. Kupchan, S. M., Karim, A., and Marcks, C., Taxodione and taxodone, two novel diterpenoid quinone methide tumor inhibitors from *Taxodium distichum*, *J. Am. Chem. Soc.*, 90, 5923, 1968.
16. Harring, S. R., and Livinghouse, T., Terminator regulated mechanistic divergence in $BF_3 \bullet MeNO_2$ promoted cascade annulations of geometrically defined trienoate derivatives, *J. Chem. Soc., Chem. Commun.*, 503, 1992.
17. Kosugi, H., Hoshino, K., and Uda, H., Cyclisation of polyenic β-keto sulphoxides. A novel route to optically active intermediates for cyclic terpenoid synthesis, *J. Chem. Soc., Chem. Commun.*, 1173, 1991.
18. Battiste, M. A., Rocca, J. R., Wydra, R. L., Tumlinson, J. H., III, and Chuman, T., Total synthesis and structure proof of (3*E*,8*E*)-suspensolide, *Tetrahedron Lett.*, 29, 6565, 1988.
19. Saito, A., Matsushita, H., and Kaneko, H., Synthesis of (±)-anastrephin and (±)-epianastrephin, *Chem. Lett.*, 729, 1984.
20. Battiste, M. A., Strekowski, L., Coxon, J. M., Wydra, R. L., and Harden, D. B., Highly stereoselective rearrangement of (*E*,*E*)-suspensolide to anastrephin and epianastrephin, *Tetrahedron Lett.*, 32, 5303, 1991.
21. Aldrich Chemical Company, Inc., unpublished information.
22. Majetich, G., Song, J. S., Ringold, C., and Nemeth, G. A., A direct stereoselective synthesis of (±)-14-deoxyisoamijiol, *Tetrahedron Lett.*, 31, 2239, 1990.
23. Majetich, G., and Khetani, V., Intramolecular reactions of unactivated alkenes with conjugated dienones, *Tetrahedron Lett.*, 31, 2243, 1990.
24. Santelli-Rouvier, C., and Santelli, M., The Nazarov cyclisation, *Synthesis*, 429, 1983.
25. Marino, J. P., and Linderman, R. J., Chemistry of substituted [α-(carboethoxy)vinyl]cuprates and their synthetic application to cyclopentenone annulations, *J. Org. Chem.*, 46, 3696, 1981.
26. Andrews, J. F. P., and Regan, A. C., Nazarov cyclisation of dienone–esters and tetrahydropyrones using trimethylsilyltriflate, *Tetrahedron Lett.*, 32, 7731, 1991.
27. Gassman, P. G., Ray, J. A., Wenthold, P. G., and Mickelson, J. W., Synthesis of perfluoroalkylated indenes, *J. Org. Chem.*, 56, 5143, 1991.
28. Royer, J., and Husson, H.-P., Asymmetric synthesis. XI. A short synthesis of the chiral pyrrolidine synthon, 2-cyano-5-oxazolopyrrolidine, *Tetrahedron Lett.*, 28, 6175, 1987.
29. Pandey, G., and Kumaraswamy, G., A novel and selective procedure for generating iminium cation by photosensitized (SET) oxidative reaction, *Tetrahedron Lett.*, 29, 4153, 1988.
30. Pandey, G., Kumaraswamy, G., and Krishna, A., Photosensitized single electron transfer oxidation of *N*-hydroxylamines: a convenient synthesis of cyclic nitrones, *Tetrahedron Lett.*, 28, 2649, 1987.
31. Chen, C.-K., Hortmann, A. G., and Marzabadi, M. R., ClO_2 oxidation of amines: synthetic utility and a biomimetic synthesis of elaeocarpidine, *J. Am. Chem. Soc.*, 110, 4829, 1988.
32. Leonard, N. J., and Musker, W. K., Unsaturated amines. XVI. An oxidative cyclization route to oxazolidines and tetrahydro-1,3-oxazines, *J. Am. Chem. Soc.*, 82, 5148, 1960.
33. Ahn, K. H., and Lee, S. J., Syntheses of 1-oxaquinolizidines *via* reductive cyclization of hydroxy–lactams, *Tetrahedron Lett.*, 33, 507, 1992.
34. Guerrier, L., Royer, J., Grierson, D. S., and Husson, H.-P., Chiral 1,4-dihydropyridine equivalents: a new approach to the asymmetric synthesis of alkaloids. The enantiospecific synthesis of (+)- and (–)-coniine and -dihydropinidine, *J. Am. Chem. Soc.*, 105, 7754, 1983.
35. Royer, J., and Husson, H.-P., Asymmetric synthesis. II. Practical method for the asymmetric synthesis of indolizidine alkaloids: total synthesis of (–)-monomorine I, *J. Org. Chem.*, 50, 670, 1985.

36. Yue, C., Royer, J., and Husson, H.-P., Asymmetric synthesis. XXI. The first enantioselective synthesis of natural (+)-tetraponerine-8: a new extention of the CN(*R*,*S*) method to an uncommon skeleton, *J. Org. Chem.*, 55, 1140, 1990.
37. Deslongchamps, P., *Stereoelectronic Effects in Organic Chemistry*, Baldwin, J. E., Ed., Pergamon Press, Oxford, 1984.
38. Stevens, R. V., and Lee, A. W. M., On the stereochemistry of the Robinson–Schöpf reaction. A stereospecific total synthesis of the ladybug defense alkaloids precoccinelline and coccinelline, *J. Am. Chem. Soc.*, 101, 7032, 1979.
39. Soerens, D., Sandrin, J., Ungemach, F., Mokry, P., Wu, G. S., Yamanaka, E., Hutchins, L., DiPierro, M., and Cook, J. M., Study of the Pictet–Spengler reaction in aprotic media: synthesis of the β-galactosidase inhibitor, pyridindolol, *J. Org. Chem.*, 44, 535, 1979.
40. Urban, F. J., New synthesis of spiro-benzopyran amino acids by intramolecular amidoalkylation, *Tetrahedron Lett.*, 29, 5493, 1988.
41. Massa, S., Mai, A., and Corelli, F., Synthesis of a new tetracyclic system related to aptazapine(CGS 7525A) by one-pot double annelation, *Tetrahedron Lett.*, 29, 6471, 1988.
42. Comins, D. L., and Badawi, M. M., Asymmetric Pictet–Spengler synthesis of tetrahydroisoquinolines. An enantioselective synthesis of (–)-laudanosine, *Tetrahedron Lett.*, 32, 2995, 1991.
43. Brown, R. T., Dauda, B. E. N. and Santos, C. A. M., Hydrolysis and selective reduction with yeast: enantiospecific synthesis of antirhine from secologanin, *J. Chem. Soc., Chem. Commun.*, 825, 1991.
44. Overman, L. E., and Sharp, M. J., Nucleophile-promoted electrophilic cyclization reactions of alkynes, *J. Am. Chem. Soc.*, 110, 612, 1988.
45. Bohlmann, F., and Winterfeldt, E., Lupinen-alkaloide. XVI. Die konstitution des angustifolins, *Chem. Ber.*, 93, 1956, 1960.
46. Ahmad, V. U., Feuerherd, K.-H., and Winterfeldt, E., Reaktionen an indolderivaten. XXXIV. Stereoselektive cyclisierungen in der tetrahydro-β-carbolin–reihe, *Chem. Ber.*, 110, 3624, 1977.
47. Meyers, A. I., Miller, D. B., and White, F. H., Chiral and achiral formamidines in synthesis. The first asymmetric route to (–)-yohimbone and an efficient total synthesis of (±)-yohimbone, *J. Am. Chem. Soc.*, 110, 4778, 1988.
48. Larsen, S. D., Grieco, P. A., and Fobare, W. F., Reactions of allylsilanes with simple iminium salts in water: a facile route to piperidines *via* an aminomethano desilylation–cyclization process, *J. Am. Chem. Soc.*, 108, 3512, 1986.
49. Murahashi, S.-I., Naota, T., and Yonemura, K., Ruthenium-catalyzed cytochrome P-450 type oxidation of tertiary amines with alkyl hydroperoxides, *J. Am. Chem. Soc.*, 110, 8256, 1988.
50. Agami, C., Couty, F., Poursoulis, M., and Vaissermann, J., Asymmetric synthesis of pipecolic acid derivatives, *Tetrahedron*, 48, 431, 1992.
51. Agami, C., Couty, F., Prince, B., and Puchot, C., Asymmetric synthesis of *N*-methyl-α-amino esters from a glyoxal derived chiral heterocycle, *Tetrahedron*, 47, 4343, 1991.
52. Overman, L. E., Robinson, L. A., and Zablocki, J., First total synthesis of (+)-allopumiliotoxin 339A. A practical entry to dendrobatid alkaloids of the allopumiliotoxin class, *J. Am. Chem. Soc.*, 114, 368, 1992.
53. Overman, L. E., and Kakimoto, M., Carbon–carbon bond formation *via* directed 2-azonia-[3,3]-sigmatropic rearrangements. A new pyrrolidine synthesis, *J. Am. Chem. Soc.*, 101, 1310, 1979.
54. Overman, L. E., Jacobsen, E. J., and Doedens, R. J., Synthesis applications of cationic aza-Cope rearrangements. Stereoselective synthesis of *cis*- and *trans*-3a-aryl-4-oxodecahydrocyclohepta[*b*]pyrroles, *J. Org. Chem.*, 48, 3393, 1983.
55. Overman, L. E., Mendelson, L. T., and Jacobsen, E. J., Applications of cationic aza-Cope rearrangements for alkaloid synthesis. Stereoselective preparation of *cis*-3a-aryloctahydroindoles and a new short route to *Amaryllidaceae* alkaloids, *J. Am. Chem. Soc.*, 105, 6629, 1983.

56. Overman, L. E., and Sugai, S., Total synthesis of (–)-crinine. Use of tandem cationic aza-Cope rearrangement / Mannich cyclizations for the synthesis of enantiomerically pure *Amaryllidaceae* alkaloids, *Helv. Chim. Acta*, 68, 745, 1985.
57. Overman, L. E., Sworin, M., and Burk, R. M., A new approach for the total synthesis of pentacyclic *Aspidosperma* alkaloids. Total synthesis of *dl*-16-methoxytabersonine, *J. Org. Chem.*, 48, 2685, 1983.
58. Overman, L. E., Kakimoto, M., Okazaki, M. E., and Meier, G. P., Carbon–carbon bond formation under mild conditions *via* tandem cationic aza-Cope rearrangement–Mannich reactions. A convenient synthesis of polysubstituted pyrrolidines, *J. Am. Chem. Soc.*, 105, 6622, 1983.
59. McCann, S. F., and Overman, L. E., Medium effects and the nature of the rate-determining step in Mannich-type cyclizations, *J. Am. Chem. Soc.*, 109, 6107, 1987.
60. Jacobsen, E. J., Levin, J. , and Overman, L. E., Scope and mechanism of tandem cationic aza-Cope rearrangement–Mannich cyclization reactions, *J. Am. Chem. Soc.*, 110, 4329, 1988.
61. Fevig, J. M., Marquis, R. W., Jr., and Overman, L. E., New approach to *Strychnos* alkaloids. Stereocontrolled total synthesis of (±)-dehydrotubifoline, *J. Am. Chem. Soc.*, 113, 5085, 1991.
62. Overman, L. E., and Shim, J., First total synthesis of *Amaryllidaceae* alkaloids of the 5,11-methanomorphanthridine type. An efficient total synthesis of (±)-pancracine, *J. Org. Chem.*, 56, 5005, 1991.
63. Overman, L. E., Robertson, G. M., and Robichaud, A. J., Use of aza-Cope rearrangement–Mannich cyclization reactions to achieve a general entry to *Melodinus* and *Aspidosperma* alkaloids. Stereocontrolled total syntheses of (±)-deoxoapodine, (±)-meloscine, and (±)-epimeloscine and a formal synthesis of (±)-1-acetylaspidoalbidine, *J. Am. Chem. Soc.*, 113, 2598, 1991.
64. Winkler, J. D., Muller, C. L., and Scott, R. D., A new method for the formation of nitrogen-containing ring systems *via* the intramolecular photocycloaddition of vinylogous amides. A synthesis of mesembrine, *J. Am. Chem. Soc.*, 110, 4831, 1988.
65. Earley, W. G., Jacobsen, E. J., Meier, G. P., Oh, T., and Overman, L. E., Synthesis studies directed toward gelsemine. A new synthesis of highly functionalized *cis*-hydroisoquinolines, *Tetrahedron Lett.*, 29, 3781, 1988.
66. Evans, D. A., Baillargeon, D. J., and Nelson, J. V., A general approach to the synthesis of 1,6-dicarbonyl substrates. New applications of base-accelerated oxy-Cope rearrangements, *J. Am. Chem. Soc.*, 100, 2242, 1978.
67. Earley, W. G., Oh, T., and Overman, L. E., Synthesis studies directed toward gelsemine. Preparation of an advanced pentacyclic intermediate, *Tetrahedron Lett.*, 29, 3785, 1988.
68. Hiemstra, H., Vijn, R. J., and Speckamp, W. N., Synthetic studies toward gelsemine. II. Preparation of the tetracyclic skeletal part by way of a highly stereospecific intramolecular reaction of a silyl enol ether with an *N*-acyliminium ion, *J. Org. Chem.*, 53, 3882, 1988.
69. Heathcock, C. H., Davidsen, S. K., Mills, S. G., and Sanner, M. A., *Daphniphyllum* alkaloids. X. Classical total synthesis of methyl homodaphniphyllate, *J. Org. Chem.*, 57, 2531, 1992.
70. Heathcock, C. H., and Stafford, J. A., *Daphniphyllum* alkaloids. XIII. Asymmetric total synthesis of (–)-secodaphniphylline, *J. Org. Chem.*, 57, 2566, 1992.
71. Heathcock, C. H., Ruggeri, R. B., and McClure, K. F., *Daphniphyllum* alkaloids. XV. Total syntheses of (±)-methyl homodaphniphyllate and (±)-daphnilactone A, *J. Org. Chem.*, 57, 2585, 1992.
72. Ruggeri, R. B., and Heathcock, C. H., Biomimetic total synthesis of (±)-methyl homodaphniphyllate, *J. Org. Chem.*, 55, 3714, 1990.
73. Mukaiyama, T., The directed aldol reaction, *Org. React.*, 28, 203, 1982.
74. Alexakis, A., Chapdelaine, M. J., Posner, G. H., and Runquist, A. W., A new synthetic approach to pseudoguaianes, *Tetrahedron Lett.*, 4205, 1978.
75. Cockerill, G. S., Kocienski, P., and Treadgold, R., Silicon-mediated annulation. I. A synthesis of tetrahydropyran-4-ones, oxepan-4-ones, and oxocan-4-ones *via* intramolecular directed aldol reactions, *J. Chem. Soc., Perkin Trans. 1*, 2093, 1985.

76. Cockerill, G. S., Kocienski, P., and Treadgold, R., Silicon-mediated annulation. II. A synthesis of β-alkoxy cyclo-octanones *via* intramolecular directed aldol reactions, *J. Chem. Soc., Perkin Trans. 1*, 2101, 1985.
77. Baldwin, J. E., Rules for ring closure, *J. Chem. Soc., Chem. Commun.*, 734, 1976.
78. Blumenkopf, T. A., Bratz, M., Castañeda, A., Look, G. C., Overman, L. E., Rodriguez, D., and Thompson, A. S., Preparation of eight-membered cyclic ethers by Lewis acid promoted acetal–alkene cyclizations, *J. Am. Chem. Soc.*, 112, 4386, 1990.
79. Craig, D., and Munasinghe, V. R. N., Stereoselective template-directed *C*-glycosidation. Silver(I)-mediated intramolecular reactions of (2-pyridylthio)glycosidic silyl enol ethers, *Tetrahedron Lett.*, 33, 663, 1992.
80. Hopkins, M. H., and Overman, L. E., Stereocontrolled preparation of tetrahydrofurans by acid-catalyzed rearrangement of allylic acetals, *J. Am. Chem. Soc.*, 109, 4748, 1987.
81. Hopkins, M. H., Overman, L. E., and Rishton, G. M., Stereocontrolled preparation of tetrahydrofurans from acid-promoted rearrangements of allylic acetals, *J. Am. Chem. Soc.*, 113, 5354, 1991.
82. Herrinton, P. M., Hopkins, M. H., Mishra, P., Brown, M. J., and Overman, L. E., Ring-enlarging furan annulations, *J. Org. Chem.*, 52, 3711, 1987.
83. Brown, M. J., Harrison, T., Herrinton, P. M., Hopkins, M. H., Hutchinson, K. D., Mishra, P., and Overman, L. E., Acid-promoted reaction of cyclic allylic diols with carbonyl compounds. Stereoselective ring-enlarging tetrahydrofuran annulations, *J. Am. Chem. Soc.*, 113, 5365, 1991.
84. Hirst, G. C., Howard, P. N., and Overman, L. E., Stereocontrolled construction of carbocyclic rings by sequential cationic cyclization–pinacol rearrangements, *J. Am. Chem. Soc.*, 111, 1514, 1989.
85. Johnson, T. O., and Overman, L. E., Ring-enlarging cyclopentene annulations, *Tetrahedron Lett.*, 32, 7361, 1991.
86. Trost, B. M., and Lee, D. C., Cyclization terminators. Vinylcyclopropanol as a composite functional group, *J. Am. Chem. Soc.*, 110, 6556, 1988.
87. Sworin, M., and Neumann, W. L., Cyclopentanoid synthesis *via* directed cationic cyclizations. Efficient generation and rearrangement of the intermediate cyclohexyl cation, *J. Org. Chem.*, 53, 4894, 1988.
88. Gassman, P. G., and Lottes, A. C., $2\pi+4\pi$ cycloadditions of carbocation activated allenes, *Tetrahedron Lett.*, 32, 6473, 1991.
89. Gassman, P. G., and Lottes, A. C., Cyclobutane formation in the $2\pi+2\pi$ cycloaddition of allyl and related cations to unactivated olefins. Evidence for the second step in the proposed mechanism of the ionic Diels–Alder reaction, *Tetrahedron Lett.*, 33, 157, 1992.
90. Mander, L. N., and Mundill, P. H. C., Olefinic cyclisations initiated by Pummerer rearrangement products; synthesis of cyclic ketones, *Synthesis*, 620, 1981.
91. Ishibashi, H., Ozeki, H., and Ikeda, M., Synthesis of optically active (–)-trachelanthamidine from L-prolinol, *J. Chem. Soc., Chem. Commun.*, 654, 1986.
92. Gallagher, T., Magnus, P., and Huffman, J. C., Pentacyclic systems for indole alkaloids. Formation of the C_{11}–C_{12} bond. Two syntheses of (±)-aspidospermidine, *J. Am. Chem. Soc.*, 105, 4750, 1983.
93. Magnus, P., and Cairns, P. M., Methods for indole alkaloid synthesis. Enantiospecific synthesis of pentacyclic *Desethylaspidosperma*-type alkaloids using an exceptionally mild retro-Diels–Alder reaction, *J. Am. Chem. Soc.*, 108, 217, 1986.
94. Cardwell, K., Hewitt, B., Ladlow, M., and Magnus, P., Methods for indole alkaloid synthesis: a study of the compatibility of the indole–2,3-quinodimethane strategy for the synthesis of 16-methoxy-substituted *Aspidosperma*-type alkaloids. Synthesis of (+)- and (–)-16-methoxytabersonine, *J. Am. Chem. Soc.*, 110, 2242, 1988.
95. Kaneko, T., Possible biomimetic synthesis of β-lactams, *J. Am. Chem. Soc.*, 107, 5490, 1985.
96. Kita, Y., Tamura, O., Yasuda, H., Itoh, F., and Tamura, Y., Chemistry of *O*-silylated ketene acetals: preparation of α-siloxy phenyl sulfides and methyl 3-(phenylthio)butyrates from alkyl phenyl sulfoxides, *Chem. Pharm. Bull.*, 33, 4235, 1985.

97. Kita, Y., Tamura, O., Miki, T., and Tamura, Y., Chemistry of *O*-silylated ketene acetals: an efficient synthesis of α-thio-*N*-heterocycles from ω-amidosulfoxides by a novel intramolecular Pummerer-type rearrangement, *Tetrahedron Lett.*, 28, 6479, 1987.
98. Fleming, I., Some uses of silicon compounds in organic synthesis, *Chem. Soc. Rev.*, 10, 83, 1981.
99. Blumenkopf, T. A., and Overman, L. E., Vinylsilane- and alkynylsilane-terminated cyclization reactions, *Chem. Rev.*, 86, 857, 1986.
100. Organosilicon chemistry in organic synthesis, Fleming, I., Ed., Tetrahedron Symposia-in-Print number 32, *Tetrahedron*, 44, 3761, 1988.
101. Fleming, I., Dunoguès, J., and Smithers, R., The electrophilic substitution of allylsilanes and vinylsilanes, *Org. React.*, 37, 57, 1989.
102. Lambert, J. B., and Chelius, E. C., The β effect of silicon in the synperiplanar geometry, *J. Am. Chem. Soc.*, 112, 8120, 1990.
103. Lambert, J. B., The interaction of silicon with positively charged carbon, *Tetrahedron*, 46, 2677, 1990.
104. Flann, C. J., and Overman, L. E., Enantioselective total synthesis of streptazolin. The tandem use of iminium ion–vinylsilane cyclizations and intramolecular acylations, *J. Am. Chem. Soc.*, 109, 6115, 1987.
105. Overman, L. E., and Robichaud, A. J., Total syntheses of (+)-geissoschizine, (±)-geissoschizine, and (±)-(*Z*)-isositsirikine. Stereocontrolled synthesis of exocyclic double bonds by stereospecific iminium ion–vinylsilane cyclizations, *J. Am. Chem. Soc.*, 111, 300, 1989.
106. Overman, L. E., and Thompson, A. S., Total synthesis of (–)-laurenyne. Use of acetal-initiated cyclizations to prepare functionalized eight-membered cyclic ethers, *J. Am. Chem. Soc.*, 110, 2248, 1988.
107. Nakamura, E., Fukuzaki, K., and Kuwajima, I., Cyclopentenones by internal acylation of vinylsilanes. Rapid construction of trichothecane-type carbon frameworks, *J. Chem. Soc., Chem. Commun.*, 499, 1983.
108. Kishi, N., Mikami, K., and Nakai, T., Vinylsilane-terminated cycloacylation: a general synthetic approach to four- to six-membered cyclic ketones and its regiochemical features, *Tetrahedron*, 47, 8111, 1991.
109. Yamazaki, S., Katoh, S., and Yamabe, S., [2+1] cycloaddition of (*E*)-1-(phenylseleno)-2-(trimethylsilyl)ethene. Novel cyclopropane ring formation by a 1,2-shift of a trimethylsilyl group, *J. Org. Chem.*, 57, 4, 1992.
110. Barbachyn, M. R., Johnson, C. R., and Glick, M. D., Synthesis and absolute configuration of (–)-rothrockene, a non-head-to-tail monoterpene, *J. Org. Chem.*, 49, 2746, 1984.
111. Grieco, P. A., and Fobare, W. F., Intramolecular variants of aminomethano desilylation: reactions of *in situ* generated immonium ions with allylsilanes, *Tetrahedron Lett.*, 27, 5067, 1986.
112. Slosse, P., and Hootelé, C., Structure and absolute configuration of myrtine, a new quinolizidine alkaloid from *Vaccinium myrtillus*, *Tetrahedron Lett.*, 397, 1978.
113. Gelas-Mialhe, Y., Gramain, J.-C., Louvet, A., and Remuson, R., Diastereoselective cyclization of allylsilanes on α-acyliminium ions. A new approach to (±)-myrtine and (±)-epimyrtine, *Tetrahedron Lett.*, 33, 73, 1992.
114. Mooiweer, H. H., Hiemstra, H., and Speckamp, W. N., Total synthesis of (±)-α-allokainic acid *via* two allylsilane *N*-acyliminium ion reactions, *Tetrahedron*, 47, 3451, 1991.
115. Schinzer, D., and Bo, Y., Synthesis of heterocycles by tandem reactions: Beckmann rearrangements / allylsilane cyclizations, *Angew. Chem. Int. Ed. Engl.*, 30, 687, 1991.
116. Chakraborty, R., and Simpkins, N. S., Cyclisation chemistry of some functionalised allylsilanes, *Tetrahedron*, 47, 7689, 1991.
117. Mekhalfia, A., and Markó, I. E., The silyl modified Sakurai (SMS) reaction. An efficient and versatile one-pot synthesis of homoallylic ethers, *Tetrahedron Lett.*, 32, 4779, 1991.
118. Mekhalfia, A., Markó, I. E., and Adams, H., The intramolecular silyl modified Sakurai (ISMS) reaction. A novel and highly convergent synthesis of oxocenes, *Tetrahedron Lett.*, 32, 4783, 1991.

119. Markó, I. E., Mekhalfia, A., Bayston, D. J., and Adams, H., The intramolecular silyl modified Sakurai (ISMS) reaction. A simple and versatile synthesis of tetrahydropyrans, spiroethers, and spiroketals, *J. Org. Chem.*, 57, 2211, 1992.
120. Tietze, L. F., and Wünsch, J. R., Stereoselective synthesis of annelated piperidines by photochemical cycloaddition and iminium ion–allylsilane cyclization, *Angew. Chem. Int. Ed. Engl.*, 30, 1697, 1991.
121. Ullrich, J. W., Chiu, F.-T., Tiner-Harding, T., and Mariano, P. S., Electron-transfer-initiated photospirocyclization reactions of β-enaminone–derived allyliminium salts, *J. Org. Chem.*, 49, 220, 1984.
122. Chiu, F.-T., Ullrich, J. W., and Mariano, P. S., Model studies examining the application of allyliminium salt photospirocyclization methodologies in synthetic approaches to the harringtonine alkaloids, *J. Org. Chem.*, 49, 228, 1984.
123. Ohga, K., Yoon, U. C., and Mariano, P. S., Exploratory and mechanistic studies of the electron-transfer-initiated photoaddition reactions of allylsilane–iminium salt systems, *J. Org. Chem.*, 49, 213, 1984.
124. Tu, C.-L., and Mariano, P. S., Dual diradical and cation diradical cyclization mechanisms for silylalkenyl iminium salt. Electron-transfer-induced photocyclization reactions, *J. Am. Chem. Soc.*, 109, 5287, 1987.
125. Thebtaranonth, C., and Thebtaranonth, Y., Developments in cyclisation reactions, *Tetrahedron*, 46, 1385, 1990.
126. Hiemstra, H., and Speckamp, W. N., Completely regioselective α-acyliminium ion cyclizations with allyl and propargyl silanes, *Tetrahedron Lett.*, 24, 1407, 1983.
127. Klaver, W. J., Moolenaar, M. J., Hiemstra, H., and Speckamp, W. N., Intramolecular reactions of 2-propynylsilanes with *N*-acyliminium ions; cuprate S_N2'reactions of the allenic products as a route to *trans*-fused carbobicycles, *Tetrahedron*, 44, 3805, 1988.
128. Schinzer, D., Dettmer, G., Ruppelt, M., Sólyom, S., and Steffen, J., Intramolecular cyclization of allyl- and propargylsilanes. The regio- and stereochemical outcome of the cyclization, *J. Org. Chem.*, 53, 3823, 1988.
129. Esch, P. M., Hiemstra, H., and Speckamp, W. N., Oxazinone *versus* allene formation in the reaction of *N*-alkoxycarbonyliminium ions with propargyltrimethylsilane, *Tetrahedron Lett.*, 29, 367, 1988.
130. Fisher, M. J., and Overman, L. E., Intramolecular *N*-(acyloxy)iminium ion–alkyne cycloadditions. A new route to bicyclic α-amino ketones, *J. Org. Chem.*, 55, 1447, 1990.
131. Danheiser, R. L., and Becker, D. A., Application of allenylsilanes in a regiocontrolled [3+2] annulation route to substituted isoxazoles, *Heterocycles*, 25, 277, 1987.
132. Danheiser, R. L., Stoner, E. J., Koyama, H., Yamashita, D. S., and Klade, C. A., A new synthesis of substituted furans, *J. Am. Chem. Soc.*, 111, 4407, 1989.
133. Trost, B. M., and Urabe, H., Novel silicon-directed alkylative cyclization, *J. Am. Chem. Soc.*, 112, 4982, 1990.
134. Mayr, H., Bäuml, E., Cibura, G., and Koschinsky, R., Synthesis of cyclopentenes *via* [3+2]-cycloadditions of silylated propargyl $\leftrightarrow$ allenyl cations with alkenes, *J. Org. Chem.*, 57, 768, 1992.
135. Fujiwara, T., Tomaru, J., Suda, A., and Takeda, T., The ethylaluminum dichloride promoted ring enlargement reaction of 1-acylbicyclo[4.2.0]oct-3-ene. A new route to hydroindanone system, *Tetrahedron Lett.*, 33, 2583, 1992.
136. Kang, K.-T., Kim, S. S., Lee, J. C., and U, J. S., Synthesis of α-methylenecyclopentanones *via* silicon-directed Nazarov reaction of α-trimethylsilylmethyl-substituted divinyl ketones, *Tetrahedron Lett.*, 33, 3495, 1992.
137. Franck-Neumann, M., Miesch, M., and Gross, L., Synthesis of the angular triquinane (±)-pentalenene *via* small ring intermediates, *Tetrahedron Lett.*, 33, 3879, 1992.
138. Xiao, X., and Prestwich, G. D., Enzymatic cyclizations of 26- and 29-hydroxy-2,3-oxidosqualenes give 19- and 21-hydroxylanosterols, *Tetrahedron Lett.*, 32, 6843, 1991.
139. Bohlmann, R., The folding of squalene; an old problem has new results, *Angew. Chem. Int. Ed. Engl.*, 31, 582, 1992.

140. Lee, A. W. M., Chan, W. H., and Lee, Y.-K., Chiral acetylenic sulfoxide in alkaloid synthesis. Total synthesis of (*R*)-(+)-carnegine, *Tetrahedron Lett.*, 32, 6861, 1991.
141. Gesson, J. P., Jacquesy, J. C., and Rambaud, D., 1-Bromoalkynes as π-nucleophiles in acyliminium ion cyclizations. A formal synthesis of lupinine and epilupinine, *Tetrahedron Lett.*, 33, 3633, 1992.
142. Yue, C., Royer, J., and Husson, H.-P., Asymmetric synthesis. XXVI. An expeditious enantioselective synthesis of the defense alkaloids (–)-euphococcinine and (–)-adaline *via* the *CN*(*R*,*S*) method, *J. Org. Chem.*, 57, 4211, 1992.

Chapter 3

RADICAL CYCLIZATION

I. INTRODUCTION

It is surely true, as we look through literature, that the last decade has been an era of radical chemistry. Information on various kinds of radicals, their properties, chemistry, and uses in organic synthesis simply erupted in publications. New reagents and new precursors for the generation of radicals were routinely reported. But in spite of all this, radical chemistry, particularly its most useful reaction, the radical cyclization, is still a long way from perfection. Leading research groups worldwide are currently studying factors that control radical reactions, characteristics of radical acceptors that determine regio- and chemoselectivities, and features that direct the stereochemistry of products, with the ultimate aim of application to the synthesis of complex molecules. In the midst of all this, the discovery of bioactive natural products exhibiting both antibiotic and anticancer properties, such as neocarzinostatin, the calicheamicins, esperamicins and dynemicins, has added further fuel to the study of radical chemistry. The general excitement surrounding these molecules, apart from their novel molecular architecture, lies in their ability to cleave DNA by the generation of sp^2 carbon-centered radicals upon activation.[1] Even nature, so it would seem, is co-operating in opening up new windows and extending visibility in the study of radical chemistry!

Since numerous recent reviews of radical chemistry, from theoretical studies[2] to reactions and applications in organic synthesis,[3–10] have been published, the aim of this chapter, under these circumstances, is therefore to provide a general understanding of radical cyclizations through the use, as much as possible, of the latest published results, in an endeavor to avoid tiresome repetitions.

II. CYCLIZATION TO UNACTIVATED ALKENES AND ALKYNES

The most basic radical cyclization is the intramolecular cyclization of alkenyl radicals, in particular the 5-hexenyl radical, which cyclizes predominantly in the 5-*exo*-trig fashion to yield five-membered rings. Theoretical studies[8,11,12] suggest a "chair-like" transition state from which we may predict the stereochemical outcome of products as demonstrated in Scheme 1 (substituents A, B and C directing the reaction to yield products **5**, **6** and **7** respectively).

Scheme 1

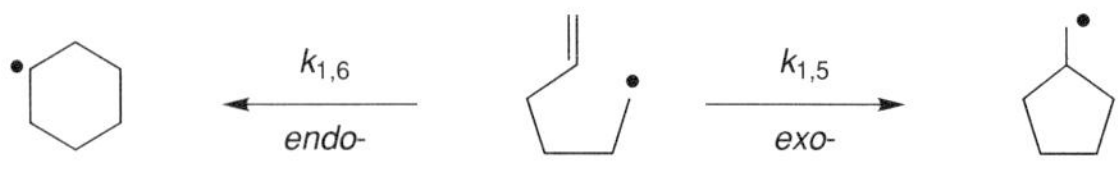

3 1 2

at 60° $k_{1,5} / k_{1,6} \approx 50$

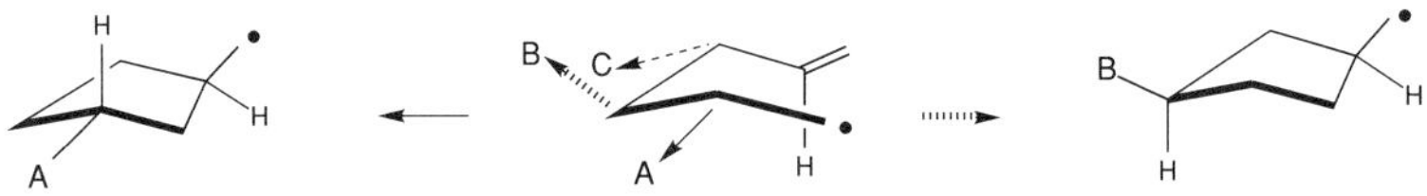

5 4 6

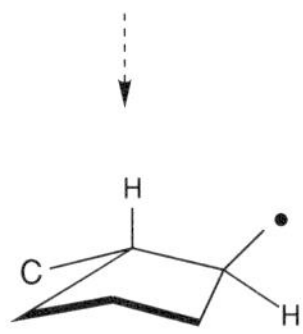

7

A host of radical centers, from carbon radicals (*e.g.*, alkyl, vinyl, aryl, allyl, acyl, imidoyl radicals, *etc.*) to heteroatom radicals (*e.g.*, aminyl, iminyl, alkoxy, and sulfide radicals) can be efficiently generated. As far as radical generation is concerned, the use of alkyltin hydride[13] is currently the method of choice, and is complemented by other methodologies such as irradiation,[14] the electrochemical process,[15] as well as the newly introduced, less toxic free radical mediators such as tris(trimethylsilyl)silane.[16]

A. ALKYL RADICALS

Much use is already made of the carbon–carbon bond formation between alkyl radicals and unactivated alkenes and alkynes, the more recent examples of which are shown in Equations 1 to 4 below. The examples demonstrate the use of various radical precursors in the formation of carbocycles and constitute key steps in the synthesis of important molecules, *i.e.*, Corey lactone **12** (Equation 1),[17] spirocompound **15**, a synthetic model for fredericamycin A (Equation 2),[18] cyclopentyl ether and sulfide **18** (Equation 3, an interesting method of radical generation),[19] and modhephene **23** (Equation 4).[20]

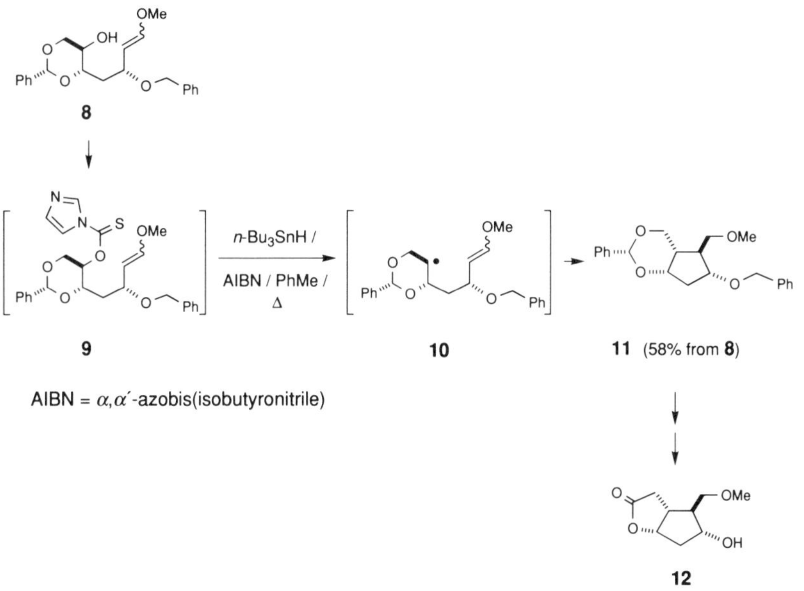

Equation 1

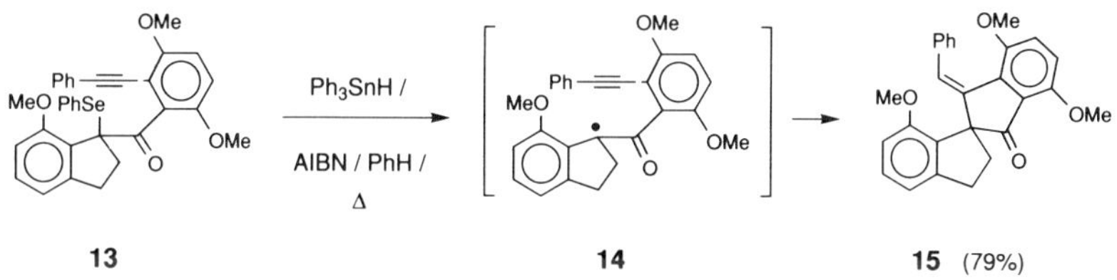

Equation 2

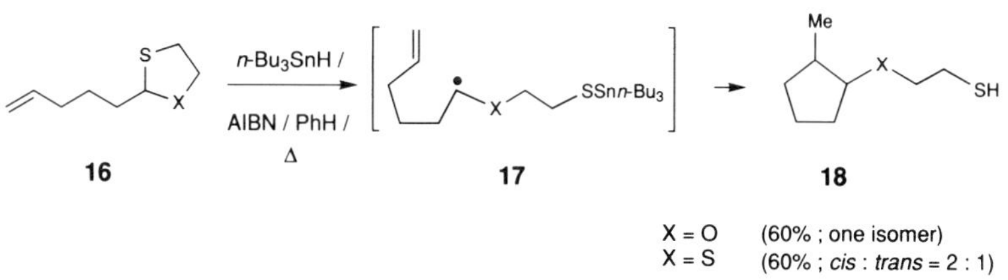

Equation 3

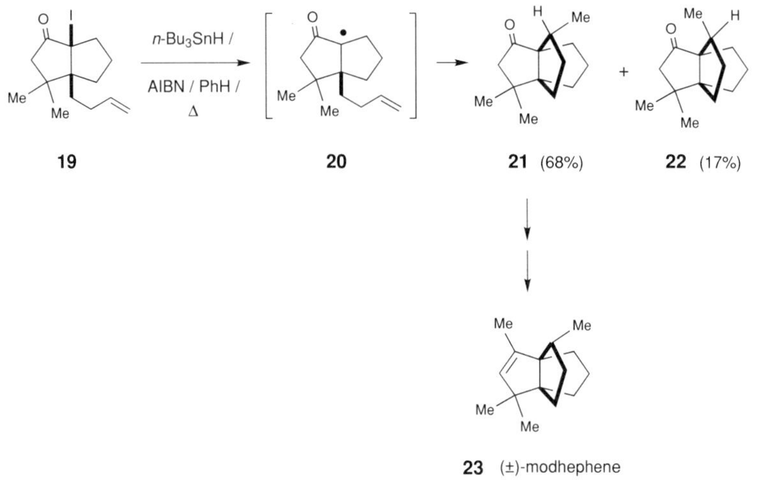

Equation 4

The use of nitrogen-containing precursors to generate alkyl radicals for cyclization is a good way to synthesize nitrogen heterocycles such as the proline derivatives **26** and **30** (Equations 5 and 6)[21] and pyrrolizidinone **33** (Equation 7).[22] Also related to Equation 7 is the rather unusual reaction shown in the following Equation 8, in which the intermediate **35** (generated from **34**), which can cyclize in both *endo* and *exo* fashions *via* **36** and **38**, respectively, yields predominantly product **37** from *endo* cyclization. This behavior has been attributed to the steric bulk and radical-stabilizing property of the bromine atom, together with the inherent strain in the pyrrolizidinone ring system.[22]

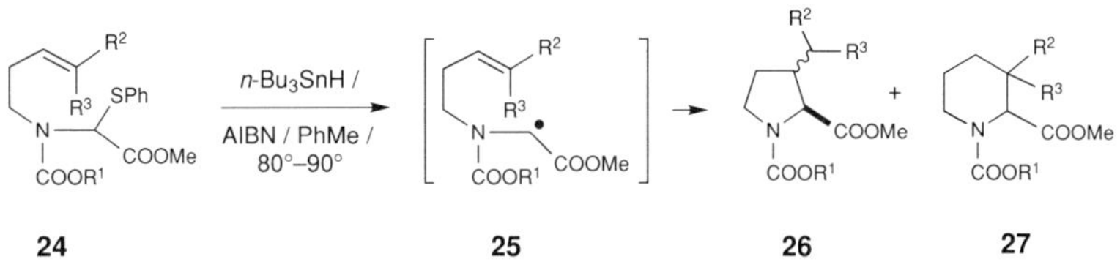

	26	27
R^1 = Et , R^2 = R^3 = H	60% (*cis* : *trans* = 35 : 65)	30%
R^1 = Me , R^2 = Et , R^3 = H	93% (*cis* : *trans* = 35 : 65)	–
R^1 = Me , R^2 = H , R^3 = Et	91% (*cis* : *trans* = 30 : 70)	–

Equation 5

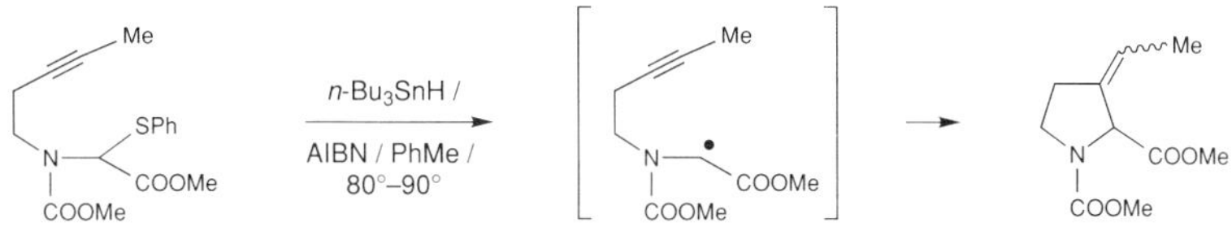

28 **29** **30**

(75% ; $E:Z$ = 59 : 41)

Equation 6

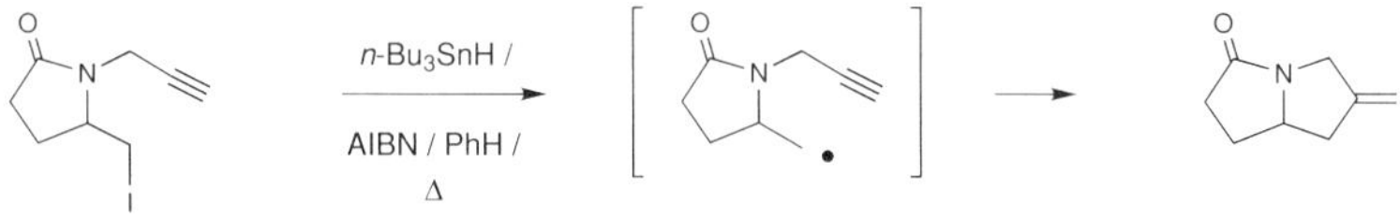

31 **32** **33** (84%)

Equation 7

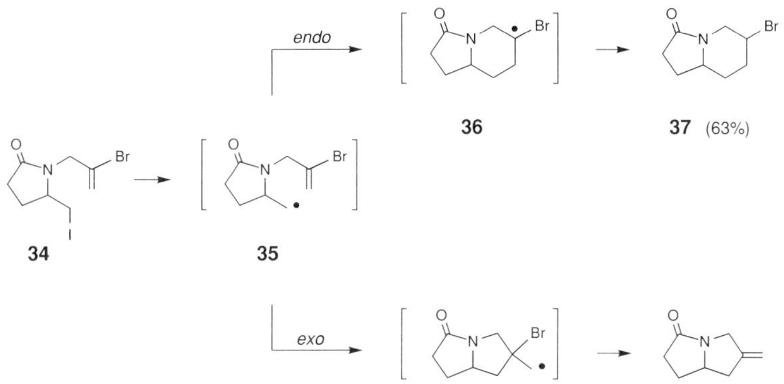

34 **35** **36** **37** (63%) **38** **33** (15%)

Equation 8

Alkyl radical cyclization has also been applied to the synthesis of numerous oxygen heterocycles, for example, methylenefurobenzopyran **41** (Equation 9),[23] the antifungal mold metabolite isoavenaciolide **45** (Equation 10),[24] (–)-mintlactone **48** (Equation 11),[25] tetrahydropyran **50** (Equation 12),[26] spirolactone **53**, and spirobutenolide **54** (Equation 13),[27] the last two *via* the cyclization of intermediate **52** generated by a photo-induced electron transfer process.

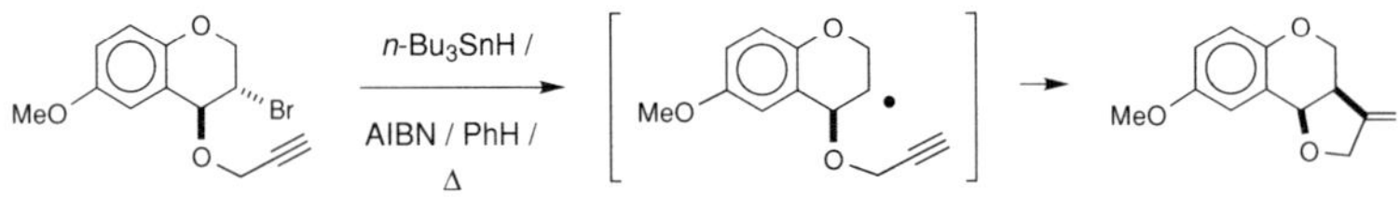

39 **40** **41** (85%)

Equation 9

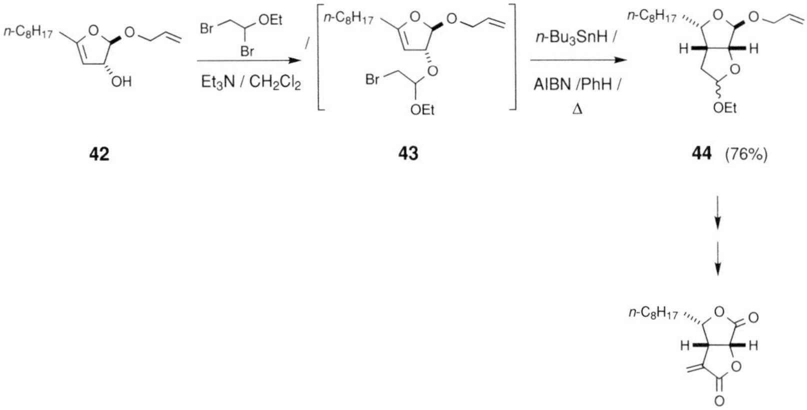

42 **43** **44** (76%)

45

Equation 10

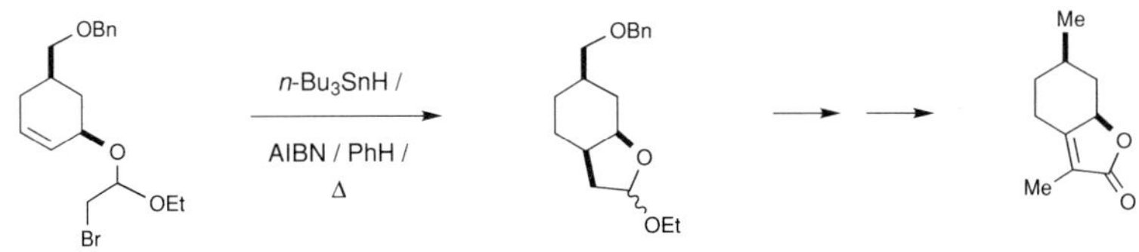

46 **47** (93%) **48**

Equation 11

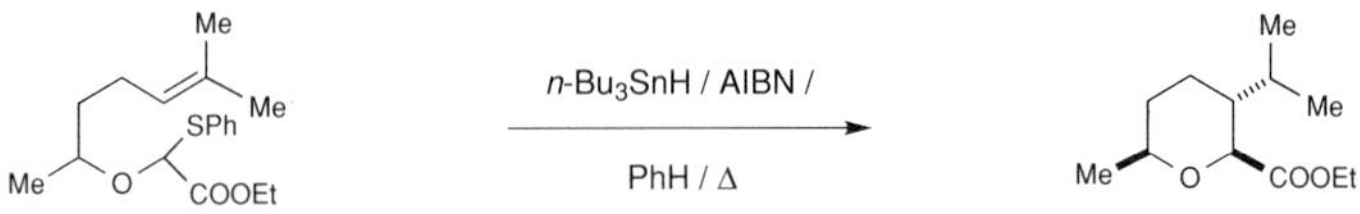

49 **50** (62%)

Equation 12

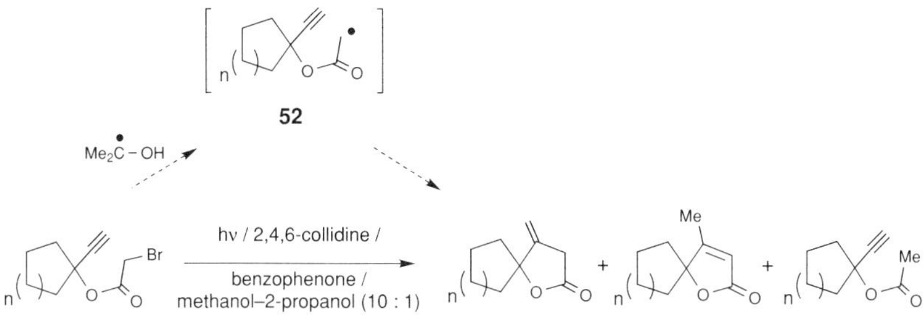

	53	54	55
n = 1	6%	88%	6%
n = 2	2%	69%	4%
n = 3	1%	71%	6%

Equation 13

Recent reports on the properties of tris(trimethylsilyl)silane (TTMSS) as a free radical mediator have shown that it rivals tributyltin hydride, the reagent of choice for radical formation, and is even preferable from the ecological and practical points of view. The Si–H bond strength of TTMSS is weak (79 kcal / mol), fairly close to the Sn–H bond strength in tributyltin hydride (74 kcal / mol),[16] and the reagent is reported[28] to be effective in reducing organic halides *via* a free radical mechanism (Scheme 2). Giese *et al.*[29] have carried out a comparative study of the two reagents, studying the generation and subsequent cyclization of the hexenyl radical from the corresponding bromide. Not much difference was found between them, although TTMSS is slightly more efficient (Equation 14).

Scheme 2

$$RX + (Me_3Si)_3Si^{\bullet} \longrightarrow R^{\bullet} + (Me_3Si)_3Si\text{-}X$$

$$R^{\bullet} + (Me_3Si)_3SiH \longrightarrow RH + (Me_3Si)_3Si^{\bullet}$$

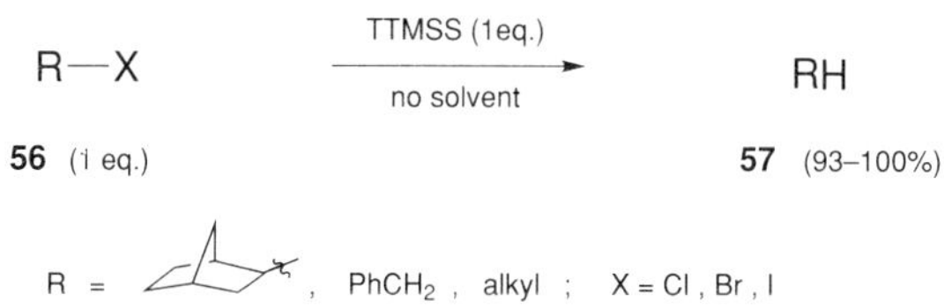

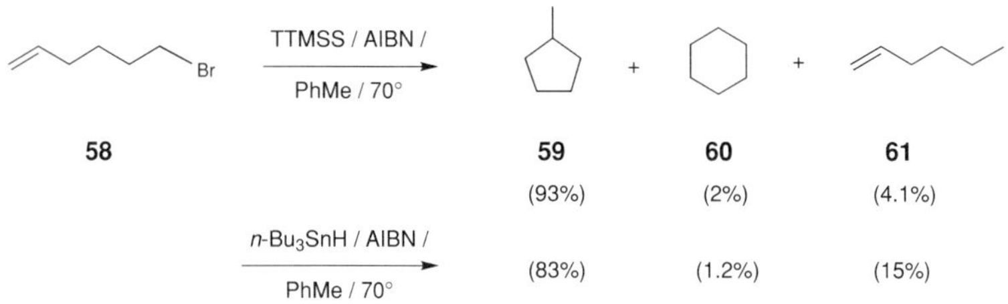

58 59 60 61

(93%) (2%) (4.1%)

(83%) (1.2%) (15%)

Equation 14

Apart from alkyl halides TTMSS also reacts with sulfur heterocycles containing a heteroatom 1,3- to sulfur (**62**) to produce alkyl radicals by ring cleavage similar to the reaction shown in Equation 3 for tin hydride. Thus, upon treatment with TTMSS, 1,3-oxathiolane **62a**, *N*(methylcarbonyl)-1,3-thiazolidine **62b**, and 1,3-dithianes **62c** and **62d** all give the corresponding acyclic analogs **64a–d** in good yields *via* the radical intermediate **63** (Equation 15).[16,30] This type of intermediate is amenable to intramolecular cyclization, as shown in Equation 16,[16,31] where cyclized product **67** is obtained along with the by-product from reduction.

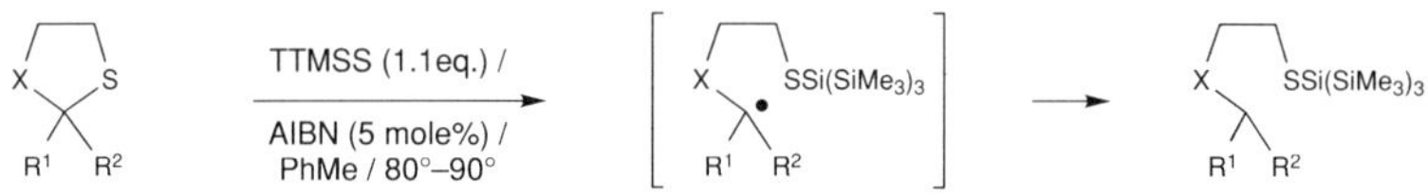

62 63 64

a) X = O, R^1 = H, R^2 = n-C_9H_{19} (82%)
b) X = NCOMe, R^1 = H, R^2 = n-C_9H_{19} (70%)
c) X = S, R^1–R^2 = n-C_4H_8 (91%)
d) X = S, R^1–R^2 = n-C_5H_{10} (90%)

Equation 15

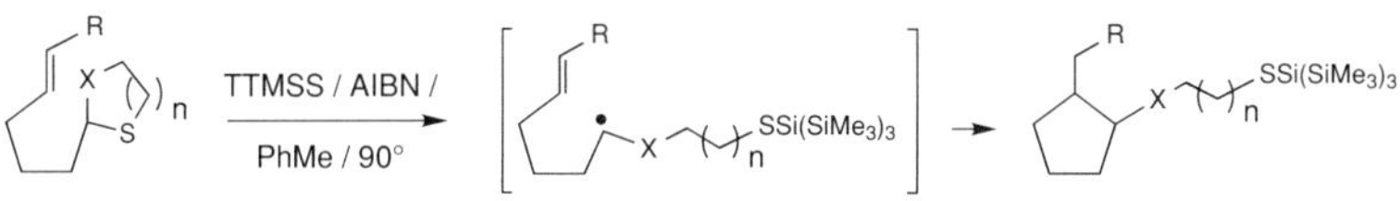

65 66 67

n = 2, X = S, R = Me (82% ; *cis* : *trans* = 1 : 1)
n = 1, X = NCOMe, R = H (41% ; *cis* : *trans* = 2.5 : 1)
n = 1, X = NCOOEt, R = H (42% ; *cis* : *trans* = 2.3 : 1)

Equation 16

Since its first report by Barton and co-workers,[32] the generation of alkyl radicals by chain decomposition of *N*-acyloxypyridine-2-thione **68** has become a very popular reaction with wide applications in organic synthesis.[33] The thione **68** decomposes *via* elimination of acyloxy radical **72**, whose decarboxylation leads to alkyl radical **74**, which may undergo reduction (to X–H) or cyclization, given the appropriate presence of a group with affinity for radicals in the molecule. Similarly the related carbamate **69** suffers chain decomposition to aminyl radical (**74**, X$^{\bullet}$ = $^{\bullet}$NR^{1}R^2) which can cyclize to give cyclic amines such as **76**.[34,35] On the other hand, the carbonate radical **77** (or **72**, X = OR), from radical-initiated (*e.g.*, $Cl_3C^{\bullet}$, *t*-BuS$^{\bullet}$, PhSe$^{\bullet}$) decomposition of thione **70**, is quite reluctant to decarboxylate, the reaction being so slow as to be uncompetitive with other intermolecular processes. However, this stability of the carbonate radical can be turned to advantage by application to the synthesis of various five- and six-membered cyclic carbonates **79** to **82**, which are obtained in moderate yields.[36]

Scheme 3

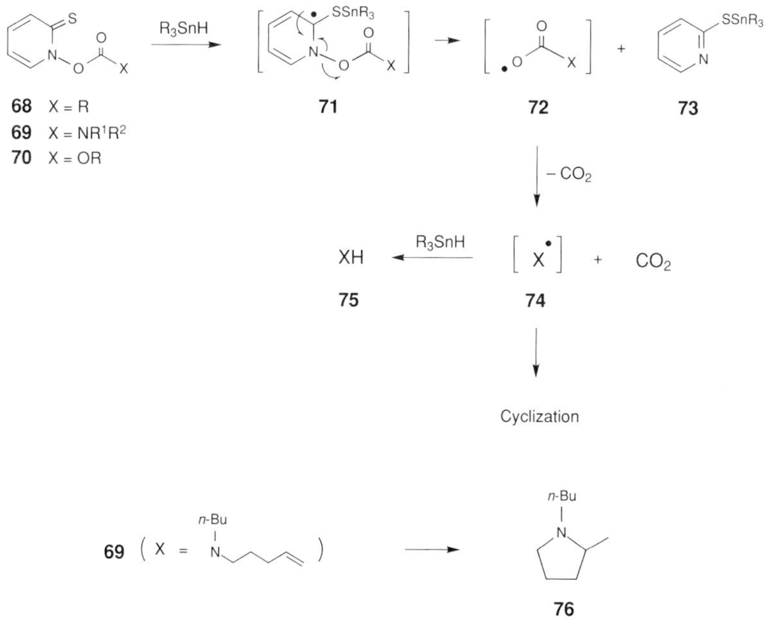

Scheme 4

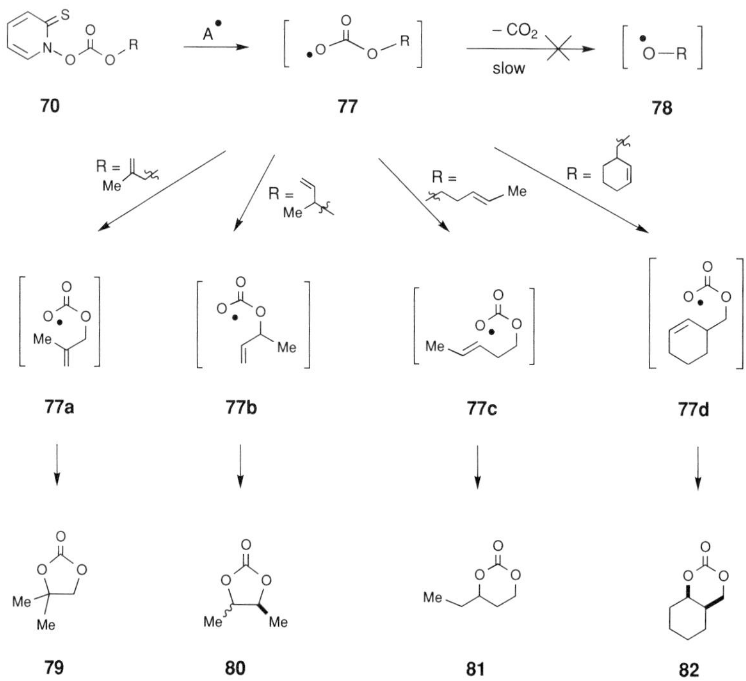

While on the subject of carbonate radical, attention should also be drawn to the alkoxycarbonyl radical **86**, obtainable from *S*-alkoxycarbonyl xanthates **85** by irradiation with a tungsten filament lamp according to the path shown in Scheme 5.[37] Here again we see an example of sluggish decarboxylation (of **86** to alkyl radical **87**) as manifested in the irradiation of xanthate **88** which, depending upon the value of n, can yield either product **91** from direct cyclization of the first-formed intermediate **89**, or product **94** from cyclization of the resultant hexenyl radical **92** following the decarboxylation of **89**.

Scheme 5

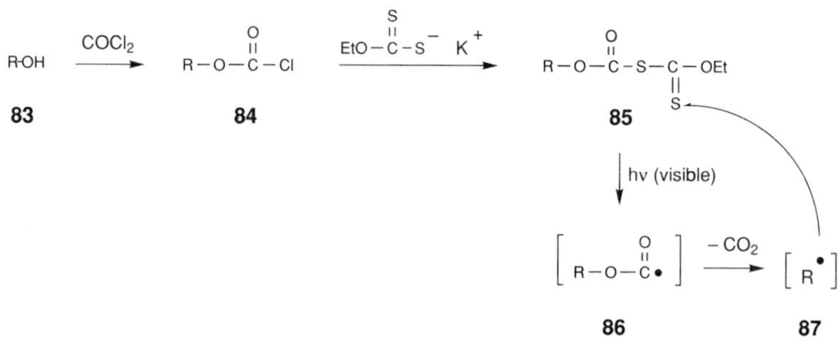

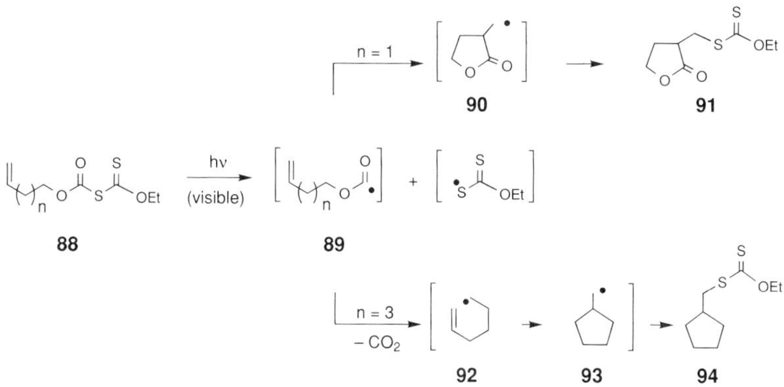

An important factor that controls radical cyclization appears to be the proximity of the reacting centers. Translated into chain length in allenyl radical **96** (generated from iodide **95**), it transpires that varying the value of n can dictate cyclization either at the terminal or central allene carbon to yield cyclized products **97** or **98**, respectively. For values of n = 1 and 4, the first type of cyclization is observed, yielding vinylcyclopropane or cyclohexane **97**. When n = 2, the second type of cyclization prevails to give cyclopentene **98**. In the case of n = 3, products from both types of cyclization may be obtained, subject to the steric interference of R,[38,39] as shown in Scheme 6. However, this pattern of behavior of the unactivated allene is disrupted upon the incorporation of an electron-withdrawing group into the allene moiety, as, for example, in radical intermediate **100**. Regardless of the value of n, no cyclization takes place at the terminal allene carbon, and radical **100** yields products **101** from reduction, **106** and **107** from cyclization *via* intermediate **103**, and **108** from hydrogen shift *via* intermediate **104** (see accompanying table in Scheme 6).[40] The failure to cyclize at the terminal allene carbon is ascribed to the steric bulk of the $-P(O)Ph_2$ group, in conjunction with its superb stabilization of the intermediate radical **103**. Therefore, when n = 2 to 4, cycloalkene **106** is obtained as the major product, and no cycloalkane with exocyclic double bond **105** is observed even when n = 1. Notable also is the not-so-common formation of a seven-membered ring **106** when n = 4.

Interest in the "naked sugars" such as optically active 7-oxabicyclo[2.2.1]hept-5-en-2-one **109** or its derivative **110** has lately grown with their use as precursors in the syntheses of rare carbohydrates,[41] C-nucleosides,[42] and biologically active natural products such as (+)-castanospermine.[43] It comes as no surprise, therefore, that compound **109** should be subjected to study in terms of radical cyclization reactions and other synthetic possibilities as shown in Scheme 7.[44]

Scheme 6

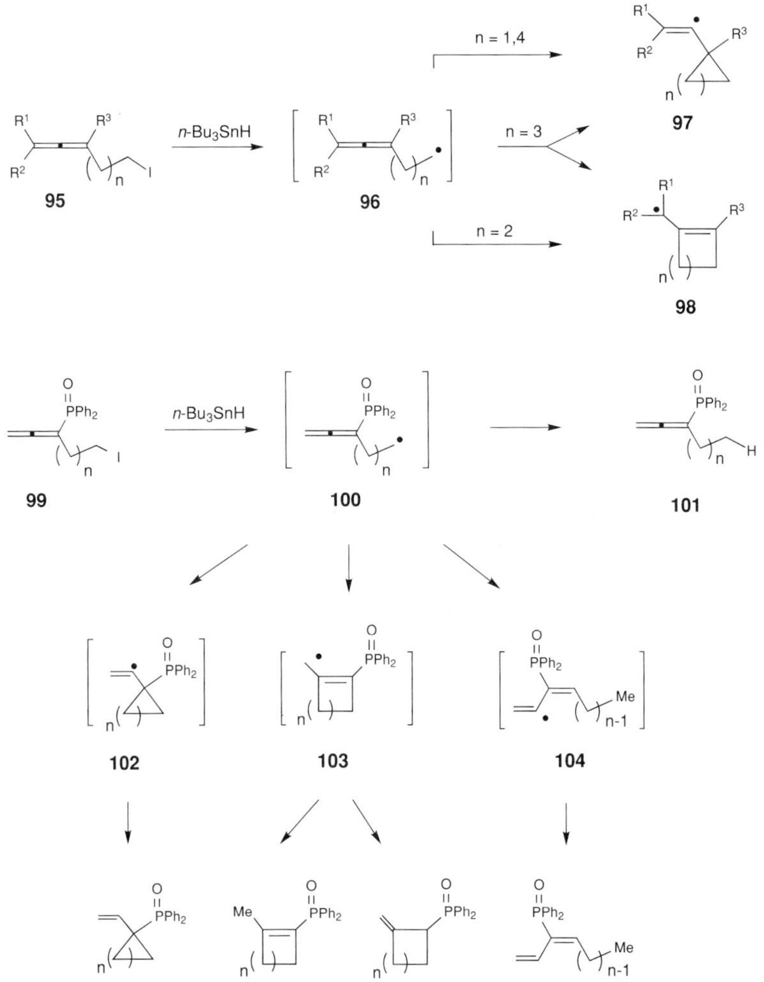

	105	106	107	108	101
n = 1	–	–	–	–	50%
n = 2	–	79%	3%	–	–
n = 3	–	85%	–	–	–
n = 4	–	39%	–	9%	35%
n = 5	–	–	–	3%	51%

Conversion of the sugar **109** to its silyl ketene acetal **112** followed by reaction with the radical generated from chloromethyl phenyl sulfone under reducing conditions results in addition to form intermediate **113**, which, from its vantage geometry cyclizes spontaneously to give **114**, thence the final product **115a**. The reaction represents the first example of radical cyclization employing initial radical addition to silyl ketene acetal to provide the dialkoxyalkyl radical for cyclization. Apart from reduction to product **115a**, the possibility of trapping the intermediate radical **114** with various groups also exists, which reaction is also expected to be stereospecific, demanding an exclusively *exo* approach of the incoming group. This extended potential has been realized: upon substituting allyl tri-*n*-butyltin for tri-*n*-butyltin hydride as the radical initiator, intermediate **114** is trapped by allyl radical to furnish the stereospecific product **115b**, albeit in low yield (Scheme 7).[44]

Scheme 7

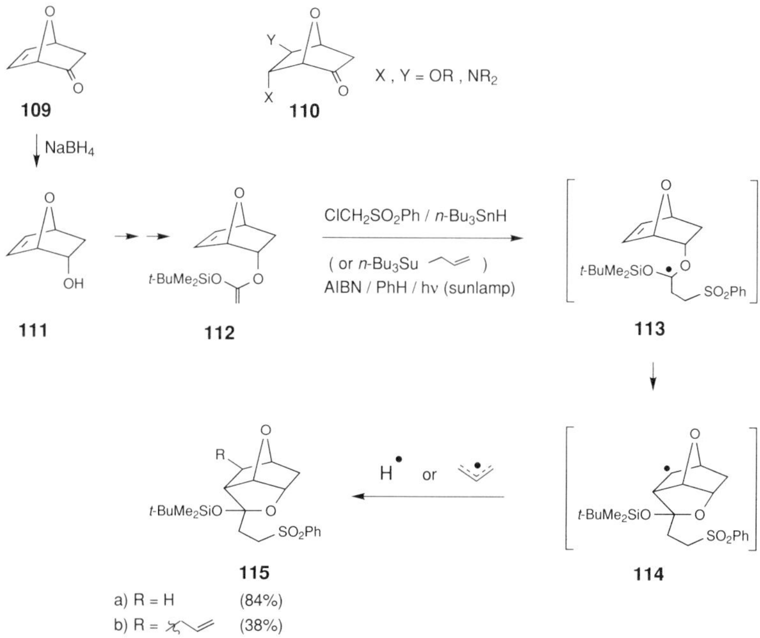

The use of the sugar molecule as a building block in organic synthesis is further exemplified in the asymmetric synthesis of bis-lactone **119**, a class of antifungal metabolites widely found in nature. Starting from D-apio-L-furanose **116**, the xanthate **117** is prepared and cyclized to afford **118** in moderate yield. Compound **118** is then readily converted to the target bis-lactone **119** (Scheme 8).[45]

Scheme 8

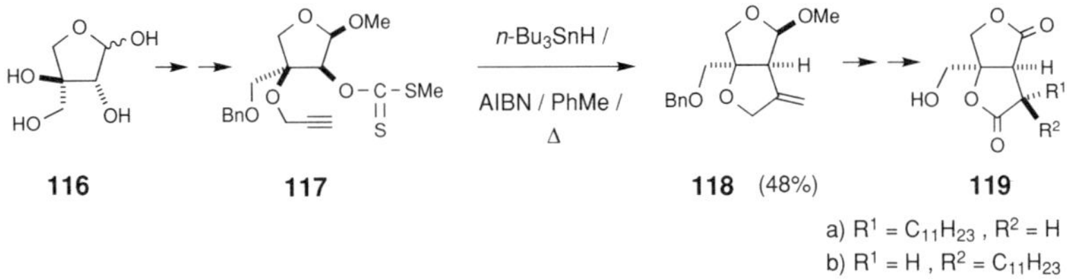

B. VINYL AND ARYL RADICALS

It is probably true that the report by Stork and Baine in 1982 on the cyclization of vinyl radical to the unactivated olefinic bond (Equation 17)[46] sparked renewed interest in radical reactions, in particular the radical cyclization. Since then the progress in this field has been altogether phenomenal.

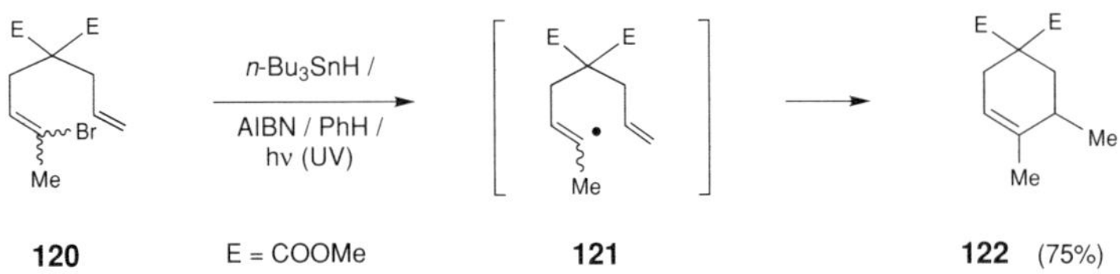

Equation 17

Straightforward, useful, and interesting examples of vinyl radical cyclization to unactivated double bonds in organic synthesis are the preparations of fused α-methylene-γ-butyrolactone (Equation 18)[47] and functionalized polycyclic carbocycles (Equation 19).[48] As for aryl radicals, there are the syntheses of fused benzofuran (Equation 20),[49] chroman (Equation 21),[50] and the construction of the indole nucleus (Equation 22).[51]

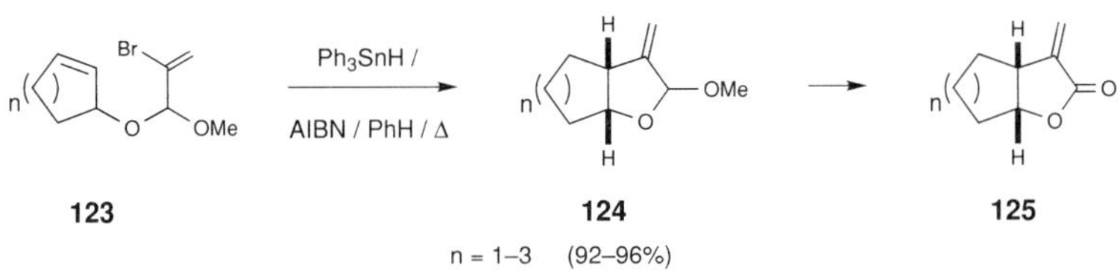

Equation 18

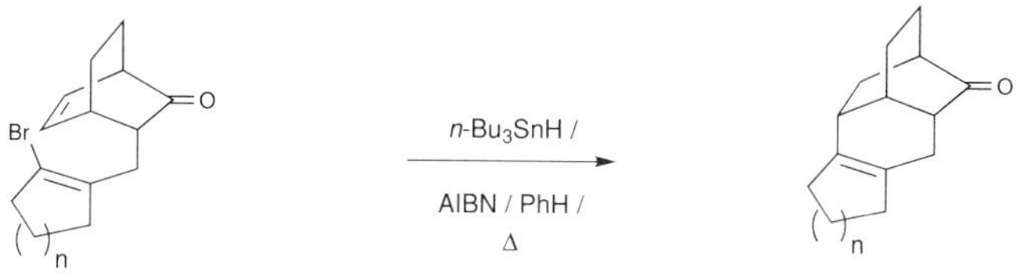

126 **127**

n = 1 (85%)
n = 2 (86%)

Equation 19

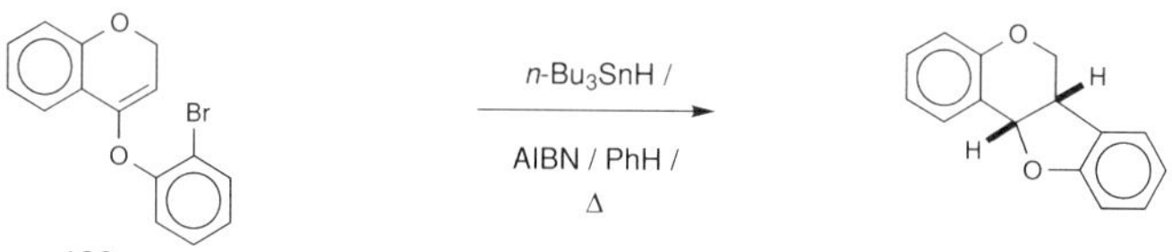

128 **129** (82%)

Equation 20

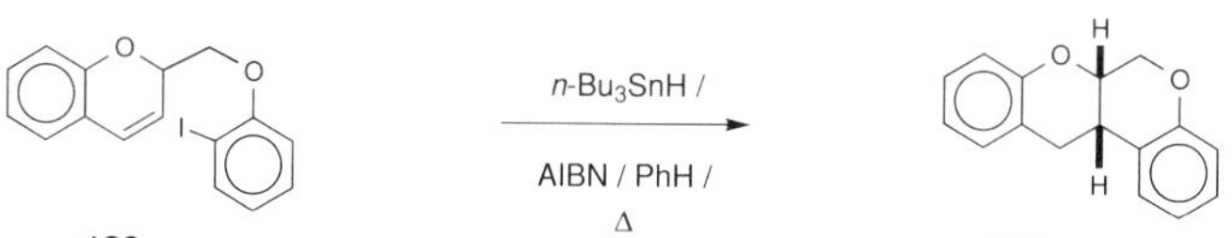

130 **131** (74%)

Equation 21

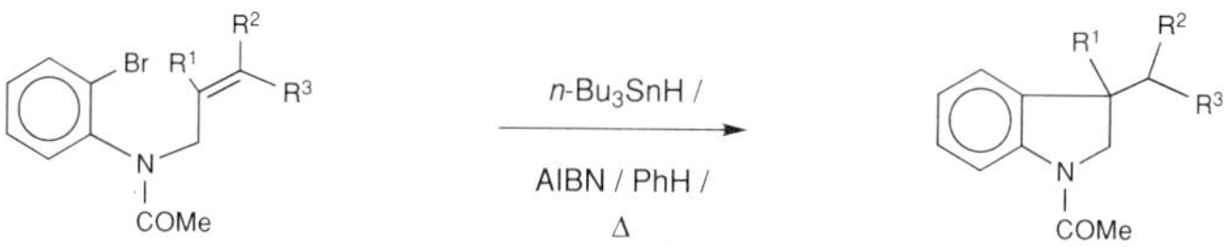

132 **133**

R1, R2, R3 = H, Me (>90%)

Equation 22

The tin-mediated radical cyclization shown in Scheme 9 constitutes a new route to spiroketals.[52] The reaction involves the generation of vinyl radical **136** from substrate **135** by the addition of tributyltin radical. Subsequent cyclization of **136** followed by removal of the tin residue furnishes methylene spiroketal **138**. The cyclization appears to be stereospecific, as evidenced in the tin-mediated cyclization of substrate **139**, which yields **140** as the sole cyclized product. The latter comprises a model study on the synthesis of phyllanthocin **142**, the aglycone of the antineoplastic agent phyllanthoside.[53]

Scheme 9

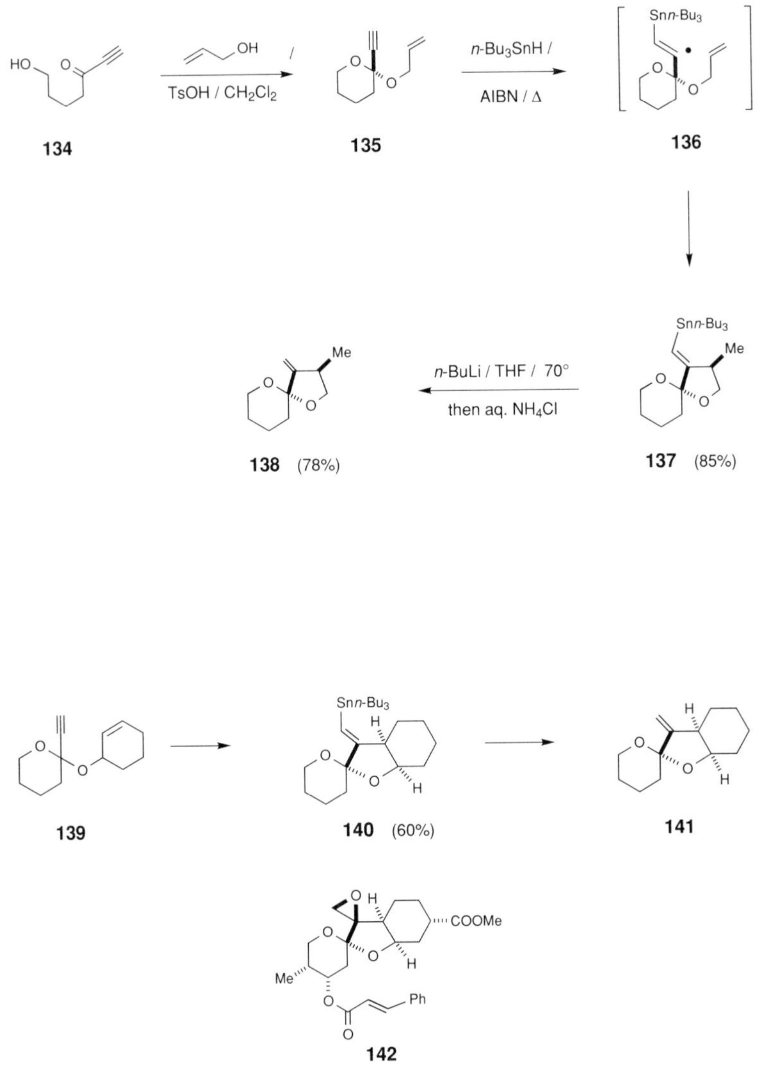

Tris(methylseleno)- and tris(phenylseleno)boranes are new reagents for the generation of RSe radicals. As illustrated in Scheme 10, treatment of acetylene **143** with tris(alkylseleno)borane **144** results in the isolation of **146a** (along with a trace amount of **146b** in certain cases), derived from radical intermediate **145**.[54] The reaction has been used to synthesize pyrrolidine derivative **149** in high yield *via* the phenylselenide-mediated cyclization of vinyl radical **148**. The high stereoselectivity associated with the cyclization is demonstrated in the use of precursor **150**, which yields *trans*-(**151**) and *cis*-(**152**) products in a 4:1 ratio (Scheme 10).[54]

Scheme 10

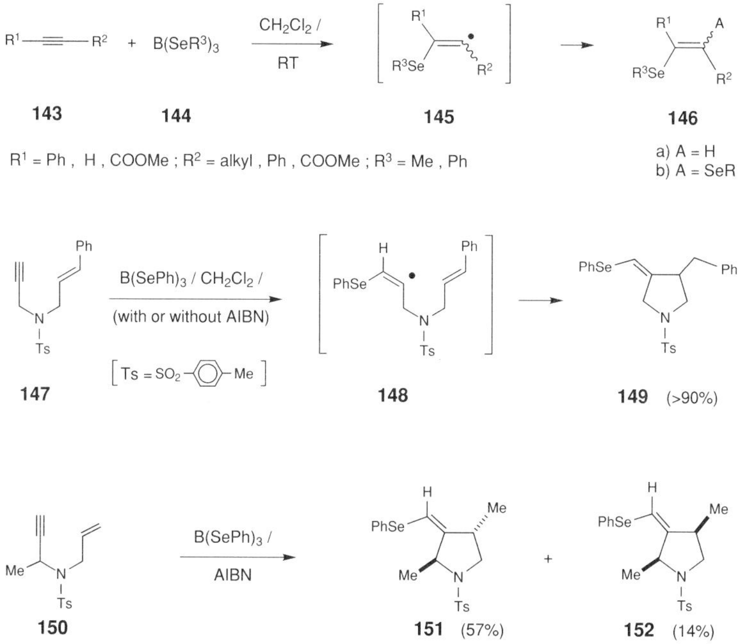

In general, the cyclization of vinyl and aryl radicals is appreciated as a very useful method of ring construction which has enjoyed successful applications in the synthesis of complex natural products.[55] However, most of these have involved the construction of five-membered rings, *i.e.*, the 5-*exo* cyclization, and reports on the 6- and 7-*exo* cyclizations to form six- and seven-membered rings have been few. The latter includes, besides the examples already shown in Equations 12, 21, and Scheme 4, the cyclization of aryl radical **154** (formed from **153** by the method of Stork: n-Bu_3SnCl / $NaBH_3CN$ / t-BuOH)[56] to afford quinoline or benzazepine derivatives **155** from 6- or 7-*exo* cyclizations, respectively, as shown in Scheme 11 and the accompanying table.[57] It should be noted that a small amount of product from 7-*endo* cyclization is observed in certain cases (**156b** and **156c**), but never the product from 8-*endo* cyclization (**156d** and **156e**). Also significant is the predominance of reduction products **157d** and **157e** over cyclized products **155d** and **155e**, a confirmation of the more favorable disposition of the 6- over the 7-*exo* mode of cyclization.

Scheme 11

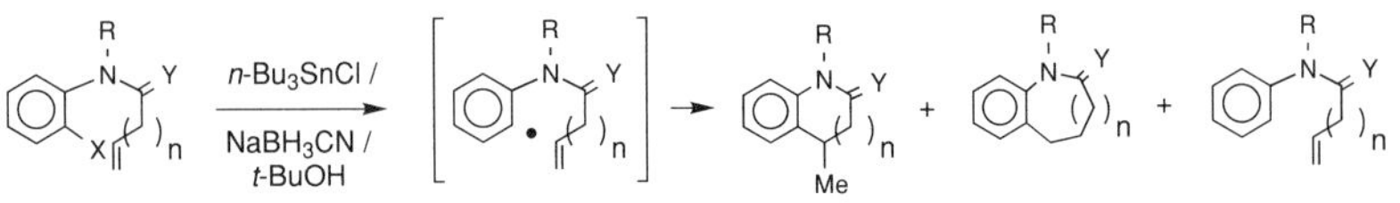

153 154 155 156 157

	155 %	156 %	157 %
a) n = 1 , X = I , R = Me , Y = O	64	–	–
b) n = 1 , X = Br , R = CH_2Ph , Y = O	58	trace	20
c) n = 1 , X = Br , R = COMe , Y = H , H	72	< 5	18
d) n = 2 , X = Br , R = CH_2Ph , Y = O	25	–	60
e) n = 2 , X = I , R = Me , Y = O	13	–	49

As regards the 6-*endo* mode of cyclization, it has already been stated in Scheme 1 that the rate of radical cyclization is much faster in the competing 5-*exo* than in the 6-*endo* mode. This is well illustrated in the simple reaction shown in Equation 14, where the cyclopentane product **59** is obtained in overwhelming ratio over cyclohexane **60**. Regardless of this fact, however, manipulations of reaction conditions can lead to a reversal in product ratio, as claimed by reports[58,59] that keeping the hydride concentration low can significantly increase the yield of 6-*endo* cyclization products, a change attributed to rearrangement of the kinetically preferred 5-*exo* cyclized radical to the thermodynamically more stable 6-*endo* counterpart (**2** → **3** in Scheme 1). This has been tested by treating radical precursors **158** and **159**, synthesized from Corey lactone,[17] with tri-*n*-butyltin hydride using either AIBN or Et_3B as the radical initiator. The results are summarized in Scheme 12 and the accompanying table. The optimum reaction condition in this case is obtained by treating vinyl bromide **159** with tri-*n*-butyltin hydride and 0.4 equivalent of Et_3B in toluene, at –30°C, for 0.5 h, which results in the 6-*endo* cyclized product **161** in 97% yield. The cyclization is completely stereocontrolled, yielding exclusively the *trans*-hydrindan **161**, unaccompanied by any product from 5-*exo* cyclization (**162**) or reduction (**163**).[60]

Further examples of aryl radical cyclization in the 6-*endo* fashion are shown in Scheme 13.[61] The *endo* cyclized **166** is obtained as the sole product in every case, although only in low to moderate yields. However, good to excellent yields can be obtained with aryl radical **168**, derived from the corresponding bromide **167**, which cyclizes in the 6-*endo* fashion to afford *trans*-octahydroanthracene **169** (Scheme 14).[62]

Scheme 12

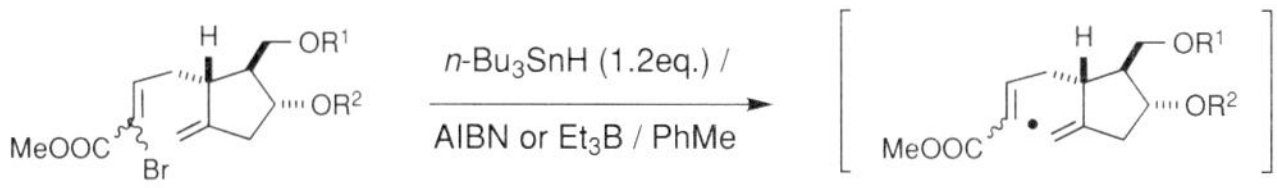

158 $R^1 = SiMe_2t\text{-}Bu$
$R^2 = THP$

159 $R^1 = SiMe_2t\text{-}Bu$
$R^2 = H$

160

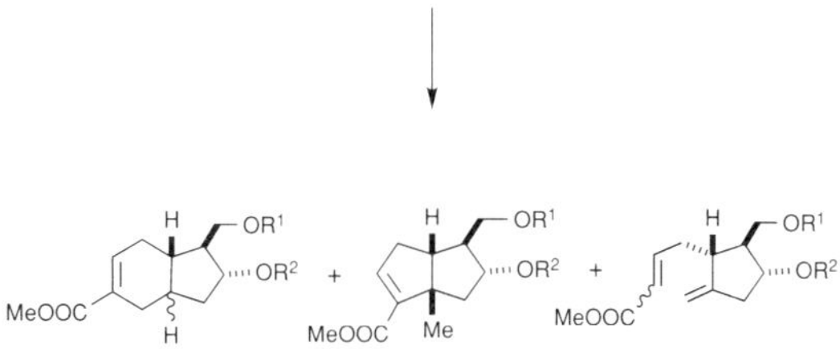

	condition	**161** (6-*endo*)	**162** (5-*exo*)	**163** (reduced)
158	AIBN (0.7 eq.) / 120° / 1 h	83% (*trans* : *cis* = 7 : 3)	–	–
	Et_3B (0.7 eq.) / – 50° / 1 h	49% (*trans* : *cis* = 8 : 2)	33%	7%
159	AIBN (0.08 eq.) / 120° / 0.5 h	72% (*trans* : *cis* = 6: 4)	–	8%
	Et_3B (0.2 eq.) / – 50° / 3 h	23% (*trans* only)	54%	8%
	Et_3B (0.4 eq.) / – 30° / 0.5 h	97% (*trans* only)	–	–

Scheme 13

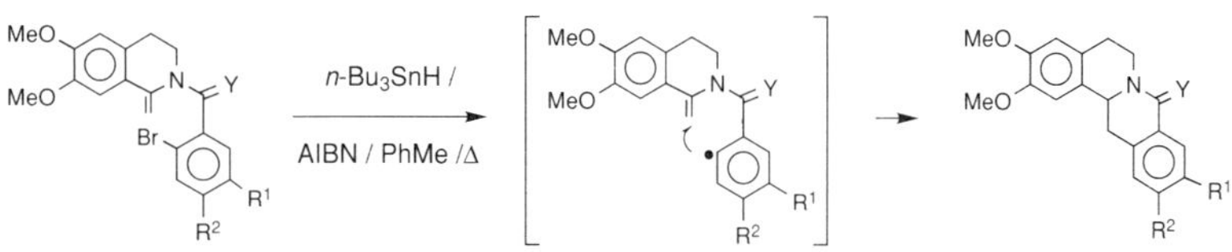

164 **165** **166**

a) Y = O , $R^1 = R^2 = H$ (61%)
b) Y = O , $R^1 = R^2 = OMe$ (62%)
c) Y = H , H , $R^1 – R^2 = – O – CH_2 – O –$ (51%)
d) Y = H , H , $R^1 = R^2 = OMe$ (32%)

Scheme 14

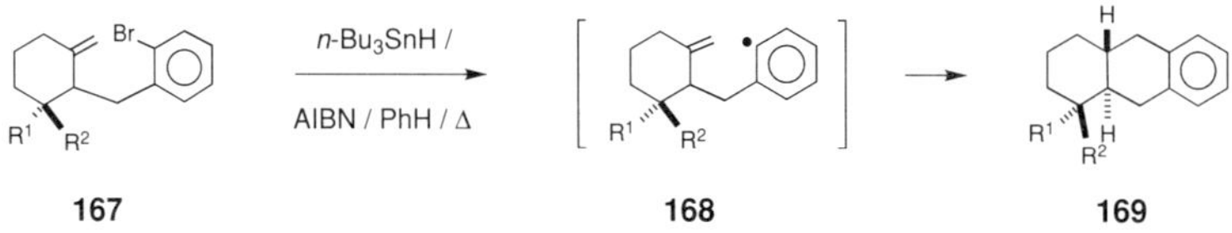

a) $R^1 = R^2 = H$ (86%)
b) $R^1 = R^2 = Me$ (95%)
c) $R^1 = Me$, $R^2 = COOMe$ (85%)

C. ACYL, THIOACYL, IMIDOYL, AND IMINYL RADICALS

A variety of methods is currently available for the generation of acyl radical **170**; for example, the reaction of acid chlorides **171**[63] or phenylseleno esters **172**[64] with tri-*n*-butyltin hydride, and the irradiation of *S*-acylxanthates **173**[65] or acylcobalt salophenpyridine complexes **174**[66] (Scheme 15).

Scheme 15

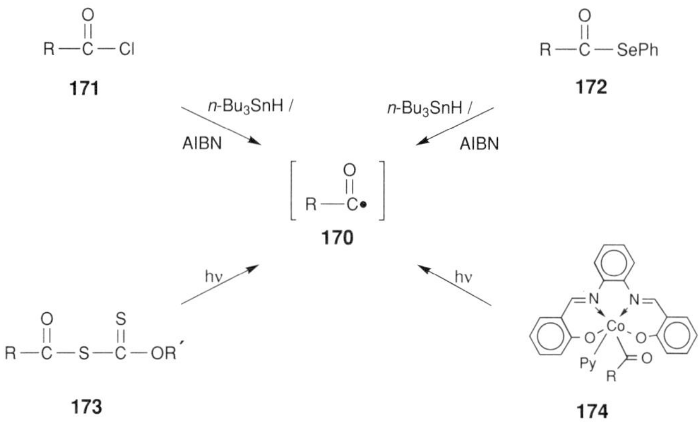

Once generated, acyl radicals can be trapped intramolecularly, provided that an appropriate group with affinity for radicals is present in suitable proximity. Equations 23 to 27[64,67,68] exemplify ring formation by intramolecular trapping of acyl radicals generated from phenylseleno esters (n-Bu_3SnH / AIBN / PhH / 80°) with unactivated double bonds, while Equation 28 is an example of ring construction employing acyl radical from the irradiation of acylcobalt salophenpyridine complex **187**.[66]

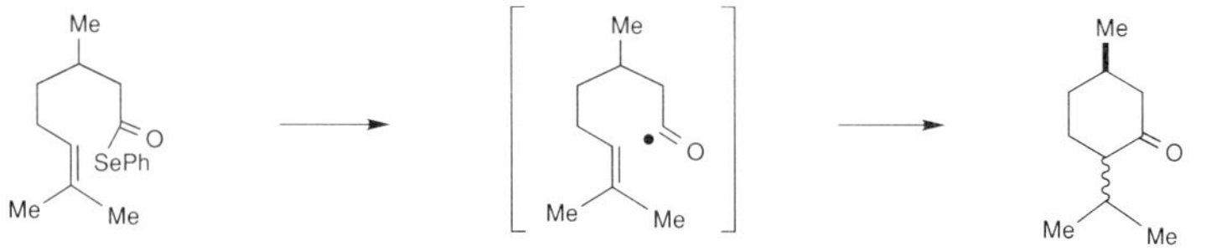

175 **176** **177** (84%)

(*trans* : *cis* = 1.3 : 1)

Equation 23

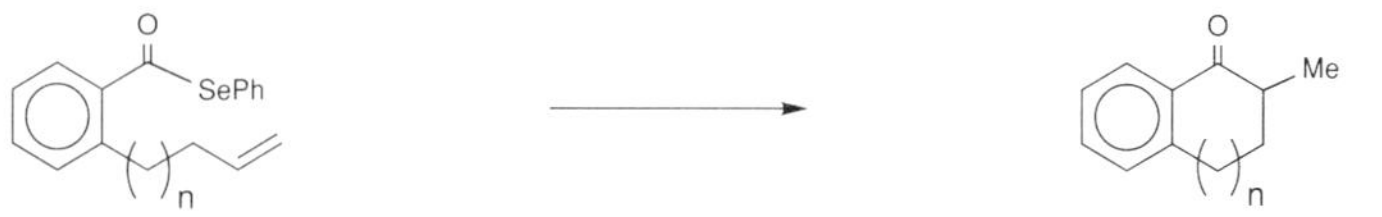

178 **179**

n = 0 (81%)
n = 1 (76%)
n = 2 (74%)

Equation 24

180 **181** (86%)

Equation 25

182 **183** (69%)

Equation 26

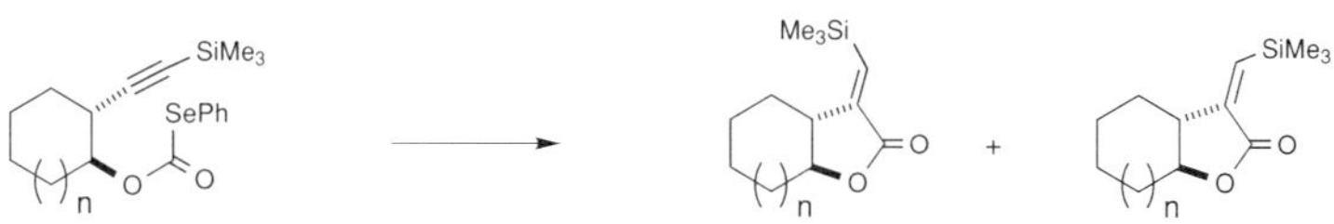

184 **185** **186**

	185	186
n = 1	(6%)	(74%)
n = 2	(8%)	(84%)

Equation 27

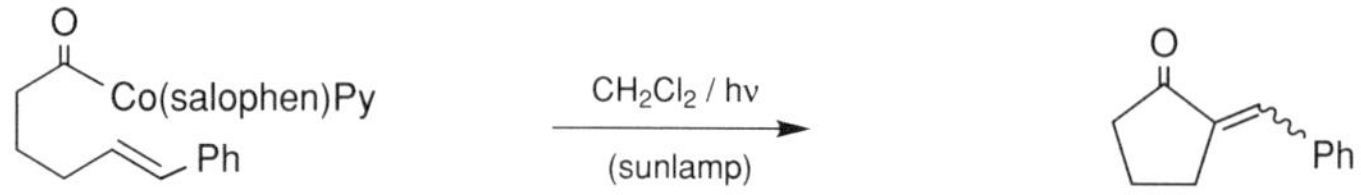

187 **188** (42%)

Equation 28

In another example the irradiation of *S*-acylxanthate **189** produces acyl radical **190** which adds to the π bond, the addition then terminating by coupling with xanthate radical to yield **193** (see Scheme 5). This coupling product has been isolated and treated with base to furnish 3-methylene-4-chromanone **194**, as depicted in Scheme 16.[65]

Scheme 16

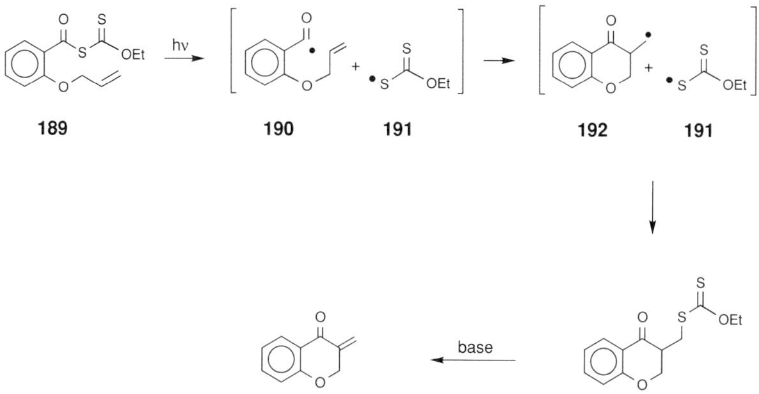

189 **190** **191** **192** **191**

194 (40%) **193** (70%)

A comparative study of the cyclization of *erythro*- and *threo*-phenylseleno esters **196** and **201** (prepared from furanose **195** and threitol **200**, respectively) shows the reaction to yield a mixture of cyclohexanones and cycloheptanones (Scheme 17).[69] From the product distribution it is clear that the *cis*-fused dioxolane ring has a distinct advantage in radical cyclization, both in terms of the overall reaction yield (quantitative product recovery from **196**) and the yield of cyclized products, here comprising **197** and **198** from 6-*exo* and 7-*endo* cyclizations, respectively. The cyclization is accompanied by only a small amount of reduction product **199**, in contrast to the reaction of the *trans*-fused substrate **201**. It is interesting to note that radical cyclization is playing an important role lately in the synthesis of cyclic carbohydrates.[70]

Scheme 17

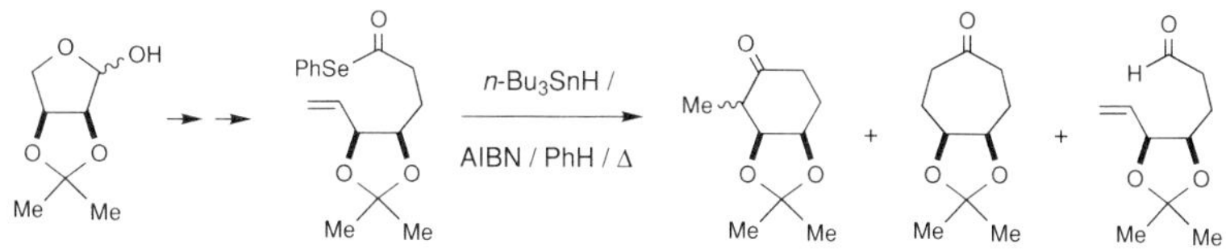

195 **196** **197** (42%) **198** (51%) **199** (7%)

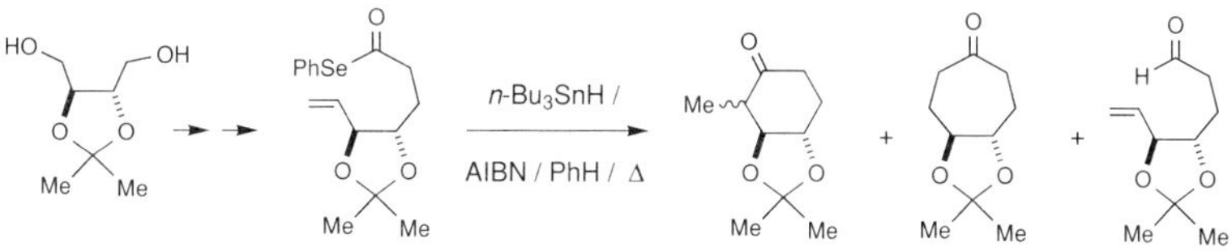

200 **201** **202** (19%) **203** (10%) **204** (44%)

Treatment of homoallylic alcohol dithiocarbonate **205** with tributyltin hydride produces a radical, which upon cyclization and subsequent hydrolysis yields γ-butyrolactone **210**. The reaction may be viewed as a cyclization of either alkyl (**206**) or thioacyl (**208**) radical, depending on whether the elimination of the *n*-Bu_3SnSMe residue occurs before or after ring formation. Whichever way it goes, the dithiocarbonate **205** can be given full credit as a thioacyl (or even acyl) radical precursor (Scheme 18).[67]

Scheme 18

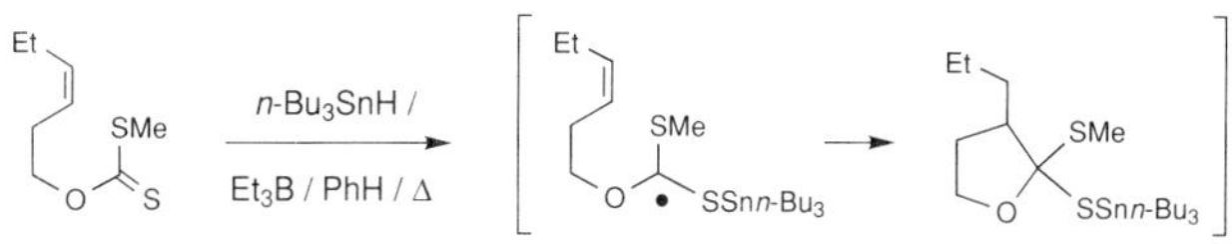

205 **206** **207**

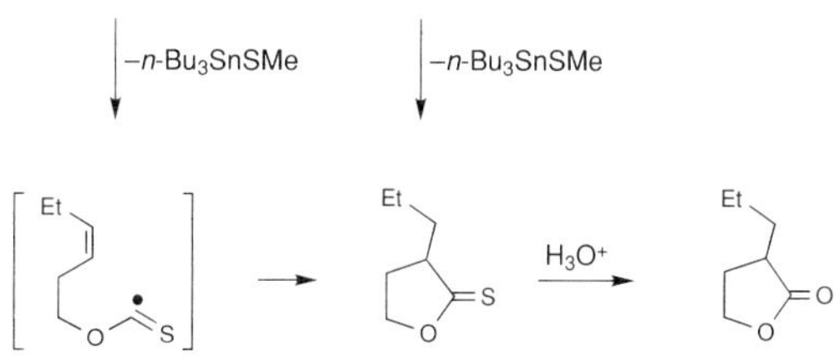

208 **209** **210** (82%)

Other dithiocarbonates have also been adopted as precursors for thioacyl radicals, as exemplified by **211** and **213** in Equations 29[71] and 30,[72] respectively.

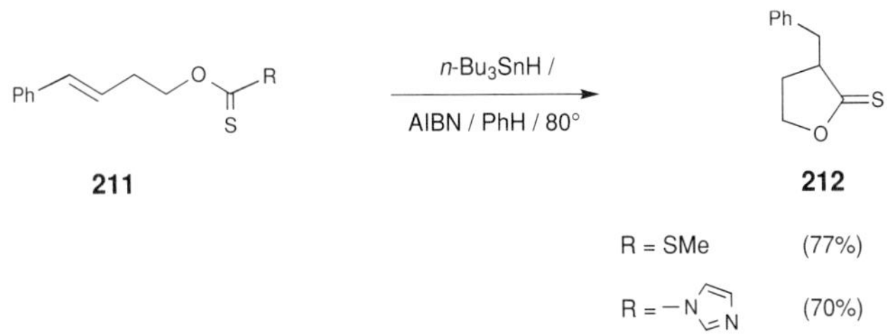

Equation 29

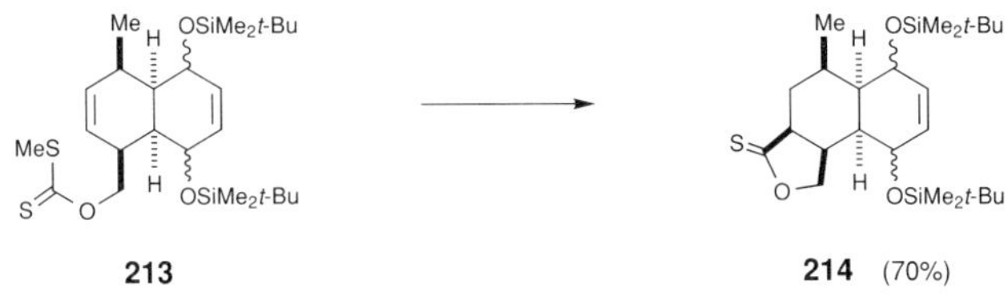

Equation 30

The generation and use of the imidoyl radical, a species isoelectronic with the vinyl, acyl, and thioacyl radicals, have also been investigated. Reduction of selenoimidate **215** with tributyltin hydride under standard conditions is used to generate the imidoyl radical **216** which, contingent upon substituent R, can react in either of two ways: fragmentation to give nitrile **217**, or cyclization *via* intermediate **218** to yield **219**, as illustrated in Scheme 19.[73]

Scheme 19

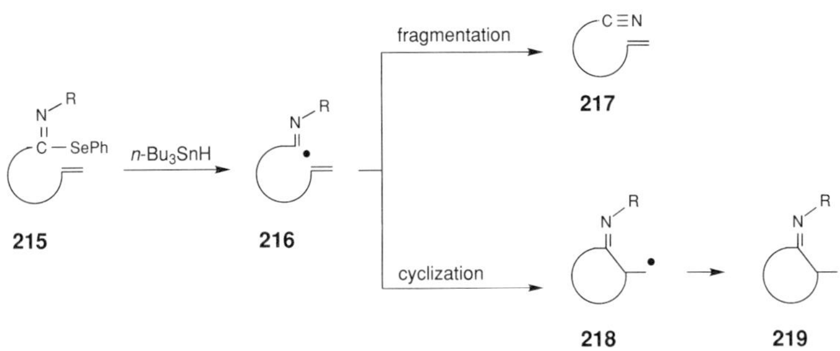

The influence of substituent R on the fate of the imidoyl radical is clearly demonstrated in the examples shown in Schemes 20 and 21. The *N*-benzylimidoyl radical **221** gives a quantitative yield of a 1:1 mixture of nitrile

222 and chromanone **224**, whereas the *N*-methyl analog **228** (generated in a manner similar to that shown earlier in Scheme 18) does not fragment but gives solely, after hydrolysis, chromanone **230**.

Scheme 20

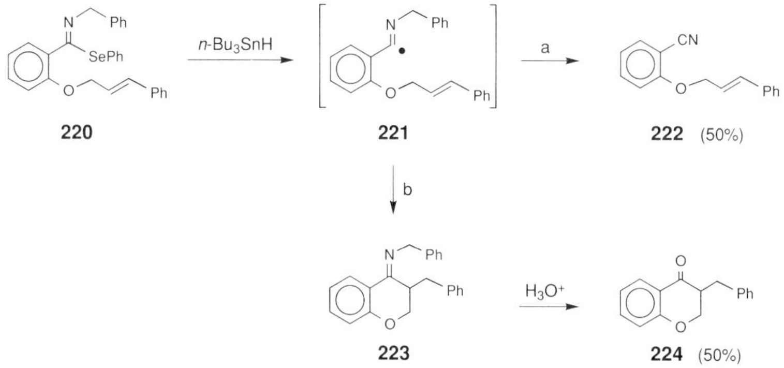

Scheme 21

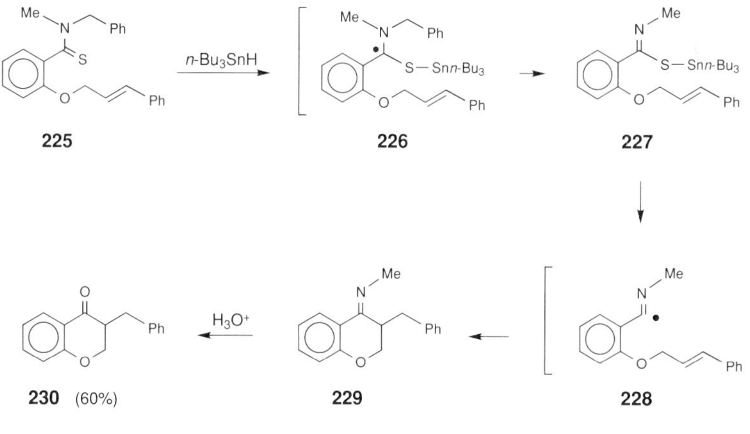

Likewise the thioimidoyl radical **232**, generated by treatment of alkenylisothiocyanate **231** with tri-*n*-butyltin hydride and AIBN, undergoes intramolecular cyclization to give, after hydrolysis, the corresponding thiolactams **235** (Scheme 22).[74]

Scheme 22

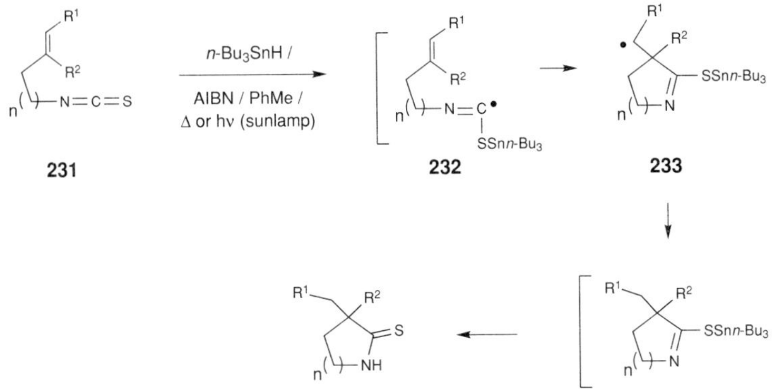

231 **232** **233** **234** **235**

n = 1 , $R^1 = R^2 = H$	(70%)
n = 1 , $R^1 = H$, $R^2 = M$	(40%)
n = 2 , $R^1 = SPh$, $R^2 = H$	(69%)

The investigation of alkenyl radical cyclizations has also been extended to aminyl[75] and iminyl[76,77] radicals. The latter can be generated conveniently by the tin-mediated fragmentation of *S*-aryl sulphenylimines. The cyclizations are exemplified in Equations 31 to 33.[76]

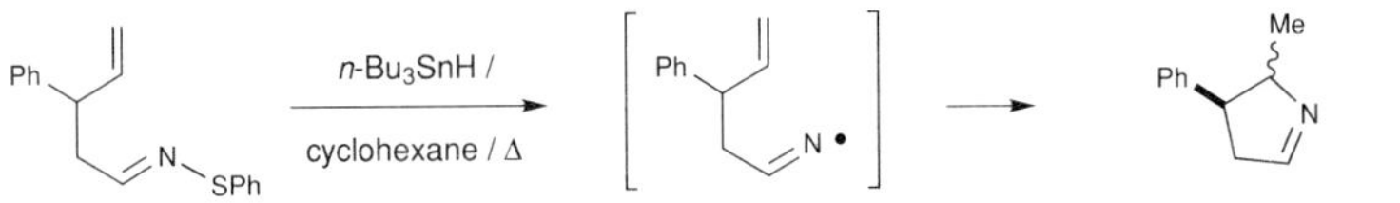

236 **237** **238** (61%)
(*trans* : *cis* = 9 : 1)

Equation 31

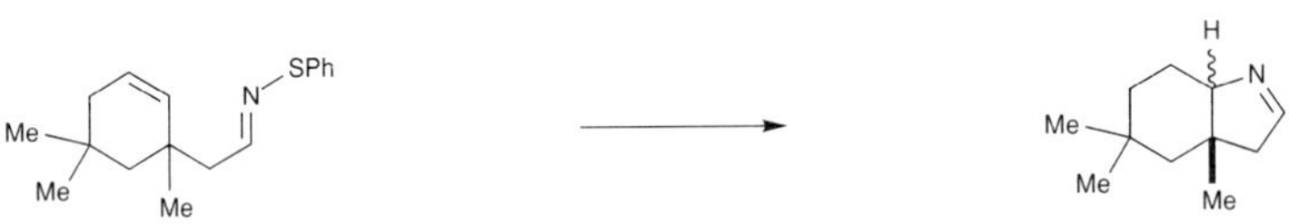

239 **240** (70%)
(*trans* : *cis* = 1 : 19)

Equation 32

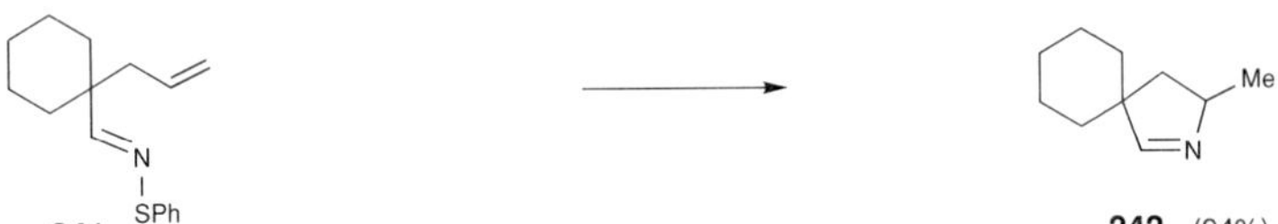

241 **242** (94%)

Equation 33

III. CYCLIZATION TO ACTIVATED OLEFINS AND CARBONYL FUNCTIONS

A. CYCLIZATION TO ACTIVATED OLEFINS

The cyclization of radical centers to activated olefinic bonds mostly involves 1,4-addition reactions in the fashion of the conjugate Michael addition. These cyclizations are much more favorable than, and thus compete successfully with, the corresponding cyclizations to unactivated π bonds, as evident in numerous recent reports. An early example of intramolecular radical conjugate addition was reported by Danishefsky *et al.*[78] and featured the acetoxymercuration of alkenylenone **243** followed by cyclization of the alkyl radical intermediate **245** formed from cleavage of the carbon–mercury bond in **244** upon treatment with $NaBH(OMe)_3$ (Scheme 23). This reaction is, in fact, a Giese reaction,[79,80] which is a method discovered by Giese for the generation of radical intermediates . Once formed, radical intermediate **245** rapidly adds to the enone system to yield bicyclic product **246**.

Scheme 23

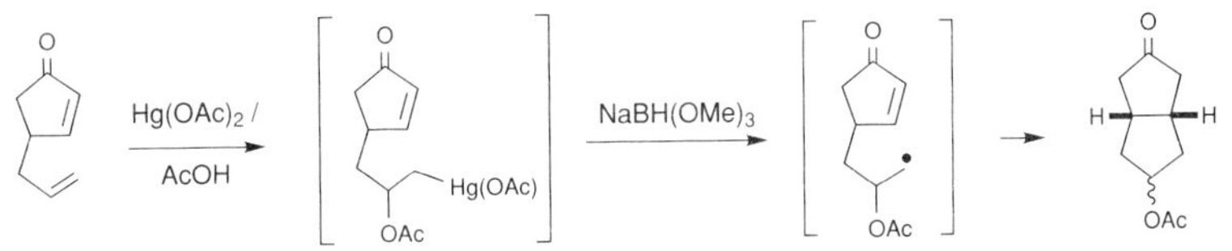

243 **244** **245** **246** (58%)

Consequent to the report by Danishefsky and co-workers and the closely following report by Stork and Baine[46] on the cyclization of vinyl radical (derived from the reaction of tin hydride / AIBN on the corresponding vinyl bromide, Equation 17) there was an explosion in activity in the area of intramolecular conjugate addition of radical centers. Reports on reactions involving all kinds of radicals and radical acceptors flooded the literature and greatly promoted the use of these reactions in organic synthesis.

Equations 34 to 43 offer cross-sectional examples of various radicals in the cyclization to α,β-unsaturated ketones and esters, namely alkyl radicals generated from the irradiation of trimethylsilylmethylamine **247** (Equation 34)[81] and dithiane **250** (Equation 35),[82] and from the treatment of thiocarbonate **253** with tributyltin hydride (Equation 36);[83] vinyl radicals **257** and **260** in the syntheses, respectively, of bicyclic ketone **258** (Equation 37)[84] and tetrahydrofuran **261** (Equation 38),[85] the latter a subunit of avermectins and milbemycins;[86] aryl radicals **263** and **267** in the construction of **264** and **268**, which are precursors of aflatoxin B_1 **265** (Equation 39)[87] and lycoramine **269** (Equation 40),[88] respectively; acyl radicals formed *via* the tin hydride treatment of phenylseleno ester **270** (Equation 41)[64] and the irradiation of acylcobalt salophenpyridine complex **273** (Equation 42),[66] the latter leading to two products **276** and **278** as shown; and lastly, the cyclization of allylic radical generated by the standard tin hydride method is represented by **280** in Equation 43.[89]

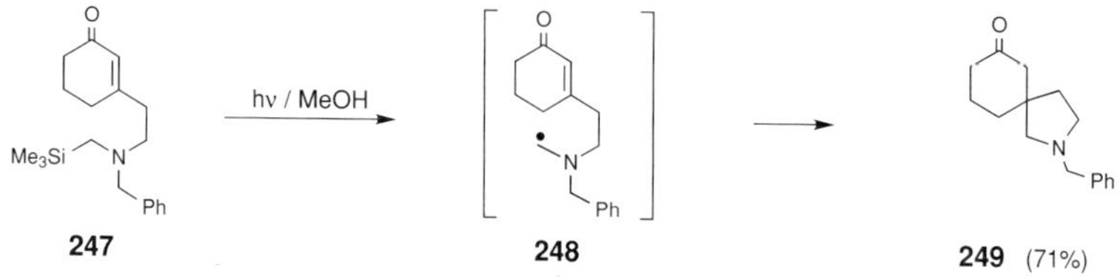

Equation 34

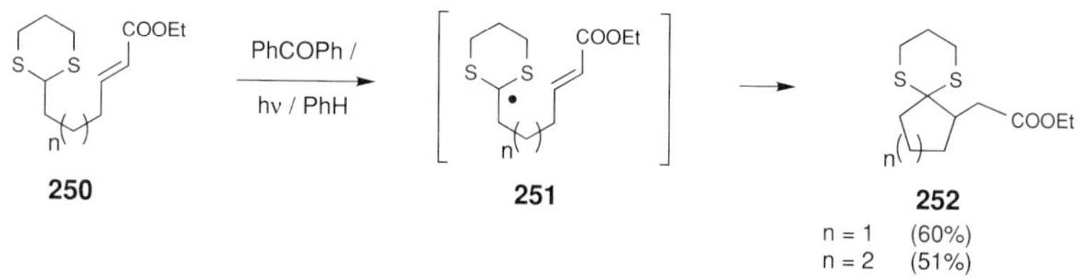

Equation 35

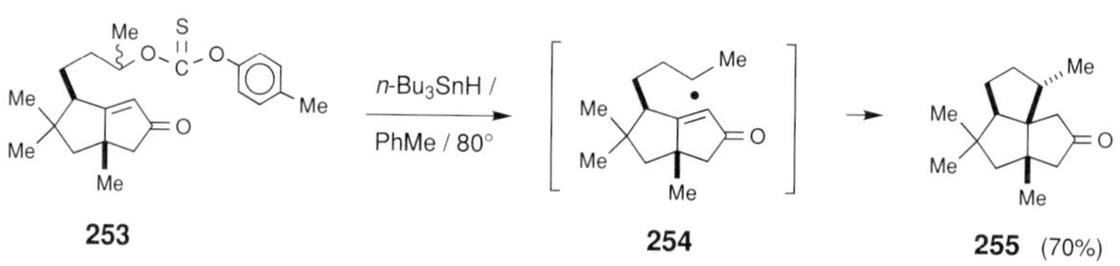

Equation 36

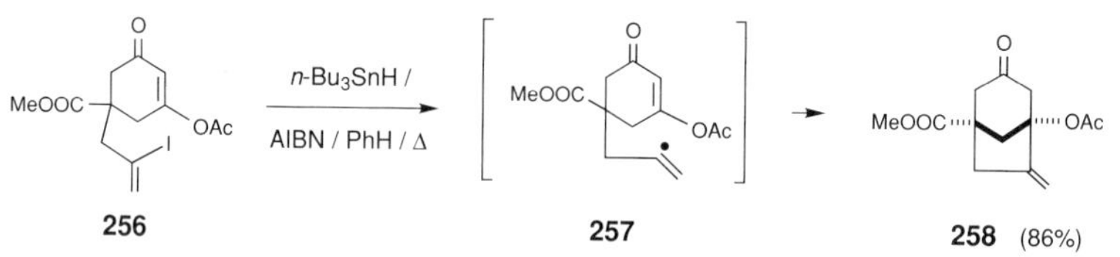

Equation 37

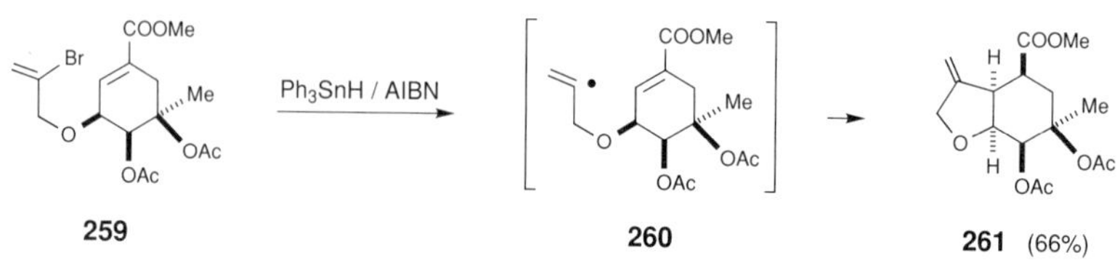

Equation 38

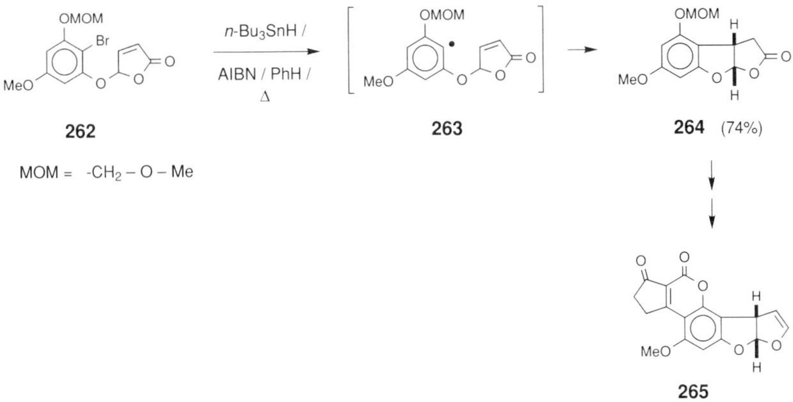

Equation 39

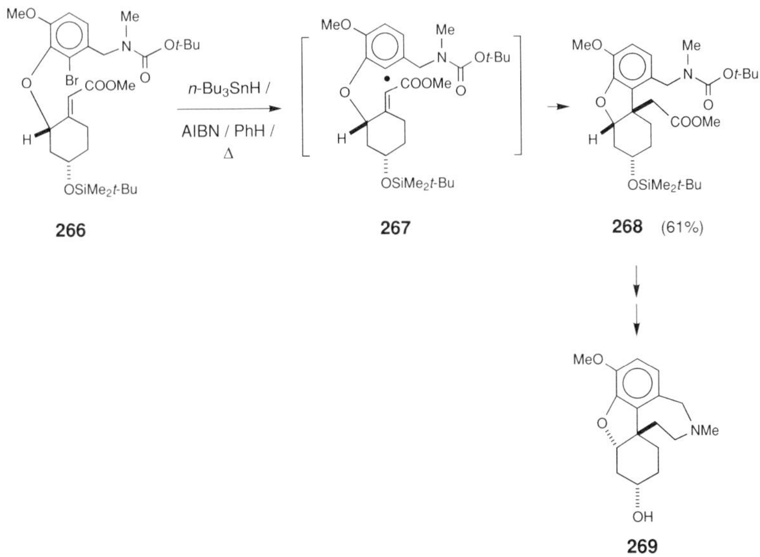

Equation 40

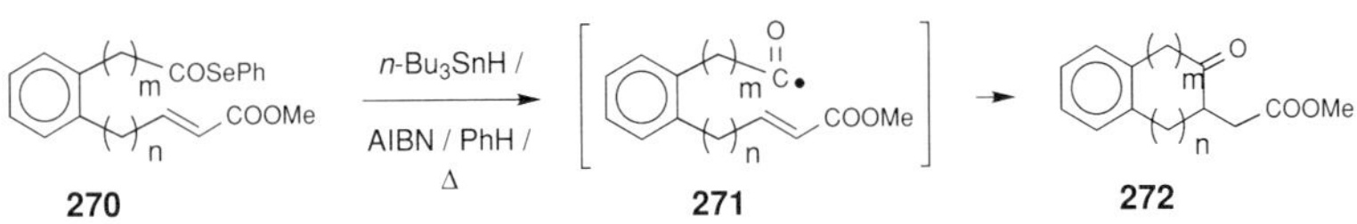

m = 0, n = 1	(88%)
m = 0, n = 2	(84%)
m = 0, n = 3	(92%)
m = 2, n = 0	(83%)
m = 2, n = 1	(71%)

Equation 41

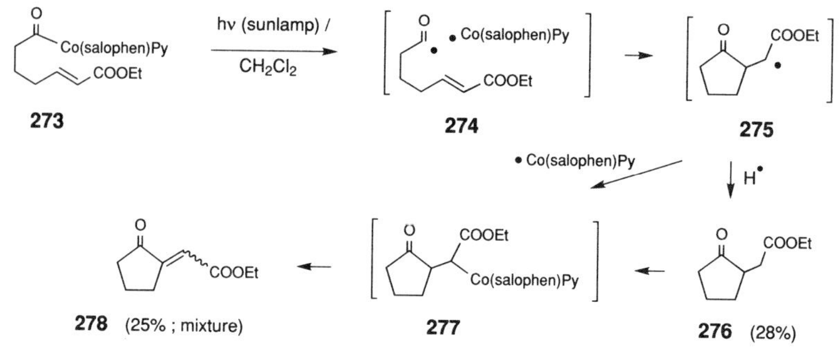

Equation 42

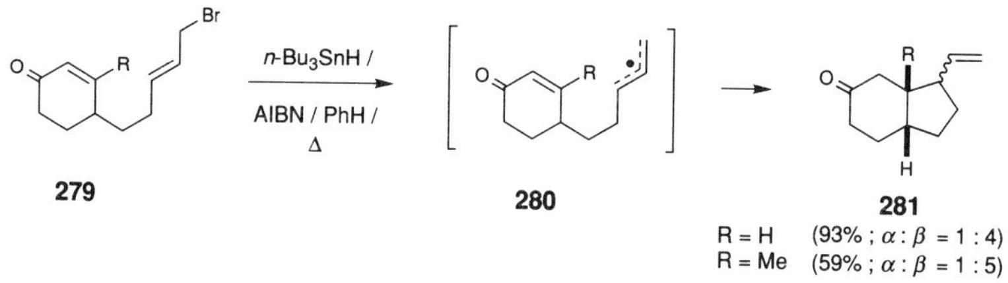

Equation 43

With the pyridinium moiety as the reactive radical acceptor,[90] it is also feasible to carry out intramolecular cyclization of a radical center generated *in situ*, as illustrated in Scheme 24.[91,92] This reaction has interesting potential as an entry to indolizidine 5,6-fused, and quinolizidine 6,6-fused alkaloids. Although regarded as an *n-endo* radical cyclization, the reaction proceeds well in the formation of 5,6-, 6,6-, and 6,7-fused heterocycles, but fails for 5,8-fused systems.[92] A notable requirement is a full equivalent of reagents, *i.e.*, tin hydride and AIBN,[91] for a complete reaction. The most plausible mechanism for the process is presented in Scheme 24, although the nature of the oxidant is, as yet, uncertain.

Scheme 24

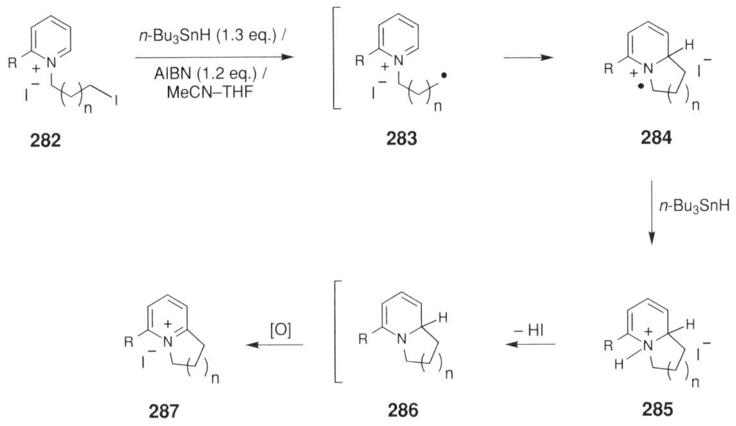

n = 1, R = H (65%)
n = 1, R = Me (65%)
n = 2, R = H (60%)
n = 3, R = H (58%)
n = 3, R = Me (58%)

Intramolecular conjugate additions of alkyl radicals generated by methods other than those given above have also been successfully utilized in synthesis, for example, in the synthesis of 5-(3-furyl)octahydro-8-methylindolizine **291**, a stereoisomeric alkaloid related to castoreum, found in the extract of the Canadian beaver scent glands (Scheme 25). The sequence demonstrates the application of known reactions, that is, the use of triphenyltin hydride to sequentially generate radical intermediates from phenylselenyl ethers **288** and **290** in high yields, the first undergoing cyclization with the enone moiety and the second, simple reduction to yield the final product **291**.[93]

O-Stannyl ketyl radicals, *e.g.*, **293** or allylic *O*-stannyl ketyl radicals, *e.g.*, **303**, produced from the reaction of the corresponding carbonyl or *α,β*-unsaturated carbonyl with trialkyltin radical, are both quite amenable to cyclization to activated alkenes, as shown in Schemes 26[94] and 27.[95] The *O*-stannyl ketyl radicals derived from aldehyde **296** and ketone **299** cyclize to yield, respectively, *trans*-alcohols **297** and **300** as major products, accompanied by lactones **298** and **301** from further cyclization of the corresponding *cis* alcoholic ester or nitrile. The isolation of the *trans*-alcohol isomers is attributed to the strain inherent in the formation of a *trans*-fused five-membered ring which inhibits further lactonization.

Scheme 25

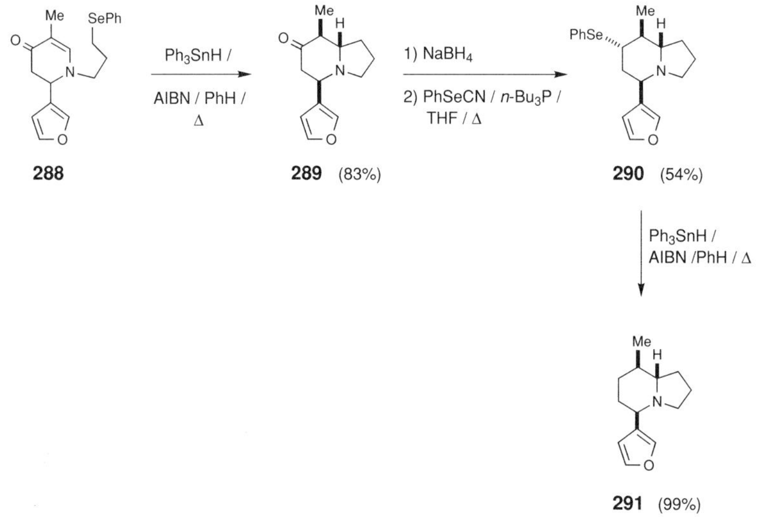

Scheme 26

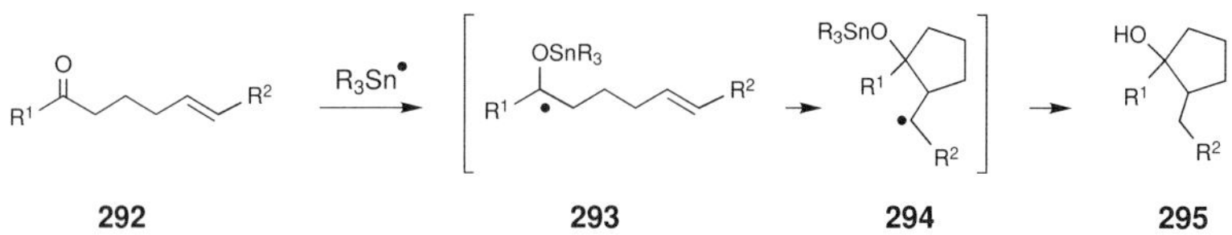

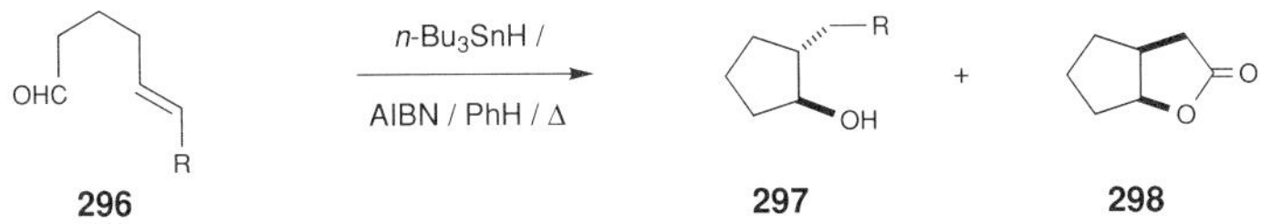

	297	298
R = COOMe	(47%)	(34%)
R = CN	(38%)	(35%)

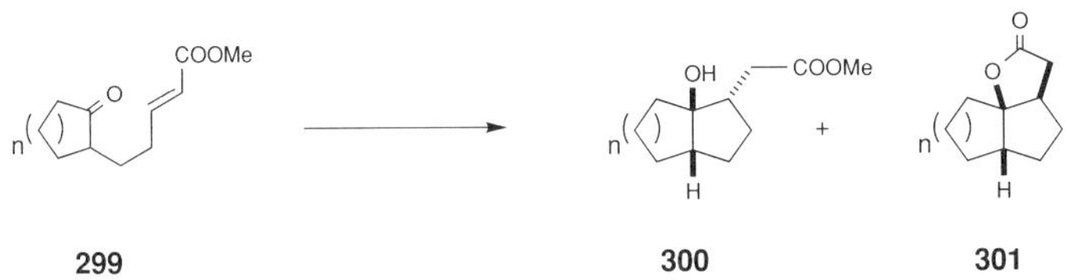

	300	301
n = 1	(52%)	(17%)
n = 2	(51%)	(37%)

In the case of the allylic *O*-stannyl ketyl radical such as **303** (Scheme 27) cyclization results primarily in product **305**, but with open options for further reactions between the trialkyltin radical and the R^2 group in intermediate **304** or the keto function in **305**, as illustrated by the use of substrates **307** and **310** in Scheme 27.[95] The significance of the activated double bond as a radical acceptor can be appreciated upon comparison with the low yield,[94] or failure to cyclize,[95] encountered upon conducting the reaction with unactivated olefins.

Scheme 27

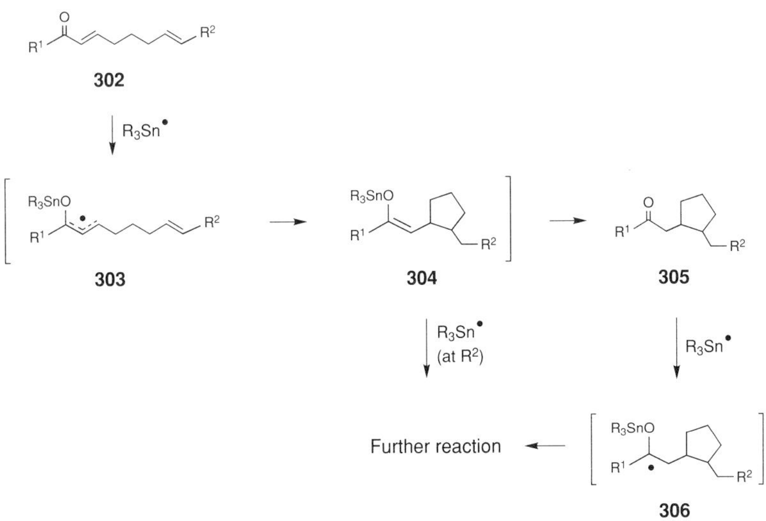

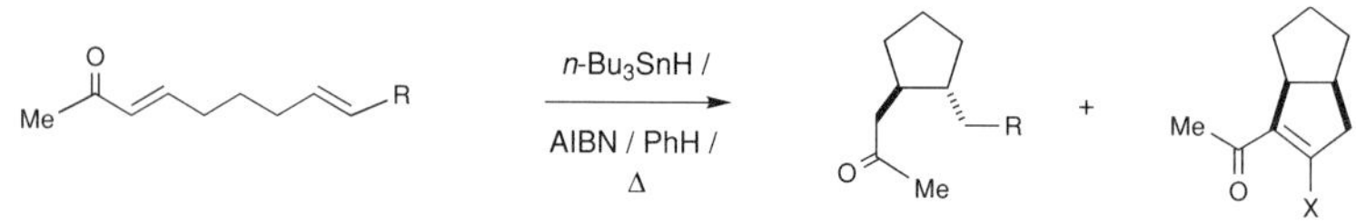

307 **308** **309**

R = COOMe (58%) X = OH (27%)
R = CN (70%) X = NH_2 (24%)

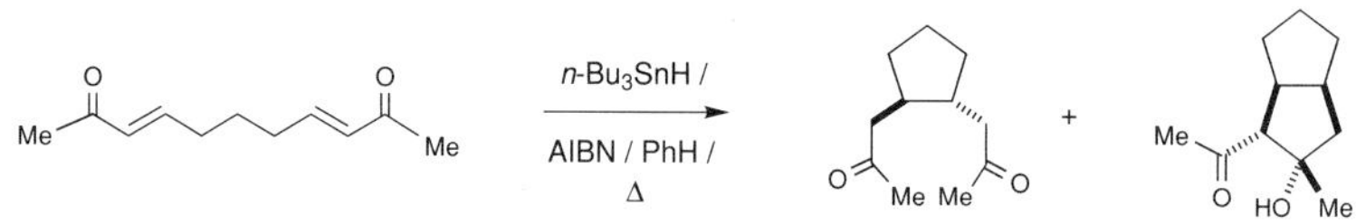

310 **311** (72%) **312** (21%)

We have already seen the use of the organosilicon radical from tris(trimethylsilyl)silane in the generation of alkyl radicals for cyclization to unactivated olefins (see Equations 14 to 16), v*ia* its reaction with alkyl halide substrates (Equation 14) or heterocyclic compounds such as the 1,3-dithianes (Equation 16). Apart from the foregoing substrates the silyl radical can also initiate radical formation from unactivated olefins and α,β-unsaturated ketones, the trichlorosilyl radical ($Cl_3Si^\bullet$) reacting selectively at the keto carbonyl while the triethylsilyl radical reacts preferably at the unactivated alkene (Scheme 28).[96] For example, treatment of substrate **313** with trichlorosilane, with di-*t*-butyl peroxide as the radical initiator and benzene as solvent, at 140°C in a sealed tube, furnishes product **316** (68%) *via* cyclization of allyl radical **314** and subsequent hydrolysis. On the other hand, similar treatment of substrate **313** (R = OMe) with triethylsilane results in the formation of product **318** in comparable yield, apparently *via* alkyl radical **317**. However, more experiments will be needed to determine the scope and limitations of the reaction before its full potential can be harnessed in organic synthesis.

Scheme 28

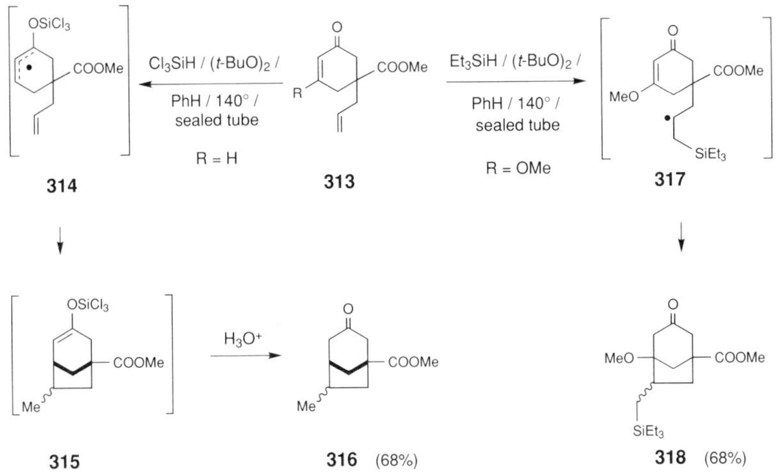

Interestingly, because the phenylthio radical (PhS•) is a good leaving group, the phenylthiomethyl substituent may permissibly be regarded as an alkene activating group. An excellent demonstration of this property is seen in the radical-initiated annulation of substrate **319** in which the radical intermediate **320**, generated by any of two methods, A or B, cyclizes spontaneously in 6-*endo* fashion, in preference over the 5-*exo* mode, to expel the phenylthio radical to yield **321**. However, the reaction shows poor stereoselectivity and only moderate yields are obtained (Scheme 29).[97]

Scheme 29

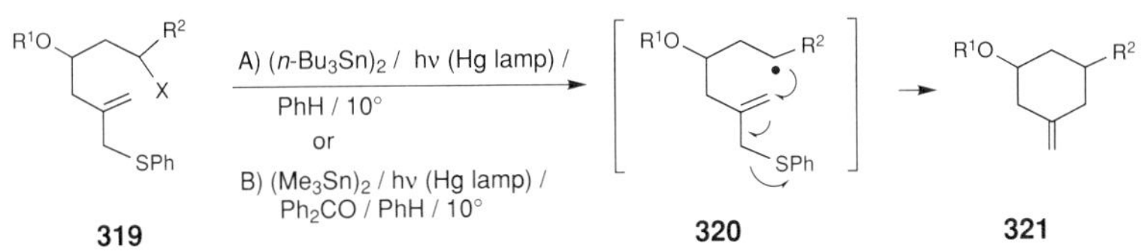

X	R^1		Method	%
Br	Me	H	A	35
Br	$-CH_2-CH_2-OH$	H	A	75
Cl	Me	Ph	A	60 (*cis* : *trans* = 1.2 : 1)
Br	Bz	Me	B	61 (*cis* : *trans* = 1.3 : 1)
Br	$SiMe_2t$-Bu	Me	B	48 (*cis* : *trans* = 1.4 : 1)

The synthetic application of intramolecular radical addition to α,β-unsaturated ketones is fast expanding, and its extension to the pursuit of macrocyclization provides excellent access to naturally occurring macrocycles. Examples include the macrocyclization of radical intermediate **323**, generated from iodoenone **322**, to afford 14-membered ring enones **324** and **325** in a 4:1 ratio (40% yield). The major product **324** is easily reduced with $LiAlH_4$ to the natural product mukulol **326** (Scheme 30).[98] Quite noteworthy is the fact that allyl radical **323**, although given the option of two enone positions (arrows in **323**) for cyclization in the *exo* or *endo* mode, reacts regiospecifically at the terminal allyl position in *endo* fashion to yield exclusively 14-membered rings **324** and **325**. In this event it is probable that steric factors play an important role in the control of cyclization. Another application, shown in Equation 44, is the synthesis of optically active (–)-zearalenone **329** *via* a similar radical macrocyclization of substrate **327**.[99]

Various 10- to 15-membered ring α-methylene macrocyclic lactones **332** have also been synthesized from the readily available tributylstannyl phenyl selenide **330**. However, attempts to construct rings smaller than 10-membered (**330**, n = 2 to 5) have been unsuccessful. For example, in the case of n = 2, substrate **330** yields the dimer **333** instead of the expected six-membered ring methylene lactone (Scheme 31).[100]

Scheme 30

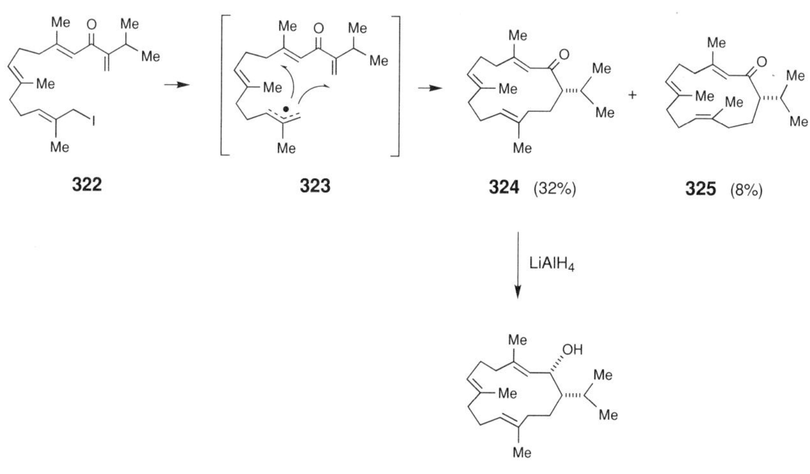

322 **323** **324** (32%) **325** (8%) **326**

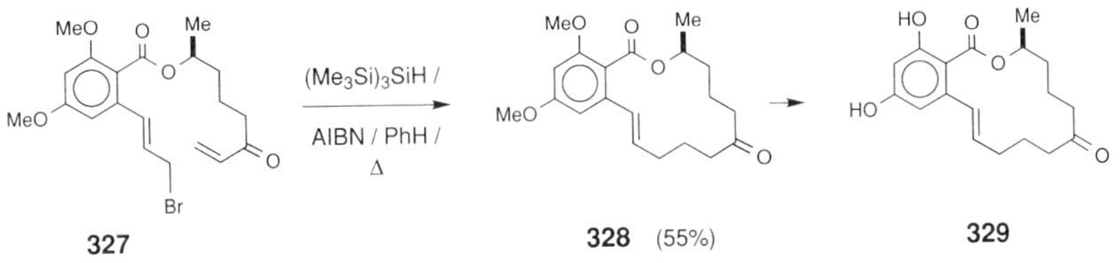

Equation 44

Scheme 31

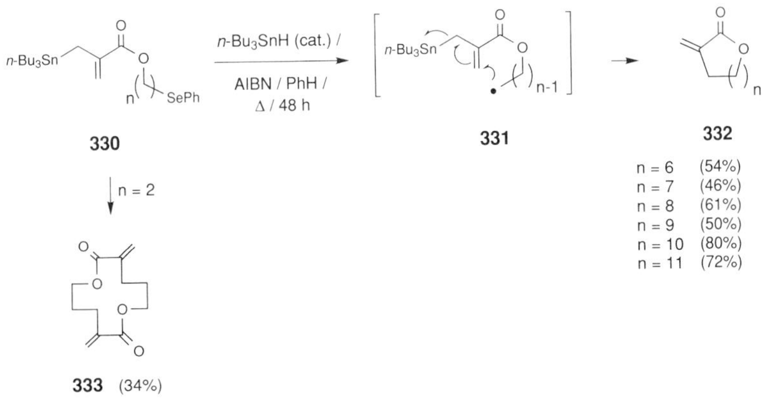

The aldehyde functionality serves well as a precursor for the *O*-alkoxymethyl radical (*e.g.*, **335**) which can undergo stereospecific cyclization with internal reactive π bonds as illustrated for the substrates 2´,3´-*O*-isopropylideneuridine 5´-aldehyde **334** and N^6-benzoyl-2´,3´-*O*-isopropylideneadenosine 5´-aldehyde **337** in Equations 45 and 46, respectively.[101]

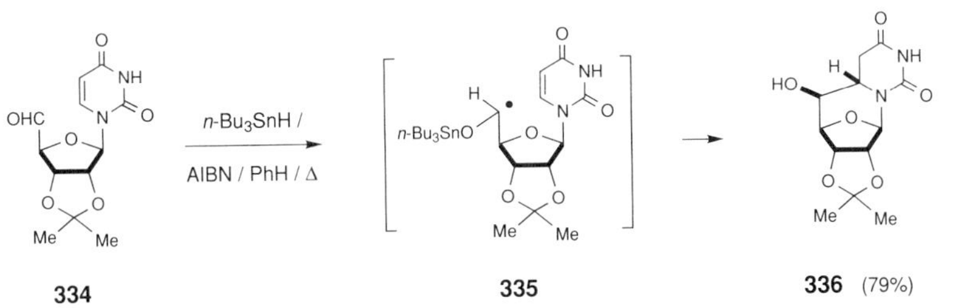

Equation 45

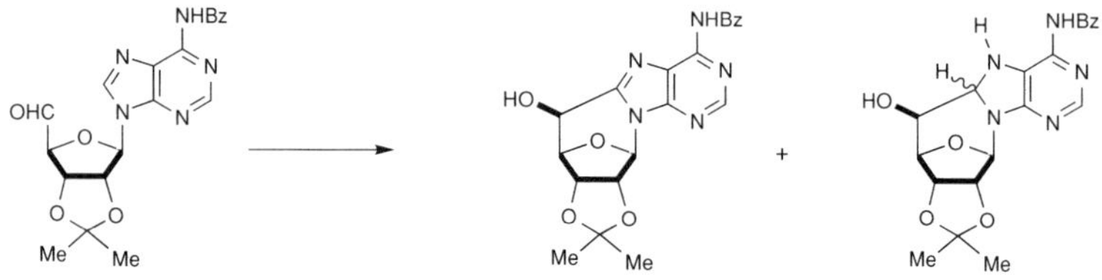

337 **338** (56%) **339** (7%)

Equation 46

An interesting aspect is the influence of geometry of the *α,β*-unsaturated ester on the stereochemistry of cyclized product. A systematic study, published since 1985,[102] reports using the *E*- and *Z*-isomers (**340** and **341**, respectively) of bromo esters as study models. Both isomers yield mixtures of *cis*- and *trans*-products, **343** and **342**, respectively, but the *Z*-olefins **341** afford consistently greater stereocontrol, as summarized in Scheme 32 and the accompanying table. Similar results are reported for the cyclization of other radical precursors (Equations 47[103] and 48[104]).

Thus the use of carbohydrate precursors in radical cyclization is fast gaining interest and a number of natural products and complex synthetic materials have already been synthesized from carbohydrate chiral templates, employing radical cyclization in the key steps.[105–108]

Scheme 32

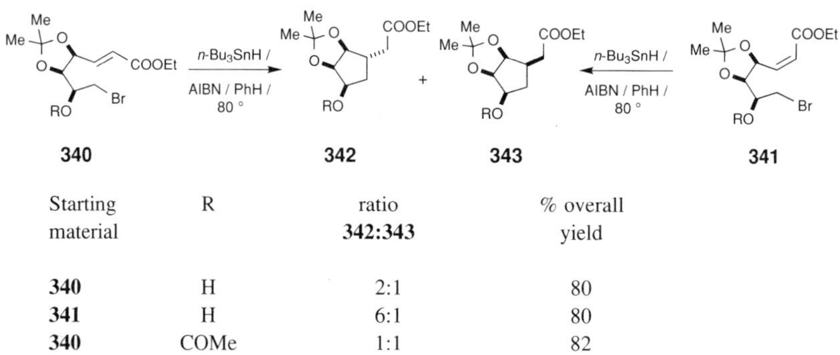

340 **342** **343** **341**

Starting material	R	ratio **342:343**	% overall yield
340	H	2:1	80
341	H	6:1	80
340	COMe	1:1	82
341	COMe	5:1	80
340	COPh	0.83:1	87
341	COPh	10:1	89

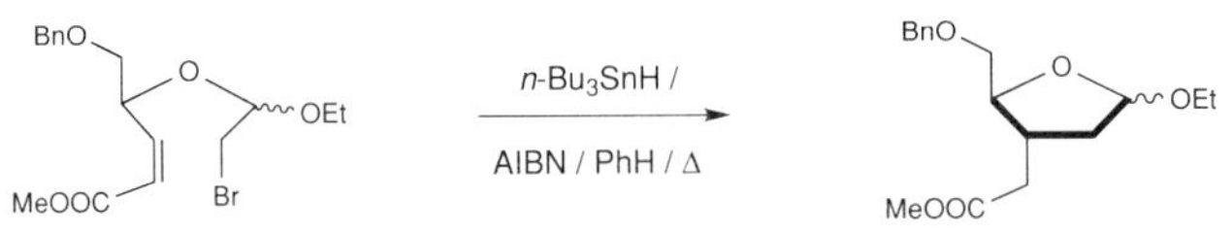

344 **345** (92%)

Equation 47

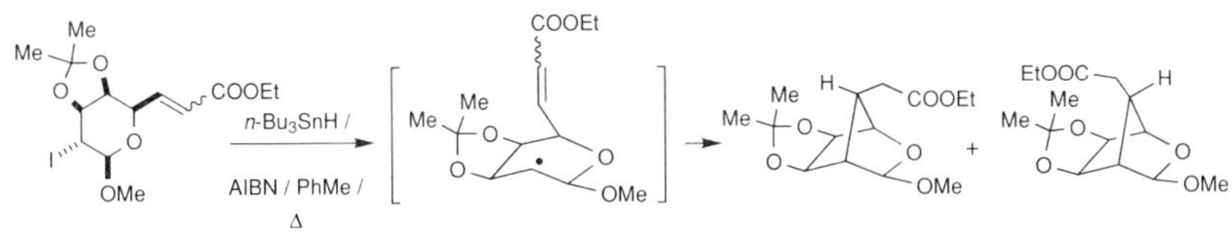

346	**347**	**348**	**349**
E		(87%)	(11%)
Z		(97%)	–

Equation 48

B. CYCLIZATION TO CARBONYL FUNCTIONS

In the study towards the synthesis of the southern moiety of avermectins and milbemycins[86] several models were investigated for the construction of the basic skeleton **350**. Drawn from retrosynthetic bond disconnection between the vinyl and alcoholic carbon atoms were two modes of radical ring closure: cyclization of an alkoxyalkyl radical such as **351**, derived from acetylenic ketone **352**, (see Scheme 26),[94,109] and cyclization of vinyl radical **353** to a carbonyl function (Scheme 33). The latter could employ as vinyl radical precursors either vinyl halide **354** (see Equation 17) or acetylene **352** (see Scheme 9).

Scheme 33

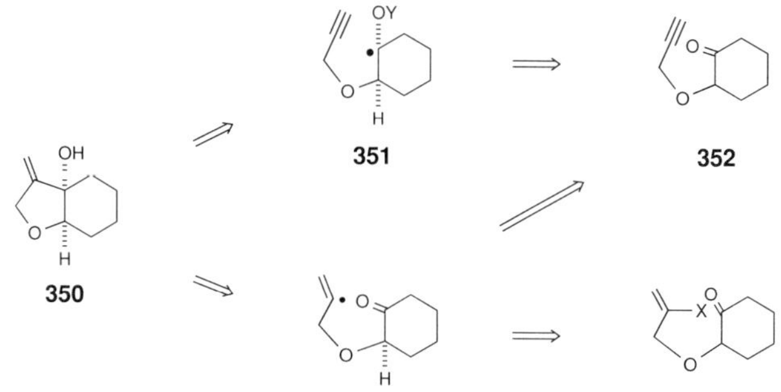

350 351 352 353 354

The cyclization of alkoxy radicals to π bonds (arrow in **351**) has already been discussed, but radical cyclization to carbonyl or carbonyl derivatives (*e.g.*, oxime ethers) has yet to be explored. Two early examples of this type of reaction are offered in Equations 49 and 50. Although the mechanism of vinyl radical (**356**) cyclization to yield **357**[110] in Equation 49 is generally acceptable, the presence of vinyl radical **359** in the reaction in Equation 50 is still somewhat arguable, since tributyltin may well induce the formation of alkoxyalkyl radical **361** instead of **359**, as both intermediates can lead to the same observed product **360**. Nevertheless, other evidence seems to favor vinyl radical cyclization to the carbonyl group to afford **360** as the sole product, but in low yield.[111]

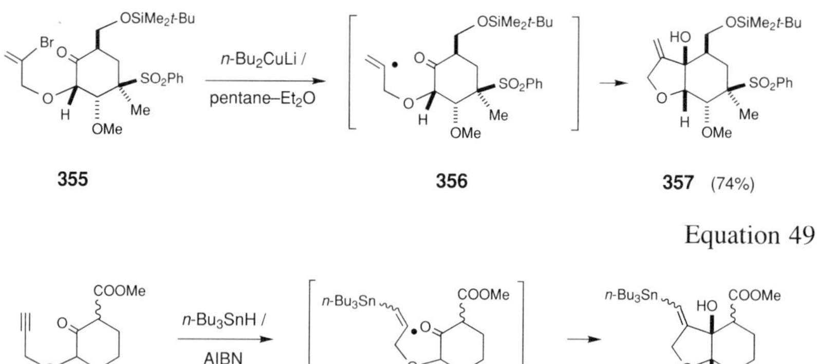

355 356 357 (74%)

Equation 49

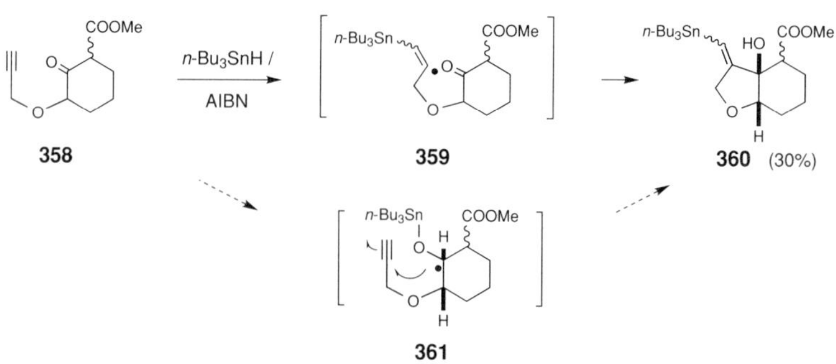

358 359 360 (30%) 361

Equation 50

Kinetic analysis of formylbutyl and formylpentyl radicals **364** and **365** (generated from bromoaldehydes **362** and **363**, respectively) indicates that both undergo fast *exo* cyclization to the carbonyl group, but the reactions are reversible with the equilibrium lying in favor of the open chain forms (Equation 51).[112] It is not surprising, therefore, that few examples of radical cyclization to carbonyl to yield the corresponding alcohols are available. In spite of this, reactions involving intramolecular radical attack on the carbonyl group followed by rearrangement of the resulting alkoxy radical to a more favorable radical are being increasingly reported, but a discussion of these is deferred to Section V of this chapter.

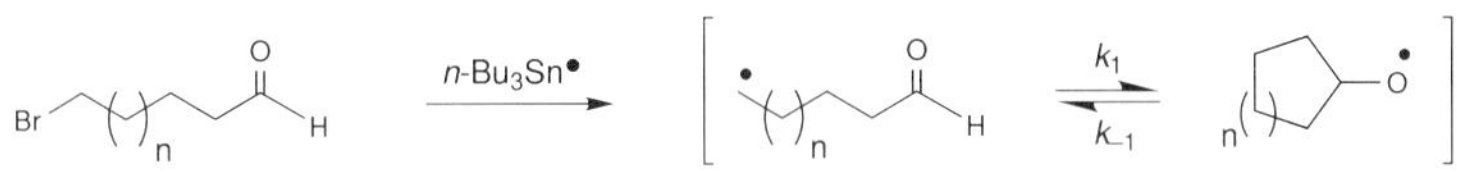

362 (n = 1) **363** (n = 2) **364** (n = 1) **365** (n = 2) **366** (n = 1) **367** (n = 2)

Equation 51

In order to drive the reaction equilibrium to the right in the radical addition to carbonyl (Equation 51) a system must be designed to trap the cyclic alkoxy radical as soon as it is formed to prevent reversal to the open chain radical intermediate. This has been maneuvered with the use of acylsilane as the trapping unit with the double function of capturing first the open chain, then the cyclized alkoxy radical, as shown in Scheme 34. Thus, standard conversion of bromoacylsilane **368** to radical **369** results in the formation of product **372**, *via* radical cyclization to carbonyl to give, initially, **370**, which then undergoes irreversible radical Brook rearrangement[113] to **371**, then the observed product **372**. Due to their high volatility, compounds **372** are characterized as their benzoate derivatives **373**, but products **375**, carrying larger silane moieties, can be isolated without derivatization.[114]

Scheme 34

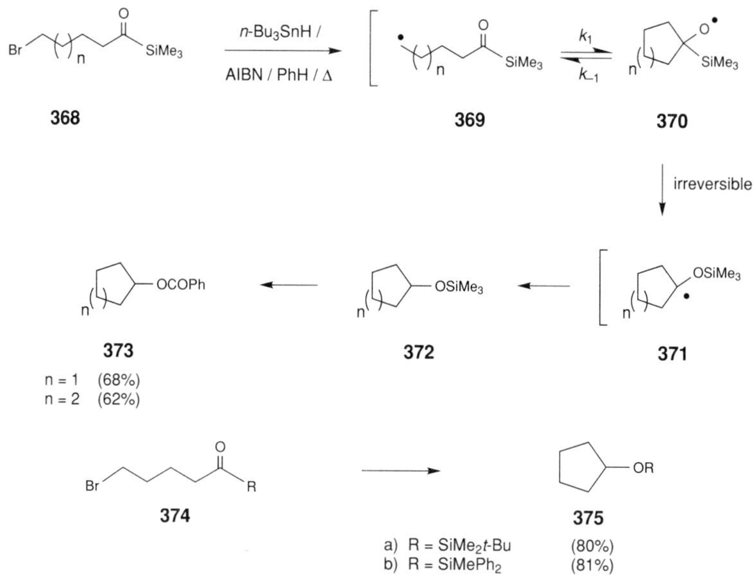

In seeking to side-step problems associated with the carbonyl group as a radical acceptor attention was shifted to derivatives of the carbonyl group, *e.g.*, the oxime ethers. Accordingly, these were found to readily undergo both inter-[115] and intramolecular[116] radical additions. One of the first examples of such cyclizations to be reported was by Corey and Pyne,[109] who generated *O*-trimethylsilylmethyl radical **377** from the reaction of keto–oxime ether **376** with Zn–$SiMe_3Cl$ in tetrahydrofuran containing 2,6-lutidine (to prevent elimination of the –$OSiMe_3$ group in the intermediate radical **377**) and observed cyclization which afforded the final product **379** in high yield as shown in Equation 52.

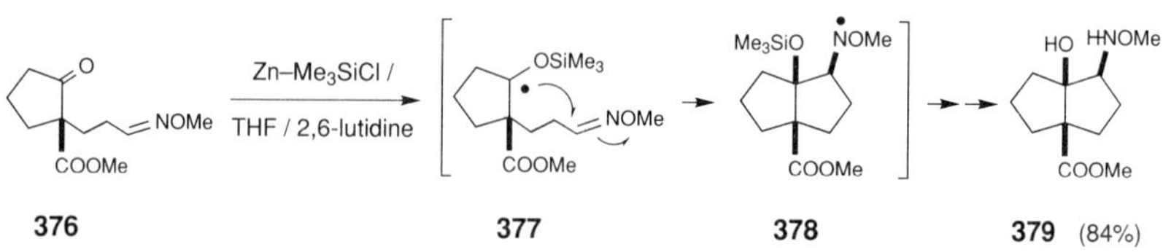

Equation 52

Many more examples of radical cyclizations to oxime ethers have since followed. They include the cyclizations of vinyl radicals (Equations 53[117] and 54[118]), aryl radicals (Equations 55 and 56),[117] and alkyl radicals from carbohydrate precursors which are used to construct polyhydroxylated cyclopentanes[119] and cyclohexanes,[120] as illustrated in Equations 57 and 58, respectively. It is interesting to note the various methodologies *i. e.*, the use of a variety of substrates for the production of radicals, the different types of cyclization performable *i.e.*, 5-, 6-, 7-*exo* cyclizations inclusive, and the stereochemistry of cyclization *e.g.*, the complete stereocontrol of radical acceptor in the 5-*exo* cyclization in Equation 57 (only β-NHOBn in products **394** and **395**) *vs.* the stereoselective 6-*exo* cyclization in Equation 58.

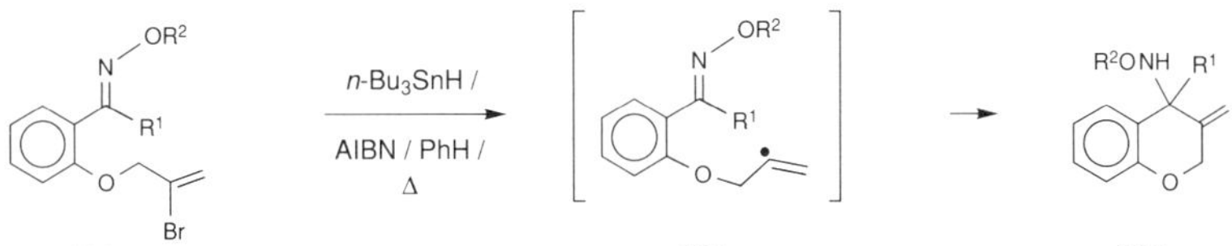

380 **381** **382**

R¹ = H , Me , R² = Me , Bn , t-Bu (71–81%)

Equation 53

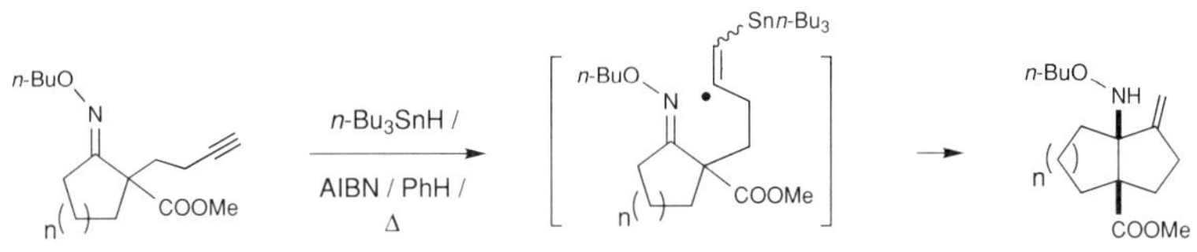

383 **384** **385**

n = 1 (90%)
n = 2 (82%)

Equation 54

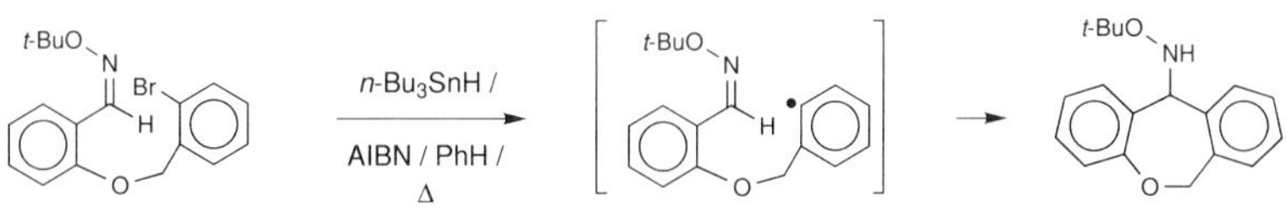

386 **387** **388** (47%)

Equation 55

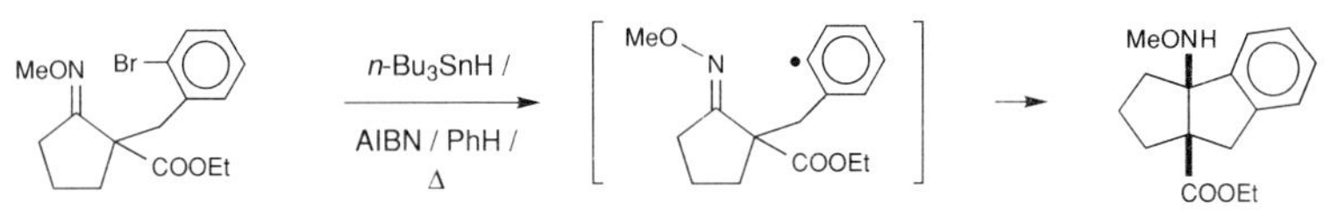

389 **390** **391** (75%)

Equation 56

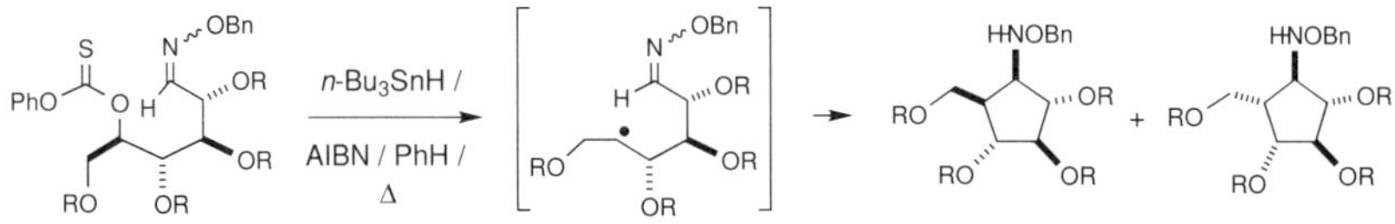

392 **393** **394** **395**

	394	395
R = Bn	(58%)	(35%)
R = Me	(56%)	(32%)

Equation 57

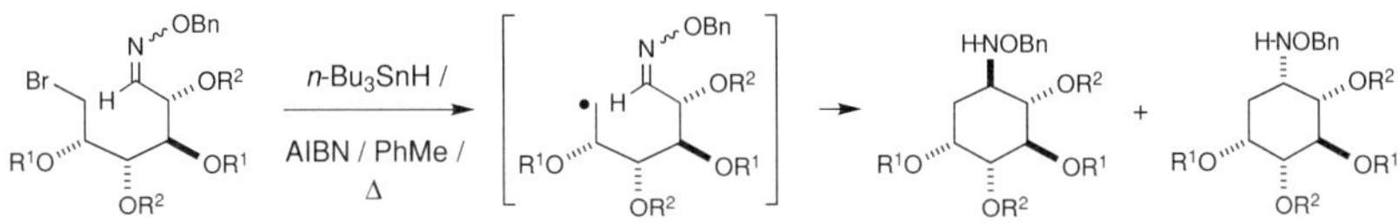

396 **397** **398** **399**

	398	399
$R^1 = R^2 = Ac$	(43%)	(9%)
$R^1 = R^2 = Bz$	(41%)	(14%)
$R^1 = Bn$, $R^2 = Ac$	(33%)	(9%)

Equation 58

IV. TANDEM CYCLIZATION

In 1964 Friedman[121] and Dowbenko[122] reported a reaction between radical $R^{\bullet}$ and cyclooctadiene **400** which yielded the 5,5-fused bicyclic compound **403** (Equation 59). Although simple and attractive, no advantage was taken of this early recorded example of radical tandem cyclization. The neglect was probably due to the then prevalent lack of information on radicals, be it the method of generation, the properties of radicals and radical acceptors, or the concept of transition states of radical cyclizations.

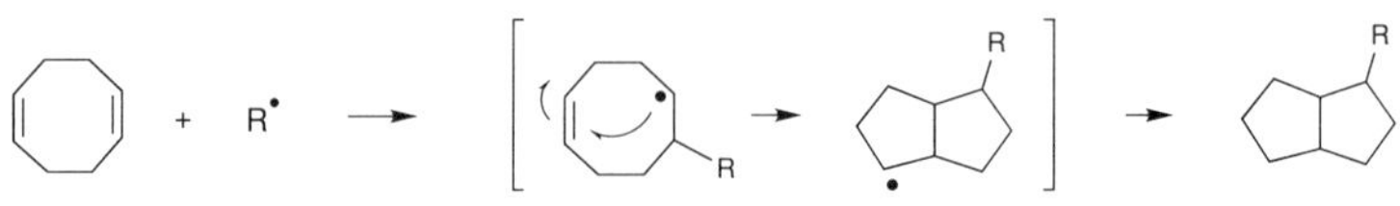

400 **401** **402** **403**

Equation 59

Twenty years elapsed before the reports by Stork and Baine[46] and Stork and Mook[123] on vinyl radical cyclization and tandem cyclization (Equation 60) stimulated activity in this area and led to the development of radical tandem cyclization, such that it now plays a prominent role in the synthesis of complex natural and unnatural molecules.[4,5,9,10]

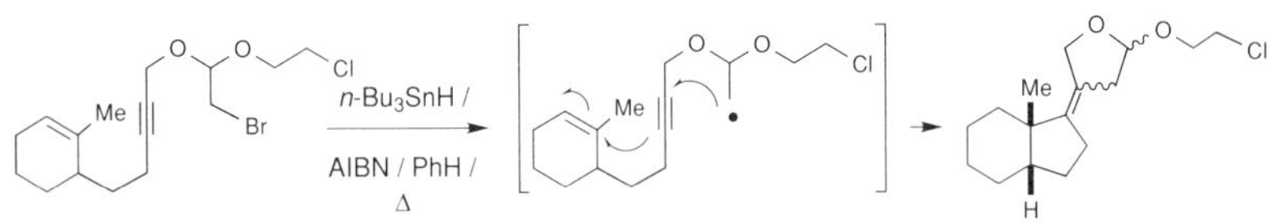

404 **405** **406**

Equation 60

A. TANDEM CYCLIZATION *AD LIB*

Endless versions of radical tandem cyclization are possible, ranging from combinations of inter- and intramolecular cyclizations (Scheme 35[14,124] and Equations 61[125] and 62[126]) to purely intramolecular tandem cyclizations, as exemplified in Equations 63, 64, and 65 for the xanthate-derived radical **423**,[127] and the alkyl and vinyl radicals from the corresponding bromides **425**[128] and **428**.[129] Shown also are the tandem cyclizations of alkyl radicals produced from the reactions of olefins with trimethyltin (Equation 66)[130] and *p*-toluenesulfonyl radical (Equation 67),[131,132] and from the tributyltin-mediated homolytic cleavage of C–NO_2 and C–Se bonds of nitrocyclohexene **439** and α-alkoxy selenide **442**, respectively (Equations 68[133,134] and 69[135]). Finally, Equation 70[136] gives an example of a Lewis acid-catalyzed tandem cyclization of aminyl radical derived from *N*-hydroxypyridine-2-thione carbamate **445**.[34,35] The abundant choice of radical precursors, the variety of methods available for the generation of radicals, and the ability to build molecules of extensive structural diversity by concomitant assembly of several cyclic structures in one operation all contribute to the broad scope, and thus the synthetic potential, of this type of reaction.

Scheme 35

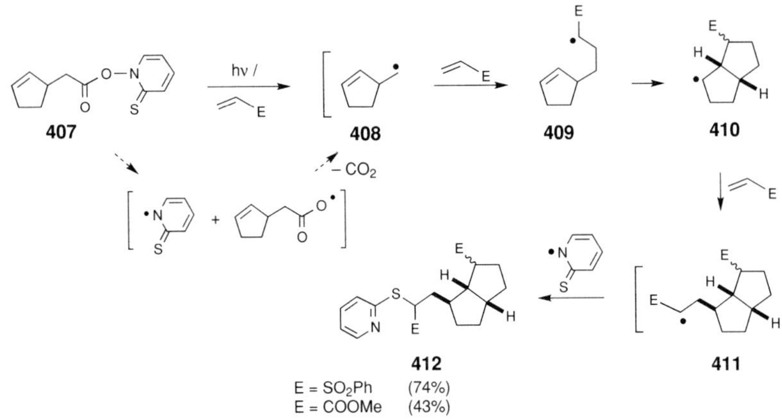

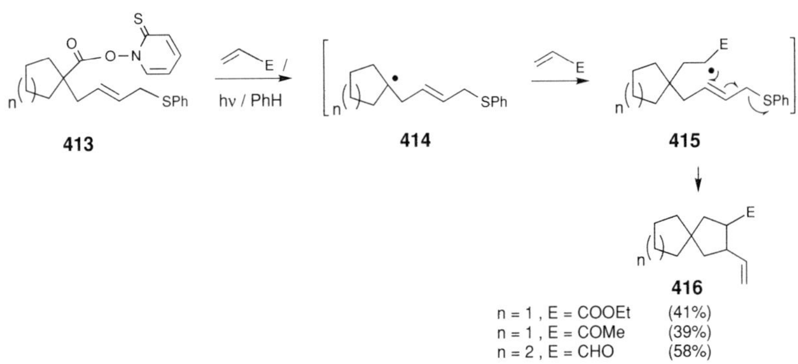

Equation 61

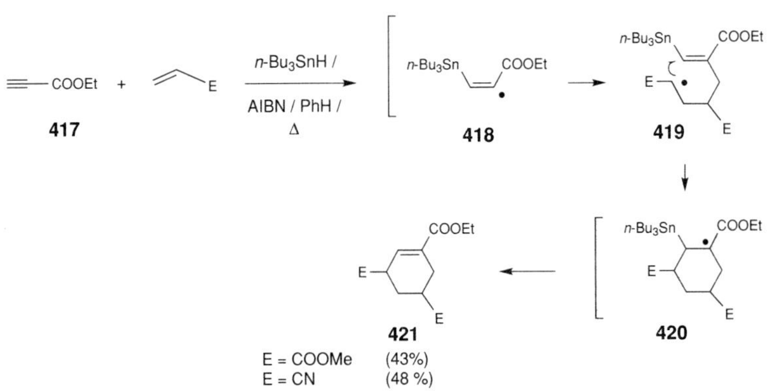

Equation 62

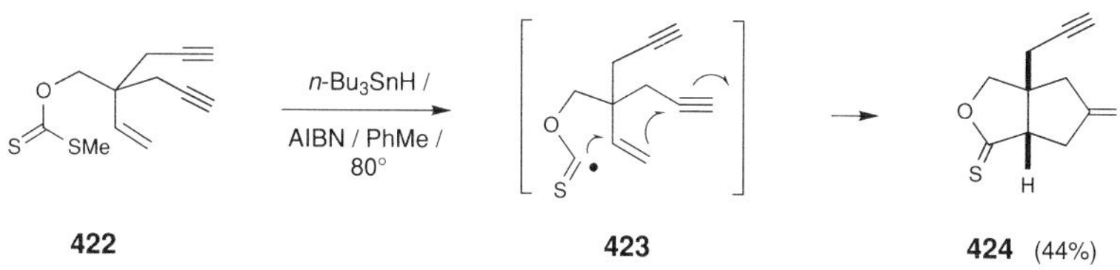

Equation 63

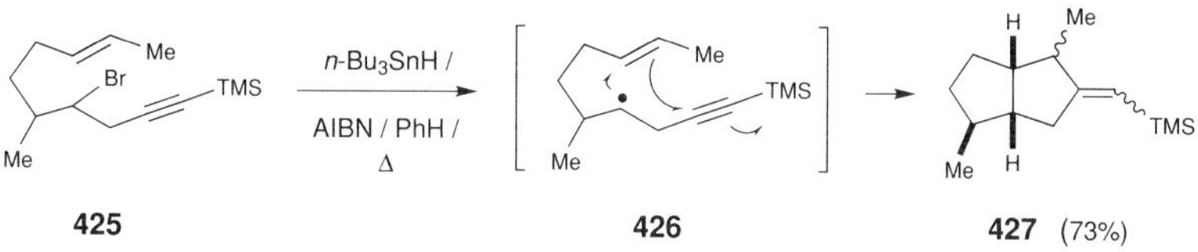

Equation 64

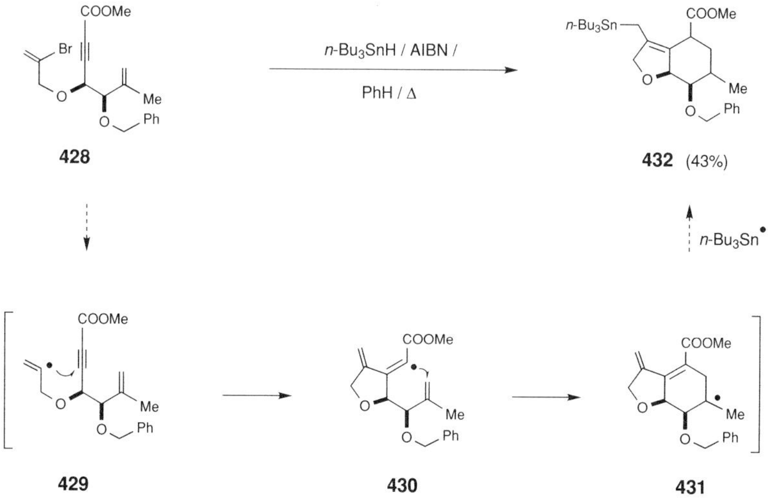

Equation 65

Equation 66

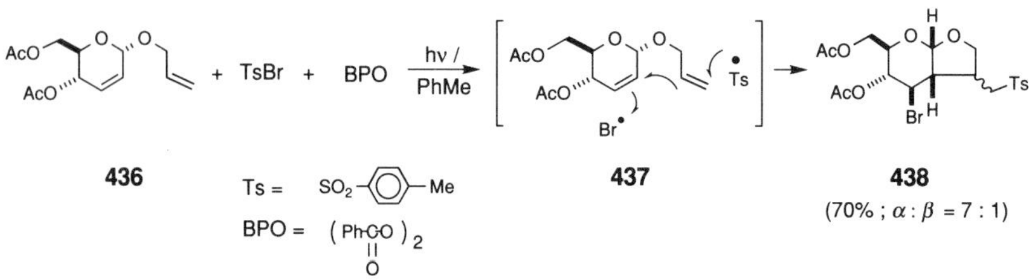

Equation 67

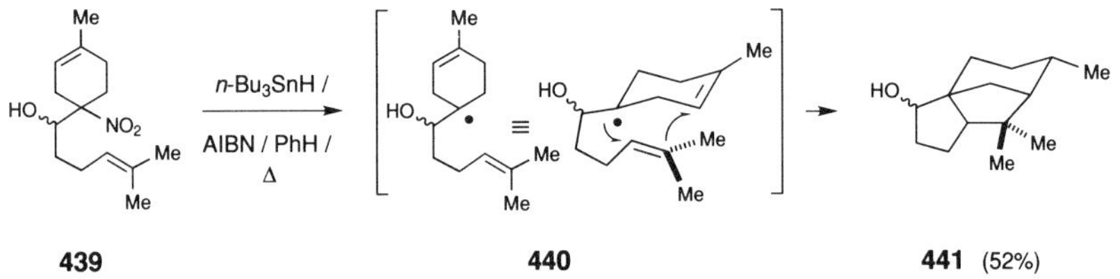

Equation 68

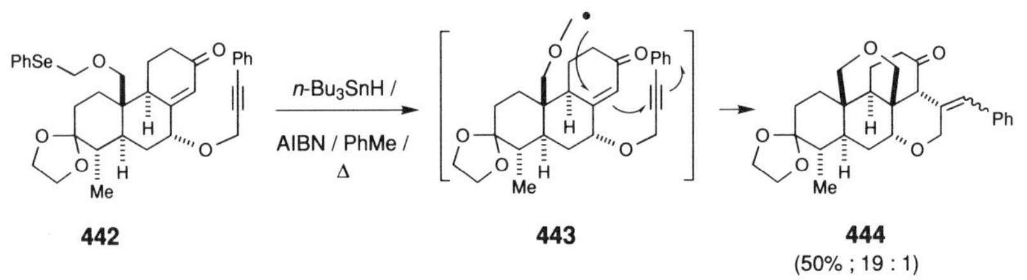

Equation 69

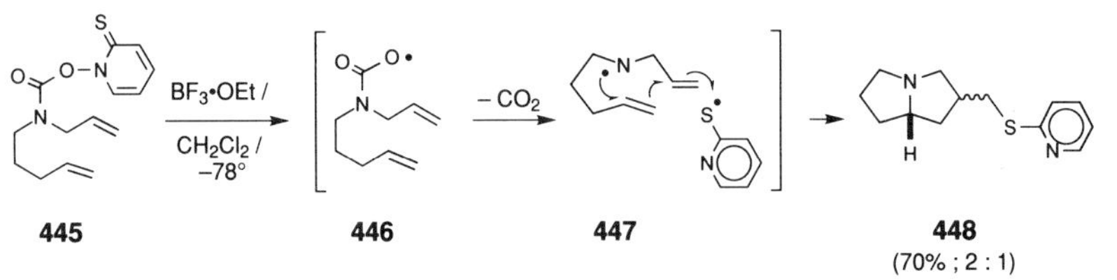

Equation 70

The reactions presented in the above scheme and equations are all 5- or 6-*exo* cyclizations except for two examples, the cyclization of intermediates **419** and **430** to the respective **420** and **431** (Equations 62 and 65) which are both 6-*endo* reactions. These are favorable since, in the first example (Equation 62), it involves radical addition to the β-position of an α,β-unsaturated ester, while in the second (Equation 65), it admits radical attack at the less hindered site to yield the thermodynamically more stable cyclized radical **431**.

A further example of tandem cyclization involving 6-*endo* ring closure is shown in Scheme 36.[137] Cyclization of acyl radical **450**, generated from the corresponding phenyl selenoester **449**, proceeds first in a 6-*endo* fashion, followed by a second 5-*exo* ring closure to finally yield product **454**. No product **452** from the alternative intermediate **451** (from initial 5-*exo* cyclization) can be detected. However, these results do not rule out the possibility of **450** undergoing 5-*exo* cyclization to **451**, then rearrangement to **453**. Such rearrangement reactions are discussed, however, in Section V.D of this chapter.

Scheme 36

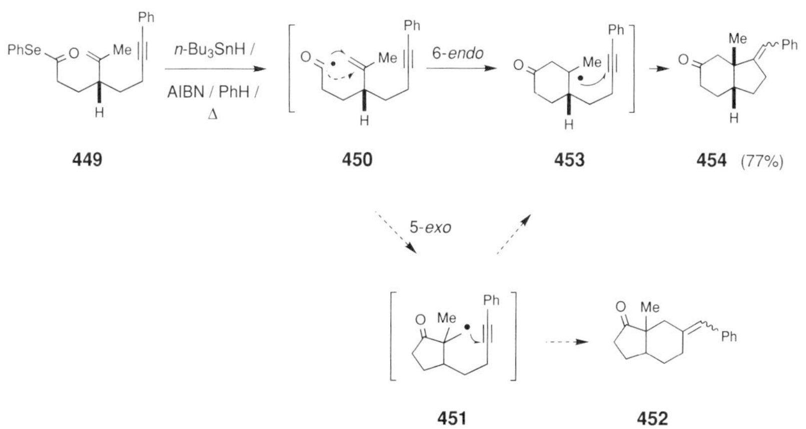

Equations 71 to 73 illustrate variations of the reaction shown in Scheme 36, the products being all formed *via* 6-*endo* ring closure in the first step of tandem cyclization.[137]

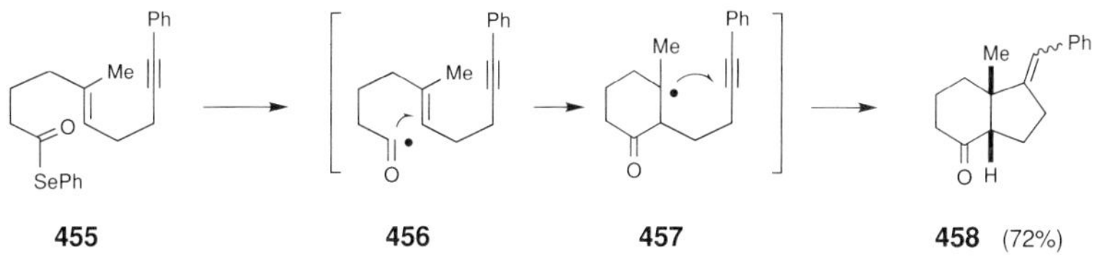

Equation 71

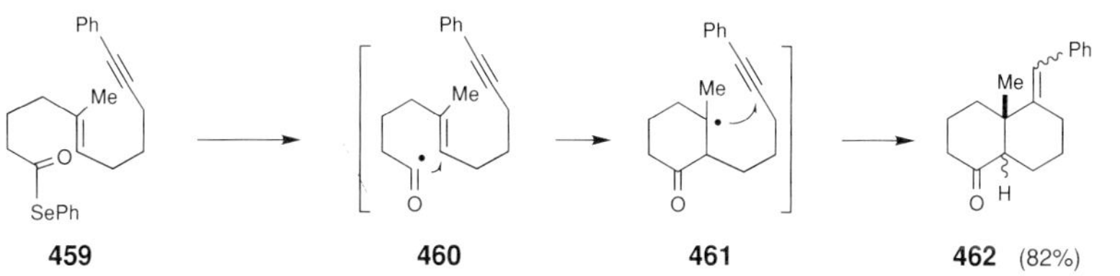

Equation 72

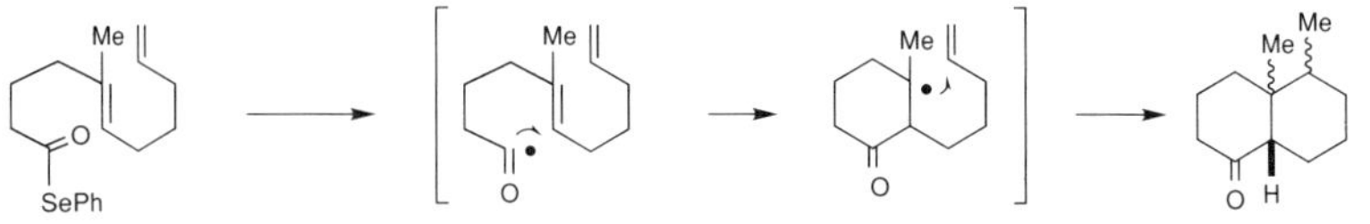

463 464 465 466 (71%)

Equation 73

Lately, there has been growing interest in the intramolecular radical cyclization of molecules "tethered" with silicon functional groups, *e.g.,* the silyl ether or ketal, for two main reasons: first, the *endo* mode of cyclization can become more competitive (competes favorably with the *exo* mode) in radical reactions that employ silicon tethers, and second, the comparatively weak O–Si or C–Si bond additionally facilitates further chemical manipulations of the cyclized products. For example, the cyclization of 2-sila-3-oxa-6-heptenyl radical **468**, derived from bromide **467**, is completely regio- and stereospecific, yielding only silaoxacycloheptane **469** (from 7-*endo* cyclization) when R = H, and only silaoxacyclohexane **470** (6-*exo* cyclization) when R = Me. Subsequent oxidations of **469** and **470** furnish diols **471** and **472**, respectively, as shown in Scheme 37.[138]

Scheme 37

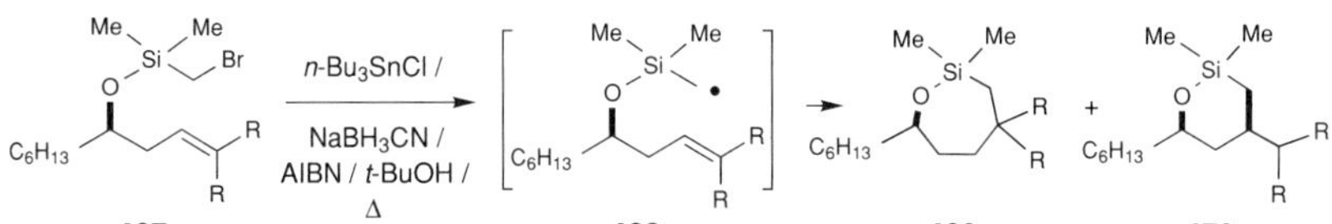

467 468 469 470

	469	470
R = H	(92%)	–
R = Me	–	(90%)

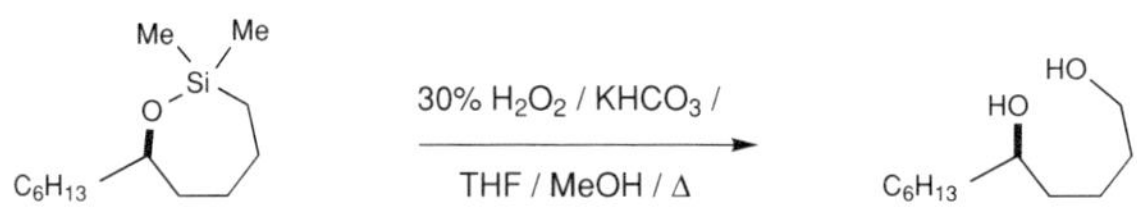

469 471

470 472

The above properties of silyl ethers have been exploited in tandem cyclizations, for example, in the stereospecific, but low-yield, cyclization of silyl ether

473 to **475** (Equation 74),[139] and in the tandem 7-*endo*, 6-*exo* cyclizations of radical **477**, derived from bromosilaketal **476**, to yield **478** which is then deprotected by tetrabutylammonium fluoride (TBAF) to yield the diol **479** (Equation 75).[140]

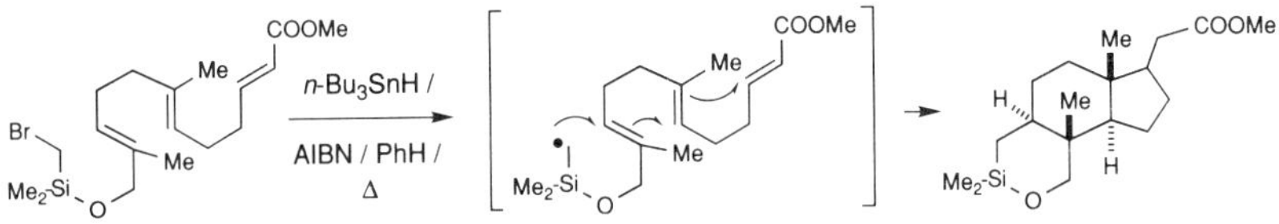

473 **474** **475**

(28% ; α : β = 2 : 1)

Equation 74

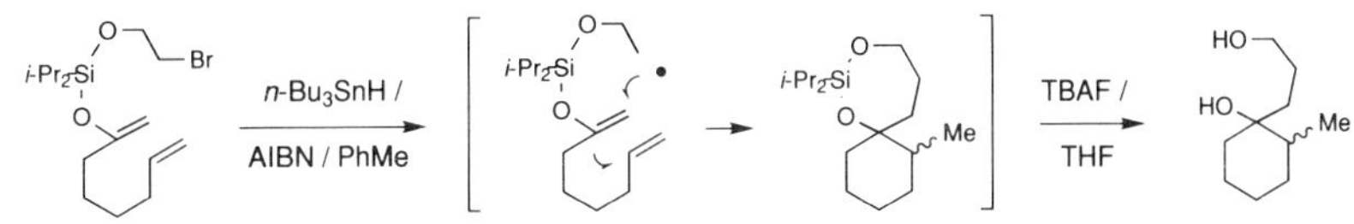

476 **477** **478** **479**

(46% ; 2 : 1)

Equation 75

A spectacular tandem radical cyclization, involving multiple radical additions in both inter- and intramolecular fashions, is the reaction of silyl ether **480** with triphenyltin hydride, as shown in Scheme 38.[141] Upon formation, the open chain methylsilyl radical **481** undergoes two consecutive cyclizations to bicyclic radical **482**, which then adds to acrylonitrile, the resulting intermediate **483** folding back in a third ring closure to form tricyclic **484**, which is eventually cleaved by oxidation to yield diol **485**. The reaction yield of 51% is reasonably high considering the number of operations involved in the overall sequence.

Scheme 38

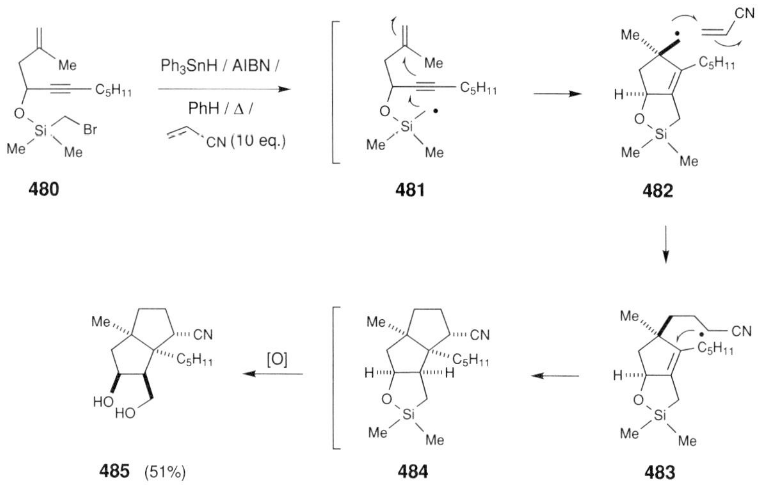

The carbonylative cyclization of 4-alkenyl halides under high pressure and free radical reaction conditions (using tin hydride) can also be categorized as a type of tandem cyclization. For example, treatment of bromide **486** with tributyltin hydride under 75 atm of carbon monoxide, with AIBN as catalyst and benzene as solvent, at 80°C, yields cyclopentanone **488** (43%), as depicted in Equation 76.[142] More preliminary results are given in Equation 77.

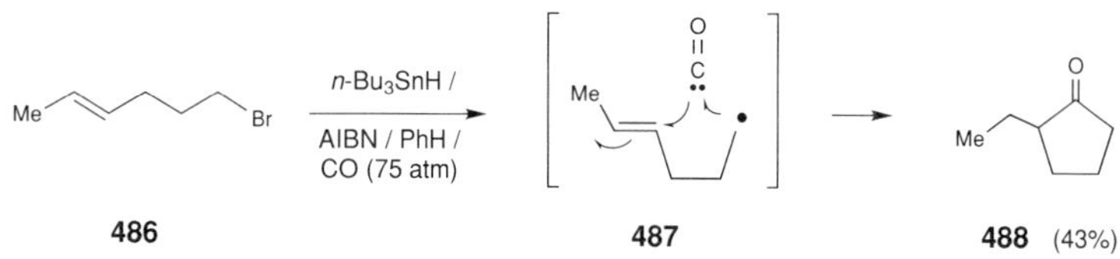

Equation 76

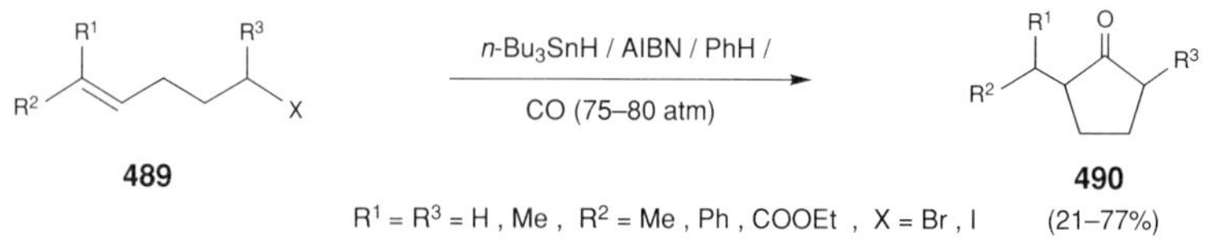

Equation 77

A very interesting reaction, related to, but not in the same category as, tandem cyclization, is the bifurcating radical cyclization. The reaction is, in other words, double radical cyclizations at one radical center, *i.e.*, two radical intermediates are formed sequentially at the same carbon site with each participating in an independent carbon–carbon bond-forming event. This pathway is exemplified in Scheme 39. Subjecting geminal dibromide **491** to radical cyclization conditions triggers the onset of bifurcating radical cyclization, which progresses, in one pot, to build products **495** and **496**. The reaction mechanism is rationalized as proceeding *via* radical intermediates **492** and **493** as shown; therefore the starting dibromide is, in effect, a source of diradical **494**.[143] Another example of bifurcating radical cyclization is shown in Scheme 40. The general concept looks promisingly useful and will no doubt continue to be explored further.

Scheme 39

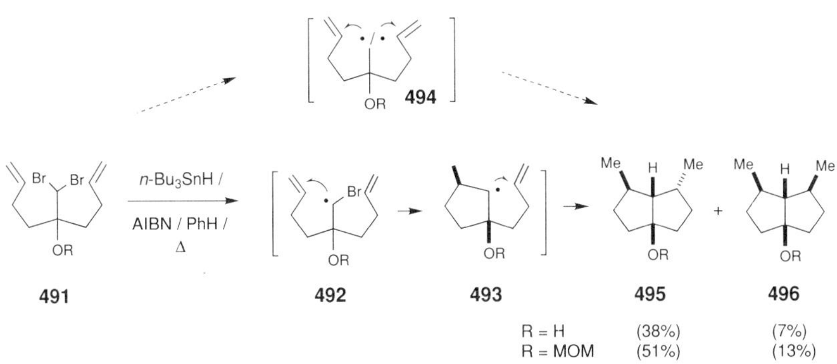

Scheme 40

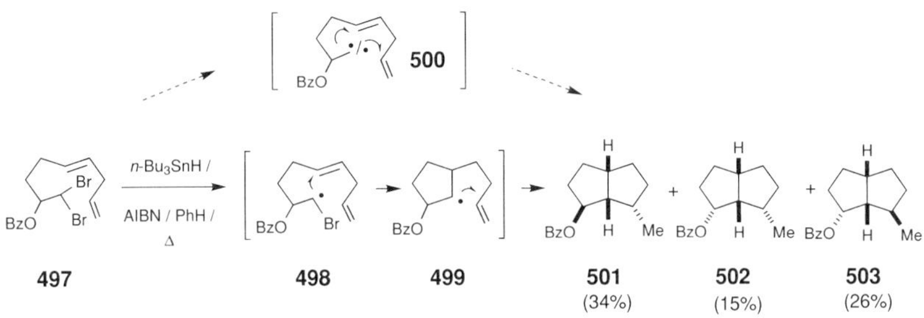

B. TANDEM CYCLIZATION TO FORM TRIQUINANE SKELETONS

The most useful application of radical tandem cyclization so far has been in the synthesis of triquinane natural products. Both linear and angular triquinane skeletons have been constructed using this approach, *e.g.*, hirsutene **504**, capnellene **505**, and hypnophilin **506** in the linear series, and silphiperfolene **507** and cripnipellin A **508** in the angular series (Scheme 41). However, the propellane triquinanes, *e.g.*, modhephene **23**, 13-acetoxy-modhephene **509**, and modhephene epoxide **510**, with their peculiar molecular architecture, do not lend themselves to construction *via* tandem cyclization but require stepwise sequential radical cyclizations to effect their synthesis. Most of the work in this area has been reported by Curran and his group and has been reviewed.[4,9]

Scheme 41

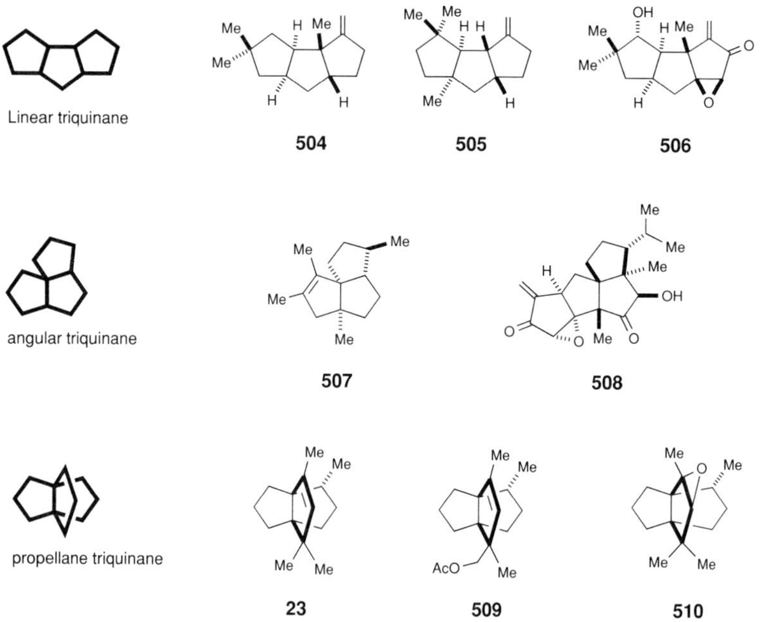

The fact that the structures of these naturally occurring triquinanes are composed of fused five-membered rings is what makes them most suitable for construction *via* the highly favorable 5-*exo* mode of radical cyclization. Depending on functional group requirements and the availability of precursors for cyclization, linear triquinanes may be constructed as, for example, outlined in Schemes 42 and 43. Racemic (±)-$\Delta^{9(12)}$-capnellene **505** can be synthesized *via* tandem (double 5-*exo*) cyclization of radical **511**, formed from the reaction of bromide **512** with tin hydride,[144] while an asymmetric synthesis of the unnatural enantiomer, (–)-$\Delta^{9(12)}$-capnellene **505** is also possible, simply by starting with the optically active bromide precursor **512**.[145]

Scheme 42

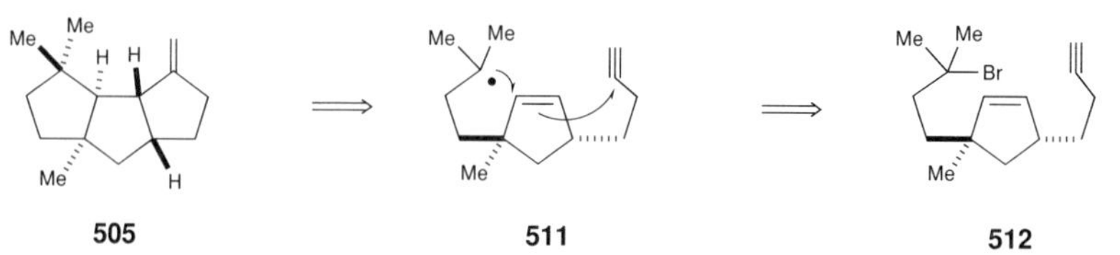

In principle, two independent paths of tandem cyclization are feasible for the synthesis of hypnophilin **506** or coriolin **520** (Scheme 43), *i.e.*, either α-oxy radical **514** or vinyl radical **515** may be employed as the radical source for tandem cyclization (route a or b, respectively). However, as discussed earlier, radical cyclization to the carbonyl group can sometimes be sluggish, therefore route a would appear to be the more attractive in this case. Accordingly, reduction of aldehyde **516** with samarium diiodide in a solvent system of THF:HMPA (20:1) starts up the process of stereospecific tandem cyclization *via* α-oxy radical **517**[146] to yield triquinane **519** (57% yield accompanied by a small amount of uncyclized reduction product).[147] Simple manipulations of **519** can lead to hypnophilin **506**[147] or coriolin **520**[148] as required (Scheme 44).

Scheme 43

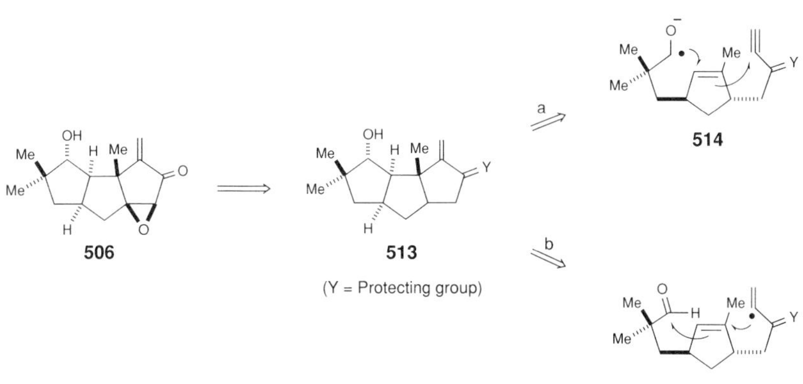

Scheme 44

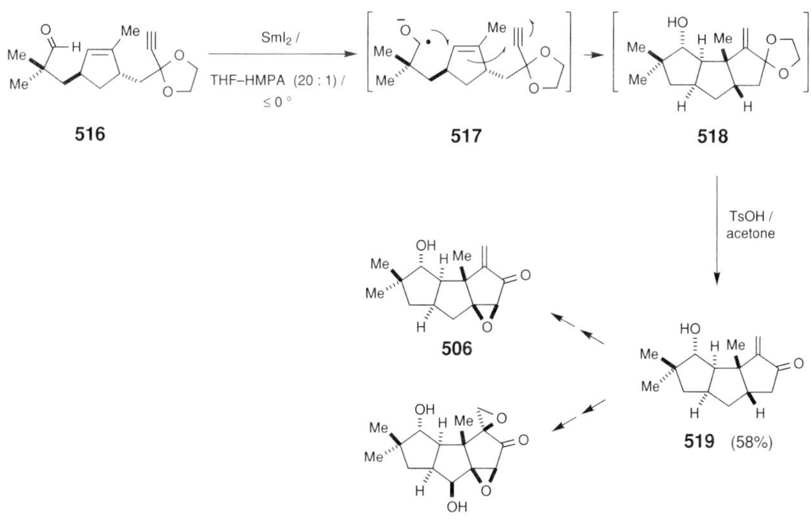

More than one mode of tandem cyclization is also possible for the construction of the angular triquinane skeleton, as depicted in Scheme 45. The two routes shown in the scheme have both been successfully applied, *e.g.*, route a, in the synthesis of silphiperfolene **507** (Scheme 46)[149] and route b in the study towards the synthesis of cripnipellin A **508**. In the latter study, tandem cyclization was experimented on iodopentadiene **529** by treating the halide with tin hydride to generate radical **530**, the cyclization of which followed the predicted route b in Scheme 45 and led to the production of triquinanes **532** and **533** in a 1:5 ratio (65% yield), apparently *via* allyl radical **531** (Scheme 47).[150] Interestingly, although given the added opportunity to cyclize in 6-*exo* fashion, allyl radical **531** undergoes exclusively 5-*exo* cyclization to yield stereoisomeric products **532** and **533**, with no trace of the alternative bridged isomer **534**.

Scheme 45

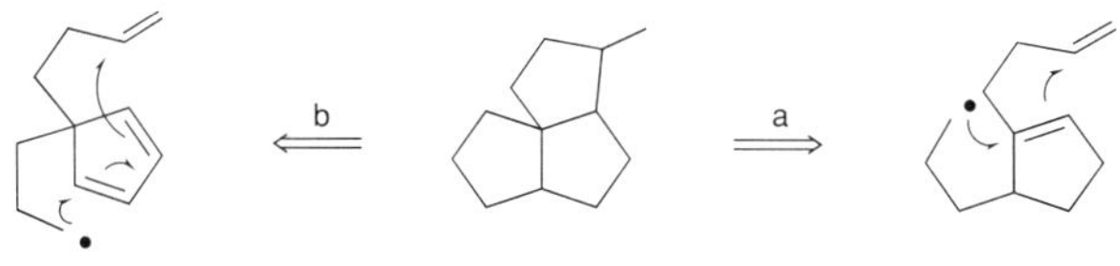

523 **521** **522**

Scheme 46

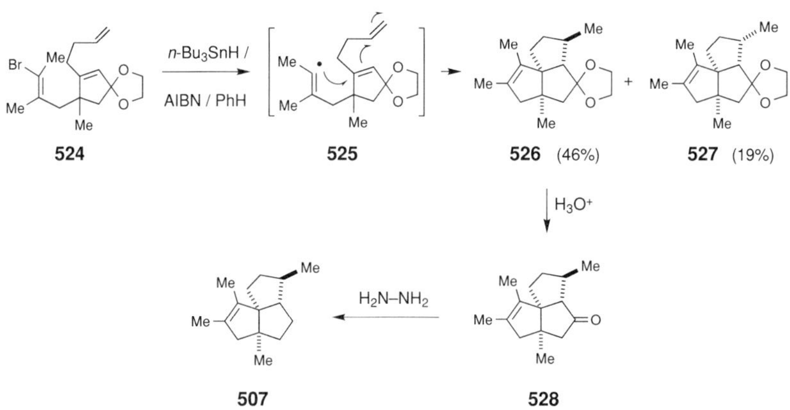

Scheme 47

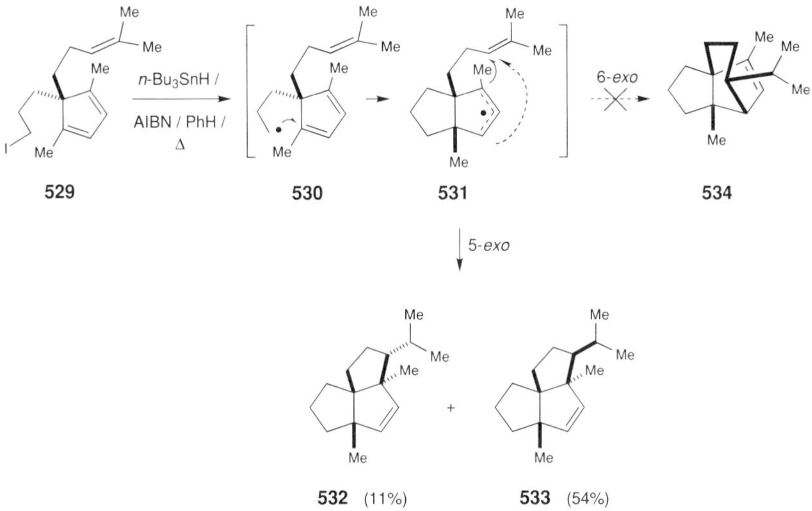

As stated earlier, the propellane skeleton is not convenient to construct by the process of tandem cyclization, and chemists have resorted to using sequential radical cyclizations for its synthesis. An example is the synthesis of modhephene **23**, in which radical cyclizations of **536** and **540** constitute the key steps, as illustrated in Scheme 48. The newly formed propellane skeleton **541** suffers spontaneous double bond isomerization to α,β-unsaturated ketone **542** which can be easily converted to modhephene **23**.[151] The synthesis is mentioned here for the sake of completeness, despite the fact that it does not comprise tandem cyclization; other methods of synthesis of modhephene having already been mentioned at the beginning of this chapter (Equation 4).[20]

Scheme 48

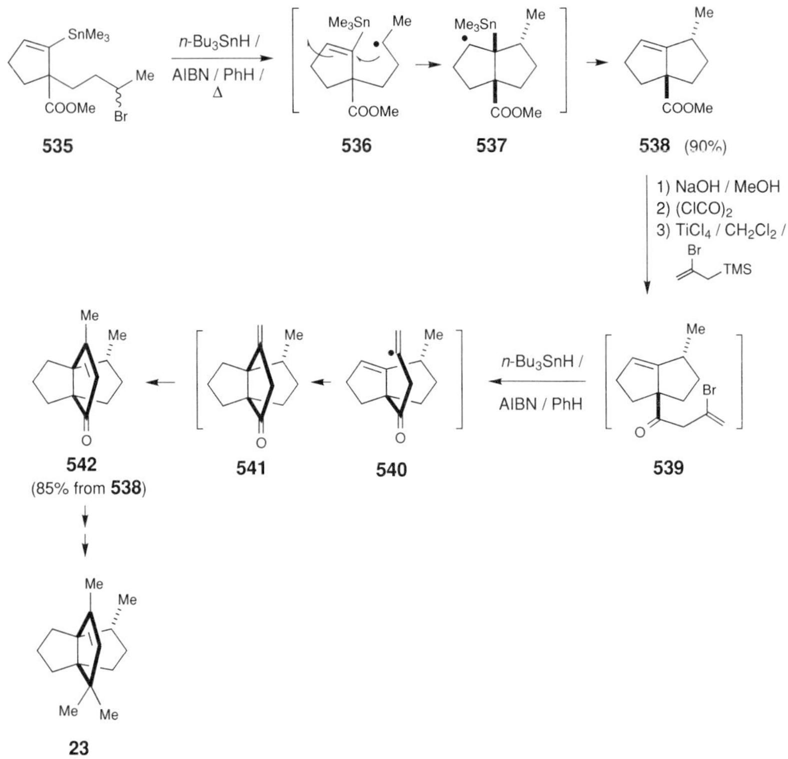

V. CYCLIZATION INVOLVING ATOM OR GROUP TRANSFER, RADICAL TRANSLOCATION, REARRANGEMENT, OR FRAGMENTATION PROCESSES

A. CYCLIZATION INVOLVING ATOM OR GROUP TRANSFER

In 1986 a report on the cyclization of 2-iodo-2-methyl-6-heptyne **543**, initiated by tri-*n*-butyltin hydride (10 mol%) and azobisisobutyronitrile (AIBN) (5 mol%) in refluxing benzene, described the formation of (iodomethylene)cyclopentane **546** in an 84% combined yield of *E*- and *Z*-isomers (15:1 respective ratio). The mechanism of the reaction was proposed as shown in Scheme 49 and the terminology "atom transfer cyclization" was assigned to this type of cyclization reaction.[152]

Scheme 49

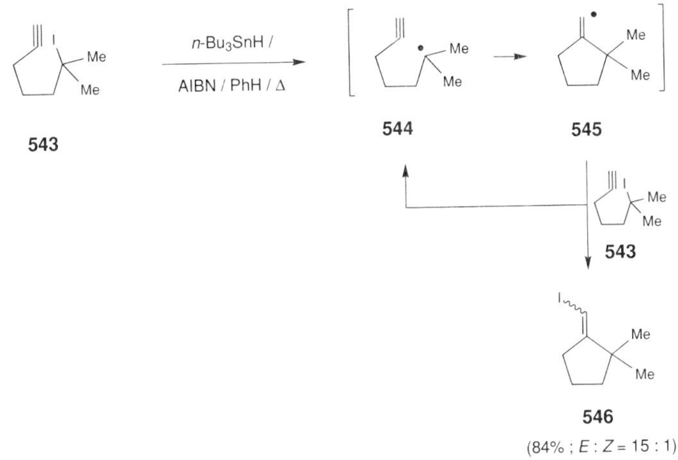

A very similar reaction, shown in Scheme 50, is the cyclization of iodide **547** to give bicyclic compound **550**.[20] Interestingly, while monitoring the reaction with nuclear magnetic resonance NMR spectroscopy, the authors found evidence of a second operating reaction pathway in their detection of iodide **551**. It stood to reason that the cyclized radical **549** could react in two ways: with hydrogen radical to yield the observed product **550** or with starting iodide **547** (by iodine abstraction) to yield the bicyclic iodide **551** with concomitant regeneration of radical **548**. The subsequent disappearance of **551** could be charged on its reversal to vinyl radical **549** (by reaction with tin radical) and thereafter final product **550**. Later results from similar studies have confirmed these mechanistic deductions.[22]

In retrospect, the reactions presented in Schemes 5[37] and 35,[14] in which radical cyclizations are terminated by the capture of xanthate and thiopyridine groups, could, just as appropriately, be designated as "group transfer cyclization".

Scheme 50

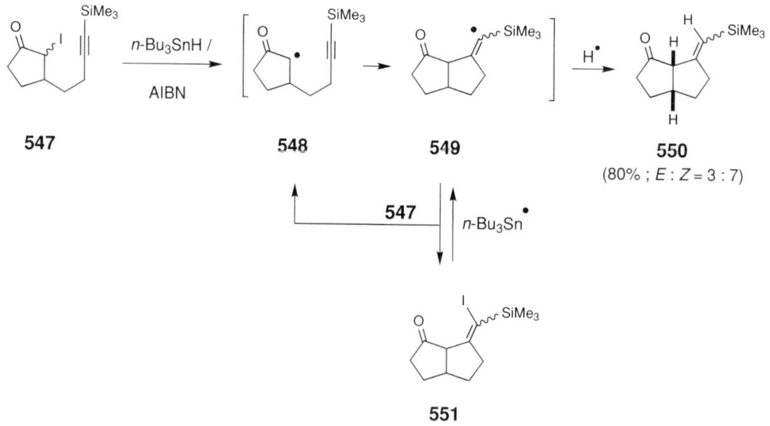

Obviously the application of atom or group transfer cyclization can be very useful in many cases. Terminating the radical center, after cyclization, with a functional group such as a halogen, xanthate, or thiopyridine, instead of the usual hydrogen atom, furnishes a product with a reactive center amenable to further chemical transformations, as exemplified in Scheme 51 and Equation 78. According to Scheme 51,[153] treatment of iodoacetamide **552** with n-Bu_3SnH / AIBN under homolytic cleavage conditions gives mainly the reduction product **554** and only a small amount of cyclization product **553**. However, the situation begins to improve with irradiation of the reaction mixture of **552** and hexabutylditin (n-Bu_3Sn–Snn-Bu_3) in benzene solution, the established reaction conditions for atom transfer cyclization,[154] resulting in some formation of the iodine-transferred cyclized product **555**, but still with **554** as the major product (1:2 respective ratio). Eventually, the addition of ethyl iodide not only improves the overall reaction yield (88%), but also dramatically alters the product ratio in favor of the cyclized product **555**. Here ethyl iodide is attributed a dual role, that of a radical sink as well as an iodine atom source. The application of this reaction to the synthesis of natural product (–)-trachelanthamidine **561** is shown in Equation 78.[153]

Further examples of group transfer cyclization, employing S-acyl xanthate dually as the alkyl radical source and the radical terminator, are shown in Scheme 52. t-Butyl peroxide, which decomposes at ~130°C to give methyl radical and acetone, is employed as the radical initiator. The cyclopentane and tetrahydrofuran **567**, as well as the bicyclic carbocycle **570**, have been prepared by this method.[155]

Scheme 51

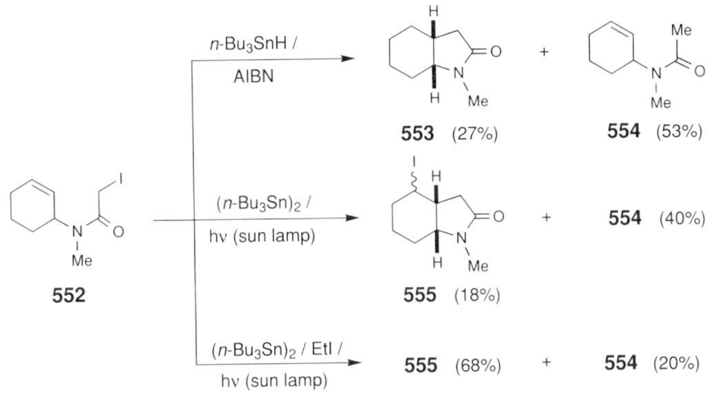

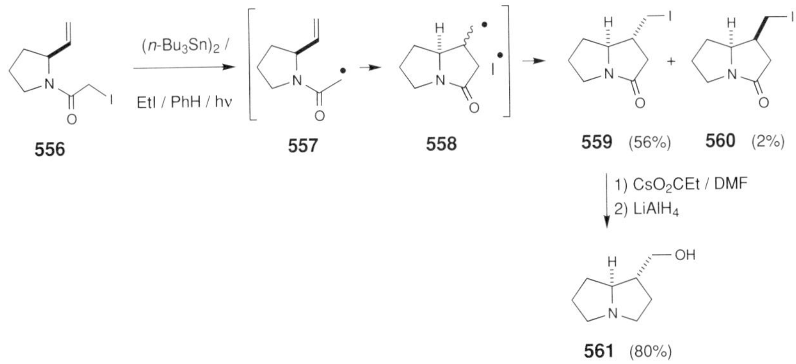

Equation 78

Scheme 52

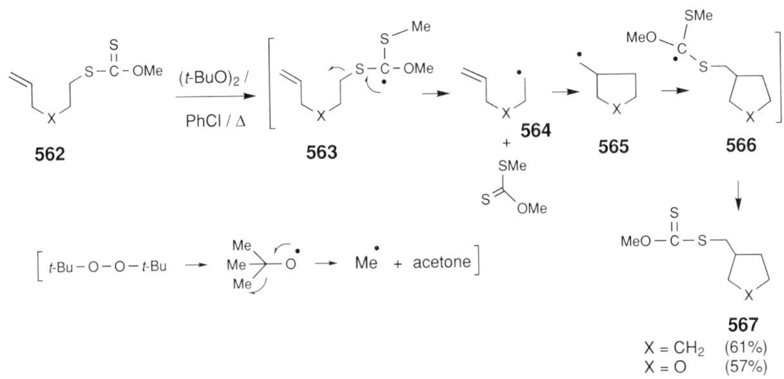

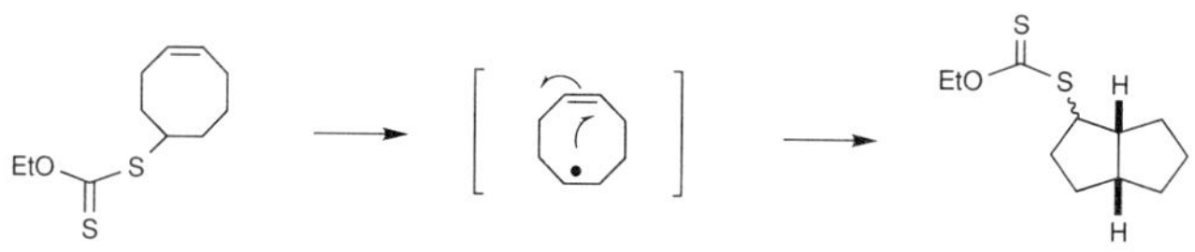

568 **569** **570** (60%)

The atom transfer reaction has also been incorporated into tandem cyclizations. For example, the reaction of iodomalonate **571** with *N-t*-butyloxycarbonyl allylamine (*N*-Boc allylamine) **572** under atom transfer cyclization conditions [(n-$Bu_3Sn)_2$ / hν (sunlamp)] yields iodoamine **575**. However, the highly sensitive amine **575** is preferably not isolated but converted straightaway to azabicycle **576** by boiling with triethylamine. The yield obtained for **576**, 43% after chromatography, is quite satisfactory considering the one-pot synthesis. The course of reaction is illustrated in Scheme 53.[156]

Scheme 53

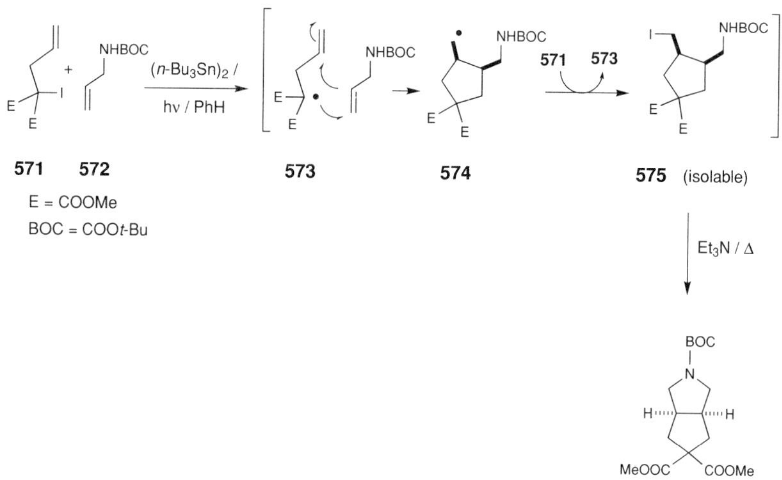

571 **572** **573** **574** **575** (isolable)

E = COOMe
BOC = COO*t*-Bu

576 (43%)

B. CYCLIZATION INVOLVING RADICAL TRANSLOCATION

Resembling the process in Scheme 53 in many respects, the reaction in Scheme 54 between propargylhalomalonate **577** and 1-hexene **578** yields cyclized products **583** and **585** accompanied by open chain by-product **581**. It is evident that methylenecyclopentane **583** is the direct result of a halogen-transfer cyclization, whereas the formation of methylenecyclopentane **585** requires a shift or transfer of hydrogen from C_1 to C_5 in the intermediate vinyl radical, *i.e.*, a translocation from **582** to **584**, before eventual termination with a halogen radical to yield the observed product.[157]

Scheme 54

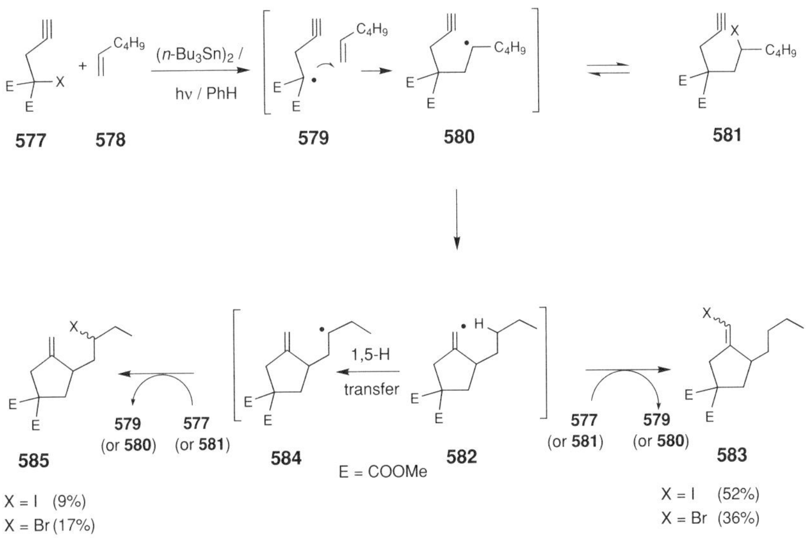

Radical translocation reactions *via* hydrogen transfer processes similar to that shown in the above Scheme (**582** → **584**) have been observed in quite a few cases. The first example is given in Equation 79. Initial generation of vinyl radical **587** (by the Stork procedure using *n*-Bu_3SnCl / $NaCNBH_3$ / AIBN / *t*-BuOH / Δ)[56] is followed by a 1,5-hydrogen atom transfer to afford the stabilized radical **588**, which subsequently cyclizes back onto the reinstated olefinic function. Substituents R^1 and R^2 can be any of various radical stabilizing groups such as alkoxyls, esters, and aryls.[158]

A second example makes use of the aryl carbon in the benzyl protecting group as the initial site of radical formation. Aryl radical **592** then proceeds through the sequence of hydrogen transfer and conjugate addition cyclization, as depicted in Equation 80.[158] In similar manner the examples in Equations 81[159] and 82[160] take advantage of the hydrogen transfer reaction to form the thermodynamically more favorable allylic radicals **597** and **601** which readily cyclize to yield the expected products.

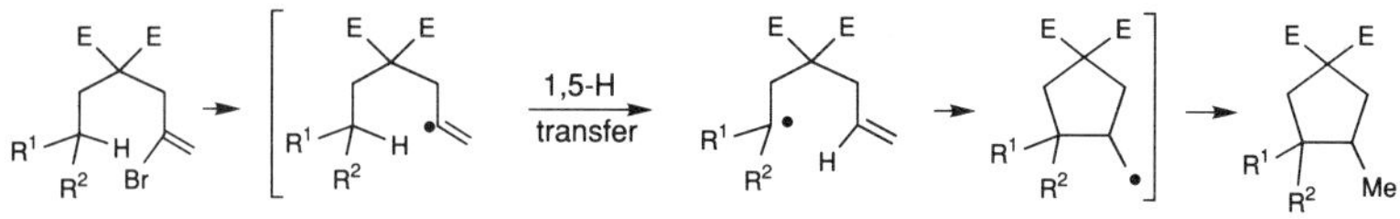

586 **587** **588** **589** **590**

E = COOMe	R^1 = H , R^2 = COOMe	(60%)
	R^1 = H , R^2 = Ph	(53%)
	R^1 = OMe , R^2 = Me	(78%)
	R^1—R^2 = O–CH_2–CH_2–O	(38%)

Equation 79

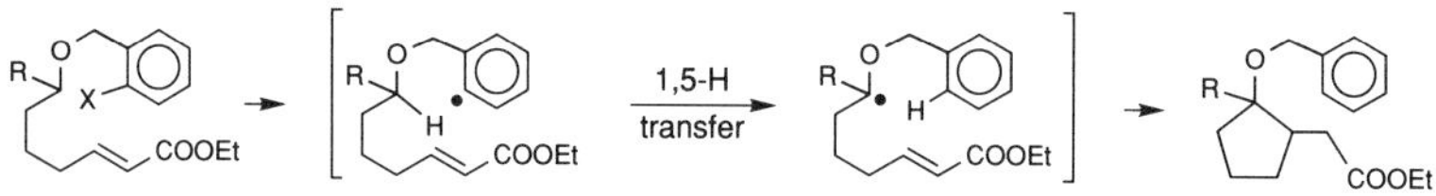

591 **592** **593** **594**

X = Br , I , R = H , Me (46–56%)

Equation 80

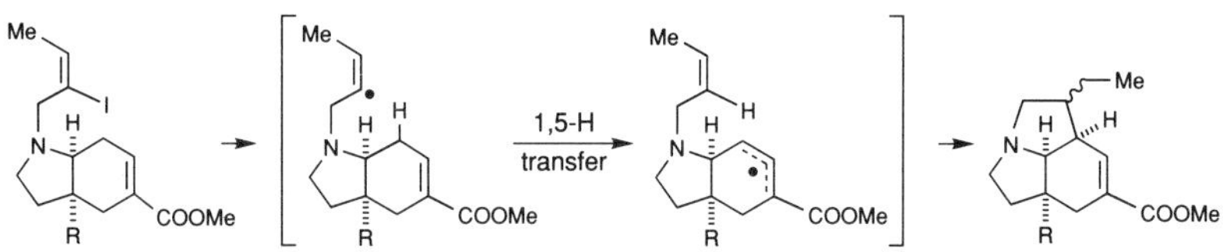

595 **596** **597** **598**

R = Ph , SPh (60–85%)
(α-ethyl : β-ethyl ≈ 4 : 1)

Equation 81

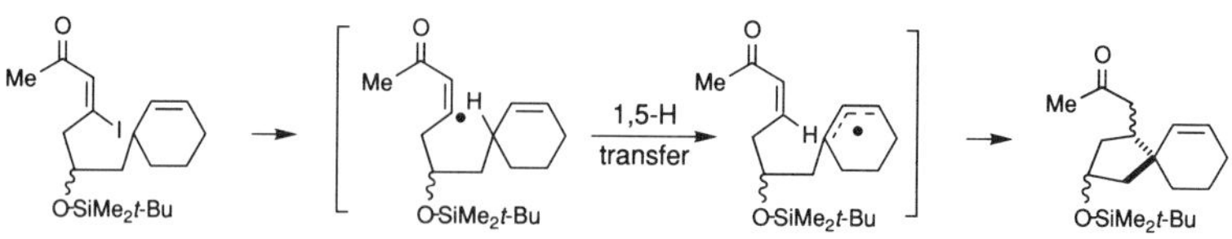

599 **600** **601** **602** (45%)

Equation 82

Radicals adjacent to the carbonyl or amino group in amide systems are also rapidly formed from C–H bonds by radical translocation reactions. Scheme 55 illustrates the process. Tributyltin radical reacts with the thermodynamically more favorable *anti* rotamer **603b** of the amide **603** to give radical **604** which spontaneously translocates to α-carbonyl alkyl radical **605**.[161] In the case of amide **606**, two α-amidoyl radicals **608** and **609** are possible from the radical translocation reaction of **607**, depending on the availability of the appropriate hydrogen at the carbon atoms attached to substituents R^1 and R^2.[162] Here the rates of 1,5-hydrogen transfer in **607a** and **607b** are found to be much faster than that of amide rotation, **607a** ⇌ **607b** (Scheme 55).

Scheme 55

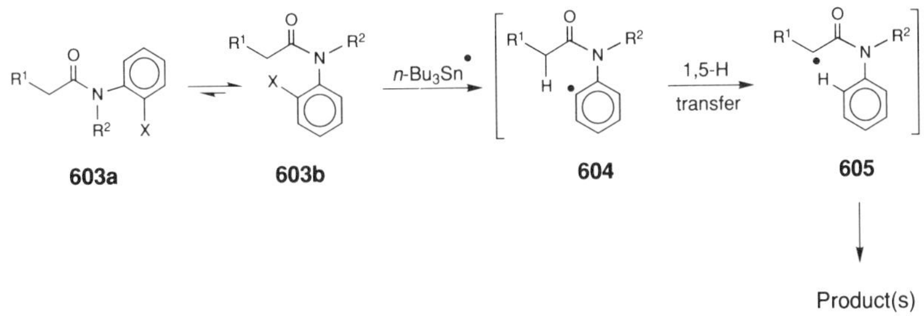

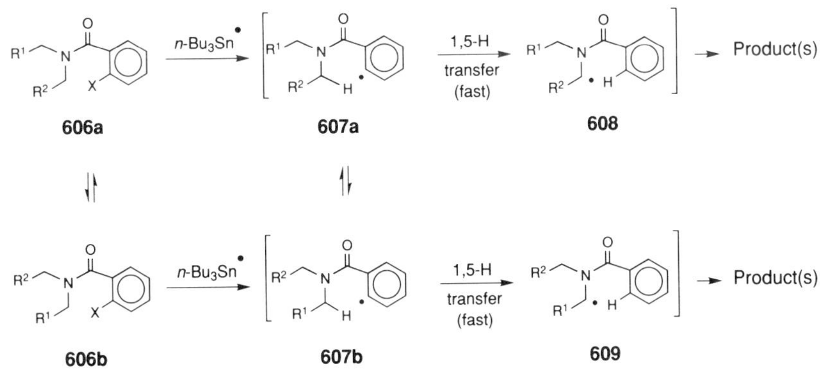

Application of the above process is nicely demonstrated in the synthesis of triquinanes,[161] in which radical translocation to α-carbonyl radicals can be used to construct both linear (Equation 83) and angular skeletons (Equation 84). Application of the similarly formed α-amidoyl radical is also illustrated in Scheme 56.[162] Starting from bromoamide **618**, standard treatment with tributyltin radical generates aryl radical **619**, which undergoes 1,5-hydrogen transfer to form α-amidoyl radicals **620** and **622**, followed by 6-*exo* cyclization to the vinyl double bond to yield angular and linear dibenzoquinolizidinones **621** and **623**, respectively. Notable are the stereospecificity of products and the minor yield of **621** compared to **623**, even though the former is formed *via* the more favorable intermediate benzylic radical **620**. This result can be explained by the earlier-mentioned slow amide rotation, *i.e.*, in the initial generation of aryl radical **619** the mixture of substrate rotamers (**618a** and **618b**) obviously gives a mixture of radical rotamers **619a** and **619b**, which, once formed, undergo radical translocation to the corresponding amidoyl radicals **620** and **622** faster than interconversion between them can occur. In other words, aryl radical **619** does not have time to entirely assume a conformation (**619a**) that would allow the more preferable transfer of benzylic hydrogen. Thus, if anything, the ratio of amidoyl radicals **620** and **622**, thence of products **621** and **623**, should reflect the starting ratio of substrate rotamers.

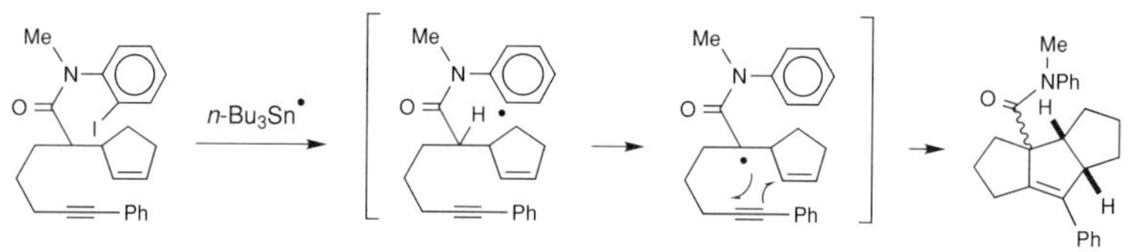

610 **611** **612** **613**

(35% ; one isomer)

Equation 83

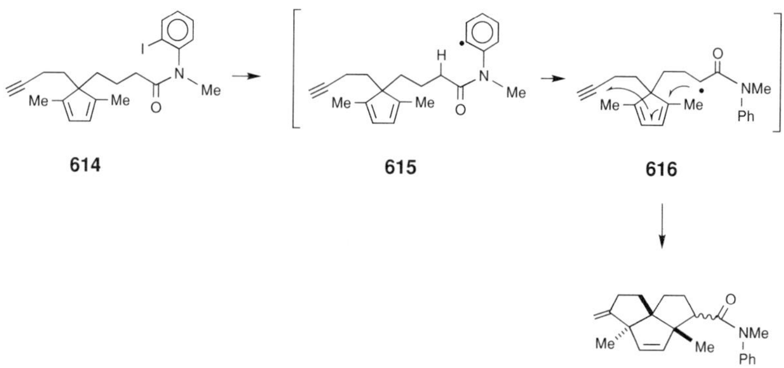

614 **615** **616**

617

(75% ; 1 : 1)

Equation 84

Scheme 56

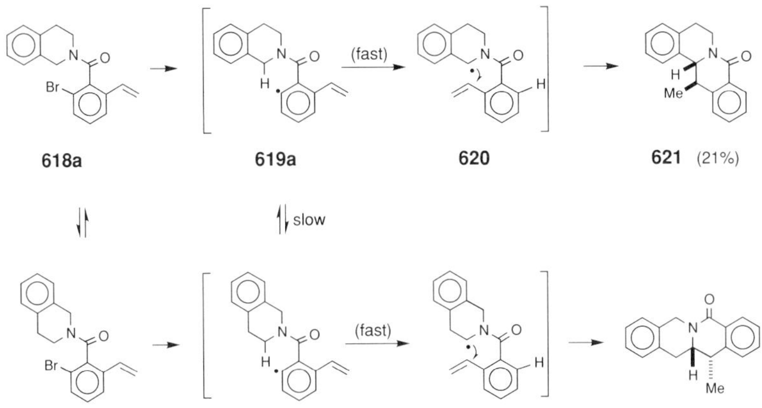

618a 619a 620 621 (21%)

618b 619b 622 623 (35%)

The foregoing examples of radical translocation have all involved 1,5-hydrogen transfer reactions, but in actual fact other types of hydrogen transfer are also possible, provided that the radical center and the C–H bond of the hydrogen to be transferred are in suitable proximity. For example, 1,4- and 1,6-hydrogen transfers are quite feasible given the appropriate molecular geometry. For instance, in the study towards the synthesis of gelsemine the cyclization of vinylogous urethane **624** was found to yield products **626** and **627** (84% yield)[163,164] but the cyclization of **628** yielded **632**. The mechanism of formation of the latter was elucidated as cyclization of aryl radical **629** to **630**, followed by a 1,4-hydrogen transfer to form the allylic radical **631** which finally terminated in the formation of product **632**. The structure of the product has been proved by X-ray crystallography, and the reaction mechanism by the use of tri-*n*-butyltin deuteride to yield deuteriated product **633** as shown in Scheme 57.[164]

In the study towards the synthesis of antitumor agent fredericamycin A, a most interesting observation was made in the cyclization of radical **635** (formed by treating **634** with $Ph_3Sn\bullet$) which resulted in only 48% yield of the requisite **637**, but with a 41% yield of an undesired product **639** (Scheme 58).[165] The proposed mechanism is that the cyclized vinyl radical **636**, besides capturing hydrogen radical (from Ph_3SnH) to form the normal product **637**, equally tends to undergo 1,6- hydrogen migration to give radical **638** which then cyclizes in the 6-*endo* mode to the observed polycyclic **639**. That the intermediate radical **638** does not abstract hydrogen from the stannane and revert to the normal product **637** has been proved with the use of triphenyltin deuteride which affords a deuteriated analog of **637** with deuterium located exclusively on the vinyl carbon. Interestingly, on using deuteriated precursor **640** for the cyclization, the yield of the desired product **641** (deuteriated) shot up to 70% while that of product **642** from 1,6-deuterium migration dropped to a mere 7%. This example constitutes an excellent application of isotope effect to control the direction of radical reaction.

Scheme 57

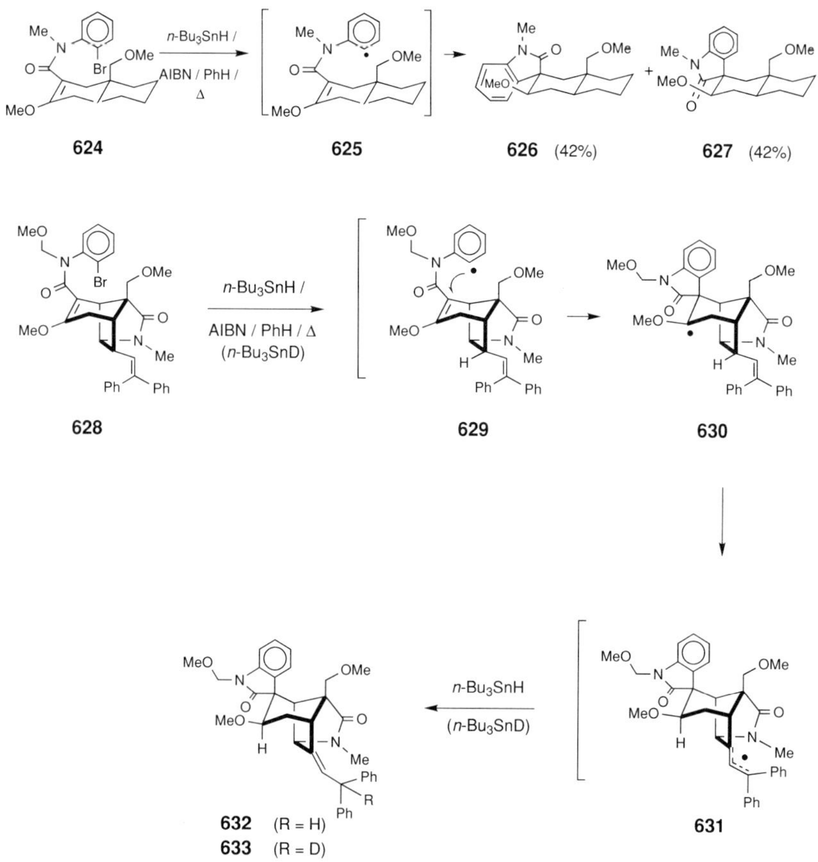

Scheme 58

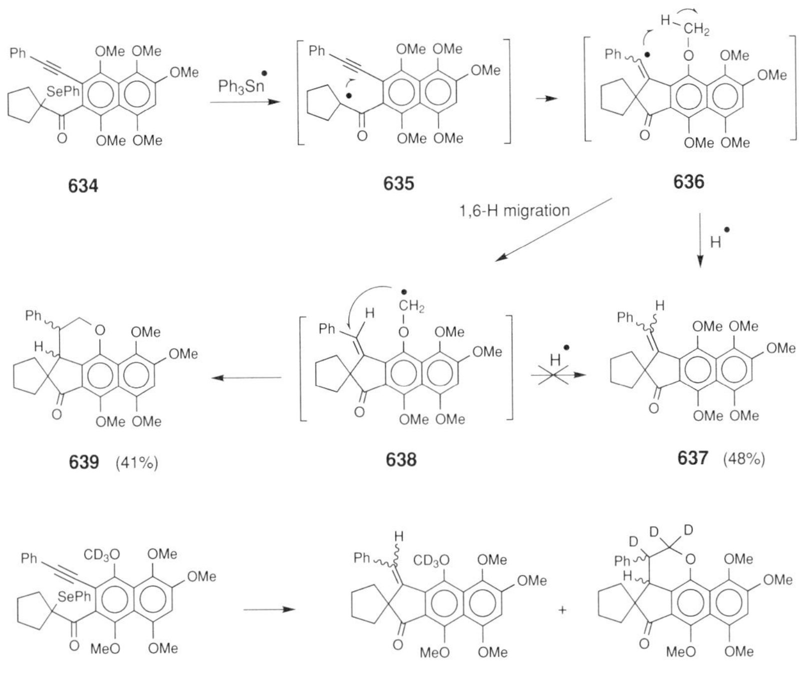

C. CYCLIZATION INVOLVING RADICAL REARRANGEMENT

Aptly regarded as a radical version of the Wharton reaction,[166] the ring-opening reaction of three-membered rings caused by an α-carbon radical as represented in Equation 85 has drawn attention through recognition of its potential in organic synthesis. Obviously, with suitably designed precursors the reaction could lead to products otherwise difficult to access. As shown in Equation 85, ring opening results in the formation of a rearranged radical **644** which, contingent upon the functional groups present in the molecule, may react further *via* simple or tandem cyclization.

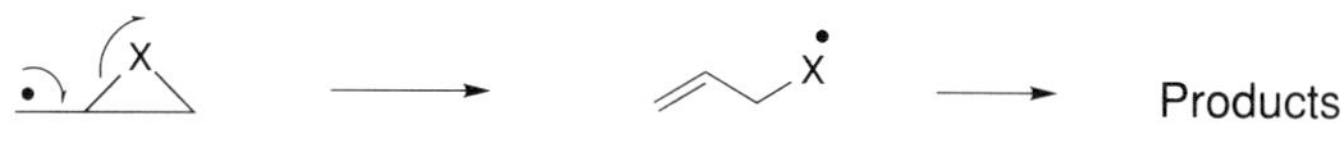

643

X = CR_2, NR, O

644

Equation 85

Such rearrangement–cyclization is observed with the α-cyclopropyl radical **647**, generated from thiocarbonyl imidazole **645**. Radical-induced cyclopropane ring opening leads to **648** which is trapped by the alkynyl group in a 5-*exo* cyclization reaction to produce spirocompound **649** (Scheme 59).[167] The reaction yields stereospecific products and can be employed as shown in Equations 86 and 87. Another mode of application is exemplified by the samarium diiodide-promoted radical ring opening of cyclopropyl ketones depicted in Equations 88 and 89,[168] while Equation 90 represents tandem aryl radical cyclization–cyclopropane ring opening in the construction of a benzofuran derivative.[169]

Scheme 59

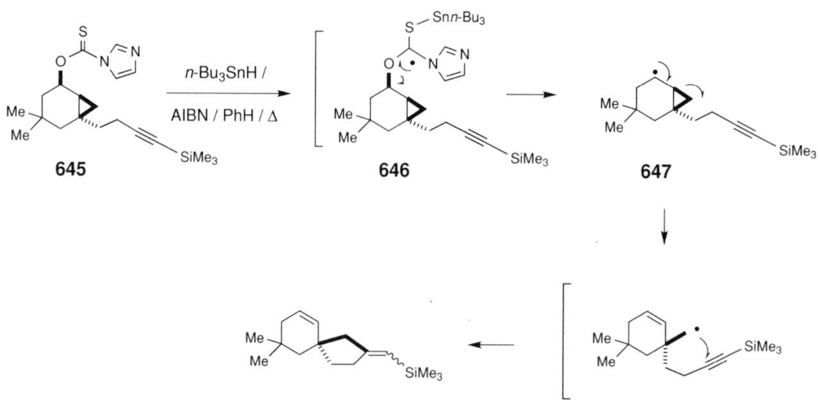

645 646 647 648 649

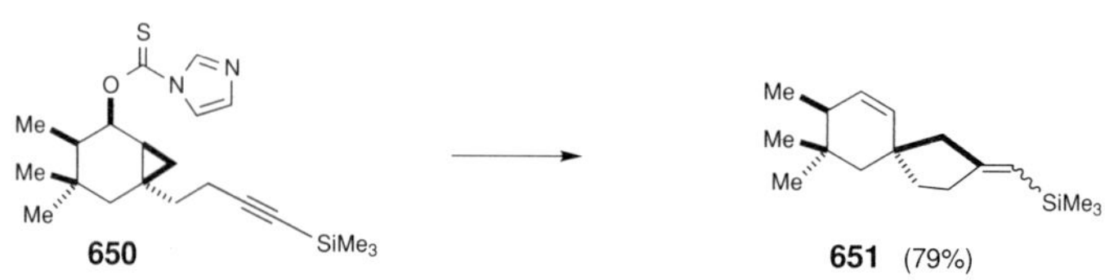

650

651 (79%)

Equation 86

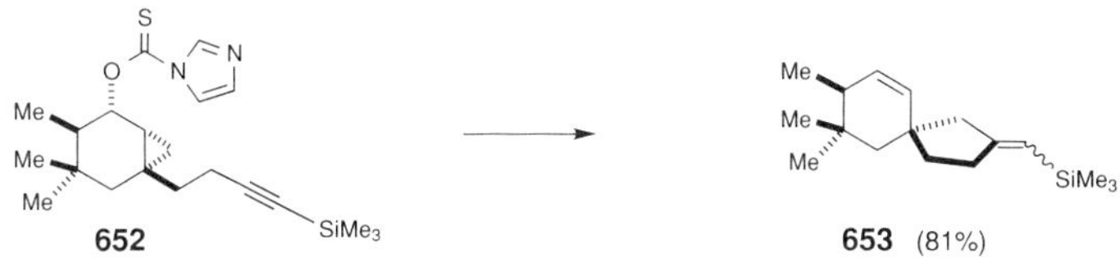

Equation 87

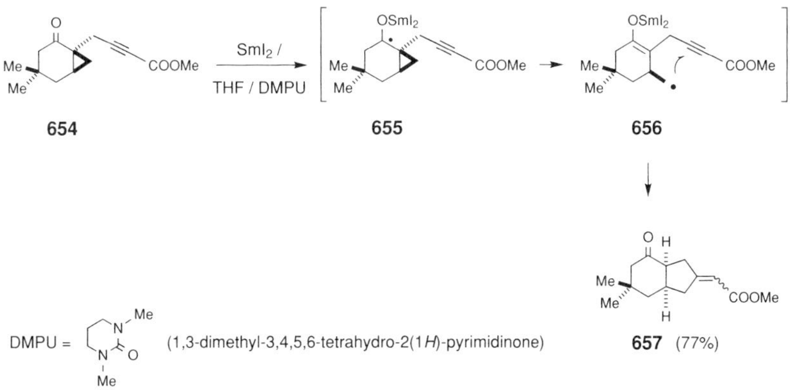

Equation 88

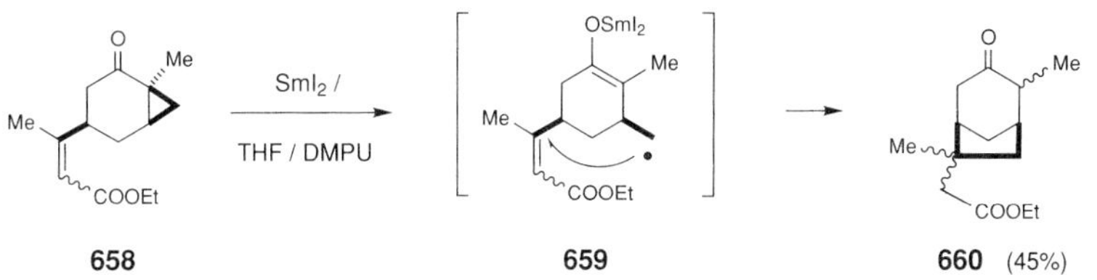

Equation 89

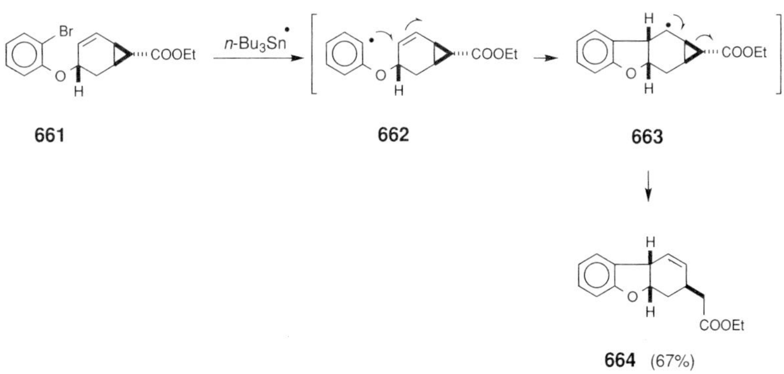

Equation 90

Radical-mediated ring opening of vinylcyclopropanes and methylenecyclopropanes have also been studied.[170,171] The example shown in Scheme 60 makes use of the reaction of the *n*-butyl sulfide radical with methylenecyclopropane **665** to generate a radical at the position α- to the ring, which intermediate suffers cyclopropane ring opening to **668** prior to intermolecular addition to olefin **666** to give intermediate **669**, before cyclizing to methylenecyclopentane **670**. The overall reaction is in fact equivalent to a [3+2]-cyclopentane annulation.[172,173]

Scheme 60

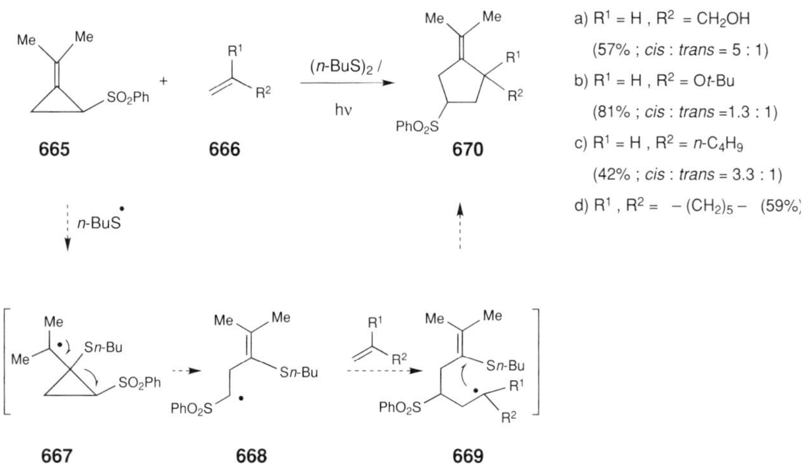

Inexplicably, the radical-induced cleavage of aziridine has not attracted the same kind of interest as has that of its analogs, cyclopropane and oxirane. Yet, from the few available examples in the literature, the reaction would seem to have quite a promising future. For example, treatment of thiocarbonylimidazole **673** (derived from aziridination[174] of geraniol **671** as shown in Scheme 61) with tin hydride / AIBN in refluxing THF, in the presence of magnesium bromide etherate, leads to the formation of pyrrolidine **676** in 70% yield. The role of $MgBr_2 \cdot Et_2O$ as Lewis acid is rather crucial for enhancing the reactivity of the cyclizing aminyl radical **675**, as evident in the drastic drop in yield of **676** to a low 30% in its absence.[175] The reaction mechanism is apparently as presented in the scheme.

A novel radical reaction of *N*-aziridinyl imines has been reported as outlined in Scheme 62 and the following Equation 91. Prepared from the corresponding carbonyl compound and 1-amino-2-phenylaziridine (caution: 1-amino-2-phenylaziridinium acetate is explosive!)[176] substrate **677** furnishes cyclopentene **682**, along with some methylenecyclopentane **683**, upon standard treatment with tin hydride. The mechanism has been elucidated as shown, the reaction producing the products stated, together with eliminated styrene and nitrogen.[176]

Scheme 61

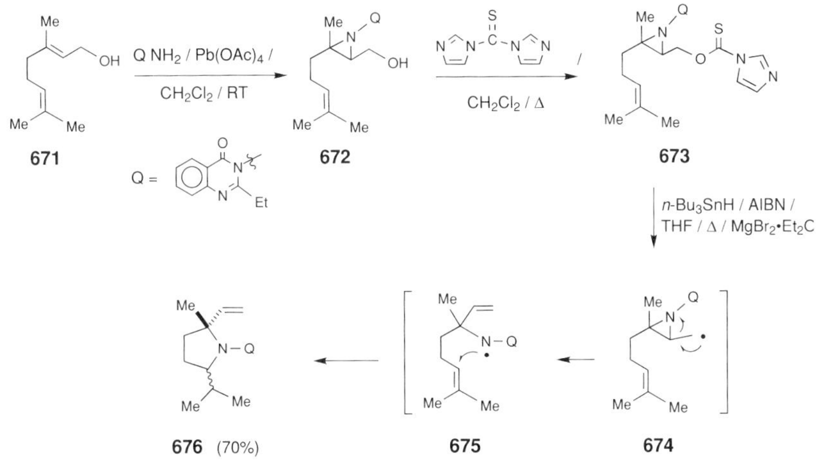

Scheme 62

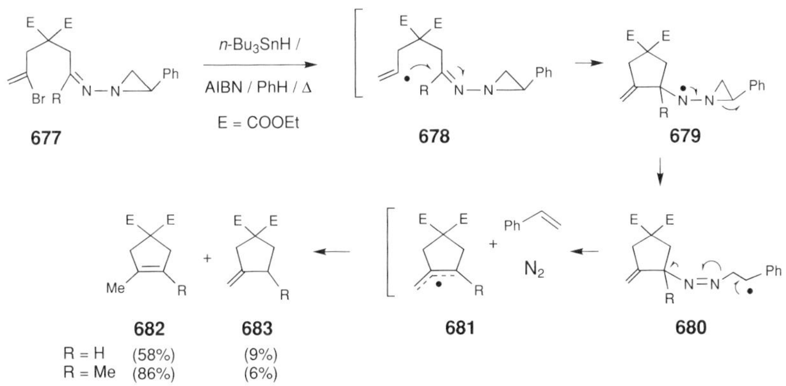

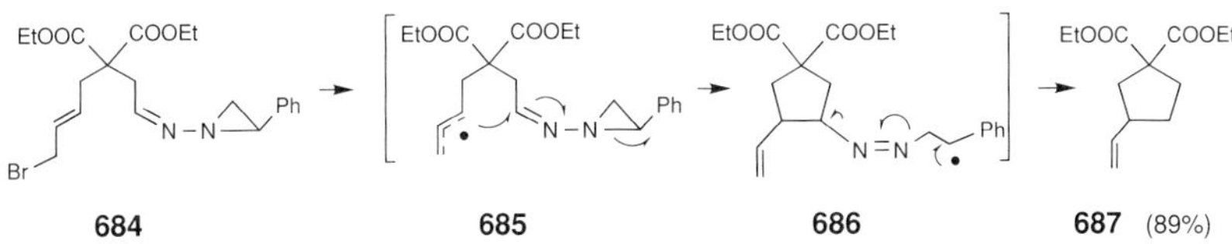

Equation 91

In looking back, oxirane ring opening induced by an α-radical center was reported as long as 30 years ago in the reaction of cyclohexene oxide with *tert*-butyloxy radical to provide a low yield (7%) of 2-cyclohexenol (Equation 92).[177] After that came the report by Barton[178] and his group on the synthesis of *cis*-carveol **695** from epoxide **692**, which synthesis also produced pinol **696** as the by-product from further cyclization of radical intermediate **694** (Equation 93). Subsequently, this latter reaction was found to be applicable to the synthesis of tetrahydrofuran derivatives.[179] As illustrated in Scheme 63,[180,181] reacting bromo epoxide **697** with tributyl tin radical gives rise to vinylpyran **701** and vinylcyclopentanol **702**, the former from direct cyclization of the first-formed alkoxide radical **699**, and the latter from an alternative pathway taken by **699**, *via* radical translocation by hydrogen transfer to afford a stabilized allylic radical **700** before cyclization to the cyclopentanol as shown.

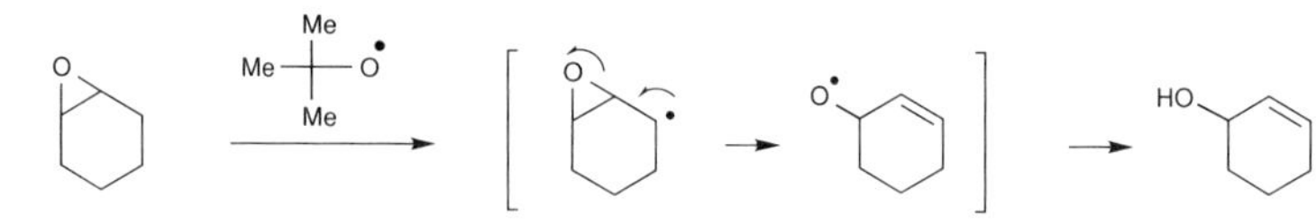

688 689 690 691

Equation 92

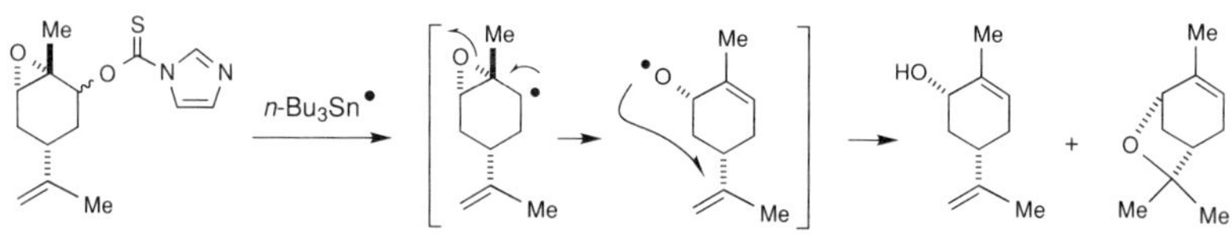

692 693 694 695 696

Equation 93

The oxygen-centered radical intermediate is highly reactive and very efficient in its abstraction of hydrogen radical, regardless of the environs of the latter. This has been demonstrated in an experiment with a series of substrates as shown in Scheme 64. Irrespective of whether group R in radical **705** is radical stabilizing or not (whether COOMe, Ph, or Me) the process of hydrogen transfer–cyclization, presumably *via* radical **706**, invariably occurs.[182] On the whole the reaction of oxirane ring opening by an α-radical center is receiving a fair amount of attention.[183–185] This is quite understandable as the reaction has considerable synthetic potential, particularly in the field of natural products, a simple example being the synthesis of α- and β-damascones from ionones.[186]

Scheme 63

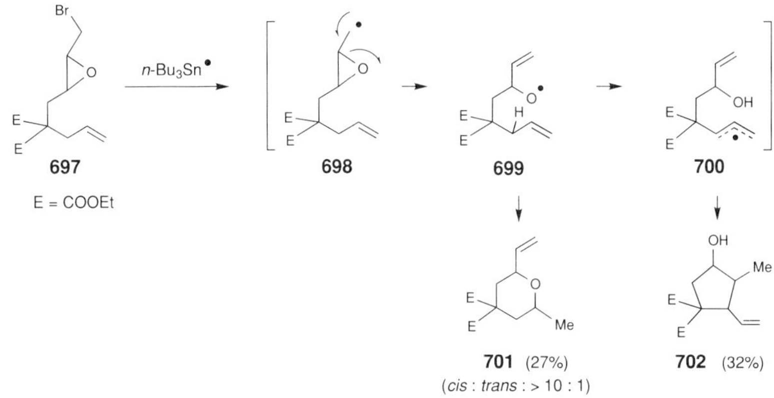

Scheme 64

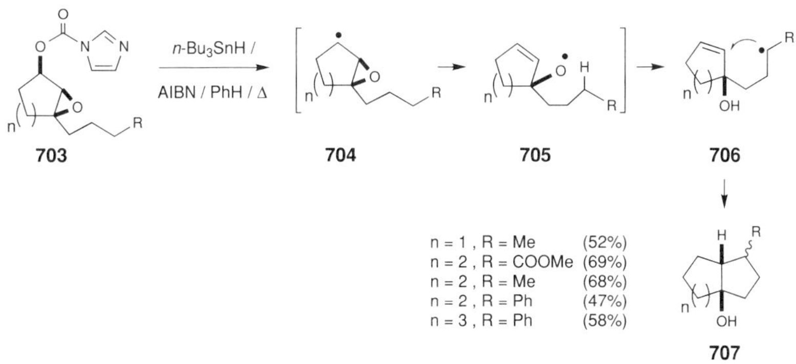

D. CYCLIZATION INVOLVING RADICAL FRAGMENTATION

Another mode of reaction of the oxygen-centered radical is by way of β-fragmentation, a process most frequently associated with ring enlargement, in which the resulting new radical may be simply terminated or further involved in cyclization. The general reaction is as presented in Scheme 65. The alkoxy radical **710**, produced either directly from the homolytic fission of the O–X bond in **708** (route a) or *via* the intramolecular attack of a preformed radical center on the carbonyl as in **709** (route b), undergoes β-fragmentation to give the new radical **711** which can either terminate or undergo further radical chain reactions, *e.g.*, cyclization, given the presence of a suitably placed radical acceptor.

Scheme 65

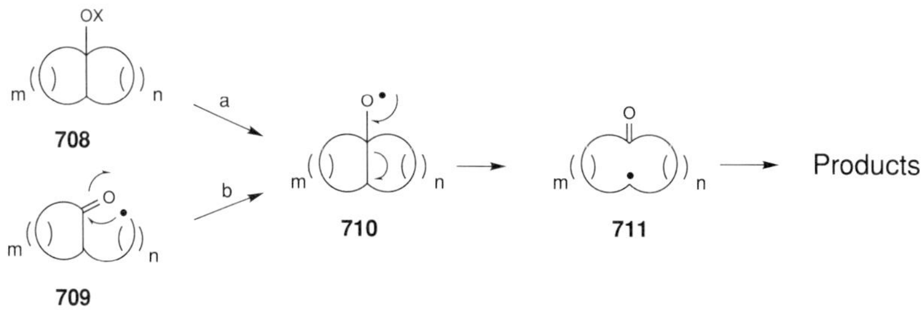

Early observations of the β-fragmentation of alkoxy radicals generated *via* route a are shown in Equations 94[187] and 95.[188] In the case of lactol **712**, upon irradiation with mercuric oxide and iodine, fragmentation of radical **713** occurs in two directions to give **714** and **715**. The same also applies to radical **717** from lactol **716**, the irradiation yielding fused- and spirolactones **718** and **719**, respectively. Remarkably, complete regiospecificity can be induced by the incorporation of a radical reactive substituent, here a SPh group, whereupon **717** fragments react in only one direction, eliminating the SPh leaving group to yield vinyl lactone **720** as the sole product.

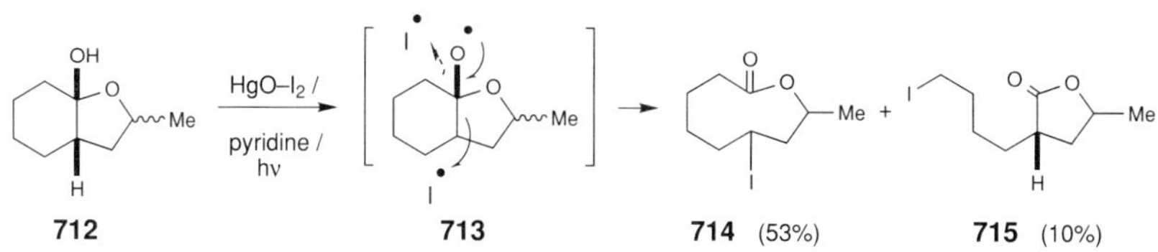

Equation 94

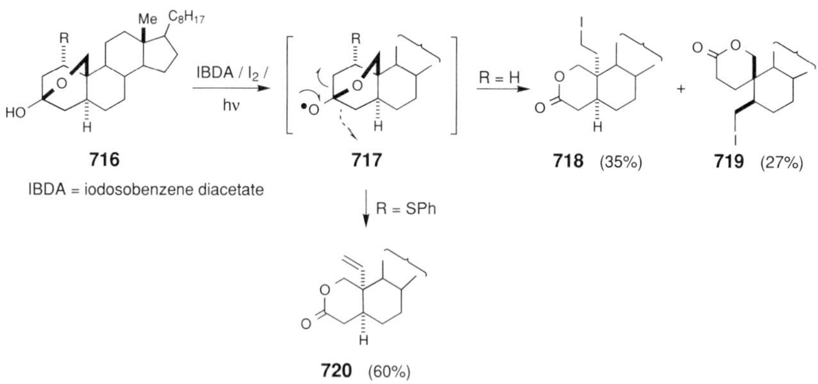

Equation 95

As for the β-fragmentation–cyclization sequence, an example is given in Equation 96. When R = Me, radical **722** exhibits regiospecific bond breaking to form tertiary radical **723**, which cyclizes to afford exclusively product **724** in a far better yield than that obtained with R = H.[189]

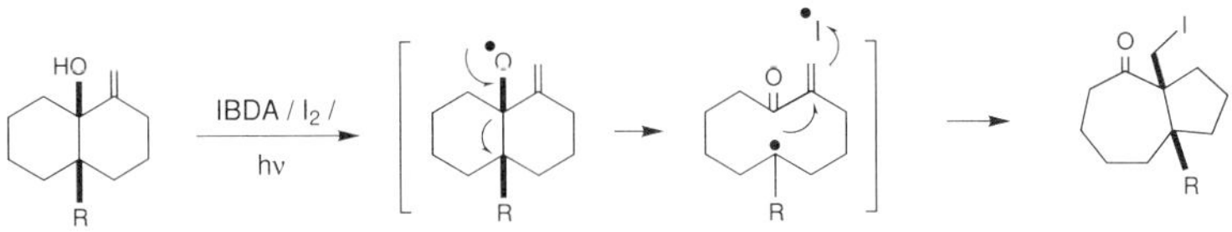

721 **722** **723** **724**

R = H (40%)
R = Me (78%)

Equation 96

The ease with which the radical center undergoes intramolecular attack onto the carbonyl group (**709**), *i.e.*, the ease of generation of alkoxy radical **710**, depends on the size of the newly formed ring. Such cyclizations to form 3-, 5-, and 6-membered rings are known, with three-membered ring formation being the most studied and the most frequently employed. For example, when iodoketone **724** (obtained from the reaction in Equation 96) is treated with tributyltin hydride, two products are obtained, one (**726**) from simple reduction and one (**728**) from ring expansion *via* radical cyclization to carbonyl to form the cyclopropylalkoxy radical **727**, which can ring expand *via* β-fragmentation as shown in Equation 97.[189] This type of ring expansion works well with cyclic β-keto ester substrates, the presence of the ester group serving to accelerate fragmentation, as summarized in Equation 98.[190,191]

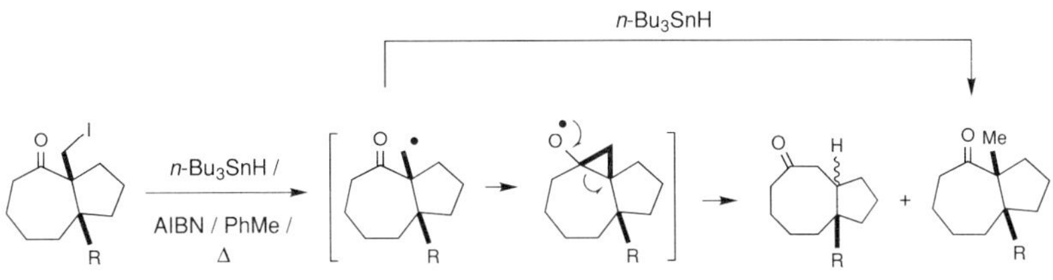

724 **725** **727** **728** **726**

	728	726
R = H	(50%)	–
R = Me	(20%)	(60%)

Equation 97

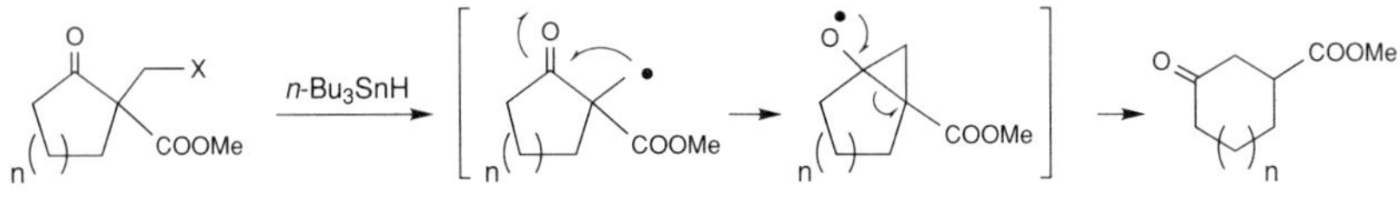

729 730 731 732

X = Br	n = 1	(75%)
	n = 2	(73%)
	n = 3	(71%)
X = I	n = 8	(79%)
	n = 10	(50%)
	n = 11	(67%)

Equation 98

The above tandem radical cyclization–alkoxy radical fragmentation process has been widely applied in synthesis, such as in carbocyclic ring expansions[192–194] and in the preparation of fused cyclooctenones from cyclohexanones. The latter in fact involves two rounds of radical cyclization–alkoxy radical fragmentation, the first, by vinyl radical **734**, proceeds as cyclization–fragmentation–recyclization to form alkyl radical **737** which then repeats the sequence of cyclization to carbonyl–*β*-fragmentation of alkoxy radical, this time terminating in elimination of the tin residue to yield fused cyclooctenone **740**. The yield from this completely tandem process is not impressive, but against the multi-step overall chemical transformation accomplished in one operation it becomes very acceptable.[195]

Scheme 66

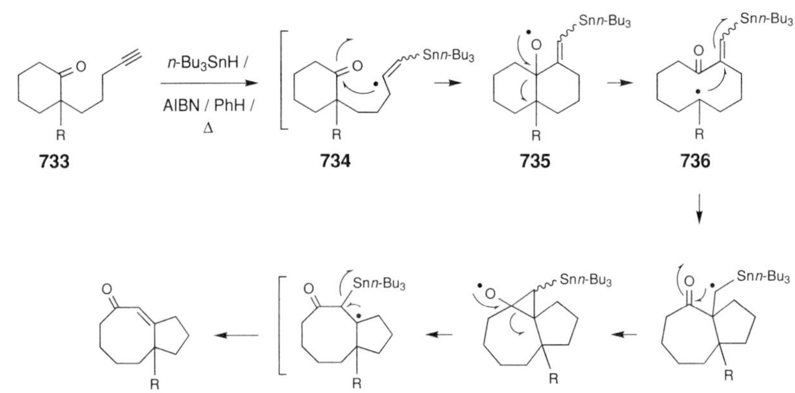

733 734 735 736

740 739 738 737

R = H (18%)
R = COOEt (22%)
R = OSiEt$_3$ (51%)

Apart from the one-carbon ring expansion, three- and four-carbon ring expansions by the same concept of radical cyclization to form five- and six-membered ring alkoxy radicals (see Scheme 65) are also feasible as shown in

Equation 99.[196] This method of three-carbon ring expansion has been applied to the synthesis of (±)-muscone **750**, a 15-membered ring carbocycle, starting from a 12-membered ring compound **745**, as illustrated in Scheme 67.[191] Although the reaction yields the reduction product **749** as the major product, access to the 15-membered macrocycle by this method is still quite useful.

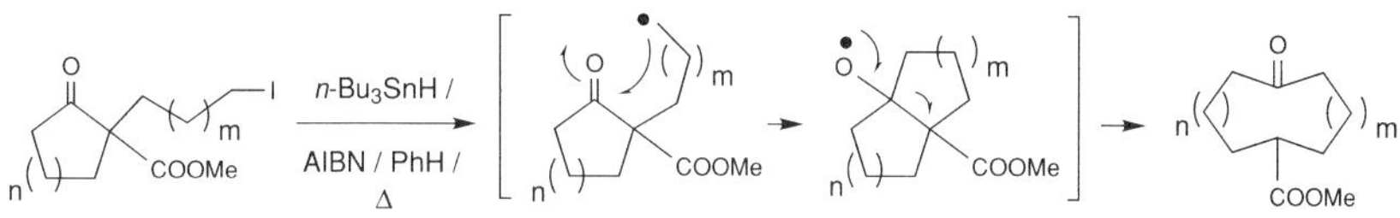

741 **742** **743** **744**

n = 1 , m = 1 (69%)
n = 1 , m = 2 (36%)
n = 2 , m = 1 (75%)
n = 2 , m = 2 (71%)
n = 3 , m = 1 (34%)
n = 3 , m = 2 (45%)

Equation 99

Scheme 67

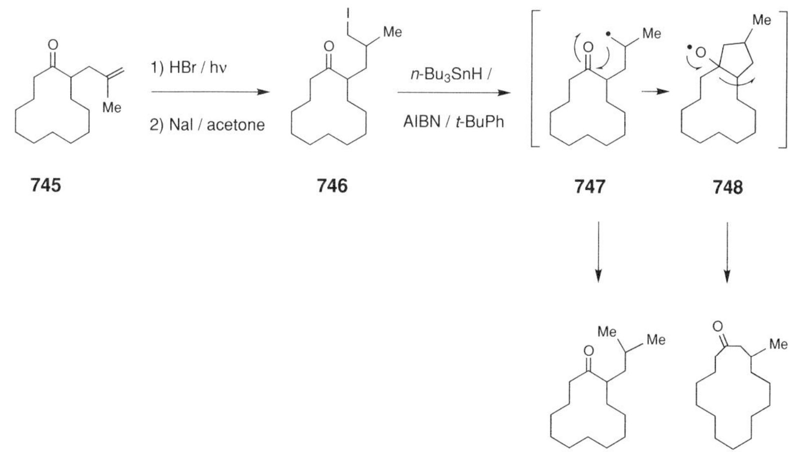

745 **746** **747** **748**

749 (63%) **750** (14%)

It will be recalled that tandem cyclizations of acyl radicals shown earlier in Scheme 36 and Equations 71 to 73 were assumed to proceed first in the 6-*endo* mode followed by either a 5- or 6-*exo* cyclization in the second stage. At this point, however, it becomes arguable that the reaction could involve the above sequential radical cyclization–alkoxy radical fragmentation–radical recyclization process. Take, for example, the reaction in Equation 71: its re-inspection (Scheme 68) reveals that acyl radical **456**, apart from 6-*endo* ring closure to give tertiary radical **457**, could well cyclize in 5-*exo* fashion to result in **752**, which is in a position to further undergo the radical cyclization–fragmentation process, leading in a roundabout way (*via* **753**) to the same intermediate **457**, hence the same product **458**, as that obtained from the *endo* cyclization process. Moreover, the direct generation of radical **752** from bromocyclopentanone **751**, also yields the bicyclic product **458**. However, in an experiment with the reaction shown formerly in Equation 72, it was found that although acyl radical **460** (from the corresponding phenyl selenide **459**) undergoes clean tandem cyclization to yield **462** (82% yield), radical **755** (from bromoketone **754**) gives not even a trace of product **462**, confirming that the latter is formed *via* 6-*endo* cyclization, which, in this case, is both kinetically and thermodynamically the preferred cyclization mode.[197]

Scheme 68

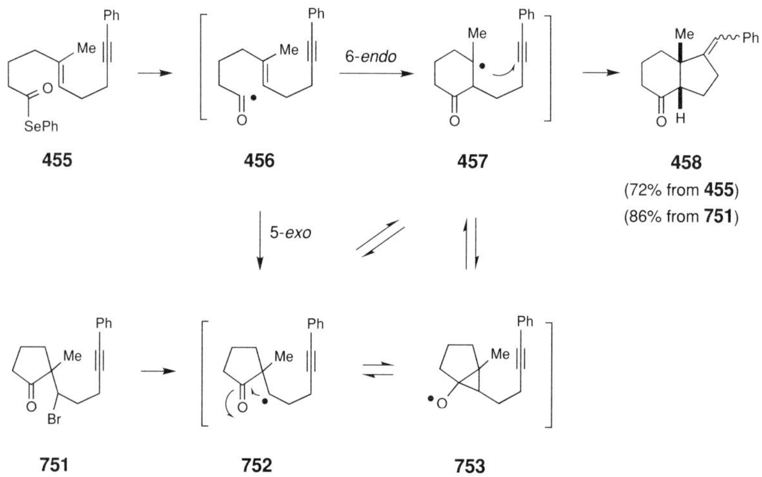

Scheme 69

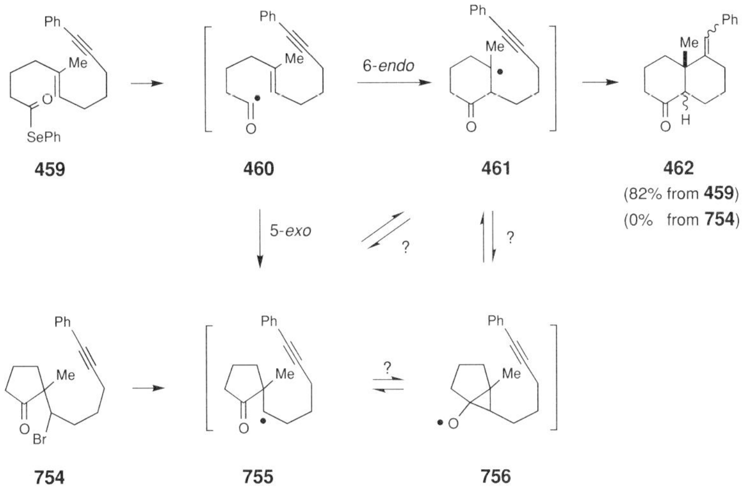

VI. CONCLUSION

Ten years, or even as recently as five years ago, one would have found it hard to believe that free radicals would be as effective a tool in synthetic chemistry as they are today, particularly in cyclization chemistry. Information on methodologies for the generation of radical centers and the utilization of radical cyclization as key synthetic steps is fast expanding, additional to the above being the generation of radicals from epoxyalkyl halides induced by sonication / zinc–copper couple,[198] the oxidative free radical cyclization using Mn(III) acetate,[199–201] the study of radical cations and anions,[202–205] and cyclization reactions involving the electron transfer process,[206–210] with atom or group transfer cyclization also making impressive progress.[211–213] Such a growing army of reactions has already led to the total synthesis of various types of natural products, most prominently the alkaloids[214–217] and terpenoids.[218,219] To date, a short and concise review by Barton[220] provides the most recent summary of radical chemistry based on the thiocarbonyl function.

In conclusion, it would indeed be difficult, even for those with deep insight into the field, to attempt to predict just where radical cyclization chemistry will be in the next decade.

REFERENCES

1. Nicolaou, K. C., and Dai, W.-M., Chemistry and biology of the enediyne anticancer antibiotics, *Angew. Chem. Int. Ed. Engl.*, 30, 1387, 1991.
2. Beckwith, A. L. J., and Zimmermann, J., Cyclization of the 3-*tert*-butylhex-5-enyl radical: a test of transition-state structure, *J. Org. Chem.*, 56, 5791, 1991.
3. Giese, B., *Radicals in Organic Synthesis: Formation of Carbon–Carbon Bonds*, Baldwins, J. E., Ed., Pergamon Press, Oxford, 1986.
4. Curran, D. P., The design and application of free radical chain reactions in organic synthesis. Parts 1 and 2, *Synthesis*, 417, 1988 and 489, 1988.
5. Stork, G., Radical cyclization in the control of regio- and stereochemistry, *Bull. Chem. Soc. Jpn.*, 61, 149, 1988.
6. Giese, B., The stereoselectivity of intermolecular free radical reactions, *Angew. Chem. Int. Ed. Engl.*, 28, 969, 1989.
7. Thebtaranonth, C., and Thebtaranonth, Y., Developments in cyclisation reactions, *Tetrahedron*, 46, 1385, 1990.
8. RajanBabu, T. V., Stereochemistry of intramolecular free-radical cyclization reactions, *Acc. Chem. Res.*, 24, 139, 1991.
9. Jasperse, C. P., Curran, D. P., and Fevig, T. L., Radical reactions in natural product synthesis, *Chem. Rev.*, 91, 1237, 1991.
10. Little, R. D., Masjedizadeh, M. R., Moeller, K. D., and Dannecker-Doerig, I., Factors affecting regioselectivity in the intramolecular diyl trapping reaction, *Synlett*, 107, 1992.
11. Beckwith, A. L. J., and Schiesser, C. H., A force-field study of alkenyl radical ring closure, *Tetrahedron Lett.*, 26, 373, 1985.
12. Spellmeyer, D. C., and Houk, K. N., A force-field model for intramolecular radical additions, *J. Org. Chem.*, 52, 959, 1987.
13. Kuivila, H. G., Reduction of organic compounds by organotin hydrides, *Synthesis*, 499, 1970.
14. Barton, D. H. R., da Silva, E., and Zard, S. Z., Multiple radical additions: an expedient entry into complex frameworks, *J. Chem. Soc., Chem. Commun.*, 285, 1988.
15. Kariv-Miller, E., Maeda, H., and Lombardo, F., Reductive tandem cyclization of allyl pentenyl ketones, *J. Org. Chem.*, 54, 4022, 1989.
16. Arya, P., Samson, C., Lesage, M., and Griller, D., Tris(trimethylsilyl)silane (TTMSS): formation of carbon-centered radicals from 1,3-dithiolanes and 1,3-dithianes, *J. Org. Chem.*, 55, 6248, 1990.
17. RajanBabu, T. V., From carbohydrates to carbocycles. II. A free radical route to Corey lactone and other prostanoid intermediates, *J. Org. Chem.*, 53, 4522, 1988.
18. Clive, D. L. J., Synthetic methodology involving radical cyclization: spiro compounds and α-oxo radicals, *Pure & Appl. Chem.*, 60, 1645, 1988.
19. Yadav, V. K., and Fallis, A. G., Cyclopentane synthesis and annulation: intramolecular radical cyclization of acetals, *Tetrahedron Lett.*, 29, 897, 1988.
20. Sha, C.-K., Jean, T.-S., and Wang, D.-C., Intramolecular radical cyclization of silylacetylenic or olefinic α-iodo ketones: application to the total synthesis of (±)-modhephene, *Tetrahedron Lett.*, 31, 3745, 1990.
21. Esch, P. M., Hiemstra, H., and Speckamp, W. N., Synthesis of cyclic α-amino acids through ring closure of glycine derived free radicals, *Tetrahedron Lett.*, 31, 759, 1990.
22. Knapp, S., Gibson, F. S., and Choe, Y. H., Radical based annulations of iodo lactams, *Tetrahedron Lett.*, 31, 5397, 1990.
23. Ariamala, G., and Balasubramanian, K. K., Radical cyclisation: synthesis of 4*H*-furo(3,2-*c*)-1-benzopyrans, *Tetrahedron Lett.*, 29, 3335, 1988.
24. McDonald, C. E., and Dugger, R. W., A formal total synthesis of (–)-isoavenaciolide, *Tetrahedron Lett.*, 29, 2413, 1988.
25. Carda, M., and Marco, J. A., Total synthesis of (–)-mintlactone, *Tetrahedron Lett.*, 32, 5191, 1991.

26. Lolkema, L. D. M., Hiemstra, H., Ghouch, A. A. A., and Speckamp, W. N., Synthesis of oxacyclic carboxylic esters through ring closure of α-alkoxy ester free radicals, *Tetrahedron Lett.*, 32, 1491, 1991.
27. Kolt, R. J., Griller, D., and Wayner, D. D. M., Unsaturated spiro-γ-lactone formation by the dissociative reduction of bromoacetates, *Tetrahedron Lett.*, 31, 7539, 1990.
28. Chatgilialoglu, C., Griller, D., and Lesage, M., Tris(trimethylsilyl)silane. A new reducing agent, *J. Org. Chem.*, 53, 3641, 1988.
29. Giese, B., Kopping, B., and Chatgilialoglu, C., Tris(trimethylsilyl)silane as mediator in organic synthesis *via* radicals, *Tetrahedron Lett.*, 30, 681, 1989.
30. Arya, P., Lesage, M., and Wayner, D. D. M., Tris(trimethylsilyl)silane (TTMSS): formation of carbon-centered radicals from oxathiolane and thiazolidine derivatives, *Tetrahedron Lett.*, 32, 2853, 1991.
31. Arya, P., and Wayner, D. D. M., Thiazolidine derivatives: a new source of α-aminoalkyl radicals for carbon–carbon bond formation in synthesis, *Tetrahedron Lett.*, 32, 6265, 1991.
32. Barton, D. H. R., Crich, D., and Motherwell, W. B., The invention of new radical chain reactions. VIII. Radical chemistry of thiohydroxamic esters; a new method for the generation of carbon radicals from carboxylic acids, *Tetrahedron*, 41, 3901, 1985.
33. Crich, D., and Quintero, L., Radical chemistry associated with the thiocarbonyl group, *Chem. Rev.*, 89, 1413, 1989.
34. Newcomb, M., Deeb, T. M., and Marquardt, D. J., *N*-Hydroxypyridine-2-thione carbamates. IV. A comparison of 5-*exo* cyclizations of an aminyl radical and an aminium cation radical, *Tetrahedron*, 46, 2317, 1990.
35. Newcomb, M., Marquardt, D. J., and Deeb, T. M., *N*-Hydroxypyridine-2-thione carbamates. V. Syntheses of alkaloid skeletons by aminium cation radical cyclizations, *Tetrahedron*, 46, 2329, 1990.
36. Beckwith, A. L. J., and Davison, I. G. E., Diastereoselective formation of cyclic carbonates by cyclization of alkenyloxycarbonyloxy radicals, *Tetrahedron Lett.*, 32, 49, 1991.
37. Forbes, J. E., and Zard, S. Z., New radical chain reaction of *S*-alkoxycarbonyl dithiocarbonates: a useful source of alkyl radicals from alcohols, *J. Am. Chem. Soc.*, 112, 2034, 1990.
38. Crandall, J. K., Tindell, G. L., and Manmade, A., Homoallenyl radicals, *Tetrahedron Lett.*, 23, 3769, 1982.
39. Apparu, M., and Crandall, J. K., Cyclizations of ω-allenyl radicals, *J. Org. Chem.*, 49, 2125, 1984.
40. Crandall, J. K., and Ayers, T. A., Radical cyclizations of functionalized allenes, *Tetrahedron Lett.*, 32, 3659, 1991.
41. Jeganathan, S., and Vogel, P., Synthesis of higher-carbon sugars *via* cross-aldolization of 7-oxanorbornan-2-one and carbohydrate aldehyde derivatives, *J. Chem. Soc., Chem. Commun.*, 993, 1989.
42. Gasparini, F., and Vogel, P., Highly stereoselective total syntheses of 2,5-anhydro-3-deoxy- and -4-deoxy-D-hexonic acids and of the related deoxyadenosines-C, *J. Org. Chem.*, 55, 2451, 1990.
43. Reymond, J.-L., and Vogel, P., A highly stereoselective total synthesis of (±) castanospermine, *Tetrahedron Lett.*, 30, 705, 1989.
44. Renaud, P., Vionnet, J.-P., and Vogel, P., Radical additions to silyl ketene acetals. Introduction of C-moieties to 7-oxabicyclo[2.2.1]hept-5-en-2-*endo*-ol, *Tetrahedron Lett.*, 32, 3491, 1991.
45. Lee, J., Teng, K., and Marquez, V. E., Synthesis of two rigid diacylglycerol analogues having a bis-butyrolactone skeleton, *Tetrahedron Lett.*, 33, 1539, 1992.
46. Stork, G., and Baine, N. H., Cyclization of vinyl radicals: a new versatile method for the construction of five- and six-membered rings, *J. Am. Chem. Soc.*, 104, 2321, 1982.
47. Dulcere, J. P., Mihoubi, M. N., and Rodriguez, J., A new synthetic route to β-bromoprop-2-ynyl mixed acetals and bromovinyl bis-allyl mixed acetals, precursors of α-methylene-γ-butyrolactones, *J. Chem. Soc., Chem. Commun.*, 237, 1988.

48. Rajamannar, T., and Balasubramanian, K. K., Enolate alkylation of bicyclo(2.2.2)oct-5-en-2-one and radical cyclisation–a potential approach for the construction of tricyclic carbocycles, *Tetrahedron Lett.*, 29, 5789, 1988.
49. Gopalsamy, A., and Balasubramanian, K. K., Radical cyclisation of 4-(*o*-bromophenoxy)-2*H*-1-benzopyrans; an efficient synthesis of pterocarpans, *J. Chem. Soc., Chem. Commun.*, 28, 1988.
50. Ahmad-Junan, S. A., and Whiting, D. A., Rotenoid synthesis *via* radical cyclisation, *J. Chem. Soc., Chem. Commun.*, 1160, 1988.
51. Dittami, J. P., and Ramanathan, H., Intramolecular radical cyclization reactions. An approach to the indole alkaloids, *Tetrahedron Lett.*, 29, 45, 1988.
52. Biggs, K. R., Parsons, P. J., Tapolzcay, D. J., and Underwood, J. M., A new route to spiroacetals and the construction of a model for the synthesis of phyllanthocin, *Tetrahedron Lett.*, 30, 7115, 1989.
53. Kupchan, S. M., LaVoie, E. J., Branfman, A. R., Fei, B. Y., Bright, W. M., and Bryan, R. F., Phyllanthocin, a novel bisabolane aglycone from the antileukemic glycoside, phyllanthoside, *J. Am. Chem. Soc.*, 99, 3199, 1977.
54. Kataoka, T., Yoshimatsu, M., Shimizu, H., and Hori, M., Vinyl radical generation with selenoborane and its application to cyclization reaction of enynes, *Tetrahedron Lett.*, 31, 5927, 1990.
55. Boger, D. L., Ishizaki, T., Kitos, P. A., and Suntornwat, O., Synthesis of *N*-(*tert*-butyloxycarbonyl)-CBI, CBI, CBI-CDPI$_1$, and CBI-CDPI$_2$: enhanced functional analogues of CC-1065 incorporating the 1,2,9,9a-tetrahydrocyclopropa[*c*]benz[*e*]indol-4-one (CBI) left-hand subunit, *J. Org. Chem.*, 55, 5823, 1990.
56. Stork, G., and Sher, P. M., A catalytic tin system for trapping of radicals from cyclization reactions. Regio- and stereocontrolled formation of two adjacent chiral centers, *J. Am. Chem. Soc.*, 108, 303, 1986.
57. Clark, A. J., Jones, K., McCarthy, C., and Storey, J. M. D., Aryl radical cyclisations: quinoline, isoquinolone, and 1-benzazepin-2-one rings *via* 6- and 7-*exo* cyclisations, *Tetrahedron Lett.*, 32, 2829, 1991.
58. Beckwith, A. L. J., and O'Shea, D. M., Kinetics and mechanism of some vinyl radical cyclisations, *Tetrahedron Lett.*, 27, 4525, 1986.
59. Stork, G., and Mook, R., Jr., Five *vs* six membered ring formation in the vinyl radical cyclization, *Tetrahedron Lett.*, 27, 4529, 1986.
60. Satoh, S., Sodeoka, M., Sasai, H., and Shibasaki, M., Control of ring-junction stereochemistry *via* radical cyclization. A new construction of *trans*-hydrindans, *J. Org. Chem.*, 56, 2278, 1991.
61. Takano, S., Suzuki, M., Kijima, A., and Ogasawara, K., Isoquinoline frameworks *via* aryl radical-initiated 1,6-cyclization, *Tetrahedron Lett.*, 31, 2315, 1990.
62. Pal, S., Mukherjee, M., Podder, D., Mukherjee, A. K., and Ghatak, U. R., A highly regioselective 6-*endo*-aryl radical cyclisation: stereocontrolled synthesis of *trans*-octahydroanthracenes, *J. Chem. Soc., Chem. Commun.*, 1591, 1991.
63. Čeković, Z., Intramolecular cyclization of unsaturated acyl chlorides by tributyltin hydride, *Tetrahedron Lett.*, 749, 1972.
64. Boger, D. L., and Mathvink, R. J., Acyl radicals: intermolecular and intramolecular alkene addition reactions, *J. Org. Chem.*, 57, 1429, 1992.
65. Delduc, P., Tailhan, C., and Zard, S. Z., A convenient source of alkyl and acyl radicals, *J. Chem. Soc., Chem. Commun.*, 308, 1988.
66. Coveney, D. J., Patel, V. F., and Pattenden, G., Acylcobalt salophen reagents. Precursors to acyl radical intermediates for inter- and intra-molecular oxidative Michael addition reactions, *Tetrahedron Lett.*, 28, 5949, 1987.
67. Nozaki, K., Oshima, K., and Utimoto, K., Synthesis of lactones by intramolecular addition of alkoxythiocarbonyl free radicals to acetylenes, *Tetrahedron Lett.*, 29, 6127, 1988.
68. Bachi, M. D., and Bosch, E., Synthesis of α-alkylidene-γ-lactones by intramolecular addition of alkoxycarbonyl free-radicals to acetylenes, *Tetrahedron Lett.*, 27, 641, 1986.

69. Batty, D., and Crich, D., An acyl radical initiated tandem 7-*endo* / 5-*exo* radical cyclization approach to enantiomerically pure bicyclo[5.3.0]decan-2-ones, *Tetrahedron Lett.*, 33, 875, 1992.
70. Gaudino, J. J., and Wilcox, C. S., A concise approach to enantiomerically pure carbocyclic ribose analogues. Synthesis of (4*S*,5*R*,6*R*,7*R*)-7-(hydroxymethyl)spiro[2.4]heptane-4,5,6-triol 7-*O*-(dihydrogen phosphate), *J. Am. Chem. Soc.*, 112, 4374, 1990.
71. Bachi, M. D., and Bosch, E., Free radical cyclization of thionocarbonic acid derivatives of 4-phenyl-3-butenol. A new route to thionolactones, *J. Org. Chem.*, 54, 1234, 1989.
72. Yamamoto, M., Uruma, T., Iwasa, S., Kohmoto, S., and Yamada, K., Tri-*n*-butyltin hydride assisted highly stereoselective lactonisation of homoallylic xanthates, *J. Chem. Soc., Chem. Commun.*, 1265, 1989.
73. Bachi, M. D., and Denenmark, D., Cyclizations of ene radicals. Imidoyl radicals as intermediates in the synthesis of heterocyclic compounds, *J. Am. Chem. Soc.*, 111, 1886, 1989.
74. Bachi, M. D., and Denenmark, D., Intramolecular addition of carbon-centered tinthioimidoyl radicals to carbon–carbon double bonds. Synthesis of γ- and δ-thiolactams, *J. Org. Chem.*, 55, 3442, 1990.
75. Newcomb, M., and Deeb, T. M., *N*-Hydroxypyridine-2-thione carbamates as aminyl and aminium radical precursors. Cyclizations for synthesis of the pyrrolidine nucleus, *J. Am. Chem. Soc.*, 109, 3163, 1987.
76. Boivin, J., Fouquet, E. and Zard, S. Z., Cyclisation of imine radicals derived from sulphenylimines: a simple access to Δ^1-pyrrolines, *Tetrahedron Lett.*, 31, 85, 1990.
77. Boivin, J., Fouquet, E., and Zard, S. Z., Multiple radical additions starting with iminyl radicals derived from sulphenylimines, *Tetrahedron Lett.*, 31, 3545, 1990.
78. Danishefsky, S., Chackalamannil, S., and Uang, B.-J., Reductive cyclization of mercurial enones, *J. Org. Chem.*, 47, 2231, 1982.
79. Giese, B., and Meister, J., Die addition von kohlenwasserstoffen an olefine eine neue synthetische methode, *Chem. Ber.*, 110, 2588, 1977.
80. Giese, B., and Heuck, K., Carbon–carbon bond formation between alkylated alkenes and acrylic ester *via* 2-methoxyalkyl radicals, *Tetrahedron Lett.*, 21, 1829, 1980.
81. Xu, W., Jeon, Y. T., Hasegawa, E., Yoon, U. C., and Mariano, P. S., Novel electron-transfer photocyclization reactions of α-silyl amine α,β-unsaturated ketone and ester systems, *J. Am. Chem. Soc.*, 111, 406, 1989.
82. Nishida, A., Nishida, M., and Yonemitsu, O., Generation and intramolecular cyclization of free radicals at the carbon between two heteroatoms under non-reductive conditions, *Tetrahedron Lett.*, 31, 7035, 1990.
83. Rao, Y. K., and Nagarajan, M., Synthesis of (±) silphinene, *Tetrahedron Lett.*, 29, 107, 1988.
84. Marinovic, N. N., and Ramanathan, H., The synthesis of fused and bridged ring systems by free radical carbocyclization. A general route to masked 1,4-diketones, *Tetrahedron Lett.*, 24, 1871, 1983.
85. Hanessian, S., Beaulieu, P., and Dubé, D., A novel synthetic route to the hexahydrobenzofuran subunit of the avermectins and milbemycins, *Tetrahedron Lett.*, 27, 5071, 1986.
86. Davies, H. G., and Green, R. H., Avermectins and milbemycins. I and II, *Chem. Soc. Rev.*, 20, 211, 1991 and 20, 271, 1991.
87. Sloan, C. P., Cuevas, J. C., Quesnelle, C., and Snieckus, V., Aryl radical-induced cyclization routes to furo[2,3-*b*]benzofurans. Abbreviated formal syntheses of aflatoxins B_1 and B_2, *Tetrahedron Lett.*, 29, 4685, 1988.
88. Parker, K. A., and Kim, H.-J., A short stereospecific synthesis of *dl*-lycoramine. Control of relative stereochemistry by dipole effects, *J. Org. Chem.*, 57, 752, 1992.
89. Stork, G., and Reynolds, M. E., Allylic radicals in cyclization reactions, *J. Am. Chem. Soc.*, 110, 6911, 1988.
90. Minisci, F., Vismara, E., and Fontana, F., Recent developments of free-radical substitutions of heteroaromatic bases, *Heterocycles*, 28, 489, 1989.
91. Murphy, J. A., and Sherburn, M. S., Intramolecular addition of free radicals to quaternised heterocyclic rings, *Tetrahedron Lett.*, 31, 1625, 1990.

92. Murphy, J. A., and Sherburn, M. S., Intramolecular free-radical substitution reactions of pyridinium rings: efficient formation of [5,6]- and [6,7]-fused ring systems, *Tetrahedron Lett.*, 31, 3495, 1990.
93. Clive, D. L. J., and Bergstra, R. J., Short synthesis of (±)-5-(3-furyl)octahydro-8-methylindolizines, alkaloids related to a component of castoreum. Use of radical cyclization, *J. Org. Chem.*, 56, 4976, 1991.
94. Enholm, E. J., and Prasad, G., Tributyltin hydride-induced *O*-stannyl ketyls in the cyclization of aldehydes and ketones with alkenes, *Tetrahedron Lett.*, 30, 4939, 1989.
95. Enholm, E. J., and Kinter, K. S., Free-radical cyclizations promoted by allylic *O*-stannyl ketyls: the intramolecular coupling of the *β*-carbons of activated alkenes, *J. Am. Chem. Soc.*, 113, 7784, 1991.
96. Kraus, G. A., and Liras, S., Organosilicon radical-induced cyclization reactions, *Tetrahedron Lett.*, 31, 5265, 1990.
97. Ward, D. E., and Kaller, B. F., [3+3] Annulation by sequential two electron and one electron allylation, *Tetrahedron Lett.*, 32, 843, 1991.
98. Cox, N. J. G., Pattenden, G., and Mills, S. D., Radical macrocyclisations in synthesis. A new approach to mukulol and marine cembranolide lactones, *Tetrahedron Lett.*, 30, 621, 1989.
99. Hitchcock, S. A., and Pattenden, G., Synthesis of macrocycles *via* allylic radical intermediates. A total synthesis of (–)-zearalenone, *Tetrahedron Lett.*, 31, 3641, 1990.
100. Baldwin, J. E., Adlington, R. M., Mitchell, M. B., and Robertson, J., Intramolecular S_N2' macrocyclisations, *Tetrahedron*, 47, 5901, 1991.
101. Sugawara, T., Otter, B. A., and Ueda, T., A novel stereospecific radical cyclization of 2′,3′-*O*-isopropylideneuridine and -adenosine 5′-aldehyde to the corresponding 6,5′-cyclodihydrouridine and 8,5′-cycloadenosine derivatives, *Tetrahedron Lett.*, 29, 75, 1988.
102. Wilcox, C. S., and Thomasco, L. M., New syntheses of carbocycles from carbohydrates. Cyclization of radicals derived from unsaturated halo sugars, *J. Org. Chem.*, 50, 546, 1985.
103. Lavallée, J.-F., and Just, G., Asymmetric synthesis of 3′-carbomethoxymethyl-3′-deoxythymidine *via* radical cyclization, *Tetrahedron Lett.*, 32, 3469, 1991.
104. Alonso, R. A., Vite, G. D., McDevitt, R. E., and Fraser-Reid, B., Radical cyclization routes to bridged pyranosides as precursors of densely functionalized cycloalkanes, *J. Org. Chem.*, 57, 573, 1992.
105. Sharma, G. V. M., and Vepachedu, S. R., Stereoselective synthesis of (–)-canadensolide from D-glucose, *Tetrahedron*, 47, 519, 1991.
106. Wu, J.-C., Xi, Z., Gioeli, C., and Chattopadhyaya, J., Intramolecular cyclization–trapping of carbon radicals by olefins as means to functionalize 2′- and 3′-carbons in *β*-D-nucleosides, *Tetrahedron*, 47, 2237, 1991.
107. Velázquez, S., Huss, S., and Camarasa, M.-J., Stereoselective synthesis of [3.3.0] fused lactones (*γ*-butyrolactones) of sugars and nucleosides by free radical intramolecular cyclization, *J. Chem. Soc., Chem. Commun.*, 1263, 1991.
108. Rashid, A., Taylor, G. M., and Wood, W. W., The 'off-template' problem: towards a general solution, *J. Chem. Soc., Perkin Trans. 1*, 1289, 1990.
109. Corey, E. J., and Pyne, S. G., Conversion of ketones having *δ,ε*-π-functions to cyclopentanols by zinc–trimethylchlorosilane, *Tetrahedron Lett.*, 24, 2821, 1983.
110. Kozikowski, A. P., and MaloneyHuss, K. E., A model study for the synthesis of the hexahydrobenzofuran fragment of the milbemycins and avermectins, *Tetrahedron Lett.*, 26, 5759, 1985.
111. Ardisson, J., Férézou, J. P., Julia, M., and Pancrazi, A., Free radical cyclisation approach to the southern moiety of avermectins, *Tetrahedron Lett.*, 28, 2001, 1987.
112. Beckwith, A. L. J., and Hay, B. P., Kinetics and mechanism of the *exo* cyclizations of *ω*-formylalkyl radicals, *J. Am. Chem. Soc.*, 111, 2674, 1989.
113. Brook, A. G., Some molecular rearrangements of organosilicon compounds, *Acc. Chem. Res.*, 7, 77, 1974.
114. Tsai, Y.-M., and Cherng, C.-D., Intramolecular free radical cyclizations using acylsilanes as radicalphiles, *Tetrahedron Lett.*, 32, 3515, 1991.

115. Hart, D. J., and Seely, F. L., Bis(trimethylstannyl)benzopinacolate-mediated intermolecular free-radical carbon–carbon bond-forming reactions: a new one-carbon homologation, *J. Am. Chem. Soc.*, 110, 1631, 1988.
116. Marco-Contelles, J., Martinez-Grau, A., Bernabe, M., Martin, N., and Seoane, C., Synthesis of annulated furanoses *via* stereoselective radical cyclization of haloalkenes derived from diacetone glucose, *Synlett*, 165, 1991.
117. Booth, S. E., Jenkins, P. R., and Swain, C. J., The synthesis of bicyclic and tricyclic ring systems by radical cyclisation reactions of oxime ethers, *J. Chem. Soc., Chem. Commun.*, 1248, 1991.
118. Enholm, E. J., Burroff, J. A., and Jaramillo, L. M., Free radical cyclizations of terminal alkynes with oxime ethers, *Tetrahedron Lett.*, 31, 3727, 1990.
119. Bartlett, P. A., McLaren, K. L., and Ting, P. C., Radical cyclization of oxime ethers, *J. Am. Chem. Soc.*, 110, 1633, 1988.
120. Marco-Contelles, J., Pozuelo, C., Jimeno, M. L., Martínez, L., and Martínez-Grau A., 6-*Exo* free radical cyclization of acyclic carbohydrate intermediates: a new synthetic route to enantiomerically pure polyhydroxylated cyclohexane derivatives, *J. Org. Chem.*, 57, 2625, 1992.
121. Friedman, L., 2-*Exo,cis*-bicyclo[3.3.0]octane derivatives *via* free-radical additions to *cis,cis*-1,5-cyclooctadiene, *J. Am. Chem. Soc.*, 86, 1885, 1964.
122. Dowbenko, R., Free-radical transannular cycloadditions to *cis-cis*-1,5-cyclooctadiene, *Tetrahedron*, 20, 1843, 1964.
123. Stork, G., and Mook, R., Jr., Vinyl radical cyclization. II. Dicyclization *via* selective formation of unsaturated vinyl radicals by intramolecular addition to triple bonds. Applications to the synthesis of butenolides and furans, *J. Am. Chem. Soc.*, 105, 3720, 1983.
124. Barton, D. H. R., and Zard, S. Z., Invention of new reactions useful in the chemistry of natural products, *Pure Appl. Chem.*, 58, 675, 1986.
125. Saičić, R. N., and Čeković, Z., Radical annulation methodology. 2-Vinylcyclopentane derivative formation by a 3+2 cycloaddition reaction, *Tetrahedron Lett.*, 31, 4203, 1990.
126. Lee, E., and Hur, C. U., Intermolecular addition of α-alkoxycarbonyl-β-stannylvinyl radicals to electron-deficient alkenes. Simple addition and [2+2+2] annulation reactions, *Tetrahedron Lett.*, 32, 5101, 1991.
127. Iwasa, S., Yamamoto, M., Kohmoto, S., and Yamada, K., Tandem radical cyclization of acyclic homoallylic xanthates: cyclopentannulated γ-thionolactone and γ-lactones, *J. Org. Chem.*, 56, 2849, 1991.
128. Kilburn, J. D., Stereoselective tandem radical cyclisations: synthesis of isoiridomyrmecin, *Tetrahedron Lett.*, 31, 2193, 1990.
129. Parsons, P. J., Willis, P. A., and Eyley, S. C., A tandem radical cyclisation approach to the hexahydrobenzofuran skeleton for avermectin synthesis, *J. Chem. Soc., Chem. Commun.*, 283, 1988.
130. Hanessian, S., and Léger, R., Expedient assembly of carbocyclic, heterocyclic, and polycyclic compounds by Me_3Sn radical mediated carbocyclizations of dienes and trienes: a novel oxidative cleavage of the C–Sn bond, *J. Am. Chem. Soc.*, 114, 3115, 1992.
131. Nouguier, R., Lesueur, C., De Riggi, E., Bertrand, M. P., and Virgili, A., Stereoselective free-radical cyclization on a sugar template. The sulphonyl radical as a synthetic tool for functionalized glycosides, *Tetrahedron Lett.*, 31, 3541, 1990.
132. Chuang, C.-P., The free radical cyclization reaction of 1,6-dienes with allylsulfones, *Tetrahedron*, 47, 5425, 1991.
133. Chen, Y.-J., and Lin, W.-Y., A facile synthesis of a Δ^2-cedrene skeleton *via* tandem radical cyclization, *Tetrahedron Lett.*, 33, 1749, 1992.
134. Ono, N., and Kaji, A., Reductive cleavage of aliphatic nitro groups in organic synthesis, *Synthesis*, 693, 1986.
135. Kim, S., and Fuchs, P. L., Intramolecular 1,4-addition of α-heteroatom-substituted radicals to β,β-disubstituted enones. Applications to tandem cyclizations, *J. Am. Chem. Soc.*, 113, 9864, 1991.

136. Newcomb, M., and Ha, C., Lewis acid activated and catalyzed cyclizations of aminyl radicals from *N*-hydroxypyridine-2-thione carbamates, *Tetrahedron Lett.*, 32, 6493, 1991.
137. Boger, D. L., and Mathvink, R. J., Tandem free-radical alkene addition reactions of acyl radicals, *J. Am. Chem. Soc.*, 112, 4003, 1990.
138. Koreeda, M., and Hamann, L. G., Chirality transmission involving a free-radical-mediated 6-*exo* cyclization process. Stereocontrolled synthesis of branched-chain 1,4-diols, *J. Am. Chem. Soc.*, 112, 8175, 1990.
139. Lee, E. R., Lakomy, I., Bigler, P., and Scheffold, R., Reductive radical cyclisations of bromo acetals and (bromomethyl)silyl ethers of terpenoid alcohols, *Helv. Chim. Acta*, 74, 146, 1991.
140. Hutchinson, J. H., Daynard, T. S., and Gillard, J. W., Silaketals as tethers in intramolecular radical cyclisations, *Tetrahedron Lett.*, 32, 573, 1991.
141. Journet, M., Smadja, W., and Malacria, M., Radical cyclization of (bromomethyl)dimethylsilyl propargyl ethers: VI. Serial radical cyclizations leading to a stereoselective synthesis of functionalized diquinane framework from an acyclic substrate, *Synlett*, 320, 1990.
142. Ryu, I., Kusano, K., Hasegawa, M., Kambe, N., and Sonoda, N., Free-radical-mediated carbonylative cyclisation of alk-4-enyl halides leading to cyclopentanones, *J. Chem. Soc., Chem. Commun.*, 1018, 1991.
143. Nagai, M., Lazor, J., and Wilcox, C. S., Syntheses of bicyclo[3.3.0]octanes *via* bifurcating radical cyclization pathways, *J. Org. Chem.*, 55, 3440, 1990.
144. Curran, D. P., and Chen, M.-H., Radical-initiated polyolefinic cyclizations in condensed cyclopentanoid synthesis. Total synthesis of (±)-$\Delta^{9(12)}$-capnellene, *Tetrahedron Lett.*, 26, 4991, 1985.
145. Meyers, A. I., and Bienz, S., Asymmetric total synthesis of (+)-$\Delta^{9(12)}$-capnellene, *J. Org. Chem.*, 55, 791, 1990.
146. Molander, G. A., and Harring, L. S., Reductive radical cyclizations of haloalkenes promoted by samarium diiodide. Sequential cyclization / intermolecular carbonyl addition reactions, *J. Org. Chem.*, 55, 6171, 1990.
147. Fevig, T. L., Elliott, R. L., and Curran, D. P., A samarium(II) iodide promoted tandem radical cyclization. The total synthesis of (±)-hypnophilin and the formal synthesis of (±)-coriolin, *J. Am. Chem. Soc.*, 110, 5064, 1988.
148. Danishefsky, S., Zamboni, R., Kahn, M., and Etheredge, S. J., Total synthesis of *dl*-coriolin, *J. Am. Chem. Soc.*, 102, 2097, 1980.
149. Curran, D. P., and Kuo, S.-C., Tandem radical cyclization approach to angular triquinanes. A short synthesis of (±)-silphiperfol-6-ene and (±)-9-episilphiperfol-6-ene, *J. Am. Chem. Soc.*, 108, 1106, 1986.
150. Schwartz, C. E., and Curran, D. P., New tandem radical cyclizations directed toward the synthesis of crinipellin A, *J. Am. Chem. Soc.*, 112, 9272, 1990.
151. Jasperse, C. P., and Curran, D. P., Sequential radical cyclization approach to propellane triquinanes. Total synthesis of (±)-modhephene, *J. Am. Chem. Soc.*, 112, 5601, 1990.
152. Curran, D. P., Chen, M.-H., and Kim, D., Atom-transfer cyclization. A novel isomerization of hex-5-ynyl iodides to (iodomethylene)cyclopentanes, *J. Am. Chem. Soc.*, 108, 2489, 1986.
153. Jolly, R. S., and Livinghouse, T., Atom-transfer annulations in heterocycle synthesis. An efficient synthesis of (–)-trachelanthamidine and related ring systems, *J. Am. Chem. Soc.*, 110, 7536, 1988.
154. Curran, D. P., and Chang, C.-T., Atom transfer cyclization reactions of α-iodo carbonyls, *Tetrahedron Lett.*, 28, 2477, 1987.
155. Forbes, J. E., Tailhan, C., and Zard, S. Z., Dithiocarbonate group transfer in radical chain reactions, *Tetrahedron Lett.*, 31, 2565, 1990.
156. Flynn, D. L., and Zabrowski, D. L., Halogen atom transfer annulations involving iodomalonates and allylamine derivatives, *J. Org. Chem.*, 55, 3673, 1990.
157. Curran, D. P., Chen, M.-H., Spletzer, E., Seong, C. M., and Chang, C.-T., Atom transfer addition and annulation reactions of iodomalonates, *J. Am. Chem. Soc.*, 111, 8872, 1989.
158. Curran, D. P., Kim, D., Liu, H. T., and Shen, W., Translocation of radical sites by intramolecular 1,5-hydrogen atom transfer, *J. Am. Chem. Soc.*, 110, 5900, 1988.

159. Lathbury, D. C., Parsons, P. J., and Pinto, I., A route to the pyrrolizidine ring system using a novel radical cyclisation, *J. Chem. Soc., Chem. Commun.*, 81, 1988.
160. Borthwick, A. D., Caddick, S., and Parsons, P. J., 1,5 Allylic abstraction, cyclisation: a new route to five membered carbocycles, *Tetrahedron Lett.*, 31, 6911, 1990.
161. Curran, D. P., Abraham, A. C., and Liu, H., Radical translocation reactions of *o*-iodoanilides: the use of carbon–hydrogen bonds as precursors of radicals adjacent to carbonyl groups, *J. Org. Chem.*, 56, 4335, 1991.
162. Snieckus, V., Cuevas, J.-C., Sloan, C. P., Liu, H., and Curran, D. P., Intramolecular α-amidoyl to aryl 1,5-hydrogen atom transfer reactions. Heteroannulation and α-nitrogen functionalization by radical translocation, *J. Am. Chem. Soc.*, 112, 896, 1990.
163. Jones, K., and McCarthy, C., Chiral induction in aryl radical cyclisations, *Tetrahedron Lett.*, 30, 2657, 1989.
164. Hart, D. J., and Wu, S. C., Intramolecular addition of aryl radicals to vinylogous urethanes: studies toward preparation of the oxindole portion of gelsemine, *Tetrahedron Lett.*, 32, 4099, 1991.
165. Clive, D. L. J., Khodabocus, A., Cantin, M., and Tao, Y., Protecting group improvement by isotopic substitution: application to the synthesis of the quinone system of fredericamycin A, *J. Chem. Soc., Chem. Commun.*, 1755, 1991.
166. Wharton, P. S., and Bohlen, D. H., Hydrazine reduction of α,β-epoxy ketones to allylic alcohols, *J. Org. Chem.*, 26, 3615, 1961.
167. Harling, J. D., and Motherwell, W. B., Generation of spirocyclic quaternary centres *via* a tandem free radical cyclopropylcarbinyl rearrangement–cyclisation strategy, *J. Chem. Soc., Chem. Commun.*, 1380, 1988.
168. Batey, R. A., and Motherwell, W. B., Samarium(II) iodide promoted radical ring opening reactions of cyclopropyl ketones, *Tetrahedron Lett.*, 32, 6211, and 6649, 1991.
169. Clive, D. L. J., and Daigneault, S., Sequential radical ring closure–radical ring opening: use in the preparation of benzofurans, *J. Org. Chem.*, 56, 5285, 1991.
170. Feldman, K. S., and Vong, A. K. K., Phenylthio radical mediated synthesis of substituted 3-vinylcyclohexanones. Interruption of the ring-opening polymerization of 1,3-dioxolan-4-yl radicals, *Tetrahedron Lett.*, 31, 823, 1990.
171. Destabel, C., and Kilburn, J. D., A novel sequence of radical rearrangements involving the 5-*exo* cyclisation of a 3-(methylenecyclopropyl)propyl radical, *J. Chem. Soc., Chem. Commun.*, 596, 1992.
172. Singleton, D. A., and Church, K. M., [3+2] Methylenecyclopentane annulations of unactivated and electron-rich olefins with 2-(phenylsulfonyl)-1-methylenecyclopropanes, *J. Org. Chem.*, 55, 4780, 1990.
173. Singleton, D. A., Church, K. M., and Lucero, M. J., Catalytic stereocontrol of a radical cyclization, *Tetrahedron Lett.*, 31, 5551, 1990.
174. Atkinson, R. S., and Kelly, B. J., Aziridination of cyclohex-2-en-1-ol and geraniol: comparison with epoxidation, *J. Chem. Soc., Chem. Commun.*, 624, 1988.
175. Dickinson, J. M., and Murphy, J. A., Pyrrolidines and allylic amines from radical-induced cleavage of aziridines, *Tetrahedron*, 48, 1317, 1992.
176. Kim, S., Kee, I. S., and Lee, S., Novel radical cyclization of *N*-aziridinyl imines, *J. Am. Chem. Soc.*, 113, 9882, 1991.
177. Sabatino, E. C., and Gritter, R. J., The free-radical chemistry of cyclic ethers. V. β-Hydrogen atom abstraction from epoxides and a thioepoxide, *J. Org. Chem.*, 28, 3437, 1963.
178. Barton, D. H. R., Motherwell, R. S. H., and Motherwell, W. B., Radical-induced ring opening of epoxides. A convenient alternative to the Wharton rearrangement, *J. Chem. Soc., Perkin Trans. 1*, 2363, 1981.
179. Johns, A., Murphy, J. A., and Sherburn, M. S., Intramolecular reactions of allyloxy radicals, *Tetrahedron*, 45, 7835, 1989.
180. Johns, A., and Murphy, J. A., Cyclisations of allyloxy radicals, *Tetrahedron Lett.*, 29, 837, 1988.

181. Begley, M. J., Housden, N., Johns, A., and Murphy, J. A., Intramolecular reactions of allyloxy radicals featuring six-centred transition states; regiochemistry and stereochemistry, *Tetrahedron*, 47, 8417, 1991.
182. Rawal, V. H., Newton, R. C., and Krishnamurthy, V., Synthesis of carbocyclic systems *via* radical-induced epoxide fragmentation, *J. Org. Chem.*, 55, 5181, 1990.
183. Bowman, W. R., Marples, B. A., and Zaidi, N. A., Reversible cleavage of 5-cholest-3-enoxyl radicals: addition of alkyl radicals to ketones, *Tetrahedron Lett.*, 30, 3343, 1989.
184. Suginome, H., and Wang, J. B., Intramolecular β,γ-addition of allylic alkoxyl radicals. A new general synthesis of α-iodoepoxides by photolysis of allylic alcohol hypoiodites in the presence of mercury(II) oxide, iodine and pyridine in benzene, *J. Chem. Soc., Chem. Commun.*, 1629, 1990.
185. Kim, S., Lee, S., and Koh, J. S., Radical reactions of vinyl epoxides *via* radical translocations by a novel 1,5-*n*-Bu_3Sn group or a 1,5-hydrogen-atom transfer, *J. Am. Chem. Soc.*, 113, 5106, 1991.
186. Sarandeses, L. A., and Luche, J.-L., A new synthesis of α- and β-damascones from the ionones, *J. Org. Chem.*, 57, 2757, 1992.
187. Suginome, S., and Yamada, S., New general synthesis of medium ring-lactones *via* a regioselective β-scission of alkoxyl radicals generated from catacondensed lactols, *Tetrahedron*, 43, 3371, 1987.
188. Francisco, C. G., Freire, R., Rodríguez, M. S., and Suárez, E., β-Fragmentation of alkoxy radicals. An approach to the synthesis of ring A of vernolepin, *Tetrahedron Lett.*, 28, 3397, 1987.
189. Ellwood, C. W., and Pattenden, G., Sequential alkoxy radical fragmentation–transannular radical cyclisation–radical ring expansion reactions. A new approach to bicyclo[5.3.0]decanones and to bicyclo[6.3.0]undecanones, *Tetrahedron Lett.*, 32, 1591, 1991.
190. Dowd, P., and Choi, S.-C., A new tri-*n*-butyltin hydride based rearrangement of bromomethyl β-keto esters. A synthetically useful ring expansion to γ-keto esters, *J. Am. Chem. Soc.*, 109, 3493, 1987.
191. Dowd, P., and Choi, S.-C., Free radical ring expansion and rearrangement of large carbocyclic rings, *Tetrahedron Lett.*, 32, 565, 1991.
192. Baldwin, J. E., Adlington, R. M., and Robertson, J., Carbocyclic ring expansion reactions *via* radical chain processes. II, *Tetrahedron*, 47, 6795, 1991.
193. Dowd, P., and Zhang, W., Free radical ring expansion of fused cyclobutanones: a new ring expansion annulation stratagem, *J. Am. Chem. Soc.*, 113, 9875, 1991.
194. Crimmins, M. T., Dudek, C. M., and Cheung, A. W.-H., A fragmentation–rearrangement sequence of cyclobutylcarbinyl radicals, *Tetrahedron Lett.*, 33, 181, 1992.
195. Nishida, A., Takahashi, H., Takeda, H., Takada, N., and Yonemitsu, O., Sequential radical cyclization, alkoxy radical fragmentation, and recyclization processes: a novel method for the synthesis of fused cycloheptanones and cyclooctenones from cyclohexanones, *J. Am. Chem. Soc.*, 112, 902, 1990.
196. Dowd, P., and Choi, S.-C., Free radical ring expansion by three and four carbons, *J. Am. Chem. Soc.*, 109, 6548, 1987.
197. Boger, D. L., and Mathvink, R. J., Tandem free-radical ring expansion and 5-*exo-dig* 5-hexynyl radical cyclization: a useful approach to fused bicyclic carbocycles, *J. Org. Chem.*, 55, 5442, 1990.
198. Sarandeses, L. A., Mouriño, A., and Luche, J.-L., Sonochemistry of epoxyalkylhalides in the presence of a zinc–copper couple, *J. Chem. Soc., Chem. Commun.*, 798, 1992.
199. Snider, B. B., Merritt, J. E., Dombroski, M. A., and Buckman, B. O., Solvent effects on manganese(III)-based oxidative free-radical cyclizations: ethanol and acetic acid, *J. Org. Chem.*, 56, 5544, 1991.

200. Curran, D. P., Morgan, T.M., Schwartz, C. E., Snider, B. B., and Dombroski, M. A., Cyclizations of unsaturated $^{\bullet}CR(COX)_2$ radicals. Manganese(III) acetate oxidative cyclizations of unsaturated acetoacetates and atom-transfer cyclizations of unsaturated haloacetoacetates give the same radicals, *J. Am. Chem. Soc.*, 113, 6607, 1991.
201. Snider, B. B., Zhang, Q., and Dombroski, M. A., Control of the regioselectivity of oxidative free-radical cyclizations by addition to haloalkenes, *J. Org. Chem.*, 57, 4195, 1992.
202. Bauld, N. L., Cation radical cycloadditions and related sigmatropic reactions, *Tetrahedron*, 45, 5307, 1989.
203. Müller, F., and Mattay, J., [3+2] Cycloadditions with azirine radical cations: a new synthesis of *N*-substituted imidazoles, *Angew. Chem. Int. Ed. Engl.*, 30, 1336, 1991.
204. Barton, D. H. R., Dalko, P. I., and Géro, S. D., Radical cation reactions associated with the thiocarbonyl group, *Tetrahedron Lett.*, 33, 1883, 1992.
205. Hobbs-Mallyon, D., and Whiting, D. A., Tricyclic spirodienones *via* intramolecular radical phenol–nitronic acid coupling, *J. Chem. Soc., Chem. Commun.*, 899, 1991.
206. Mattay, J., Photoinduced electron transfer in organic synthesis, *Synthesis*, 233, 1989.
207. Hasegawa, E., Brumfield, M. A., Mariano, P. S., and Yoon, U.-C., Photoadditions of ethers, thioethers, and amines to 9,10-dicyanoanthracene by electron transfer pathways, *J. Org. Chem.*, 53, 5435, 1988.
208. Hasegawa, E., Xu, W., Mariano, P. S., Yoon, U.-C., and Kim, J.-U., Electron-transfer-induced photoadditions of the silyl amine Et_2NCH_2TMS to α,β-unsaturated cyclohexenones. Dual reaction pathways based on ion-pair-selective cation--radical chemistry, *J. Am. Chem. Soc.*, 110, 8099, 1988.
209. Xu, W., and Mariano, P. S., Substituent effects on amine cation radical acidity. Regiocontrol of β-(aminoethyl)cyclohexenone photocyclizations, *J. Am. Chem. Soc.*, 113, 1431, 1991.
210. Kraus, G. A., and Chen, L., Intramolecular photocyclizations of amino ketones to form fused and bridged bicyclic amines, *Tetrahedron Lett.*, 32, 7151, 1991.
211. Curran, D. P., and Seong, C. M., Atom transfer addition and annulation reactions of propargyl iodomalononitrile, *Tetrahedron*, 48, 2157, 1992.
212. Curran, D. P., and Seong, C. M., Radical annulation reactions of allyl iodomalononitriles, *Tetrahedron*, 48, 2175, 1992.
213. Kim, S., and Koh, J. S., Radical cyclization of epoxyketones *via* a 1,5-$n$$Bu_3Sn$ group or a 1,5-H atom transfer, *J. Chem. Soc., Chem. Commun.*, 1377, 1992.
214. Rigby, J. H., and Qabar, M., [1+4] Cycloaddition of vinyl isocyanates with alkyl isocyanides. Formal total synthesis of erysotrine, *J. Am. Chem. Soc.*, 113, 8975, 1991.
215. Ishibashi, H., So, T. S., Okochi, K., Sato, T., Nakamura, N., Nakatani, H., and Ikeda, M., Radical cyclization of *N*-(cyclohex-2-enyl)-α,α-dichloroacetamides. Stereoselective syntheses of (±)-mesembranol and (±)-elwesine, *J. Org. Chem.*, 56, 95, 1991.
216. Knapp, S., and Gibson, F. S., Radical routes to indolizidines. Synthesis of (–)-slaframine, *J. Org. Chem.*, 57, 4802, 1992.
217. Curran, D. P., and Liu, H., New 4+1 radical annulations. A formal total synthesis of (±)-camptothecin, *J. Am. Chem. Soc.*, 114, 5863, 1992.
218. Astley, M. P., and Pattenden, G., Natural furanocembranoids. A synthetic approach to lophotoxin based on an acyl radical macrocyclisation strategy, *Synthesis*, 101, 1992.
219. Snider, B. B., and Buckman, B. O., Total synthesis of (±)-velloziolone, *J. Org. Chem.*, 57, 4883, 1992.
220. Barton, D. H. R., The invention of chemical reactions: the last five years, *Tetrahedron*, 48, 2529, 1992.

Chapter 4

ANIONIC CYCLIZATION

I. INTRODUCTION

From the past to the present, cyclizations involving anionic intermediates have constituted a common approach to ring construction. The most commonly used procedures involve the intramolecular attack of anionic centers on electrophiles in simple S_N2 and S_N2' fashions, or in the manner of 1,2-addition to various carbonyl functions, as well as to such functions as the iminium, oxonium, and thionium. Also popular is the conjugate addition (Michael addition) of anionic centers which can involve tandem cyclizations (*e.g.*, double or triple Michael additions) and create several bonds in one operation, making it particularly suitable for the construction of polycyclic compounds.[1] Lately there has evolved a growing interest in the cyclization of reactive anionic centers onto unactivated olefinic bonds,[2] and studies along the same line as for the very successful radical counterpart have been carried out, though the reaction has not yet been developed to an extent that would make it as useful in organic synthesis as other established methods. Also exhibiting anionic character is the cyclization of allylsilanes, which is included in the discussion in this chapter.[3–5]

In addition to preexisting standard methods for the generation of anion intermediates, numerous new methods have been developed, rendering accessible a variety of novel anions for the purposes of cyclization. Examples range from metalation of the aryl nucleus[6] and heteroatom-facilitated lithiation,[7] which leads to the generation of "remote anions", to the generation and use of dianion species.[8] Recently, there have appeared reviews on anionic cyclizations in general,[9] as well as on anionic cyclizations in the building of natural product molecules,[10,11] which clearly project the role of the reaction as a powerful work-horse in organic synthesis.

II. CYCLIZATION *VIA* S_N2 AND S_N2' DISPLACEMENTS

Simple intramolecular displacement reactions allow uncomplicated access to natural product molecules. As demonstrated in Equation 1, the intramolecular displacement reaction of lithioalkyne **2**, generated from **1**, with the allylic bromide portion of the molecule readily produces the 14-membered ring **3**. This reaction constitutes the key step in the total synthesis of crassin acetate **4**,[12] the marine natural product of cembrane skeleton. [Other reported syntheses of crassin involve titanium(III) chloride-induced intramolecular coupling between ketone and aldehyde functions as the key step in the formation of the required macrocyclic ring].[13,14] Meanwhile, Equation 2 illustrates the construction of angular triquinane **9** by a one-pot Michael addition-S_N2 displacement reaction between bicyclic enone **5** and chlorocuprate **6**. An interesting observation is the persistent formation of uncyclized product **10** which always accompanies the required product **9**, despite the respectable yield of the latter (73%). Moreover, treatment of **10** with a variety of bases under various conditions give unreproducible and poor yields of **9**. These results have been rationalized in terms of the initial enolate **7** (formed from Michael addition in the first step) being thermodynamically less stable than enolate **8** due to angle strain,[15,16] hence any isomerization to the latter would not, in effect, be reversible and therefore lead to uncyclized by-product **10**. This relative stability of the two enolates also accounts for the poor yield of cyclized product obtained when **10** is treated with base (Equation 2).[17]

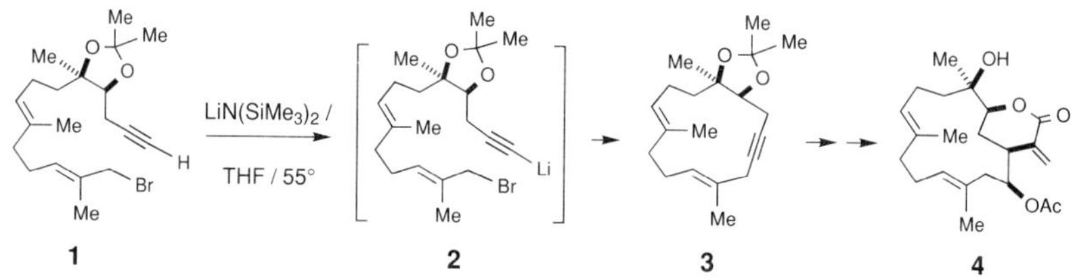

Equation 1

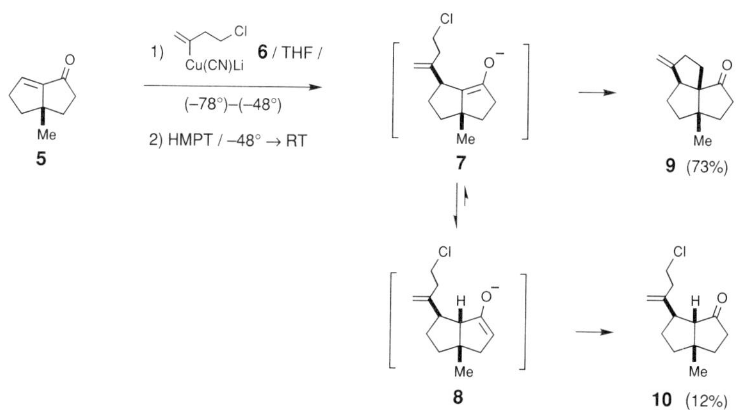

Equation 2

The thermodynamic control of enolates, though problematic in the above Equation 2, turns out to be an advantage in the synthesis of natural product α-chamigrene **17**, as shown in Equation 3. In this case the specific formation of enolates **12** and **15** allow the ready construction of spirocompound **16** from which α-chamigrene **17** is readily obtained.[18]

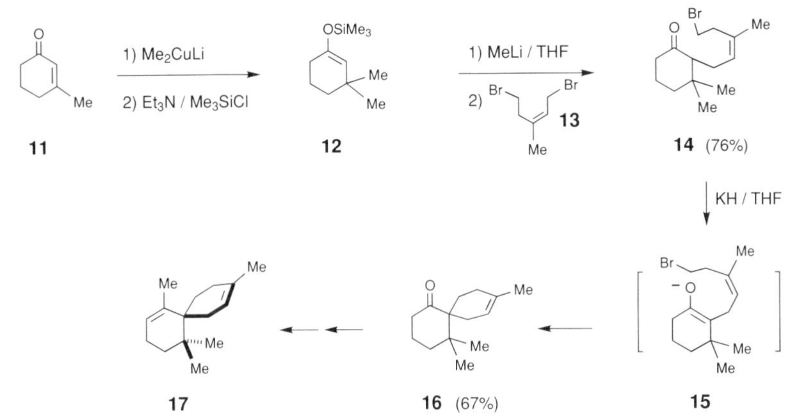

Equation 3

An example of a highly stereoselective (12:1) S_N2 cyclization is displayed in the reaction of potassium ester enolate **20** (derived from (*R*)-limonene **18** *via* **19** as in Scheme 1) which cyclizes *via* the more stable transition state geometry of **20**[19,20] to deliver methylene cyclohexane **21**,[19] a precursor for hydrindanone **22**, itself an important synthon in the synthesis of steroidal compounds.

Scheme 1

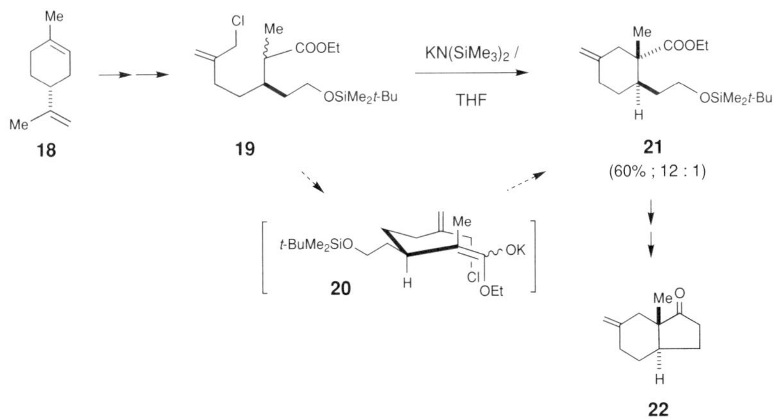

In the synthesis of *cis*-fused bicyclic compounds such as **27** *via* alkylative cyclization, a bulky group Y in substrate **23** would, in principle, dictate the stereochemistry of both the Michael addition (**23** → **24**) and cyclization (**25** → **26**) reactions, as shown in Scheme 2. If group Y were also easily removable it would enable a facile synthesis of *cis*-fused bicyclic compounds. One such "anchoring" group that meets the above requirements[21] is trimethylstannyl ($SnMe_3$) which can be cleaved easily by dissolving metal reduction. Accordingly, the synthesis of optically active *cis*-fused compound **31** was successfully carried out as depicted in Scheme 2.[22]

Scheme 2

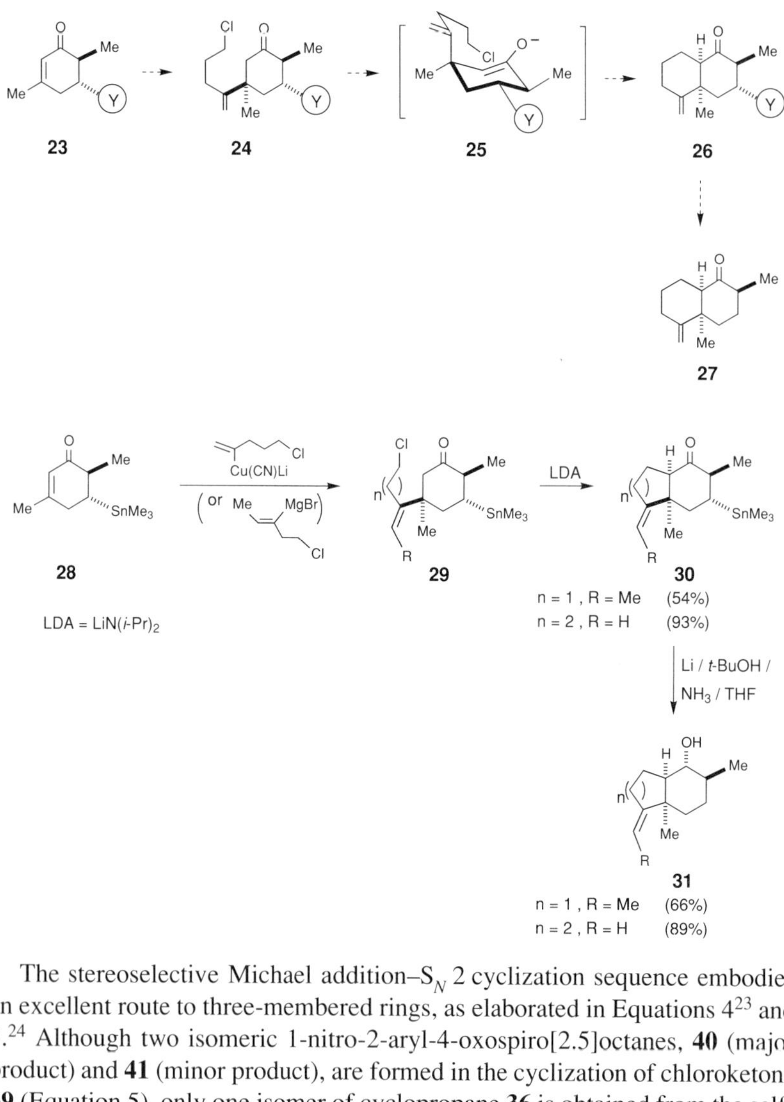

The stereoselective Michael addition–S_N2 cyclization sequence embodies an excellent route to three-membered rings, as elaborated in Equations 4[23] and 5.[24] Although two isomeric 1-nitro-2-aryl-4-oxospiro[2.5]octanes, **40** (major product) and **41** (minor product), are formed in the cyclization of chloroketone **39** (Equation 5), only one isomer of cyclopropane **36** is obtained from the self-cyclization reaction of methyl α-bromocrotonate **32** (Equation 4). In actual fact the latter experiment was originally designed to deconjugate the starting bromocrotonate **32** to **35**, which compound was obtained accompanied by a single cyclopropane product **36**. The reaction conditions were then optimized (*t*-BuOK / THF / –70°; 1 N HCl / 70°) for maximum production of **36**, eventually raising the ratio of products **36**:**35** to 7.3:1.

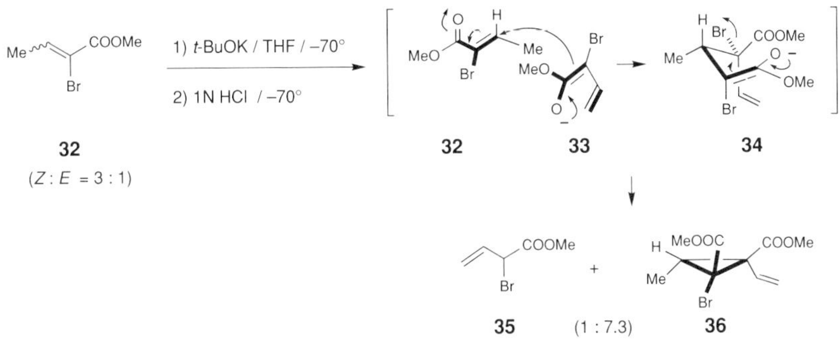

Equation 4

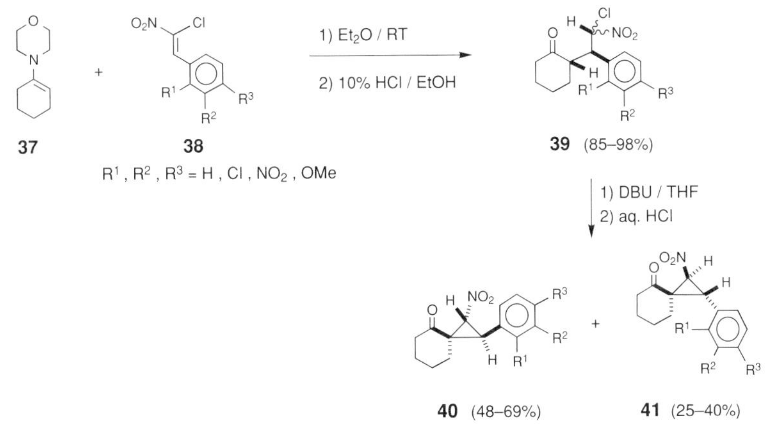

Equation 5

The preparation of cyclopenta[*b*]benzofuran **46**, the key intermediate in the synthesis of stable prostacyclin analog **47**, can be conveniently accomplished *via* S_N2' displacement. Aryl anion **45**, formed directly from a halogen–metal exchange of **44** with butyllithium in tetrahydrofuran, initiates the displacement, as shown in Scheme 3.[25]

Scheme 3

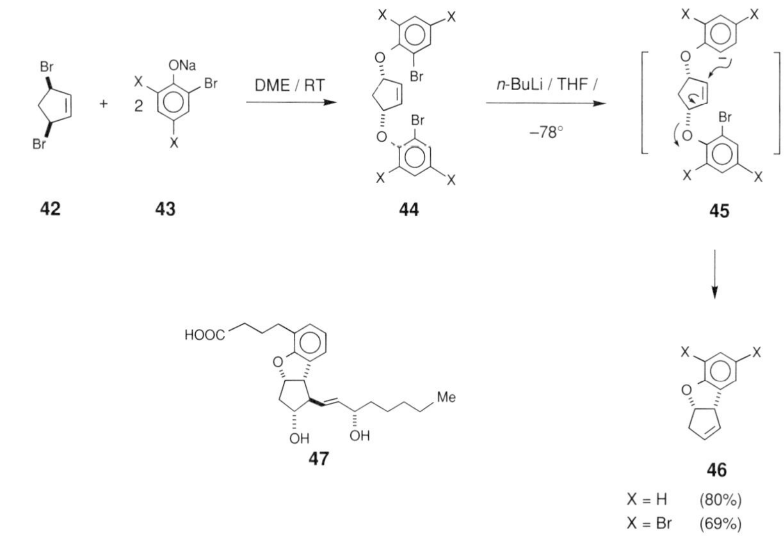

Further examples of intramolecular S_N2' enolate alkylation are shown in the annulation to form *cis*- and *trans*-fused [4.4.0] and [5.4.0]-bicyclic systems (Scheme 4), and in the synthesis of spironitrogen heterocycle depicted in Scheme 5. Scheme 4 illustrates the palladium-catalyzed cyclization of enolates[26] derived from *cis*- and *trans*-acetates **51** and **52**, themselves prepared from the chloro acetate **48**.[27] In the case of *cis*-**51** substrate, cyclization produces *cis*-**53** as the sole product, regardless of whether n = 1 or 2. The cyclization of *trans*-**52**, on the other hand, yields a 1:1 mixture of *cis*- and *trans*-bicyclic products **53** and **54** when n = 1 (six-membered ring), but when n = 2 the product stereochemistry is as expected from an S_N2' reaction, *i.e.,* only the *trans*-isomer **54** is formed.[28] Meanwhile, Scheme 5[29] depicts the S_N2' cyclization of potassium enolate **57** without the use of a catalyst, the reaction affording spiroproduct **58** which can eventually be transformed into the alkaloid isonitramine **59**.

Scheme 4

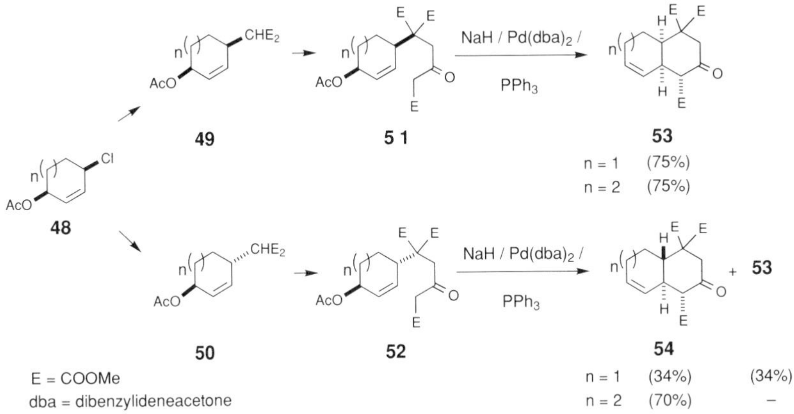

Scheme 5

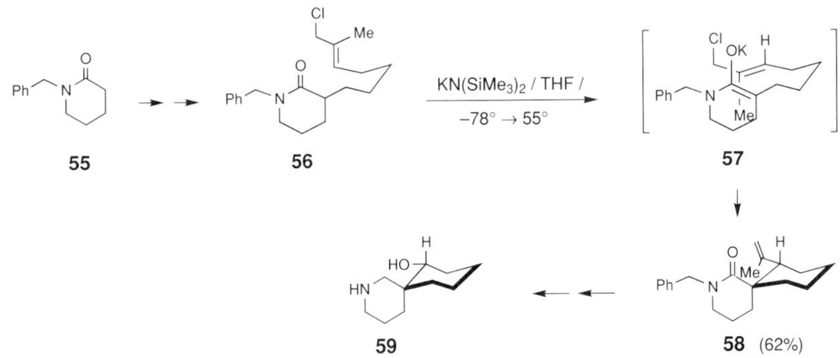

In the asymmetric, stereocontrolled total synthesis of (–)-brevianamide B **64**,[30] intramolecular S_N2' cyclization plays a prominent role in the construction of the bicyclo[2.2.2] nucleus. Such a cyclization of precursor **60** leads to products **62** and **63**, presumably *via* the **61** "*exo*" and "*endo*" transition states, respectively (Scheme 6). The dependence of product ratio on reaction conditions (solvent, base, catalyst, and temperature) is summarized in the accompanying table in Scheme 6. It can be seen that in every case the addition of crown ether effectively provokes the domination of **61** "*exo*" over **61** "*endo*" transition states, the former becoming sterically more favorable as crown ether complexes with the enolate metal ion, severely hindering the **61** "*endo*" form. Thus, the reaction can be manipulated to yield, as major product, the heterocycle **62**,[31] an advanced intermediate in the synthesis of (–)-brevianamide B **64**.

Scheme 6

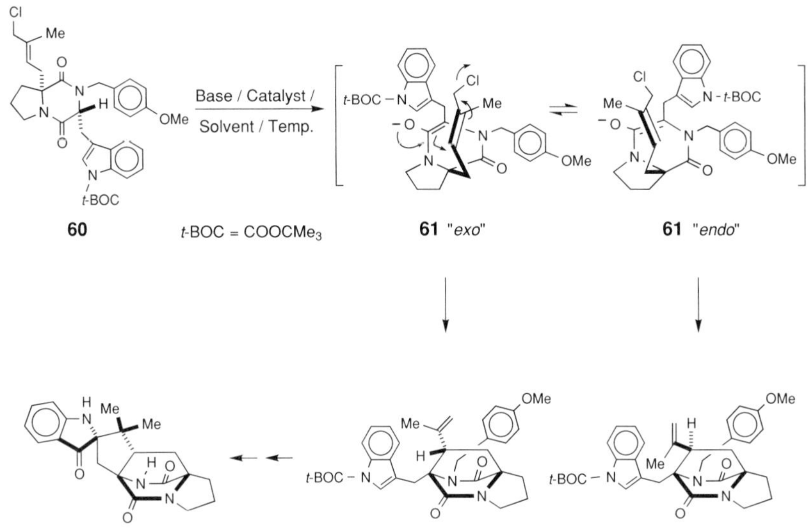

60 61 "*exo*" 61 "*endo*"

64 (−)-brevianamide B 62 63

Solvent	Temperature (°C)	Base / Catalyst	Ratio 62 : 63	% Yield
DMF	25	NaH /–	2 : 1	63
benzene	80	NaH /–	3 : 97	82
benzene	80	NaH / 18-crown-6	3.9 : 1	56
benzene	80	KH / 18-crown-6	1.1 : 1	38
benzene	80	NaH / 15-crown-5	1.42 : 1	40
benzene	25	NaH / 18-crown-6	6 : 1	14
THF	67	NaH / 18-crown-6	4.9 : 1	64

Enthusiasm over intramolecular cyclization involving epoxide ring opening escalated upon the discovery of natural polycyclic ether antibiotics, *e.g.*, the brevetoxins, and over efforts to synthesize these compounds.[32] In general, the intramolecular opening of an epoxide by a carbanion occurs at the proximal end, leading to the smaller carbocycle, as exemplified in Equation 6. Nevertheless, the reverse regiospecificity is encountered in the case of *trans*-disubstituted epoxide, in which the carbanion attacks the distal end of the epoxide part of an allylic system as shown in Equation 7.[32,33]

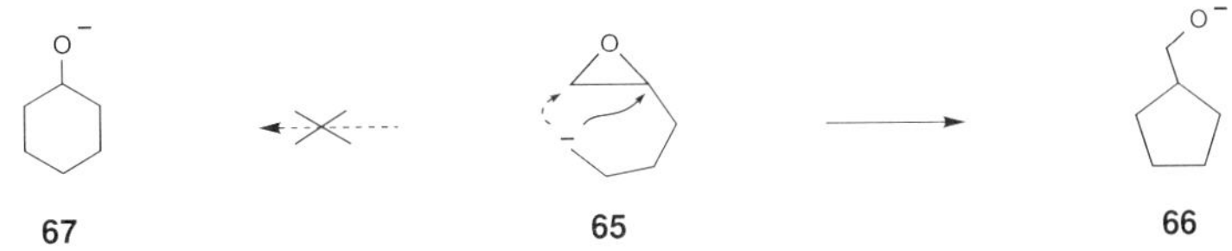

Equation 6

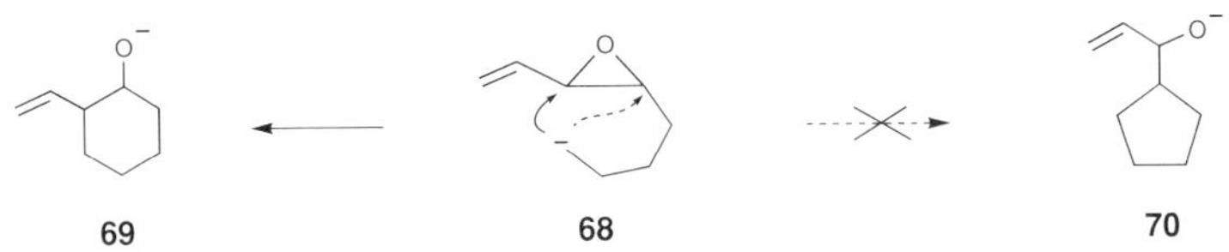

Equation 7

The reaction shown in Equation 7 has been employed successfully in the construction of tetrahydropyran derivatives ($Nu^- = O$),[32] and functionalized cyclohexanes,[33] as well as in the total synthesis of optically active (–)-histrionicotoxin 235A, **81**, and (–)-histrionicotoxin **82**[34] as illustrated in Scheme 7.

Ester enolate **71** stereospecifically cyclizes to give exclusively **73**, presumably *via* the more favorable transition state **72a**. Application of the reaction to optically active ester **75** and bromo epoxide **76** accomplishes a three-step synthesis in one pot, *i.e.*, sequential ester alkylation (**75** + **76** → **77**), cyclization *via* distal epoxide ring opening (**78** → **79**; see transition state **72a**) and lactonization to optically active **80**, which is obtained in 43% overall yield and employed as an intermediate for the synthesis of **81** and **82**.[34]

Scheme 7

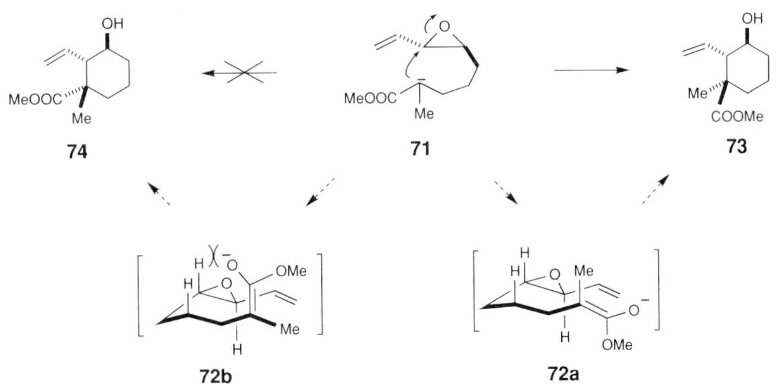

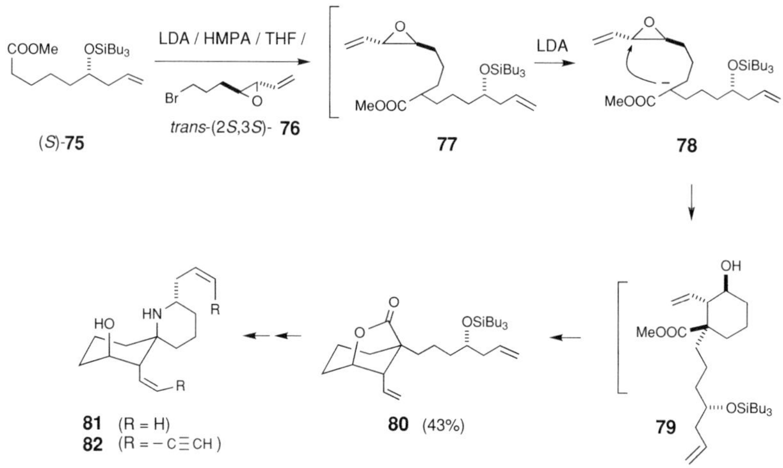

Not all epoxide ring opening reactions proceed as expected, however, and difficult-to-explain examples have been reported, such as those shown in Equation 8. Here, contingent upon the value of n, the epoxides **83** either completely fail to cyclize (n = 1 and 3), even after treatment with *tert*-butoxide/ *tert*-butanol at reflux for 2 d, or undergo *endo* cyclization (n = 2) to yield only one isomer of tricyclospirodienone **85**[35] from a starting diastereomeric mixture of **83**!

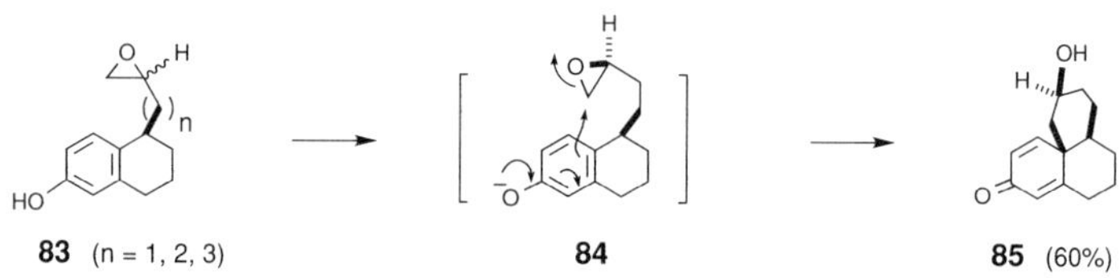

Equation 8

The first step in the total synthesis of (+)-longifolene **89**, the preparation of spiro[2.4]hepta-4,6-diene alcohol **88**, is elegantly accomplished by the addition of methyllithium to **86** to generate, *in situ*, cyclopentadienyl anion **87**, which spontaneously cyclizes *via* epoxide ring opening to the required material **88** (Equation 9).[36]

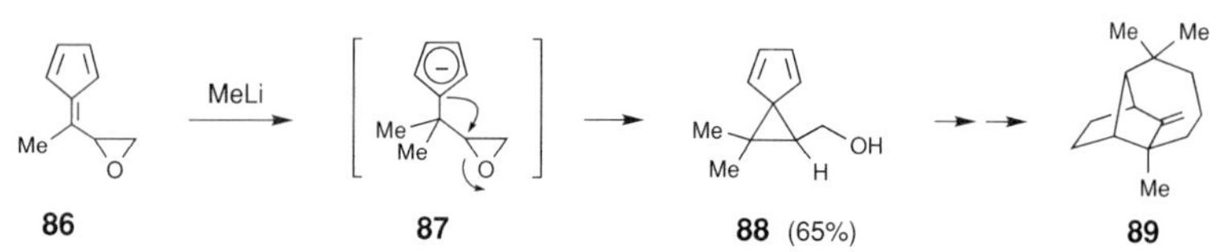

Equation 9

It is entirely true that intramolecular S_N2 cyclization offers a simple, direct, and convenient method for ring construction, as amply evidenced by the frequent reports of application, such as in the synthesis of three-[37,38] and five-membered[39–42] carbocycles of interest. The use of nonanionic nucleophiles, *e.g.*, the heteroatoms, is also effective and leads to the formation of heterocyclic compounds. Some examples are cited below.

3-Amino-1,2-diol **91**, obtainable[43] in either racemic or enantiomerically pure form from the titanium(IV)-mediated regioselective ring opening of 2,3-epoxy alcohol **90**, makes an excellent precursor for the synthesis of both 3-azetidinol **93** and aziridine **95**, as shown in Scheme 8.[44] By simply varying the amount of methanesulfonyl chloride (MsCl), the mesylation reaction of **91** under basic (triethylamine) conditions can be adjusted to yield one or the other of the three- and four-membered ring products, **95** and **93**, respectively. This is fortunately possible because the rate of S_N2 cyclization (**92** → **93**) is slower than the rate of mesylation of the secondary hydroxyl group (**92** → **94**).

Scheme 8

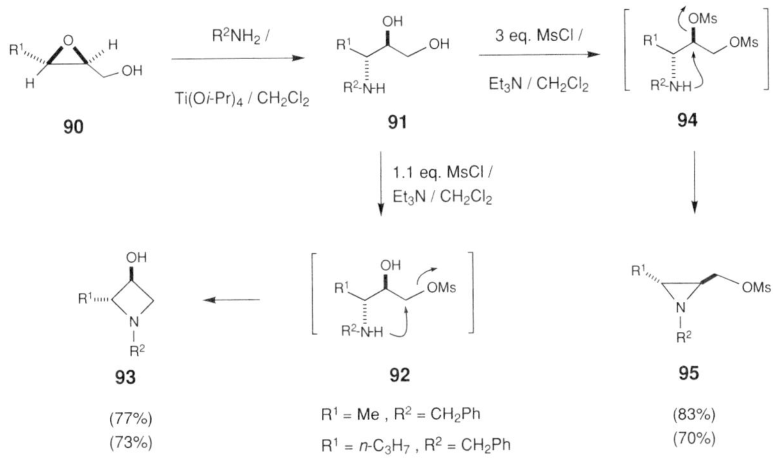

Alkaloids possessing the 1-azabicyclo[m.n.o]alkane basic skeleton, such as the pyrrolizidine (**96**: m = n = 3), indolizidine (**96**: m = 4, n = 3), and quinolizidine (**96**: m = n = 4) alkaloids, are widely found in nature and most seem to exhibit extremely interesting biological activities. Consequently, the construction of their skeleton **96** has received much attention. Scheme 9 outlines some of the approaches to the synthesis. Ring formation could involve a *trans*-annulation reaction (**99** → **98** → **97**)[45–47] in which "Y" could be halogen or phenyl selenide, or it could entail the intramolecular cycloaddition of an azide to a haloalkene (**102**) to form first the thermally labile triazoline, which would spontaneously expell N_2 and concurrently undergo a 1,2-hydrogen shift to give intermediate **101**, followed by cyclization to **100**, and subsequent reduction to **96** as presented in the scheme (**102** → **101** → **100** → **96**).[48–50] Another approach is to employ the tandem epoxide ring opening–intramolecular alkylation reaction, as represented by **104** → **103**.[51,52] Examples of applications of these approaches to the synthesis of natural products are shown in Equation 10 and Schemes 10 and 11. Respectively, these are the syntheses of the quinolizidine alkaloid, epilupinine **108**, *via* a *trans*-annular reaction;[46] of the indolizidine alkaloid, swainsonine **114**, by tandem azide cycloaddition–rearrangement-*N*-alkylation;[48,50] and of pyrrolizidine alkaloids, 7-epianstraline **119** and the unnatural 7-epialexine **120**, by the tandem epoxide ring opening–*N*-alkylation process.[52]

Scheme 9

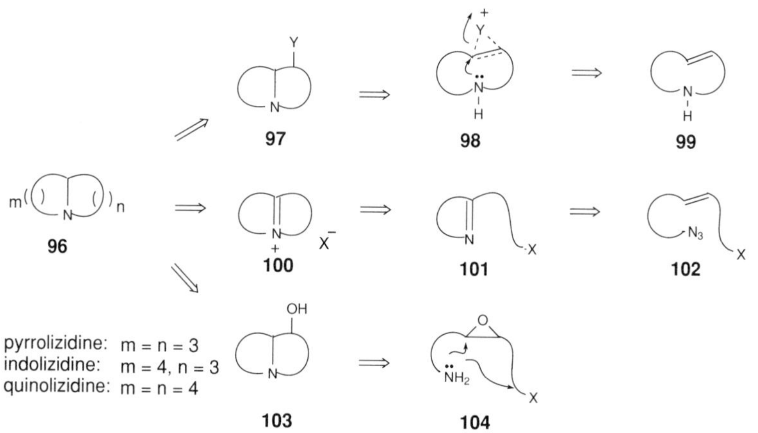

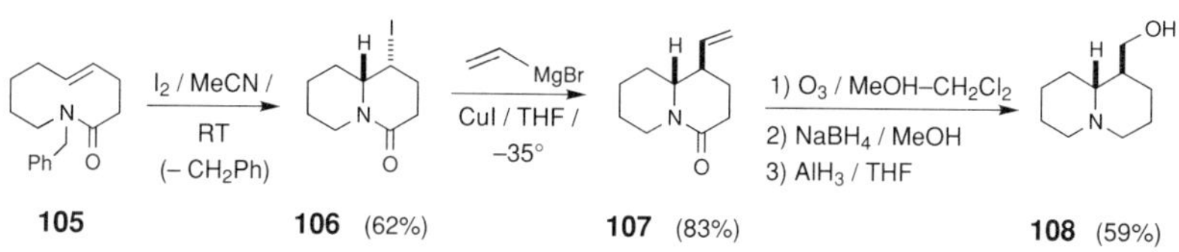

Equation 10

Scheme 10

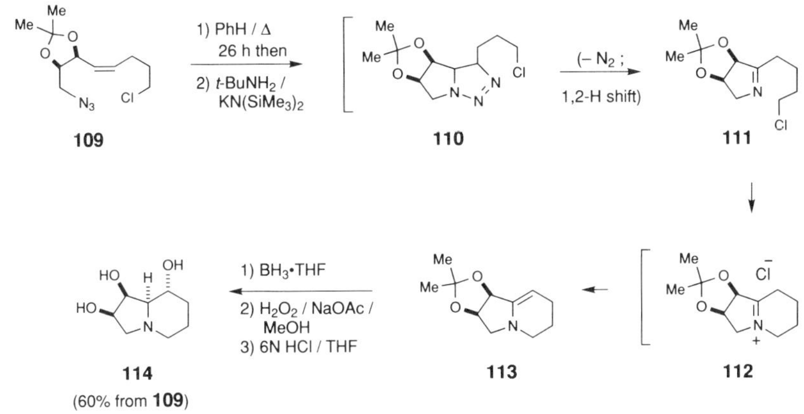

Scheme 11

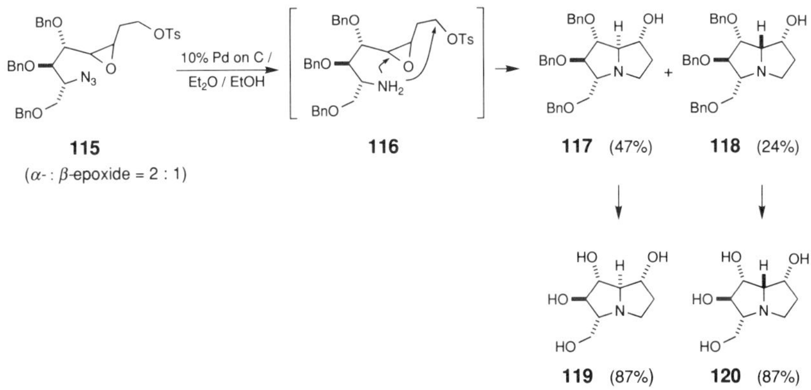

Finally, as a conclusion to intramolecular S_N2 reactions, a very interesting novel synthetic method, the ring expansion of halolactones mediated by bis(tributyltin) oxide, is shown in Scheme 12.[53] Advantage is taken of the properties of tin oxide in its use to cleave the *β*- and *γ*-lactone rings **121** to **122**, which then cyclize to **123**, thence delivering the ring-expanded products **124** in good yields. The ready cyclization (**122** → **123**) is greatly facilitated by the strong affinity of tin towards halogen. Application of the reaction to optically active starting materials **125** and **128**, respectively, yield products **127** and **130** whose stereochemistry confirms the true S_N2 nature of the cyclization by oxygen nucleophile.

Scheme 12

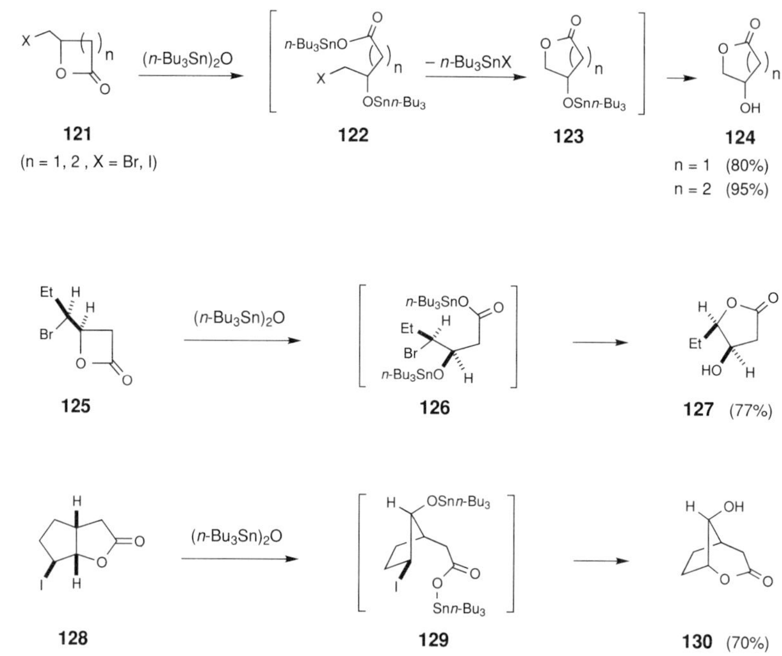

III. CYCLIZATION *VIA* ADDITION TO CARBONYL, IMINE, IMINIUM, OXONIUM, AND THIONIUM FUNCTIONS

A. ADDITION TO CARBONYL FUNCTIONS

Nucleophilic addition to sp^2 centers represents a common approach to organic synthesis in general, consequently, a multitude of new applications are constantly being reported. But while interesting and useful examples abound, the present discussion will, out of necessity, be limited to only a few selected recent results.

The aldol condensation, in all its multiformity, has been employed repeatedly as the key step in the organic synthesis of complex molecules. As shown in Equation 11,[54] the synthesis of model compound calicheamicin deoxyaglycone (an enediyne possessing potent antitumor activity[55]) employs aldol condensation between the acetylide anion and the aldehyde group. In another application, the synthesis of *Daphniphyllum* alkaloids, *e.g.*, daphnilactone A **138**, advantage is taken of the Reformatsky reaction in the first step, and S_N2 displacement of bromide in the second, to yield a single isomer of lactone–ether **137**. Here the presence of HMPA is essential for the formation of the cyclic ether (**136** → **137**), the alcohol product derived from **136** being obtained instead in its absence (Equation 12).[56] As for the process in Equation 13, it offers an example of *p*-toluenesulfonic acid-catalyzed enolization–intramolecular aldol condensation of a keto–aldehyde,[57,58] leading to an analog of benzomorphan in which the piperidine ring is replaced by a cyclohexane ring. The key step of the synthesis is the acid-catalyzed acetal deprotection–*in situ* intramolecular aldol condensation of intermediate **142**, which yields a single isomer of **143**.[58]

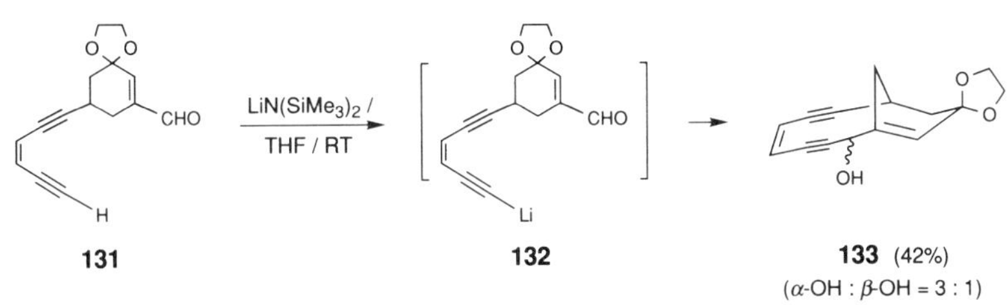

Equation 11

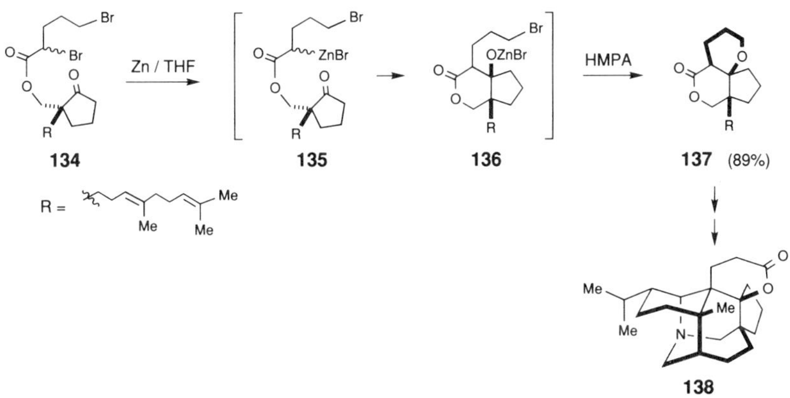

Equation 12

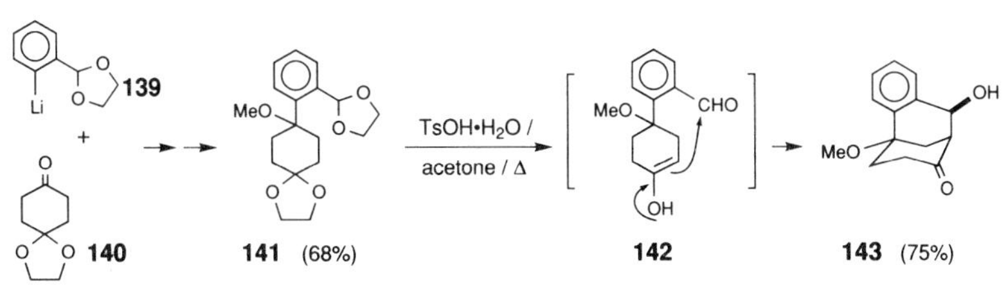

Equation 13

Silicon-directed aldol condensation is now well known and its use has become common practice in organic chemistry.[59,60] A recently reported aldol reaction of silyl enol ether **144**, derived from (*S*)-prolinol, is illustrated in Scheme 13. The reaction proceeds in the complete absence of any catalyst, yielding enantiomerically pure *anti* aldol product **147**, with only a trace of the easily separable *syn* enantiomer **148**. The isolation of enantiomerically pure **147** is indicative of transition state **145**, in which the aldehyde reacts on the concave surface of the enol by coordination to the silicon atom, with the trace of *syn* aldol **148** resulting from the less favorable transition state **146**, in which there is less overlapping between the bond-forming carbon atoms.[61]

Scheme 13

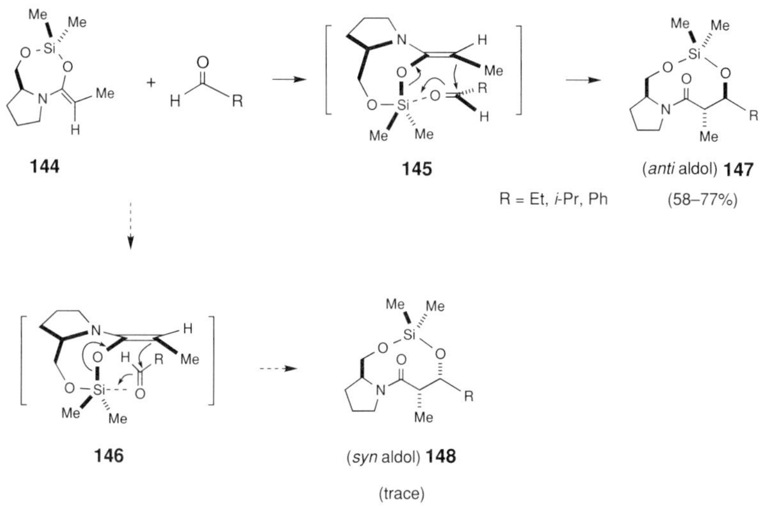

The Reformatsky-type reaction of allyl β-iodopropionate **149**, catalyzed by zinc–copper couple, with added silylating agent to trap the product, as shown in Equation 14, provides a moderate yield but convenient preparation of 1-allyloxy-1-siloxycyclopropane, a useful intermediate for further chemical elaboration.[62] An aldol cyclization is also employed in the synthesis of bicyclic compound **157**, and is effected by heating the starting ether **153** in toluene for 4 h in the presence of sodium hydride. The mechanism of this transformation, although debatable, is plausibly tandem Claisen rearrangement–aldol condensation–ester migration–elimination, as depicted in Scheme 14.[63]

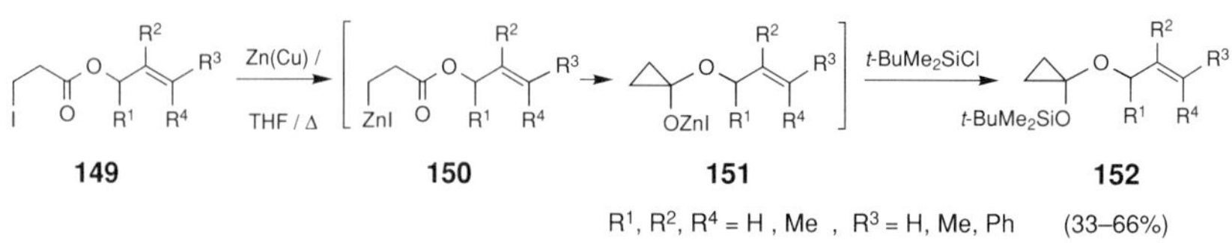

Equation 14

Scheme 14

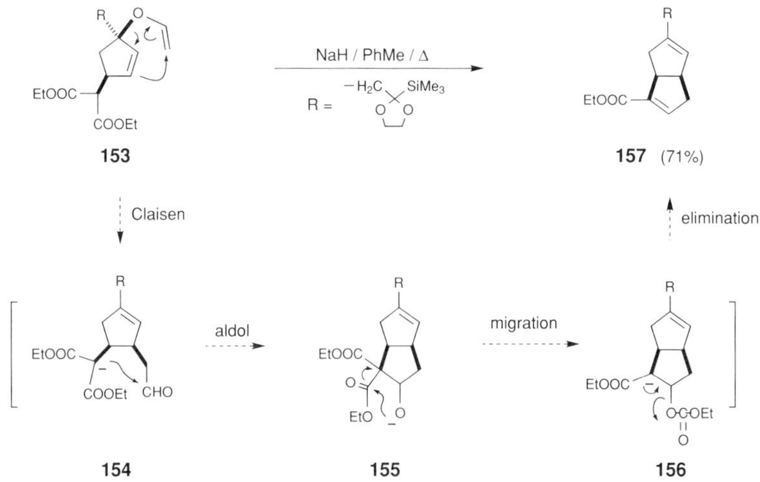

While on the subject of rearrangement–aldol condensation, special mention must be made of the tandem anionic oxy-Cope rearrangement–aldol condensation of C_2 symmetric dienediol **158**, a very interesting process illustrated in detail in Scheme 15.[64]

Theoretically, the disodium salt **159** of optically active **158** is capable of existing in two forms, **a** and **b**, depending upon the disposition of the oxyanions to dissociate or associate. In view of the strong electrostatic repulsion between the anions, the dissociated conformation **b** would seem advantageous, yet in the associated conformation **a**, both oxyanions can occupy the more favorable diequatorial positions. In any case, the oxy-Cope rearrangement of **159** would lead to **160**, which, after protonation, is expected to cyclize *via* aldol condensation to **162**, thence **163** after dehydration. It should be noted that the two enolate groups in **160** are homotopic and, accordingly, aldol condensation on either of them is at once possible and without regiochemical complications (see **161a** and **161b**), leading to the corresponding cyclopentenes **163**.

In practice, compound **158** (R = CH_2OCH_2Ph), when subjected to the above process affords a single enantiomer of formylcyclopentene (–)-**163a** in 70% yield, and with substitution of D_2O for water results in deuterium incorporation at the methylene group of (–)-**163a**. These results indicate the total association of dioxyanion **159**, *i.e.*, it exists entirely in the associated conformation **a**. The reaction significantly achieves quantitative chirality transfer from starting material **158** to the cyclopentene product (–)-**163a** in one pot.

Scheme 15

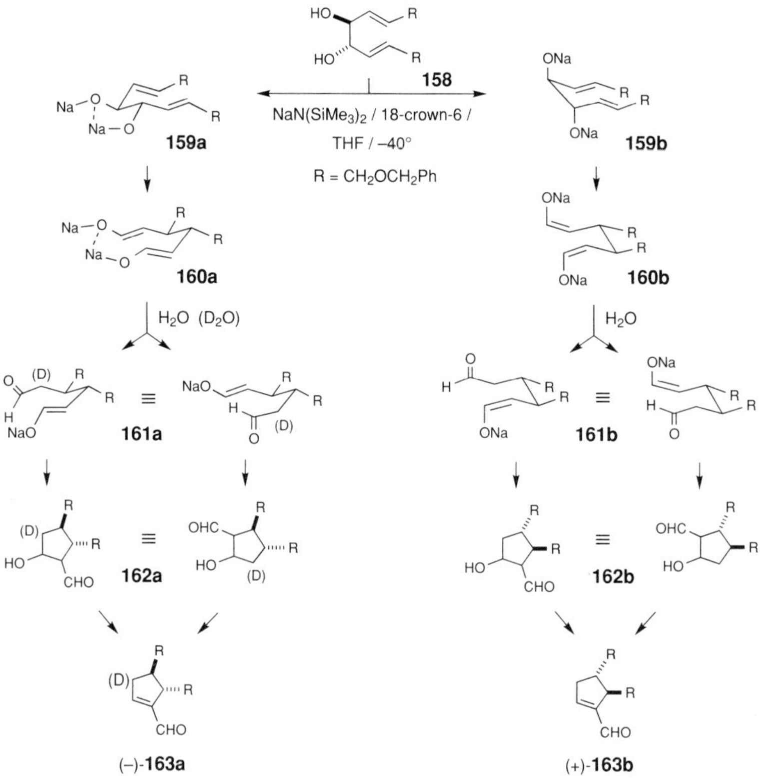

C_2 symmetric molecules have played a part in many more anionic condensation reactions, for example, in the synthesis of (+)-terrein **169** from L-tartaric acid derivative **164**, as shown in Scheme 16.[65] In addition to being an enantiomeric synthesis the process offers a much simplified approach over the former existing synthesis.[66] The sequence involves a Wittig–Horner condensation of intermediate **166**, generated *in situ* from **165**, to give **167** and **168** (Knoevenagel product) in a 4:1 ratio, followed by extension of the side chain and deprotection of the hydroxy groups to yield the requisite cyclopentenone **169**.

Related to the above condensation is the intramolecular Wittig reaction of **171**, also derived from tartaric acid, to give optically active cyclopentenone diol **172**,[67,68] an important building block in the synthesis of natural products (Equation 15). In this case epimerization takes place at the carbon atom α- to the carbonyl–ylide group, during the transformation of **171** to **172**. This is probably necessitated by the inhibition of cyclization by the strain inherent in the *trans*-fused configuration.

Scheme 16

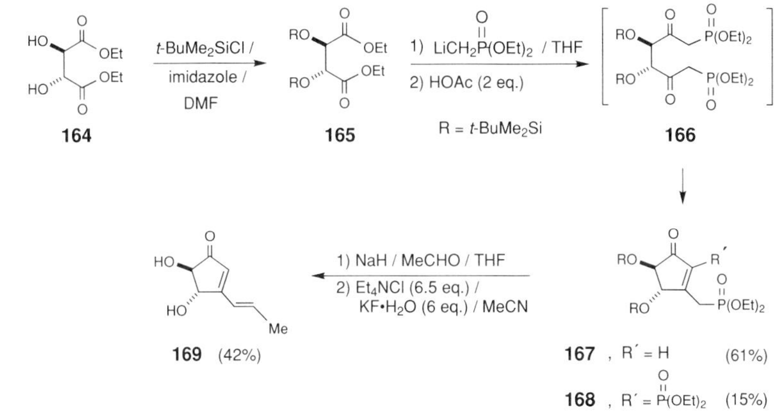

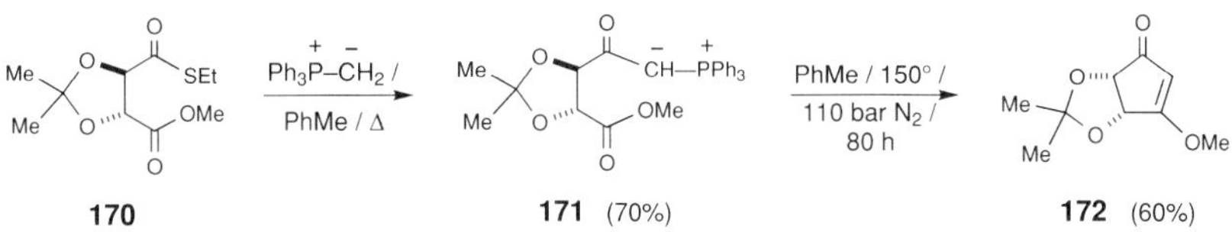

Equation 15

In the final step of the synthesis of carbacephem derivative **175**, a novel Dieckmann-type cyclization is employed, using a triacylamino group to carry out the reaction. In this way regiospecific cyclization is observed upon treatment of *N*,*N*-dibenzoyl carbonamide **173** with sodium bis(trimethylsilyl)amide at –60°C, the product from which may be isolated or further derivatized without isolation, as shown in Equation 16.[69]

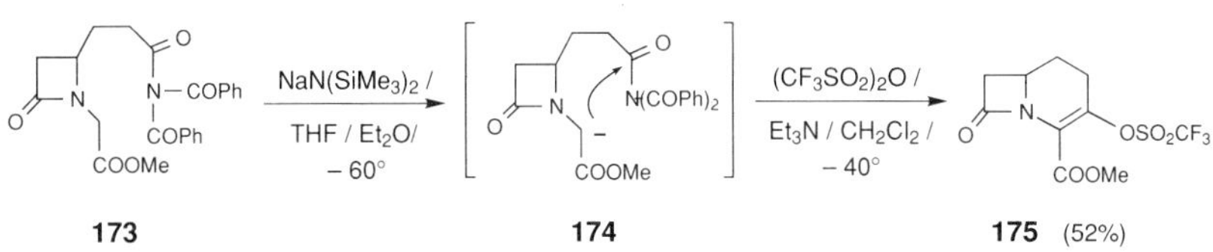

Equation 16

The more reactive carbanions, *e.g.*, vinyl and aryl carbanions, now enjoy common usage without the former hassle and inhibition associated with their generation, the techniques for which have become common practice in present day synthetic methodology. The easiest way to generate vinyl and aryl carbanions is by the direct halogen–metal exchange method,[7] in which the simple criterion to be considered is that the exchange reaction should be fast enough to permit the formation of unstabilized (or not strongly stabilized) nucleophilic centers in the presence of electrophilic functions. The first examples of the generation of vinyl and aryl anions by this method, for subsequent intramolecular 1,2-addition to the carbonyl group, are shown in Equations 17[70,71] and 18.[72] Since then these reactions have been widely applied and are in regular use today.

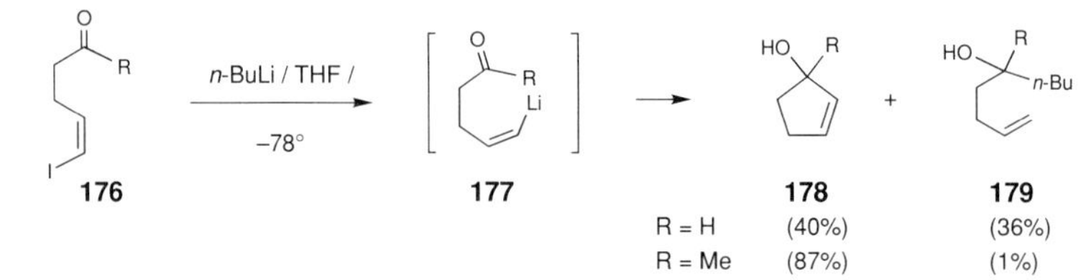

Equation 17

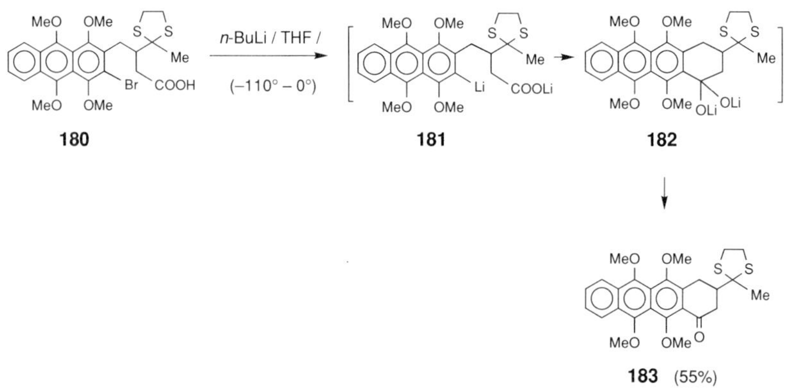

Equation 18

The application of vinyl anions in the construction of alkaloids is exemplified in the syntheses of tricyclic indoles **188** (Equation 19)[73] and **190** (Equation 20),[74] which are obtained *via* intramolecular vinyl anion addition to the carbonyl group of the corresponding oxindoles **184** and **189**, respectively, followed by lithium aluminum hydride reduction as shown in the Equations. Meanwhile, aryl anion application is shown in the synthesis of cularine alkaloid (Equation 21), the key step of which involves the cyclization of aryl anion **192**.[75]

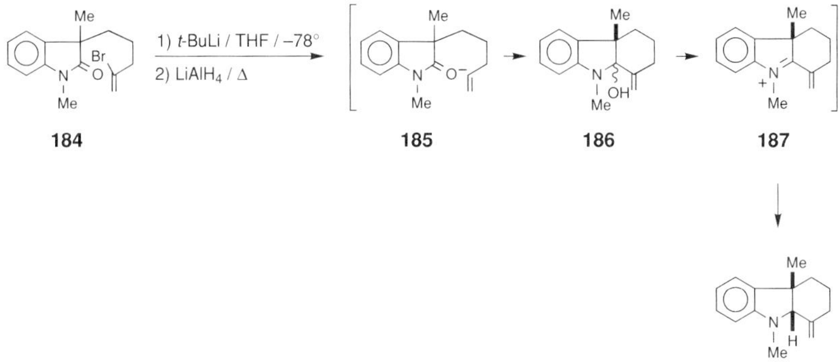

184 **185** **186** **187**

188 (85%)

Equation 19

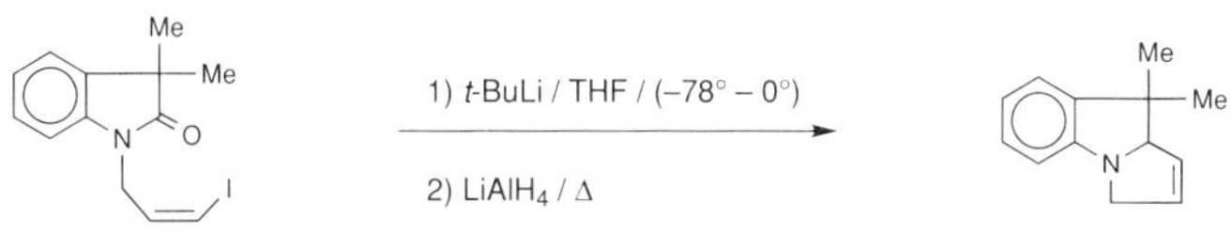

189 **190** (53%)

Equation 20

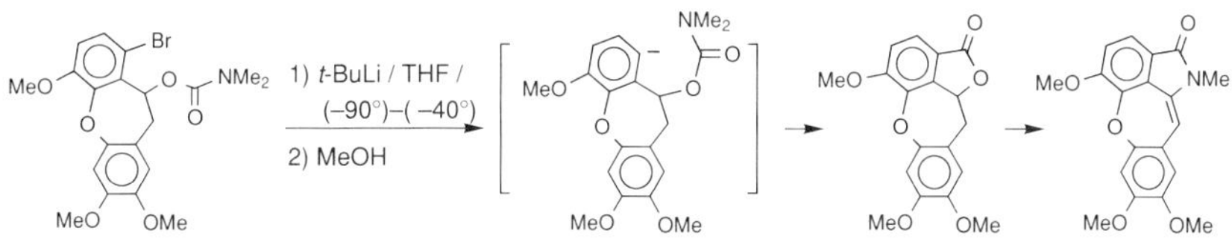

191 **192** **193** (92%) **194** (52%)

Equation 21

An excellent reagent for the generation of vinyl and aryl anions, n-Bu_3Sn–$SiMe_3$ suffers Sn–Si bond cleavage readily upon reaction with tetraalkylammonium chloride, generating n-Bu_3Sn^-, which is extremely reactive in the metal–halogen exchange reaction with vinyl and aryl halides, yielding the corresponding anions. These anions easily engage in intramolecular addition to carbonyl, *e.g.*, the ketone or ester function, as exemplified in Scheme 17.[76] When present together in the same molecule, it is observed that both keto and ester functions react to give products, *e.g.*, **199** and **200** from starting halide **197**, and products **206** and **207** from halide **205**. The most stringent geometrical requirements must also apply, as the *Z*-iodide (**205**) reacts readily while its *E*-isomer (**208**) fails to undergo a similar reaction, giving only the deiodination product **209**.

Scheme 17

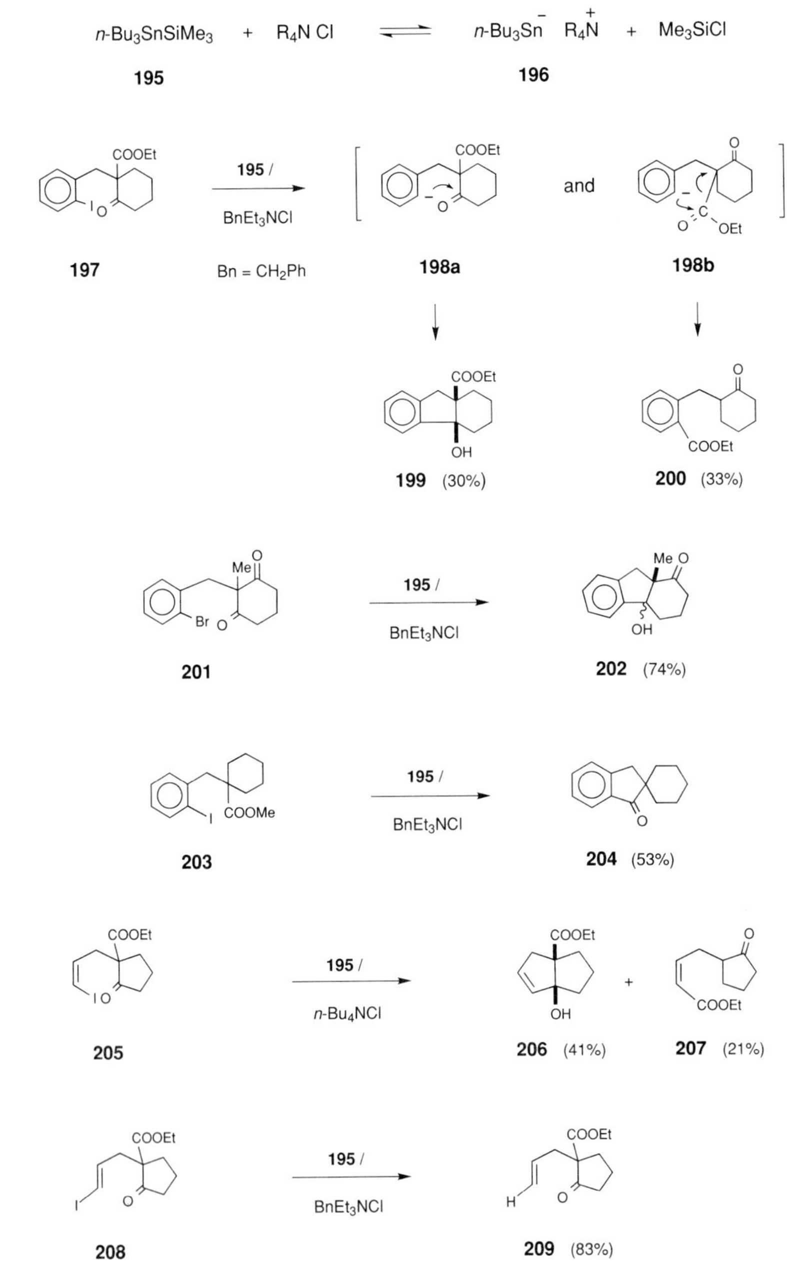

n-Bu3SnSiMe3 + R4N Cl ⇌ n-Bu3Sn⁻ R4N⁺ + Me3SiCl
195
196
COOEt
195 /
BnEt3NCl
Bn = CH2Ph
197
198a
and
198b
199 (30%)
200 (33%)
OH
OEt
Me
Br
201
202 (74%)
I
COOMe
203
204 (53%)
205
n-Bu4NCl
206 (41%)
207 (21%)
208
209 (83%)
H

Besides the discussed carbanionic intramolecular 1,2-addition to carbonyls, similar addition by nitrogen nucleophiles is also synthetically useful, particularly in alkaloid synthesis. Scheme 18 outlines the significant features of the general reaction: condensation between amino and carbonyl functions to form an iminium ion intermediate followed by *in situ* attack by a nucleophile (usually a reduction reaction, see Equations 19 and 20) delivers the corresponding cyclic amine. The most important point to consider in this reaction is the direction of nucleophile attack which essentially determines the stereochemistry of the cyclic product **212**.

Scheme 18

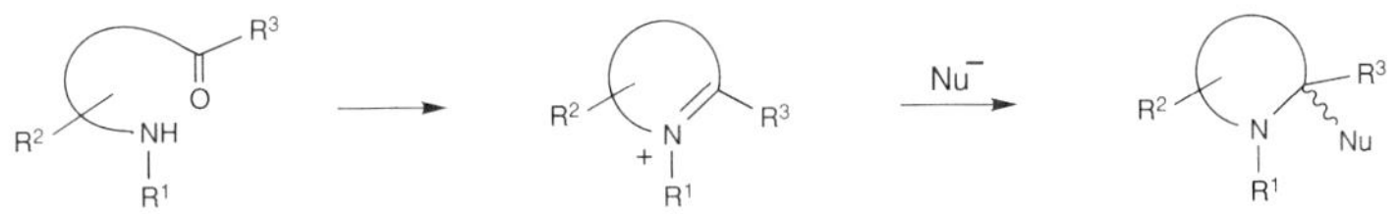

210 **211** **212**

A good example is the enantiomeric synthesis of (+)-pinidine **217**, a one-pot procedure starting from optically active keto sulfide **213**. The reaction first generates intermediate **214**, then develops through [2,3]-sigmatropic rearrangement followed by the formation of cyclic iminium ion **216** and its subsequent *in situ* reduction. Controlled by stereoelectronic effects (see Scheme 18 in Chapter 2)[77] the reduction step proceeds stereospecifically *via* transition state **216** to give exclusively (+)-pinidine **217** in 70% yield (Scheme 19).[78,79]

Scheme 19

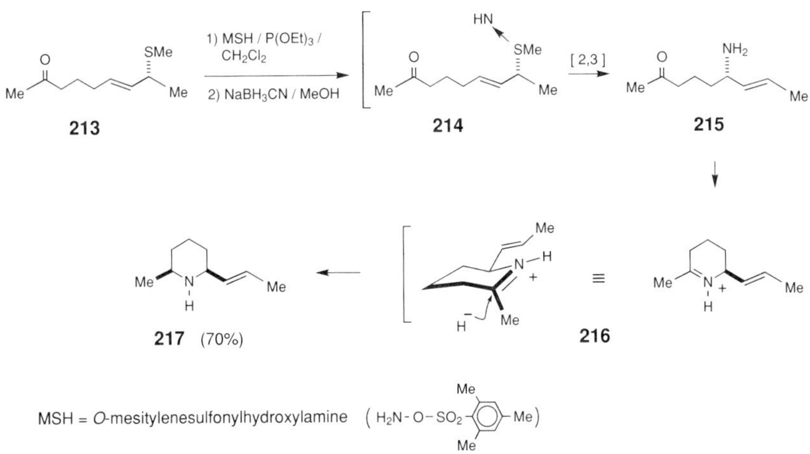

Another one-pot synthesis of octahydroindol-4-ones **222** and **223**, building blocks for the construction of indole alkaloids, is illustrated in Scheme 20.[80] The conversion of allyl dione **218** to final products **222** and **223** proceeds through successive chemical transformations, that is ozonolysis of the allyl group, aldimine formation and subsequent reduction to **220**, cyclization to cyclic iminium ion **221**, and finally reduction to the indolone products. Observed stereoselectivity varies from nil (R = CH_2COOEt) to 4 : 1 selectivity (R = α-methylbenzyl). It should be recalled here that *cis*-indolones such as **222** have previously been synthesized by the tandem rearrangement–cyclization of iminium ions and have been employed in the total synthesis of (±)-pancracine as already discussed in the foregoing chapter (Scheme 33, Chapter 2).[81]

A very interesting reductive–cyclization reaction of 5-nitropentadecane-2,8-dione **224** leading to bicyclic pyrrolizidine derivatives **225** and **226** have been reported as presented in Scheme 21.[82,83] Although producible *via* two methods A and B, the products exhibit different stereochemistry with each method. The hydrogenation of **224** according to method A yields predominantly *epi*-xenovenine **226**, most probably *via* consecutive iminium ion formation–*syn* reduction, twice repeated. Method B, on the other hand, yields xenovenine **225**, a component of ants venom, with only a trace of the **226** isomer. It has also been revealed through ^{15}N labeling experiments that the total nitrogen content of **225** is derived from NH_4OAc, stirring the speculation that intermediate **231** is probably involved in the reaction mechanism.

Scheme 20

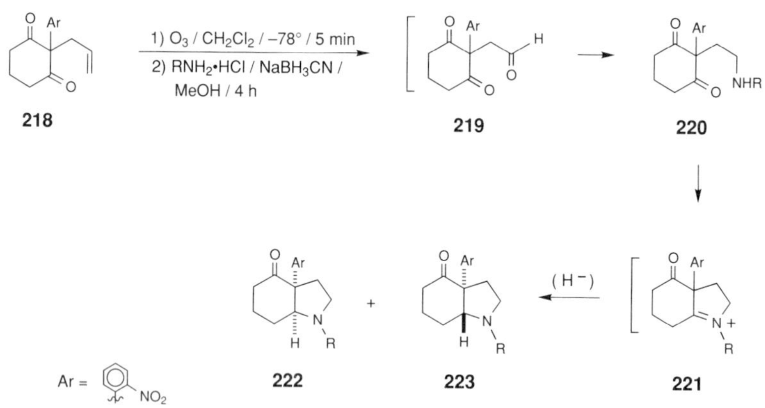

R	222	223
R = CH_2COOEt	(26%)	(26%)
= Me	(37%)	(25%)
= CH_2Ph	(38%)	(18%)
= CH_2CH_2SePh	(29%)	(12%)
= CH(Me)Ph	(28%)	(7%)

Scheme 21

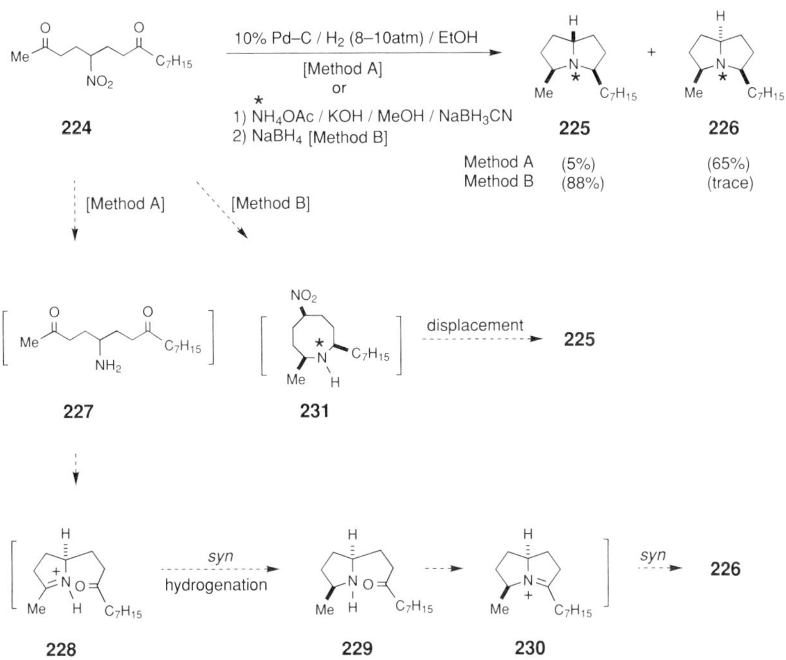

Lately, catalysis of organic reactions by high surface area inorganic reagents has become more popular. There are reported cases where these reactions proceed with great selectivity under milder conditions as compared to conventional methods. For example, high yields of pyrroles **233** are obtained when 1,4-diketone **232** and a primary amine are dispersed on alumina or clay (montmorillonite K_{10}) without solvent at room temperature.[84] Also, the method has been applied with the additional use of microwave irradiation in the synthesis of bicyclic compounds **238**.[85] The reaction and its mechanism are illustrated in Scheme 22.

Scheme 22

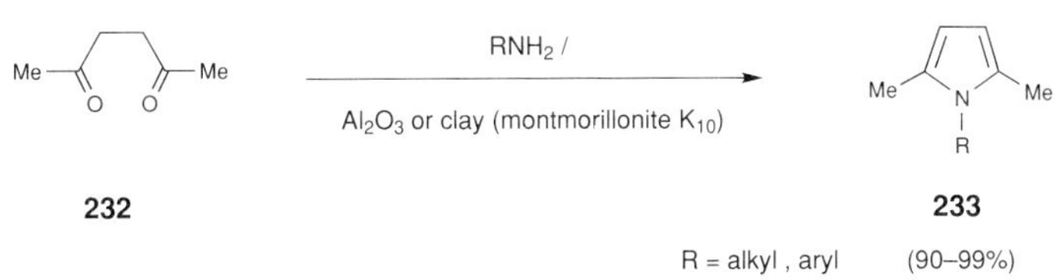

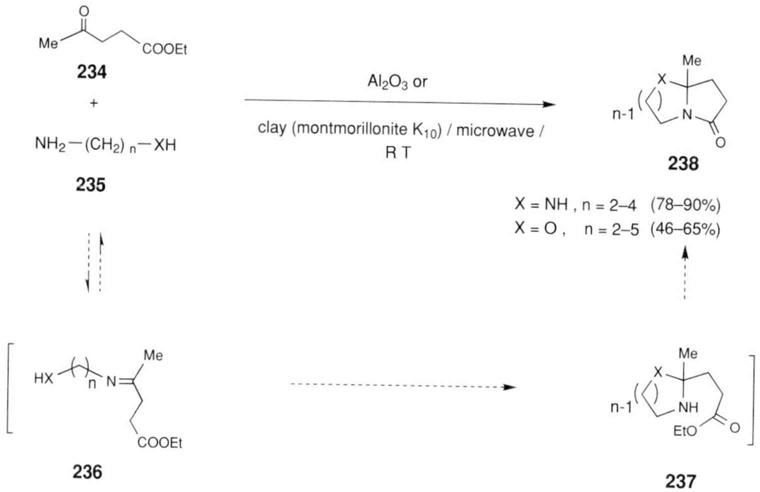

B. ADDITION TO IMINE, IMINIUM, OXONIUM, AND THIONIUM FUNCTIONS

Cyclization *via* addition to the imine group is much less frequently encountered than the analogous additions to iminium and other cationic intermediates. A recent example is the reaction of dianion **240**, derived from **239**, which readily adds to aldehydes, ketones, as well as imines **241**, to give the corresponding lactones and lactams **243** *via* intermediate **242**.[86,87] This reaction has been employed in the synthesis of γ-coniceine **244**,[88] the naturally occurring indolizidine alkaloid (Scheme 23).

Scheme 23

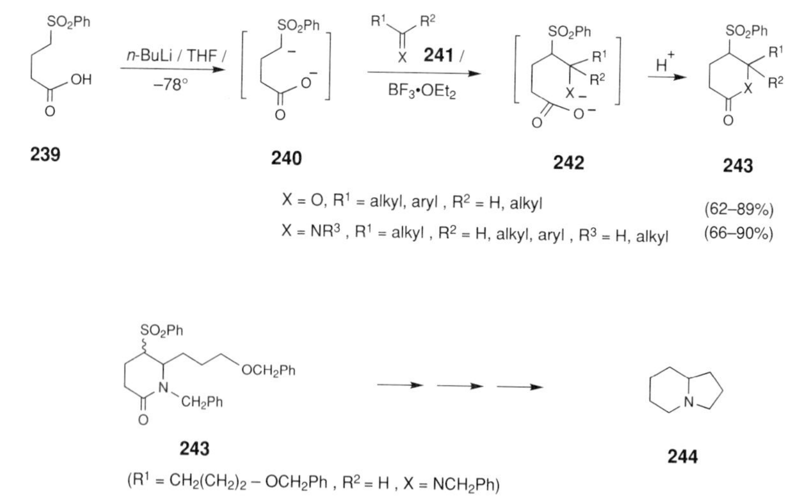

In the synthesis of strychnine, Woodward *et al.*[89] prepared spiroindole **247a** by treating imine **245a** with *p*-toluenesulfonyl chloride (TsCl), which reaction is thought to involve intramolecular cyclization of iminium ion **246a**. Continued development[90] since then has found the reaction to be viable with a variety of electrophiles and, more importantly, to exhibit excellent chiral induction. The use of starting imine **248**, for example, produces spiroproduct **249** in moderate yield but with excellent chirality transfer (97% ee),[91,92] as shown in Scheme 24.

Scheme 24

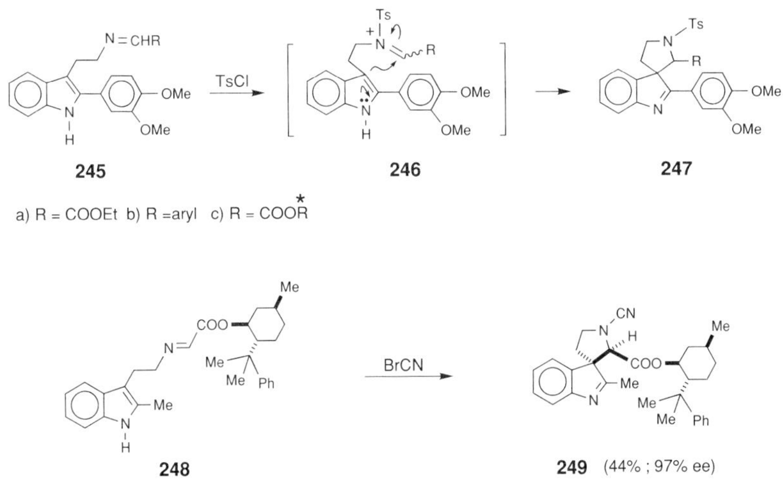

Further examples of intramolecular additions of carbanions to the iminium system are shown in Scheme 25. The condensation of cyclopentadienecarboxylic acid ester **250** with an aldehyde, *e.g.*, isobutyraldehyde **251**, at –30°C in the presence of aqueous methylamine in a methylamine–acetic acid buffer solution (pH 8 to 9) regiospecifically produces fulvene ester **252**. This latter reacts very rapidly with enamines such as **253** to generate the interesting intermediate **254** (contains both carbanion and iminium systems in the molecule) which spontaneously cyclizes with subsequent elimination of the amino group to yield the triquinane skeleton **256**.[93]

The process has been applied to the synthesis of $\Delta^{9(12)}$-capnellene **265**. Starting from **258** (obtained from the condensation of **250** with aldehyde **257**), hydrolysis yields the corresponding aldehyde **260** and its cyclic isomer **259**. Treatment of **259** with basic alumina results in condensation to yield **264**, identical with the product obtained from **260** *via* its conversion to enamine **262** which spontaneously undergoes tandem cyclization–elimination of the amino group to the observed product (**262** → **263** → **264**) in an overall yield of 52% from **258**. From tricyclic **264** the preparation of $\Delta^{9(12)}$-capnellene is then a simple matter of following known procedures.[94]

Scheme 25

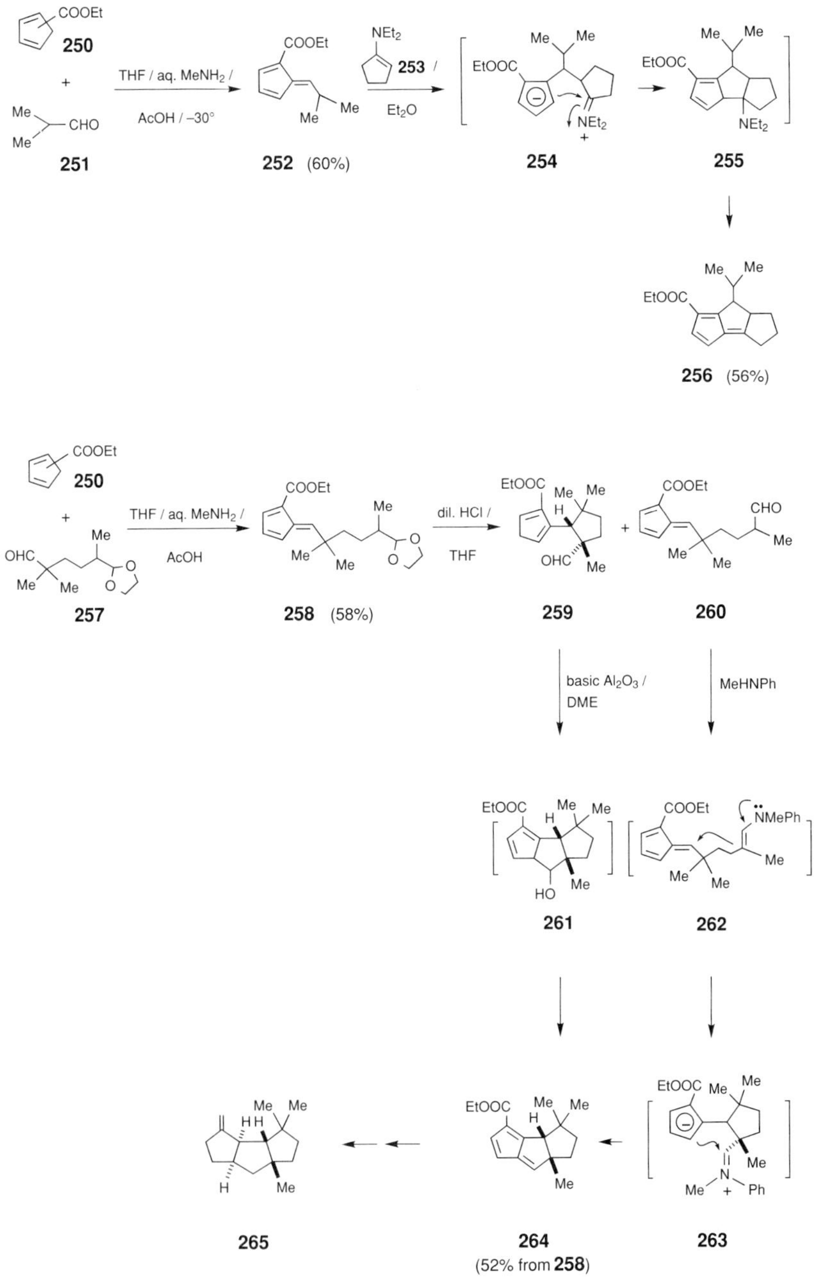
COOEt
250
251
THF / aq. MeNH2 /
AcOH / –30°
252 (60%)
NEt2
253 /
Et2O
EtOOC
254
255
256 (56%)
257
THF / aq. MeNH2 /
AcOH
258 (58%)
dil. HCl /
THF
259
260
basic Al2O3 /
DME
MeHNPh
261
262
263
264
(52% from 258)
265

Analogous anionic cyclizations to oxonium ions are exemplified in Scheme 26. Conjugative addition of organocuprate **267** to enone **266** and trapping of the resultant enolate with trimethylsilyl chloride provides substrate **268** for cyclization. Treatment of the latter with titanium tetrachloride affords the bicyclic compound **270** in respectable yields. The mechanism of the reaction is proposed to proceed *via* titanium-induced oxonium ion (**269**) formation which triggers cyclization in the *endo* mode. Hence the true role of reagent **267** is in fact that of a carbonyl ylide synthon (A).[95]

Likewise, the reagent **272** can be regarded as a synthon of the carbonyl ylide A, but, in this case, the reaction sequence is in reverse order to that of the cuprate **267**, *i.e.*, it reacts first as an electrophile, *via* intermediate oxonium ion **273**, with silyl enol ether **271** to give **274**, then functions as a nucleophile in the cyclization to the final product **275**.[96]

Scheme 26

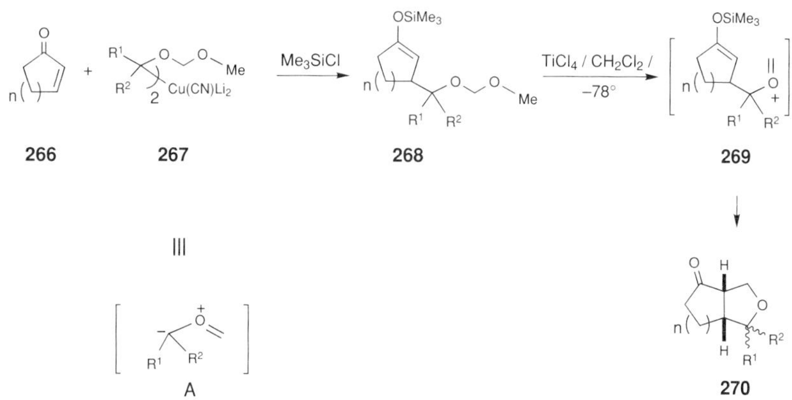

n = 1 ; R^1 = H, R^2 = $CHMe_2$ (82% ; 1 : 1)
n = 2 ; R^1 = H, R^2 = $CHMe_2$ (77% ; 7 : 3)
n = 1 ; R^1 = H, R^2 = *t*-Bu (58% ; 1 : 1)
n = 2 ; R^1 = H, R^2 = *t*-Bu (74% ; 7 : 3)
n = 1 ; R^1 = R^2 = Me (73%)
n = 2 ; R^1 = H, R^2 = *n*-C_5H_{11} (68% ; 6.5 : 4)

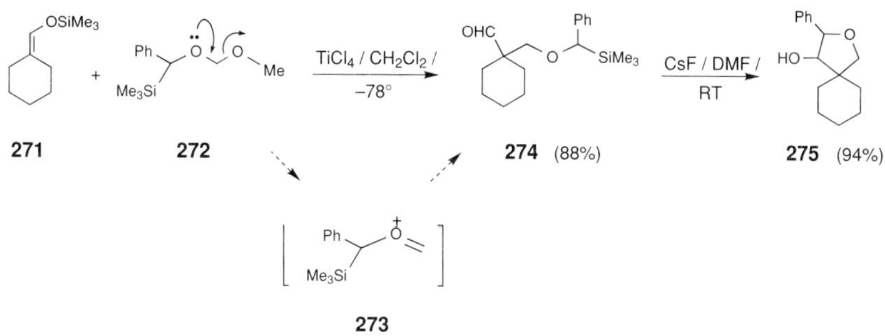

The synthetic strategy involving *in situ* formation of sulfonium ylide *via* sulfide–carbene interaction has received continuous attention.[97–99] Several pyrrolizidine alkaloids, *i.e.*, trachelanthamidine **281**, isoretronecanol **282**, and supinidine **283** have been synthesized *via* the sequence of carbene **277** → ylide **278** → cyclization of zwitterion **279**, as outlined in Scheme 27.[97] A similar approach, shown in Scheme 28, paved the way for the synthesis of penems (**289**; X = S) and carbapenems (**289**; X = CH_2).[98] Both processes (Schemes 27 and 28) exploit anion addition, to an iminium ion in the former and to a thionium ion in the latter. This type of reaction has also been employed in the synthesis of other alkaloids.[99]

Scheme 27

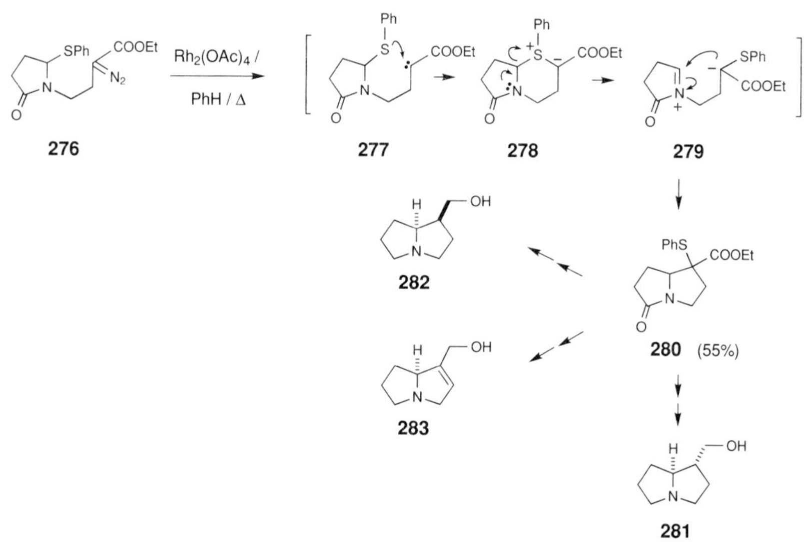

Scheme 28

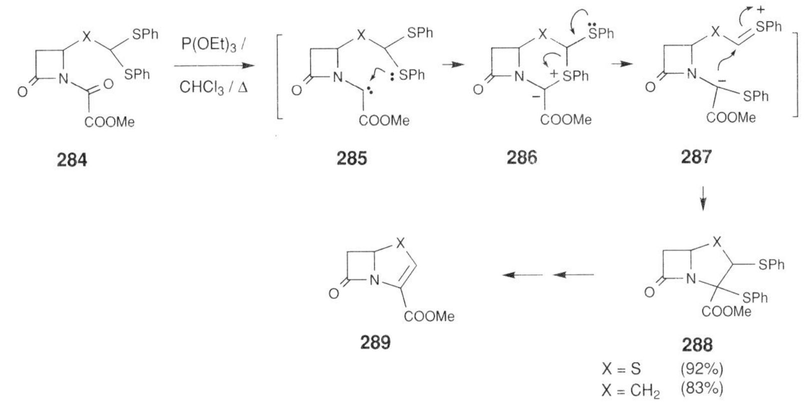

IV. CYCLIZATION *VIA* CONJUGATE ADDITION

Anionic cyclization involving the Michael addition can be regarded as one of the most important tools of organic synthesis. There are endless examples, portraying a large number of variations, from Michael-initiated reactions that are followed by reactions such as S_N2 displacement or aldol or Dieckmann condensations, to tandem Michael addition processes such as the double or triple Michael additions.

A. THE DIRECT MICHAEL ADDITION

The metal–halogen exchange reaction of iodide **290** is extremely rapid and the resulting anion **291** cyclizes smoothly to products **292** and **293**. Fearing a competing reaction of butyllithium (the base used to generate the anion) with the olefin activating group (E), the unit CO-C(PPh_3)COOR was originally employed to provide a stabilized carbonyl having an adjacent negative charge, in order to withstand nucleophilic attack.[100] However, as already expressed, the exchange reaction is so rapid that it does not warrant any precautions with the olefin activating group, which could be any ordinary ester function such as COO*t*-Bu.[101] Generally, the reaction with *E*-olefins (*e.g.*, **290**) yields only a slight excess of the *trans*-(**292**) over the *cis*-(**293**) products, but with *Z*-olefins (*e.g.*, **294**) the ratio of *trans*-(**295**) to *cis*-(**296**) products is as high as 300:1. Similar results are observed with the radical cyclization of *Z*-olefin **294**, from which the *trans*-cyclopentane **295** is produced in an overwhelming ratio of >280:1 over its *cis*-isomer **296**. These results have been rationalized in terms of the transition states depicted in Scheme 29. It is clear that transition state **298**, leading to the *trans*-product, is sterically much more favorable than its counterpart **297**, irrespective of the anionic or free radical nature of the cyclizing intermediate.

Scheme 29

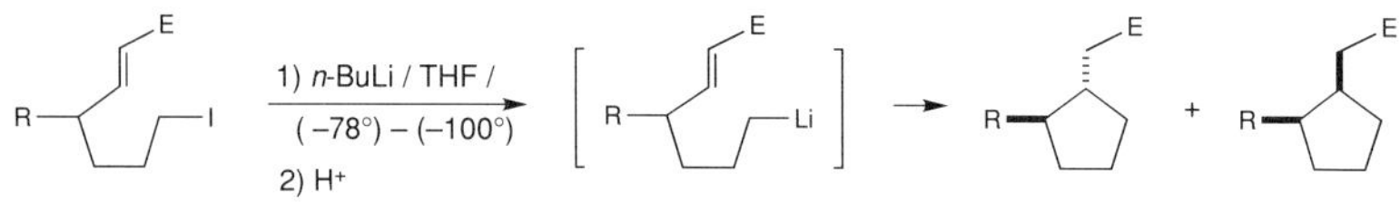

290 **291** **292** > **293**

R = Et, *t*-Bu E = C(=O)–C(=PPh$_3$)–COOEt, COO*t*-Bu (80–93%)

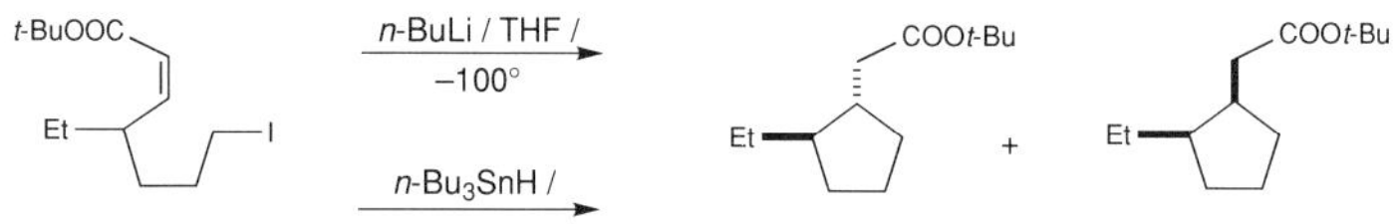

294 **295** > **296**

(~ 300 : 1)

297 X = •CH$_2$, Li–CH$_2$ **298**

Scheme 30 features the use of *S,S*′-diethyl dithiomalonate in a reaction with iodoenone **299** to yield cyclized product **303** by a consecutive alkylation–Michael addition process. The dithiomalonate function offers the added convenience of direct reduction to alcohol **304** by Raney nickel, but not without prior protection of the keto group, which otherwise would result in an extremely poor yield of reduction product. Thus, dithiomalonate **300** has, in effect, the role of a masked ethanol bis-carbanion.[102]

Scheme 30

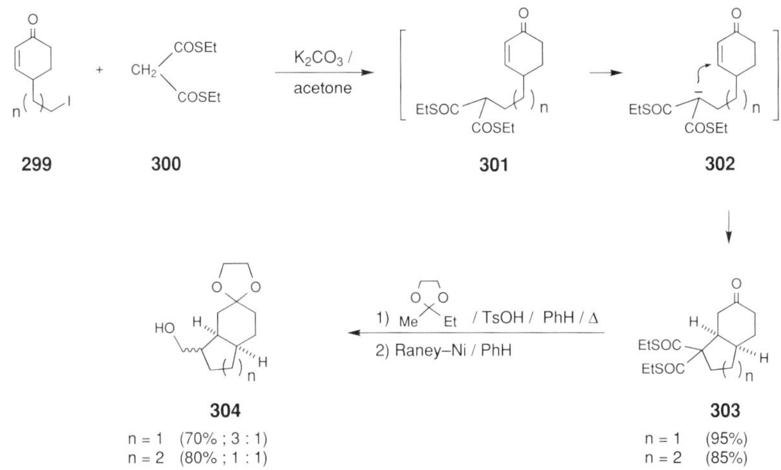

High stereoselectivity in ring closure by Michael addition has been observed in many cases, *e.g.*, the chiral induction by the optically active R group in ester enolate **306** which ring closes to cyclopentanone **307** in high enantiomeric purity (Equation 22),[103] or the use of an external chiral base to induce chirality in the product of cyclization as depicted in Scheme 31. Here the appropriate choice of either (*R*)-(+)- or (*S*)-(–)-α-methylbenzylamine to catalyze the Michael cyclization of aminoketo ester **308** determines the enantiomeric excess of either product **310** or **311**. Although the reaction yields are all good, the six-membered ring amines[104] are obtained in much higher optical purity than the five-membered ring amides.[105] Both types of products **310** and **311** are useful starting blocks in the synthesis of alkaloids.

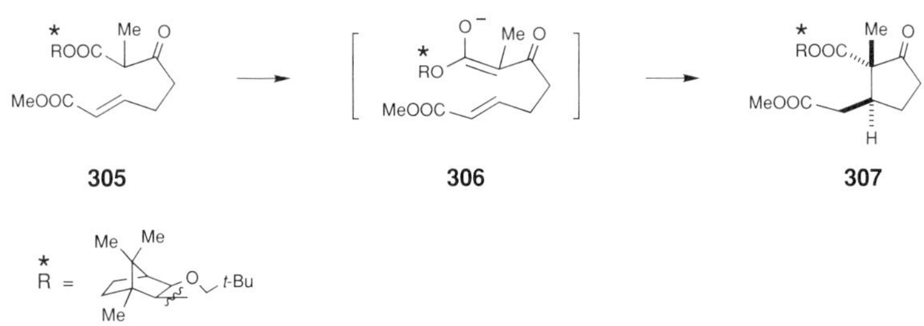

Equation 22

Scheme 31

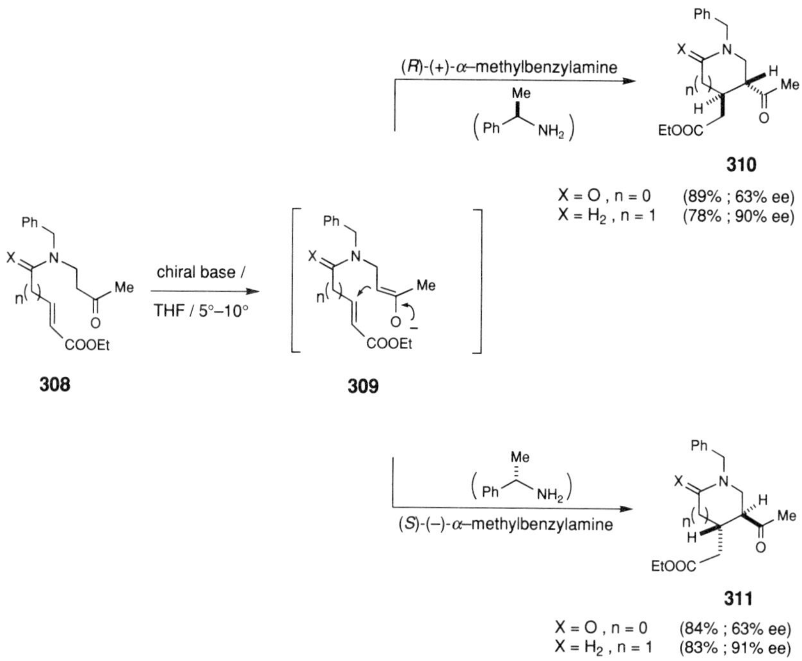

The appeal of the following simple reaction lies in its very simplicity, coupled with its versatility and usefulness. As shown in Scheme 32, the reaction between iodo-α,β-unsaturated ester **312** and benzylamine furnishes the nitrogen heterocycle **315**. It is apparent that the reaction could involve either route a or b, which differ only in the sequential order of transformations; that is, in route a S_N2 displacement would take place first, followed by Michael addition by the nitrogen nucleophile, whereas in route b the order of events would be the reverse. The actual mechanism has been confirmed to be according to the sequence shown in route a (Scheme 32).[106]

Scheme 32

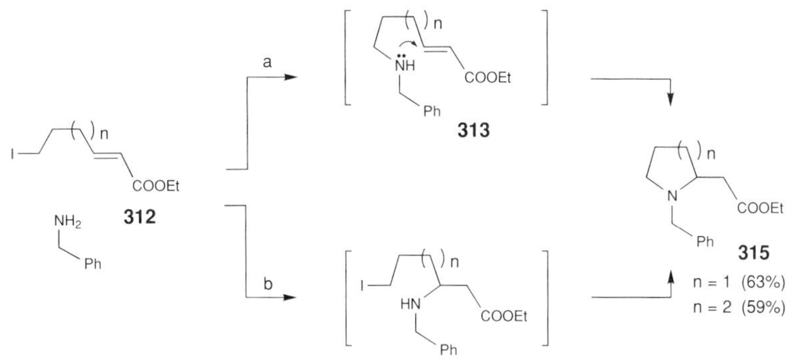

B. MICHAEL-INITIATED RING CLOSURE (MIRC)

The term MIRC (Michael-initiated ring closure)[107] in general refers to the reaction between two reactants involving nucleophilic conjugate addition to an α,β-unsaturated carbonyl compound followed by ring closure of the resulting enolate. As a matter of fact MIRC is not a novel reaction but scattered information has only recently been collected together and the scope of the reaction expanded and reviewed.[1]

The course of the MIRC reaction can be categorized as illustrated in detail in Scheme 33. The success of the reaction in general is dependent upon two factors: first, the stability of the Michael addition intermediates (**317**, **323**, **326**, and **329**) relative to the nucleophiles (as in **316**, **322**, **325**, and **328**, respectively) and second, the rate of cyclization (k_2).[107,108] As exemplified in the scheme, the second step of the reaction can be any of various reactions, *e.g.*, direct S_N2 (Michael–S_N2, **316** → **317a** → **318**), aldol condensation (Michael–aldol, **316** → **317b** → **319**), Dieckmann condensation (Michael–Dieckmann, **316** → **317c** → **320**), and Michael reaction (Michael–Michael, **316** → **317d** → **321**). These reactions are versatile and extremely useful for the construction of cyclic and polycyclic compounds of various ring sizes.

Scheme 33

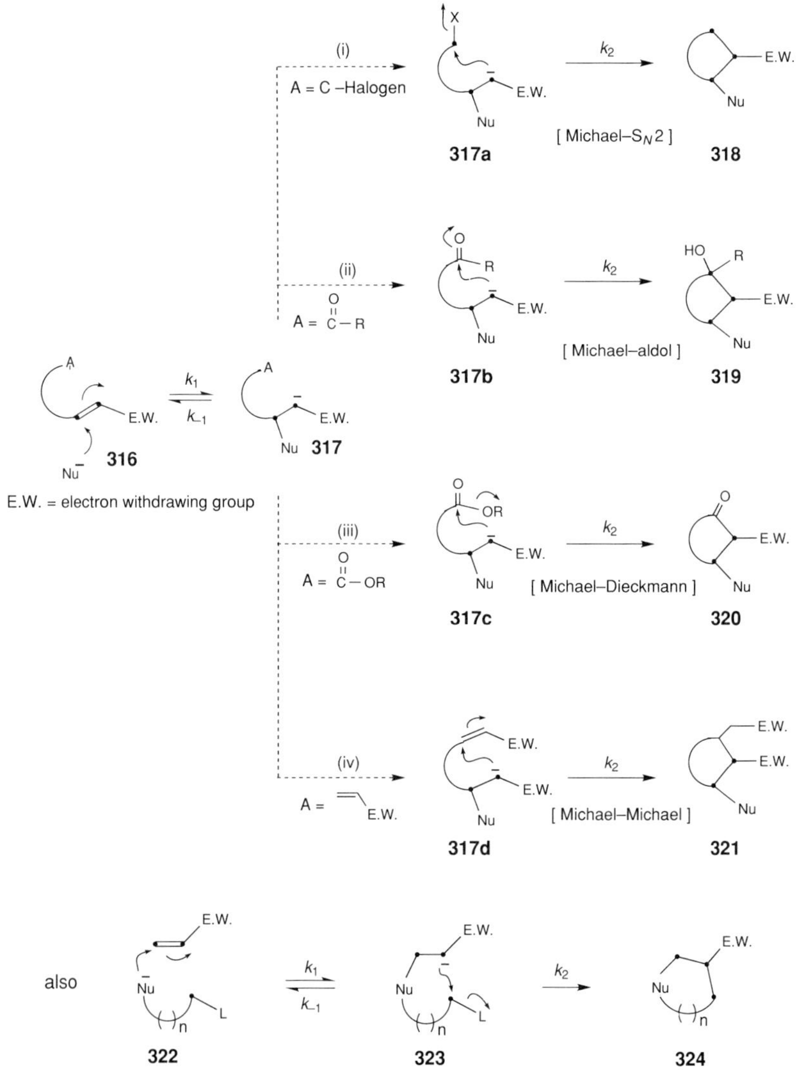
(i)
A = C –Halogen
X
E.W.
Nu
317a
k_2
[Michael–S_N2]
318
(ii)
A = C(=O)–R
R
317b
[Michael–aldol]
HO
319
A
k_1
k_{-1}
316
317
E.W. = electron withdrawing group
(iii)
A = C(=O)–OR
OR
317c
[Michael–Dieckmann]
320
(iv)
A = =\E.W.
317d
[Michael–Michael]
321
also
L
n
322
323
324

Michae l-initiated cyclopropanation

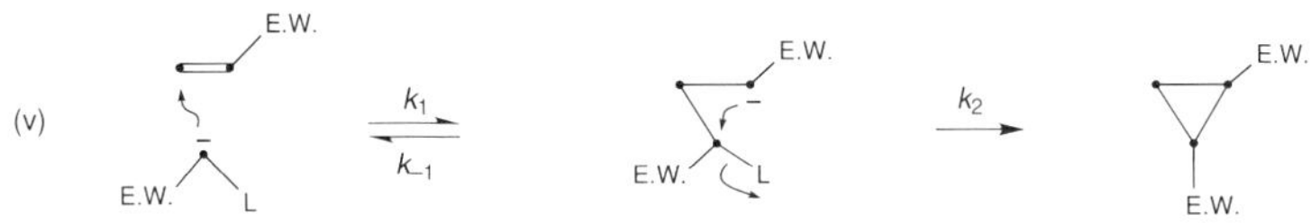

325 **326** **327**

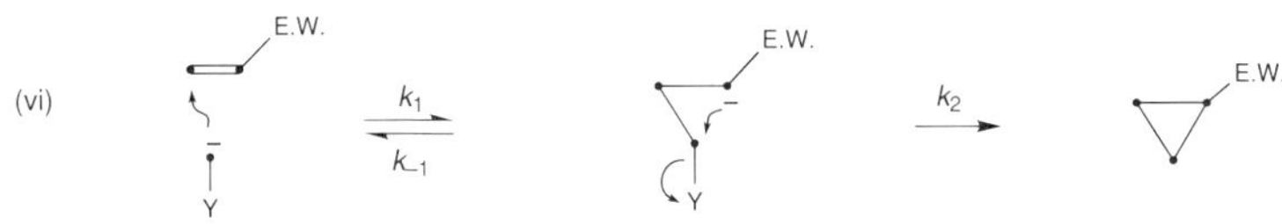

328 **329** **330**

E.W. = electron withdrawing group
L = leaving group
Y = electron withdrawing–leaving group

Lately the Michael-initiated cyclopropanation reaction, in particular, has enjoyed increasing use with innovative substrates. The reaction belongs to the Michael–S_N2 category, where the Michael acceptor reacts with a nucleophile carrying a leaving group within the same molecule. Such a coexistence does pose problems; for example, nucleophile **322** is vulnerable to intramolecular cyclization, and also elimination, which compete with the desired Michael reaction (**322** → **323**). This type of reaction is hence popularly restricted to the synthesis of cyclopropanes, where the starting material either carries both the electron-stabilizing and leaving groups on the same carbon (type v, Scheme 33) or bears a dual functional substituent "Y" (type vi, Scheme 33) which serves as both electron-stabilizing and leaving group, a strategy long known in cyclopropanation. Examples of "Y" include R_2S^+, R_2SO^+, NO_2, etc.

1. Tandem Michael Addition–S_N2 Displacement

The simplest MIRC reaction, the tandem Michael addition–S_N2 displacement process, provides a most convenient entry to carbocycles as exemplified in Equations 23[107] and 24.[109] Even optically active *trans*-substituted cyclopentanecarboxylic ester **340** can be synthesized with exceptionally high enantiomeric purity (≥95% ee) by the MIRC reaction of **331** (n = 1) with the anion derived from optically active pyrrolidine hydrazone derivative **337**, as shown in Scheme 34.[110]

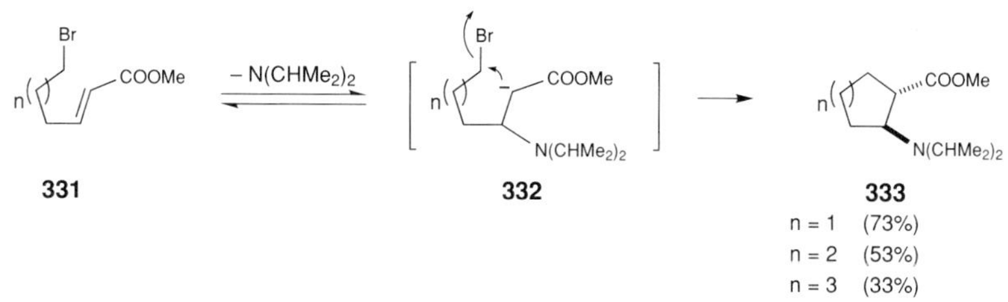

Equation 23

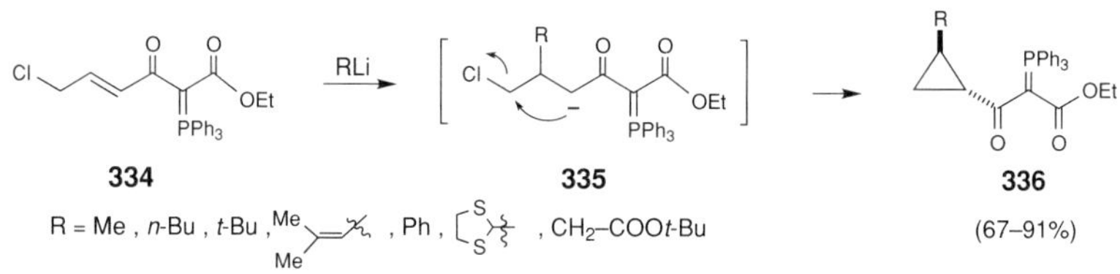

Equation 24

Scheme 34

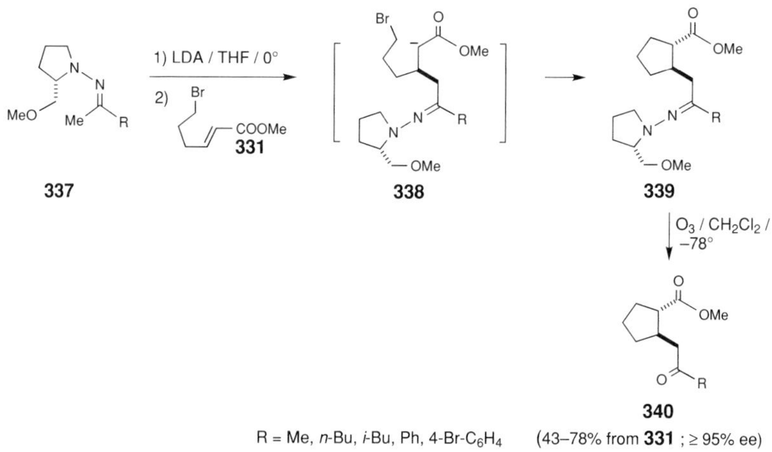

2. Tandem Michael Addition–Aldol Condensation

Tandem Michael–aldol ring closure has been applied as the key step in the total synthesis of taiwanin E **346**. By generating the highly stabilized anion **343** from starting material **341** and allowing it to react with butenolide **342**, annulation is effected, as shown in Scheme 35,[111] to furnish product **345** which can be easily converted to taiwanin E. A similar reaction is presented in Scheme 36, in which treatment of dicyano compound **347** with sodium hydride, followed by the addition of bromoacrylonitrile **348**, leads directly to dicyanonaphthylamine **354**. This product is believed to arise from a Michael addition of anion **349** to **348**, the resulting intermediate **350** then eliminating bromine to **351** before finally undergoing the aldol-like reaction *via* anion **352** as shown in the scheme.[112]

Scheme 35

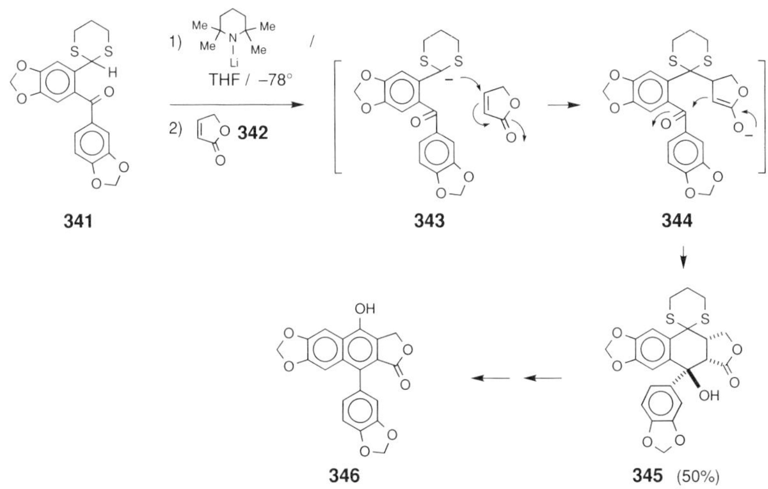

Scheme 36

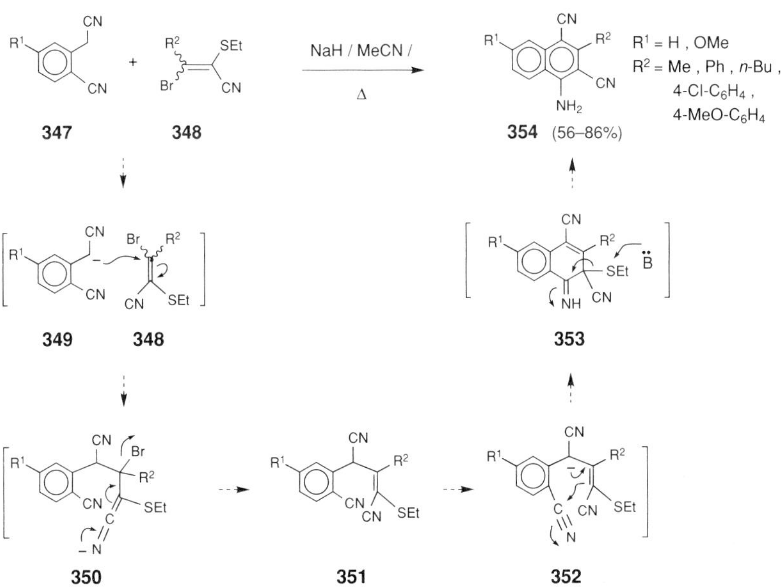

The condensation between α-hydroxy ketones **355** and di-*tert*-butylacetylenedicarboxylate has been reported to yield the corresponding 4-hydroxy-4,5-dihydrofurans, *e.g.*, **359** in Scheme 37.[113,114] The yield of product is found to be much improved by lowering the amount of methyllithium (15 to 20%) used to generate the anion in the first step, and activating the carbonyl group with lithium bromide in the aldol condensation step.[115] The advantage of minimal methyllithium stems from the high sensitivity of both starting material and product towards base. This reaction is, in fact, a tandem Michael addition–aldol condensation process involving oxygen anion as the initiating nucleophile.

Scheme 37

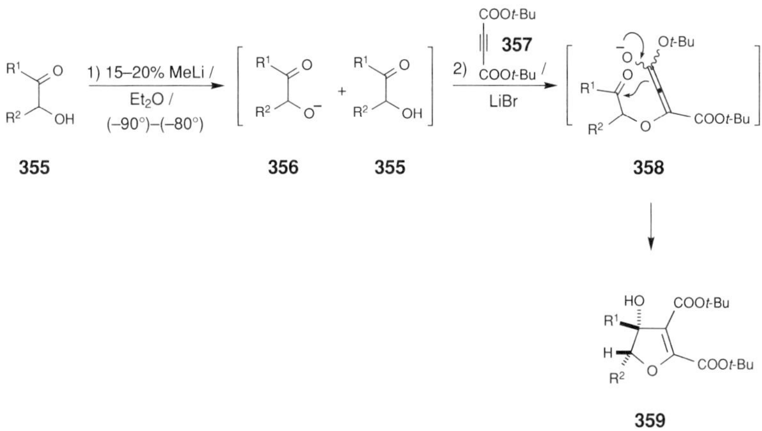

$R^1 = R^2 = Ph$ (86%)
$R^1 = R^2 = Me$ (88%)
$R^1 = Et, R^2 = H$ (83%)
$R^1—R^2 = C_4H_8$ (69%)

3. Tandem Michael Addition–Dieckmann Condensation

The tandem Michael addition–Dieckmann condensation (type iii, Scheme 33) is well suited for the synthesis of both five- and six-membered rings. Varying the nucleophile (Michael donor) and the Michael acceptor can lead to carbocycles of diverse structures. The examples in Schemes 38 and 39 demonstrate the versatility of the reaction. In brief, the addition of zinc homoenolate **360** to acetylenic ester **361** results in conjugate addition followed by intramolecular Dieckmann condensation of the silyl enolate **363** to afford 3-substituted-2-carboethoxycyclopentenones **364** in reasonable yields (Scheme 38).[116] It should be noted, however, that an electron-withdrawing substituent R^1 (*e.g.*, COMe) significantly lowers the yield of product **364**.

Scheme 38

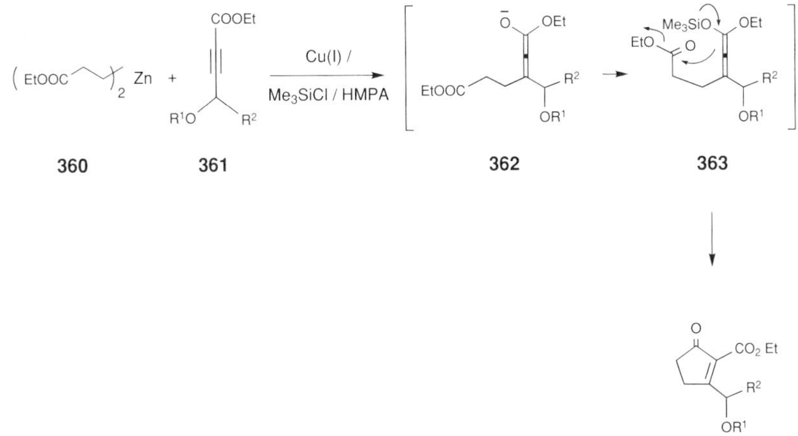

364

R^1	**R^2**	**(%)**
MOM	C_5H_{11}	71
COMe	C_5H_{11}	50
H or TMS	C_5H_{11}	70
MOM	$CHMe_2$	72
COMe	$CHMe_2$	49
H or TMS	$CHMe_2$	70

The interesting reaction shown in Scheme 39 constitutes a facile synthesis of substituted cyclohexenones **371** *via* tandem Michael addition–Dieckmann condensation of lithium dienediolate **367**, derived from **365**, with lithium salt **368**, which can be derived from the same (**365**) or a different (**366**) α,β-unsaturated acid. The remarkably simple procedure involves deprotonation of the corresponding acids with lithium diethylamide at –78°C and leaving the mixture to react at room temperature, followed by acidification which also accomplishes decarboxylation to yield cyclohexenone **371**. An important note here is that a predominantly *s*-cis population of dienediolate **367** is required in order for the Dieckmann condensation to be successful. This accounts for the use of methyl or phenyl groups as substituent R^1, since the corresponding dienediolates are known to have the required geometry.[117]

Scheme 39

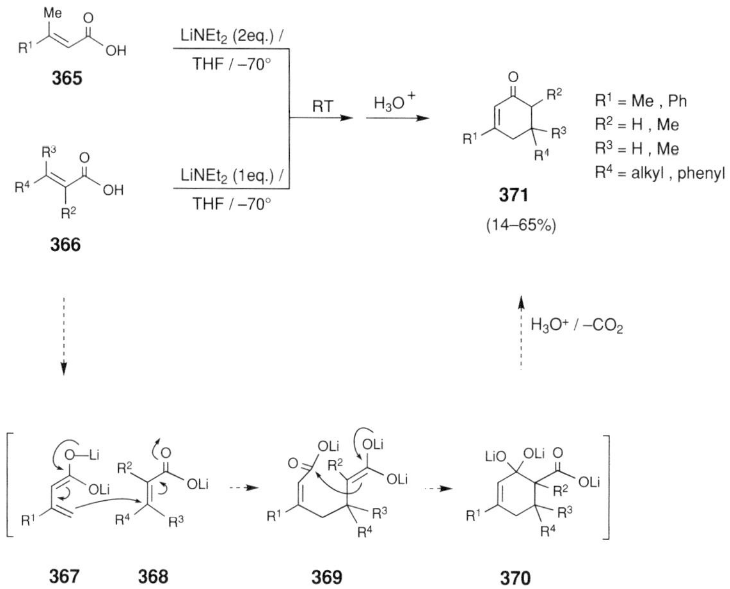

Several superb examples of the application of tandem Michael addition–Dieckmann condensation in the key steps of natural product synthesis are known, such as the reactions shown in Equations 25 to 27, which comprise the syntheses of the antibiotics sarkomycin **375** and methylenomycin A **376** (Equation 25)[118,119] anthelmintic diospyrol **381** (Equation 26),[120,121] and (–)-(9*R*)-7,11-dideoxy-13-deoxodaunomycinone **387** (Equation 27).[122–124] In the first two processes the acrylate unit conveniently serves as the Michael acceptor, reacting with the masked itaconate anion derived from **372** in the reaction in Equation 25, and with toluate anion **378**, derived from **377**, in the reaction in Equation 26. The last example (Equation 27) shows the reaction of sulfone anion **384** with hydronaphthalenone **383**, the cyclization product from which is oxidized *in situ* by oxygen gas to yield anthraquinone **386**.[124] From there, introduction of the missing hydroxy group completes the target molecule **387**.

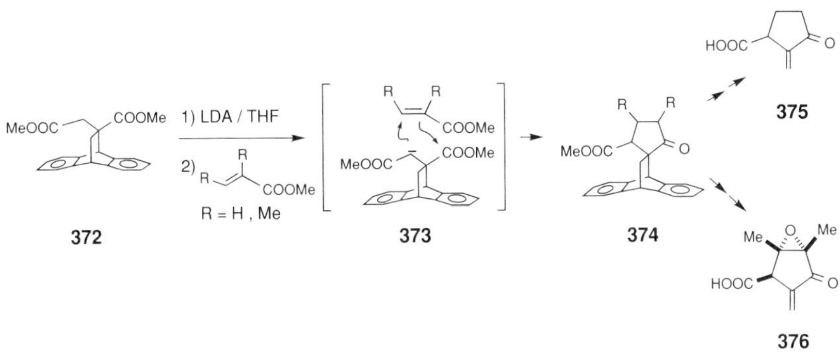

Equation 25

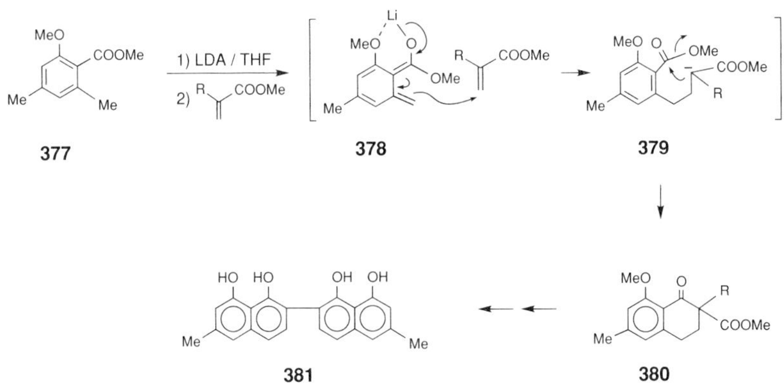

Equation 26

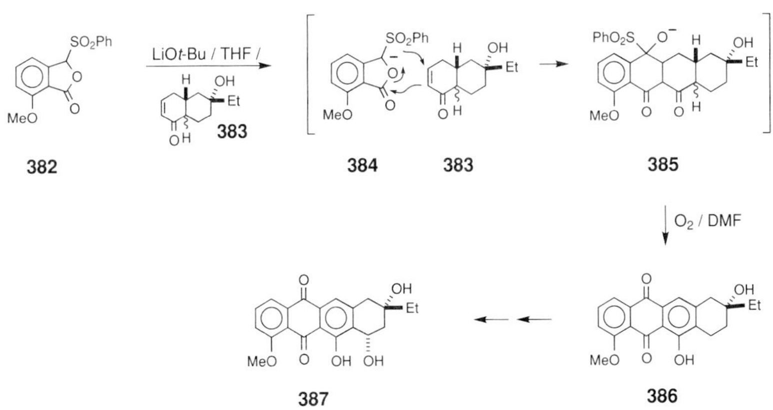

Equation 27

4. Multiple Michael Additions

Cyclizations effected by multiple Michael additions, for example the double Michael reaction shown in Scheme 33 (type iv), offer a short and simple route to polycyclic compounds. With a proper design of reactants (Michael donor and Michael acceptor) even polycycles with complex structural features, otherwise difficult to access, may be assembled.

An excellent nucleophile, lithium *N*-benzyl-*N*-(trimethylsilyl)amide (LSA) **388**, is found to add to crotonate derivatives only in Michael fashion[125] and thus is ideal for initiating ring closure reactions. Equation 28[126] exemplifies a tandem Michael addition–S_N2 cyclization process and Equation 29[127] a double Michael addition cyclization, initiated with LSA. Scheme 40 demonstrates application of the reaction in the synthesis of bioactive lactones dihydronepetalactone **399a** and isodihydronepetalactone **399b**.[126]

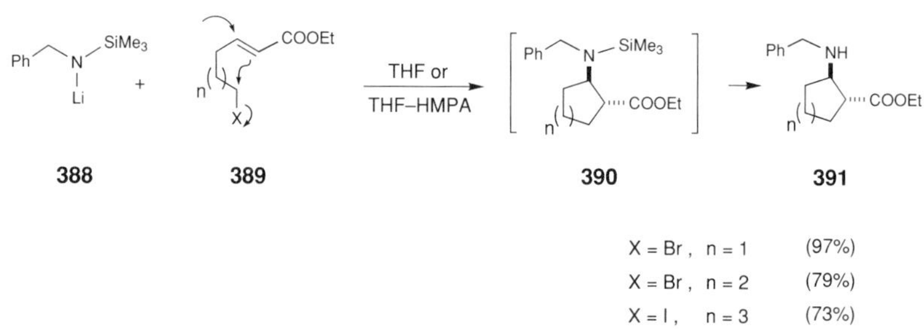

Equation 28

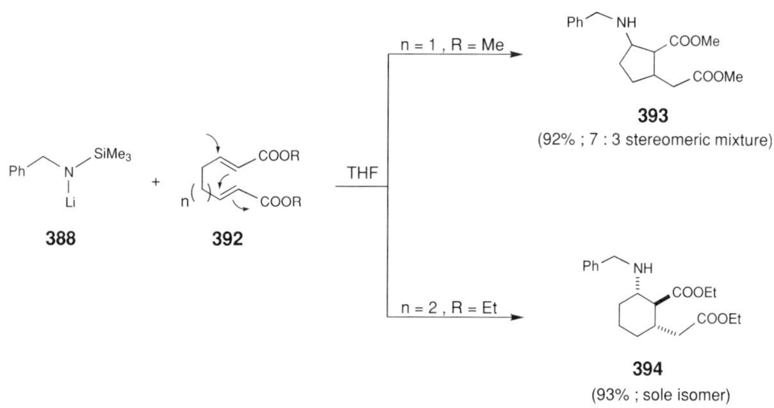

Equation 29

Scheme 40

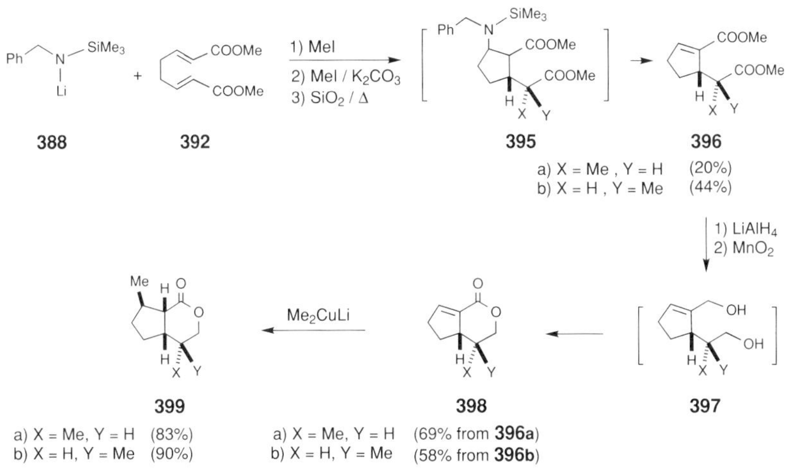

Silyl ketene acetal **401** also functions well as a Michael donor in initiating the double Michael cyclization to yield the stereo- and regiospecific product **402** (Equation 30).[128] The regiospecificity of the reaction is nicely demonstrated here, where, of the two different electron withdrawing groups on the substrate (**400**, R = Me), the nucleophile reacts solely at the β-position of the keto group. A most impressive example of stereocontrol is the double Michael reaction of optically active (*R*,*R*)-**403**, induced by PhS^- and terminated by trapping the enolate with phenylselenyl chloride, to yield product **404** (88%) with an enantiomeric excess exceeding 99% (Equation 31).[129]

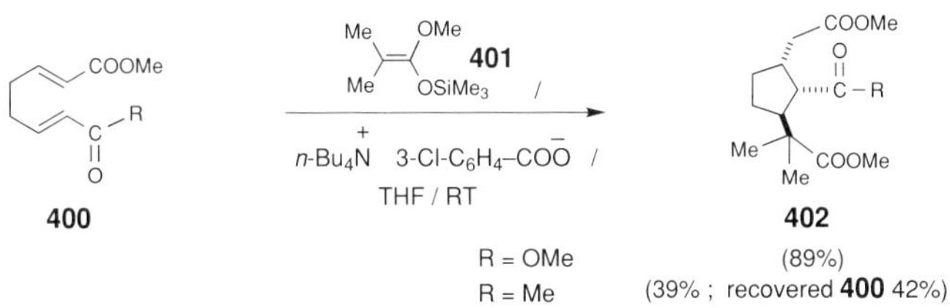

Equation 30

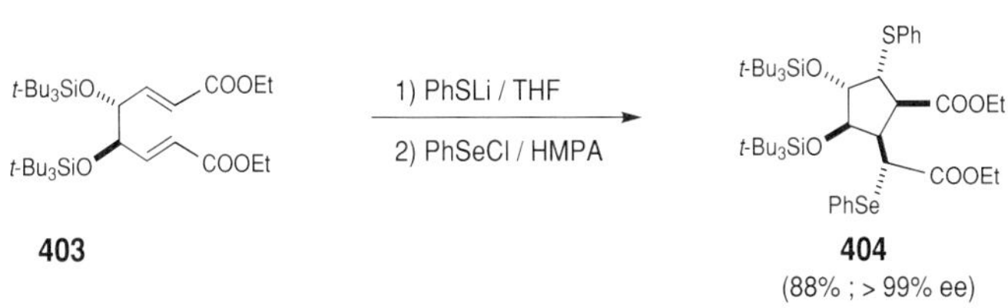

Equation 31

Variations of cyclizations *via* double Michael additions abound; for example, the reactions shown in Equations 32[130] and 33.[131] Initiating the processes are carbanion **406** in the former, and the nitrogen nucleophile **411** in the latter, both donors being similar in bearing a Michael acceptor moiety in their molecules. The nucleophiles first add intermolecularly to a Michael acceptor substrate and the resultant enolate in each case then cyclizes in a second, intramolecular, Michael addition. The *in situ* generation of nitroethylene **412**, in Equation 33, from 1-benzoyloxy-2-nitroethane **410** should also be noted.

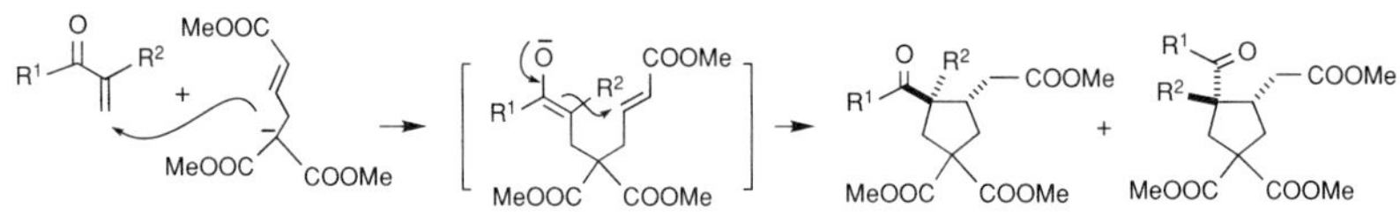

405 **406** **407** **408** **409**

R^1	R^2	%	408	:	409
Me	H	65	> 50	:	1
Me	Me	73	5	:	2
Ph	H	81	> 50	:	1
Ph	Me	83	18	:	1

Equation 32

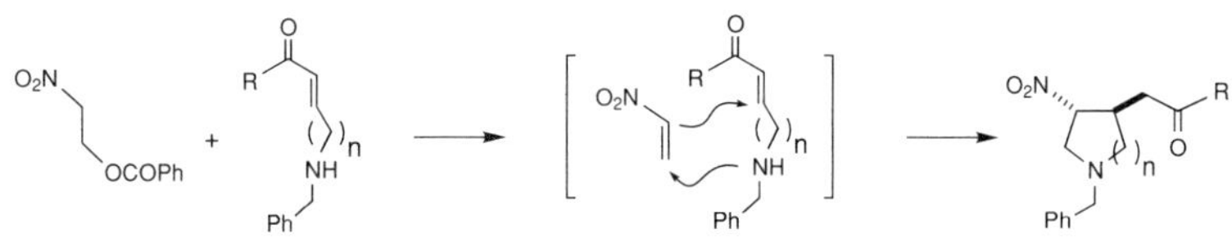

410 **411** **412** **411** **413**

n = 1 , R = OEt
n = 2 , R = Me

Equation 33

The base-catalyzed tandem double Michael addition reaction between enones **414** and **415**, which yields products **419** and **420** (after removal of E^2) as shown in Scheme 41, provides an interesting study. In principle the observed products could arise from a Diels–Alder reaction *via exo-* and (unfavorable) *endo-*transition states, respectively (**421**). The mechanism of the reaction remained debatable, although evidence was in favor of the double Michael addition process.[132] A detailed study of the reaction has now concluded that the course of reaction is influenced by several factors, that is, depending upon the substituent E^2, solvent, and base employed,[133,134] either the double Michael addition or Diels–Alder process may prevail. Varying these reaction variables leads to different ratios of products **419** and **420**. This is exemplified in Scheme 41 which tabulates results from the reaction of **414** with **415a–c**. The reaction has been applied to the synthesis of optically active 14β-hydroxy steroid **425** as shown in Equation 34.[135]

Scheme 41

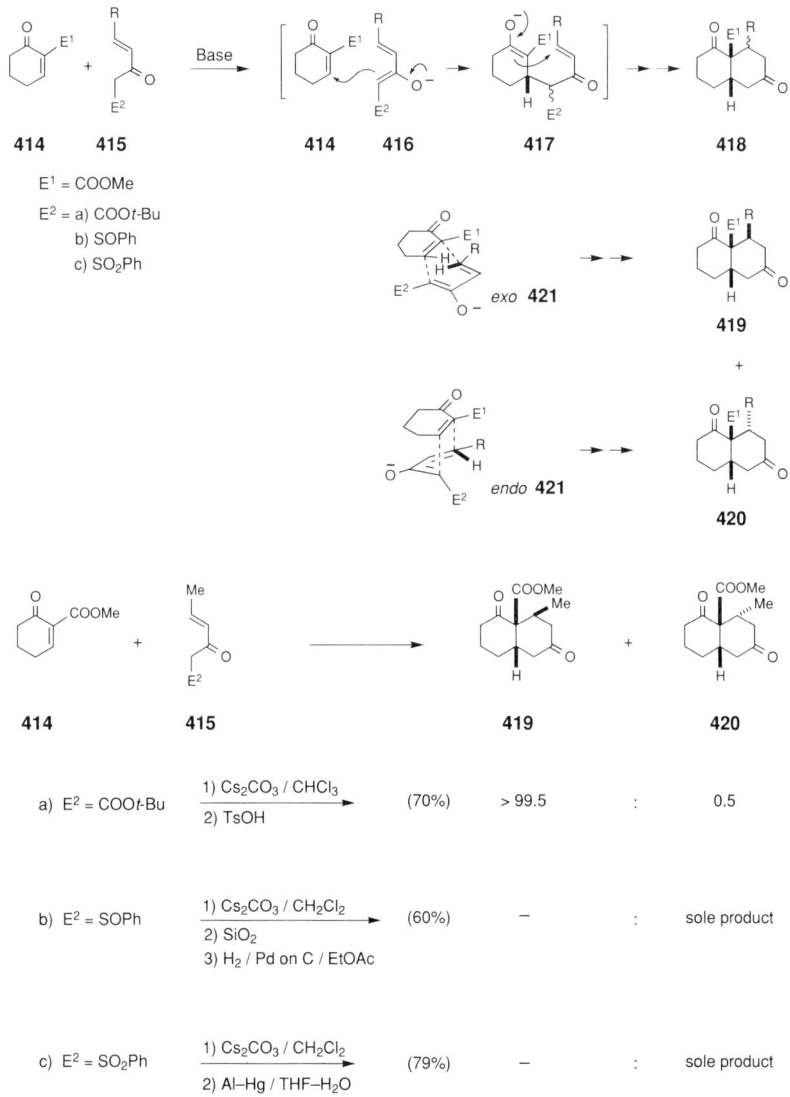

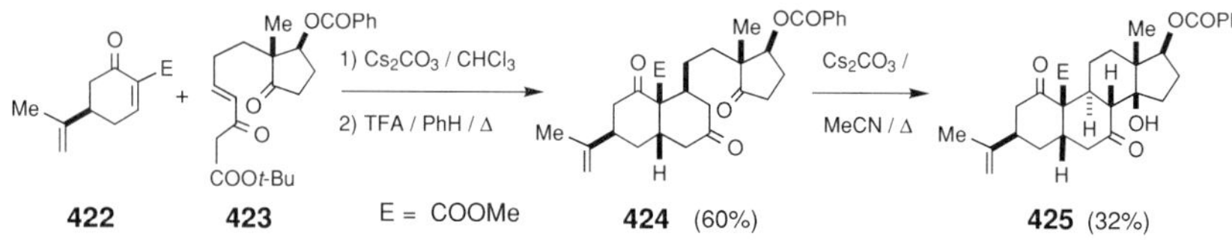

Equation 34

Bicyclic compounds, in particular, are convenient to construct with the double Michael addition reaction. Examples of such preparations of the bicyclo[2.2.2]octane system are numerous,[136–139] the one shown in Equation 35[137] being the synthesis of keto ester **430** from the reaction between (–)-carvone **426** and acrylic ester **427** with lithium hexamethyldisilylazide ($LiN(SiMe_3)_2$) as base. A similar reaction employing chiral acrylate **432** yields bicyclic product **433** with a modest diastereomeric excess, the highest chiral induction (de = 64%) being observed with the chiral acrylate derived from a camphor auxiliary (**432d**, Equation 36).[136]

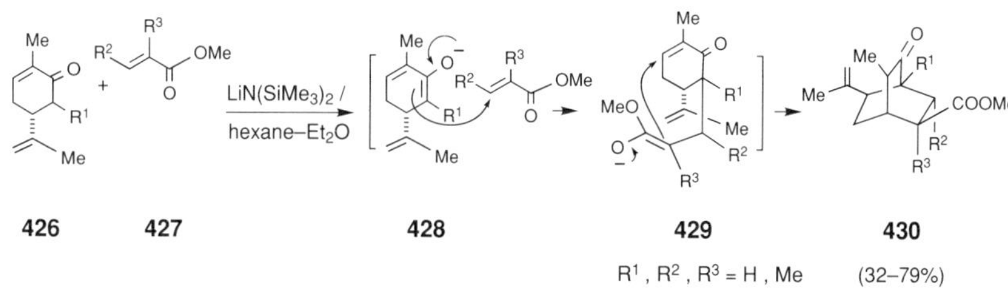

Equation 35

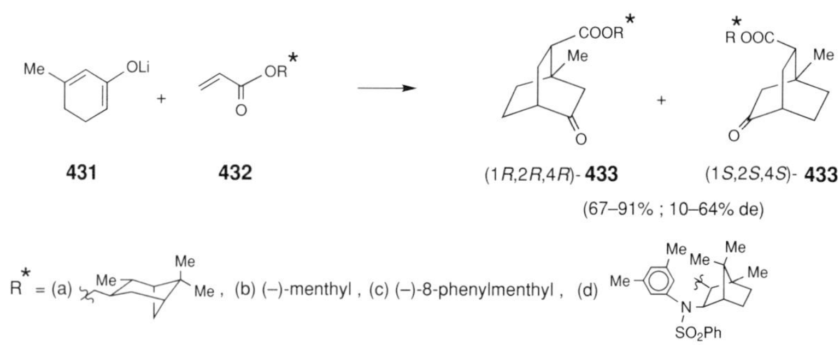

Equation 36

Elegant applications of intramolecular double Michael addition are featured in the key steps of several natural product syntheses, for example, in the synthesis of (±)-pentalenic acid **439**, the naturally occurring angular triquinane, from the key intermediate **435**, as outlined in Scheme 42,[140] and the synthesis of the estrane ring system **444** from enone ester **441**, as shown in Scheme 43.[141] An interesting observation has been made with a reaction related to that shown in Scheme 43, *i.e.*, the reaction of **445** with *tert*-butyldimethylsilyl trifluoromethanesulfonate (TBDMSOSO$_2$CF$_3$) in the presence of triethylamine in methylene chloride at room temperature, which always gives a single product **447**, as depicted in Equation 37 (~63%), regardless of the starting geometry of substrate **445**. These results are consistent with a stepwise mechanism of the double Michael addition type, mediated by sulfur, and infer as highly improbable the alternative concerted mechanism mentioned earlier. In this case the zwitterion **446** is proposed as the intermediate in the reaction.[142]

Scheme 42

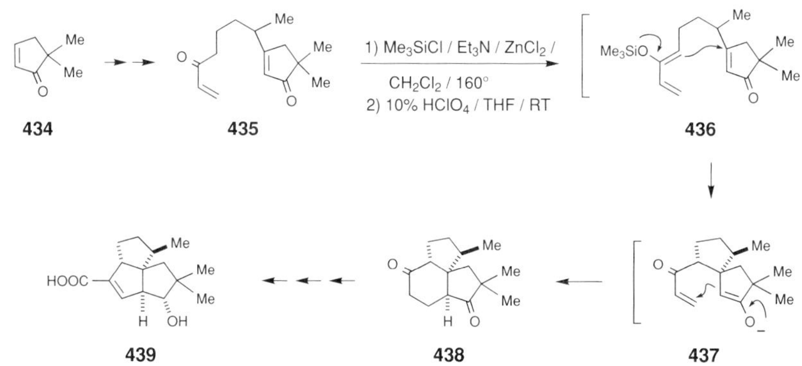

Scheme 43

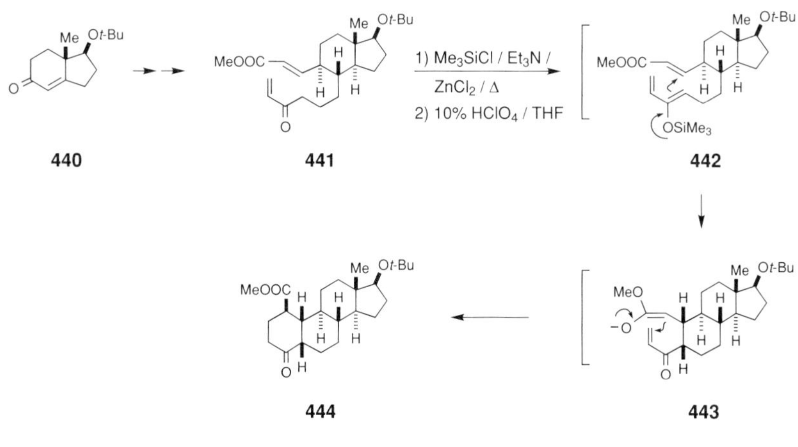

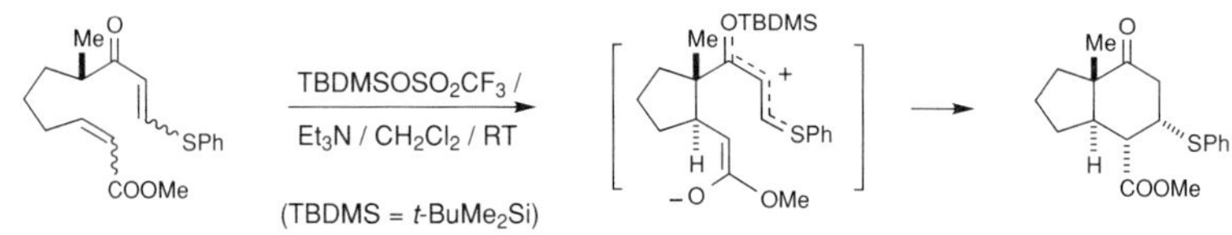

445 **446** **447** (63%)

Equation 37

A further example of the strategic use of the double Michael addition is in the construction of the complex hexacyclic alkaloid, atisine **453**, which has been synthesized both in the racemic[143] and enantiomeric forms,[144] using the reaction in the key construction steps (**449** → **450** → **451** → **452**), as illustrated in Scheme 44. The reaction plays a similarly important role in the synthesis of (–)-tylophorine **459**, in which the double addition is triggered by the nitrogen nucleophile as depicted in Scheme 45. The reaction is carried out by treating ester amide **456**, bearing a chiral auxiliary on the ester group, with standard reagents for effecting the double Michael addition [*t*-$BuMe_2SiOSO_2CF_3$ ($TBDMSOSO_2CF_3$) / Et_3N / CH_2Cl_2] at –78°C, and raising the temperature to 5° for 2 h, upon which reaction is completed (*via* transition state **457**) to give a single isomer of substituted indolizidone **458** in 78% yield.[145] (–)-Tylophorine is easily accessible from compound **458**.

Scheme 44

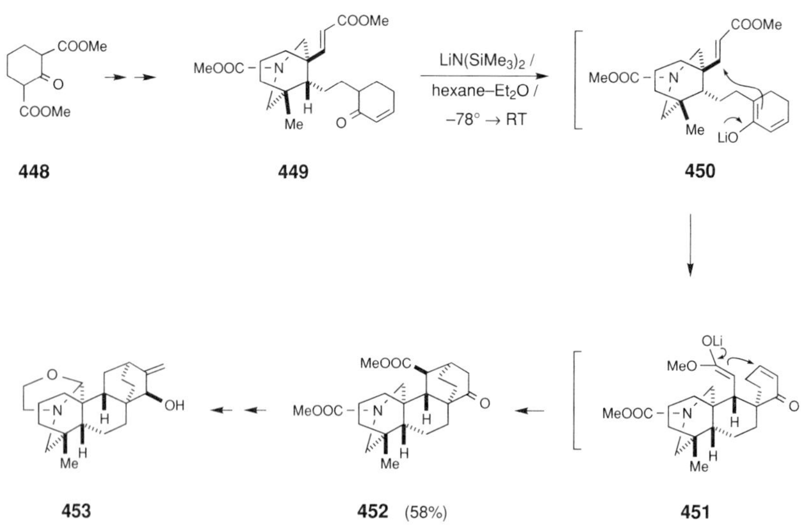

Scheme 45

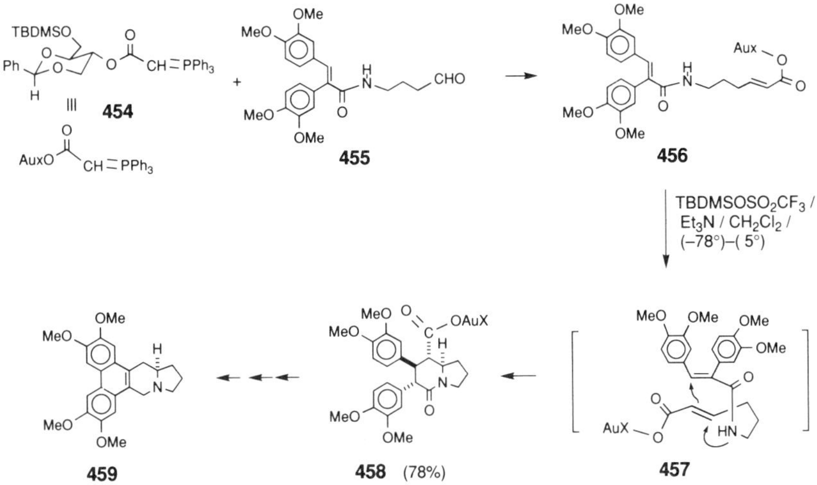

Higher orders of MIRC are also known. The examples shown in Scheme 46[146] and Equation 38[147] both represent Michael–Michael–S_N2 sequences. The former involves specific generation of enolate **462**, which reacts with bromoacrylate **461**, the sequence then terminating in bromine displacement to deliver tricyclic product **465** in moderate yield. The latter reaction in Equation 38 exemplifies a double use of the nitro group (in **467**), first as the activating group for the initial conjugate addition, then as the leaving group in the final cyclization. Such cyclizations *via* displacement of the nitro group are quite well documented.[148,149]

Scheme 46

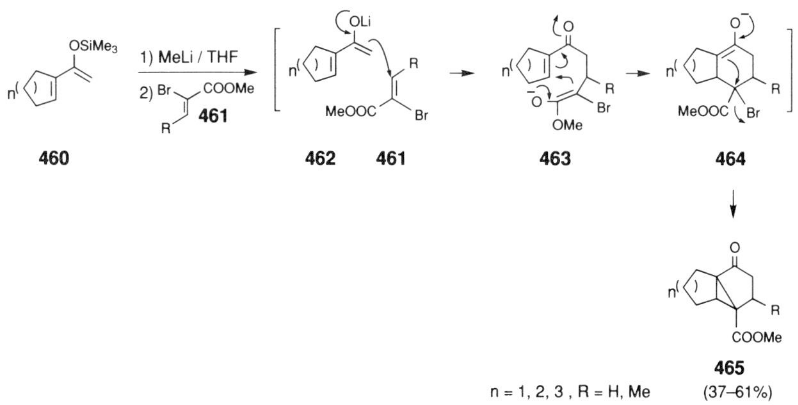

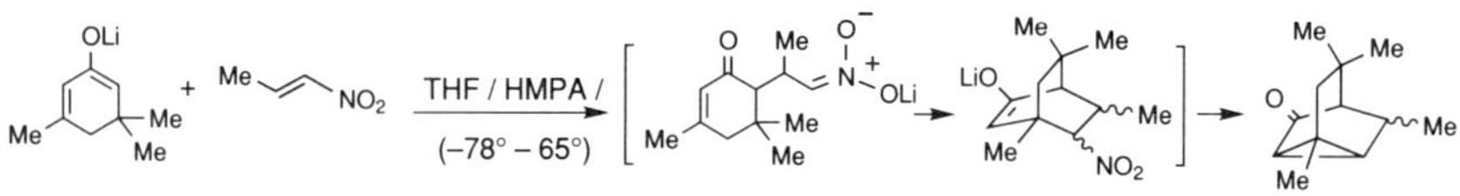

466 **467** **468** **469** **470**
(63% ; ratio 12 :1)

Equation 38

One-pot multiple Michael additions can create several bonds in one operation and therefore achieve the construction of more complex molecules. Triple and quadruple Michael additions, for example, have been employed in cyclizations that produce three and four new bonds, respectively. Apart from the reactions already documented in the earlier review,[1] recent examples include the one-pot triple Michael addition reaction to yield stereospecific product **473** which conveniently leads to the natural product seychellene **474** (Equation 39);[150,151] the three-component triple addition to yield bicyclic ketone **478** (Equation 40);[152] and the quadruple Michael addition (Equation 41) involving one molecule of nucleophile **479** and two molecules of *para*-quinone **480**.[153] The numbers depicted on the bonds in the products (Equations 39 to 41) indicate the consecutive order of their formation.

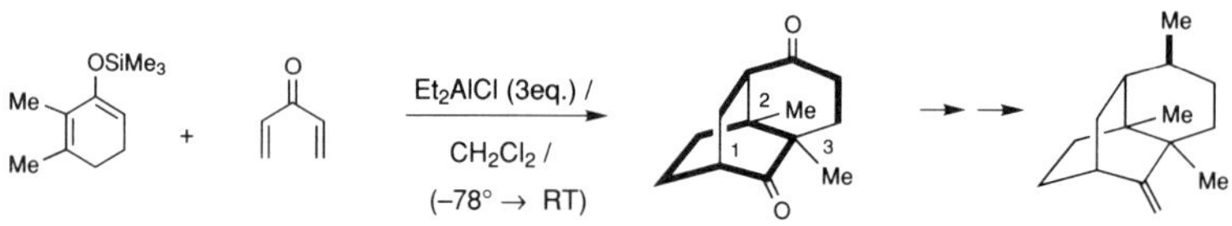

471 **472** **473** (43%) **474**

Equation 39

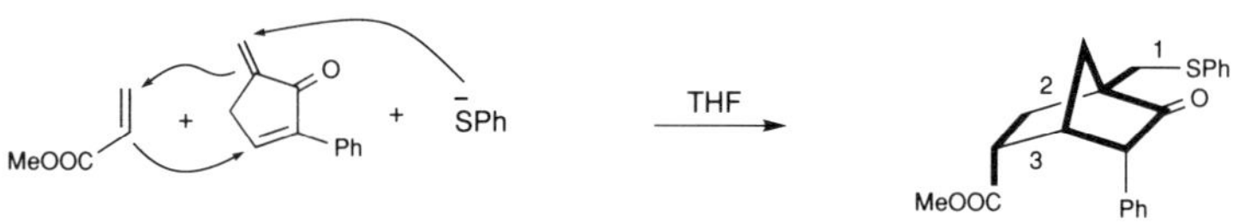

475 **476** **477** **478** (68%)

Equation 40

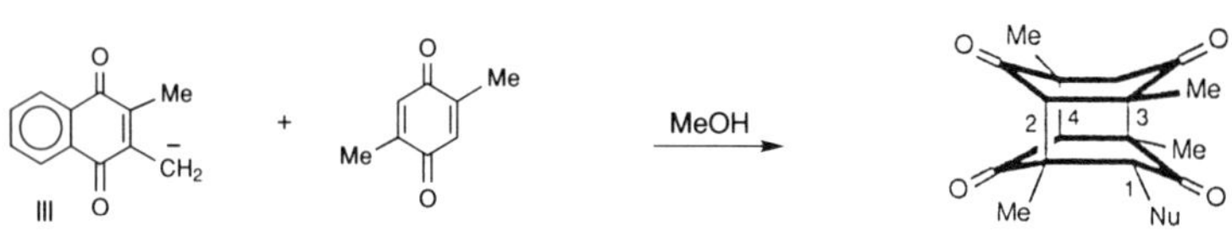

Nu⁻ **479** **480** 481 (20%)

Equation 41

5. Michael-Initiated Cyclopropanation

As outlined in Scheme 33 and discussed earlier, the incorporation of a leaving group into a nucleophile in an MIRC reaction is successful only when the leaving group is placed directly on the carbanionic center (*e.g.*, **325**) or when the carbanion stabilizing group itself doubly functions as a leaving group (*e.g.*, **328**). Such a design of nucleophile dictates the formation of a three-membered ring, as illustrated once again below.

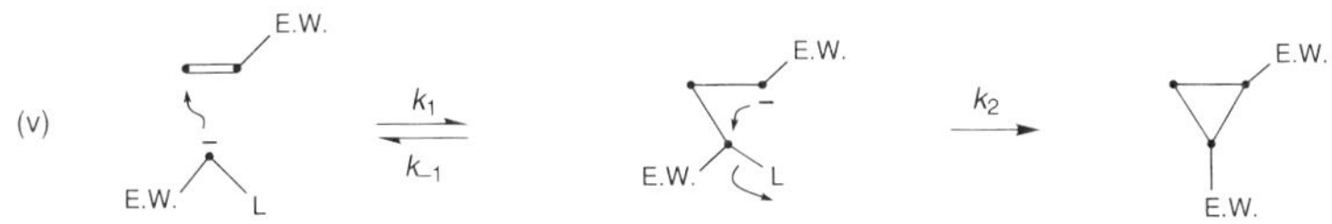

325 **326** **327**

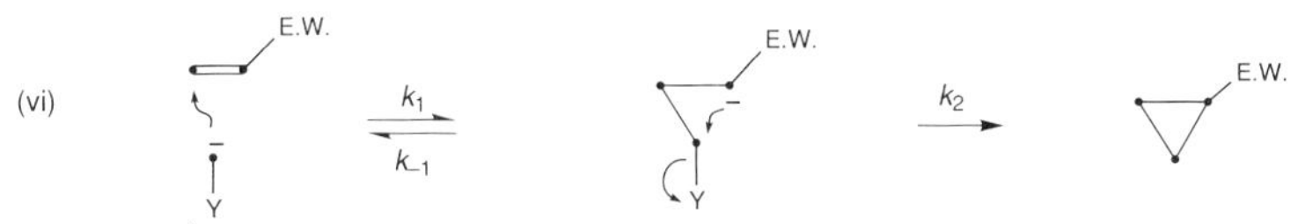

328 **329** **330**

E.W. = electron withdrawing group
L = leaving group
Y = electron withdrawing–leaving group

The Michael-initiated cyclopropanation reaction is enjoying increasing usage with the introduction of innovative substrates which allows easy preparation of cyclopropanes. Equations 42[154] and 43[155] exemplify the preparation of simple cyclopropanes, while application of the reaction to more complex substrates are shown in Equations 44[156] and 45.[157]

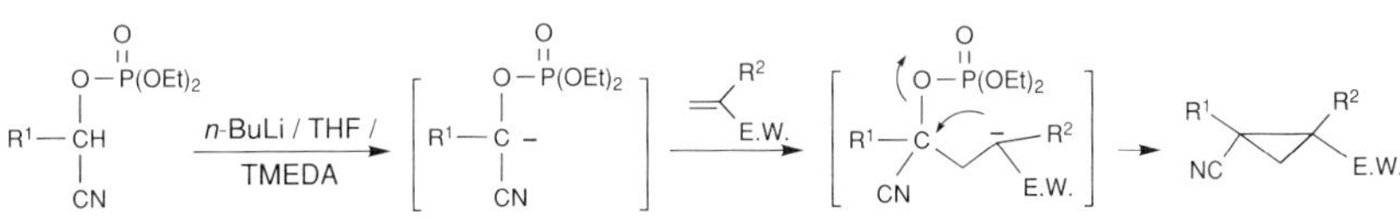

482 **483** **484** **485**

E. W. = COOEt , CN , R^1 = Ar, CH=CH–R , R^2 = H , Me (40–85%)

Equation 42

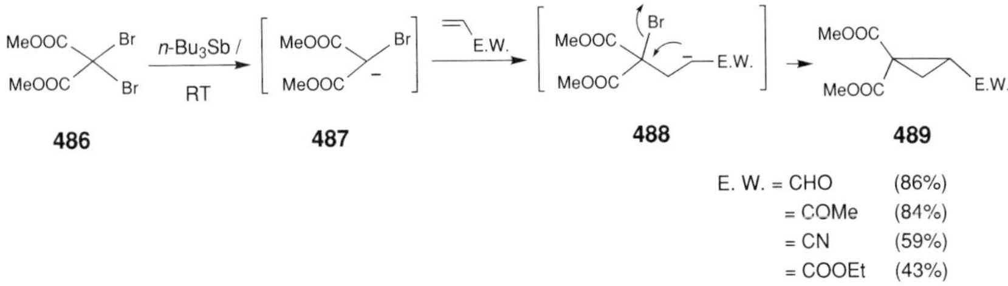

Equation 43

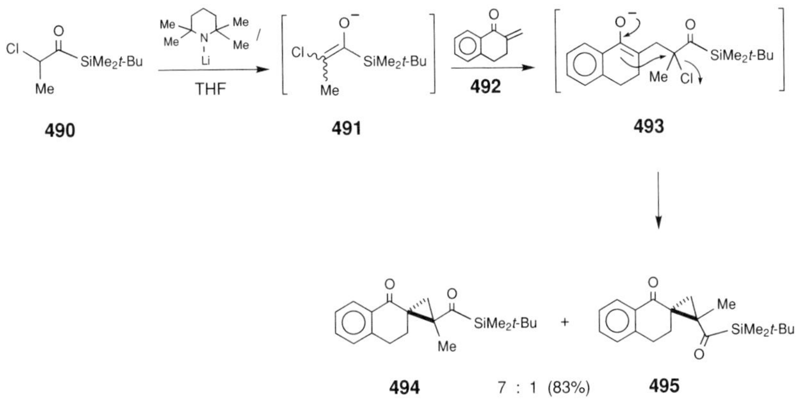

Equation 44

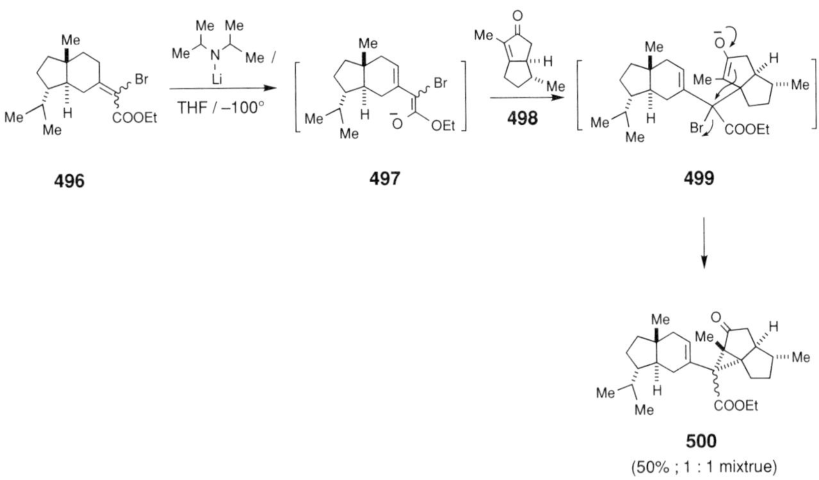

Equation 45

The use of an electron stabilizing group which also serves as a leaving group has long been practised in cyclopropanation chemistry. This tactic is the essence of the Michael-initiated cyclopropanation of type vi in Scheme 33.

Applications using the nitro group are shown in Equations 46[158] and 47.[159]

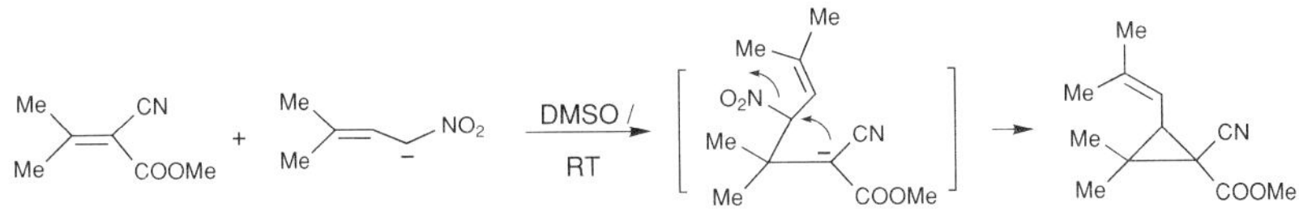

501 **502** **503** **504** (86%)

Equation 46

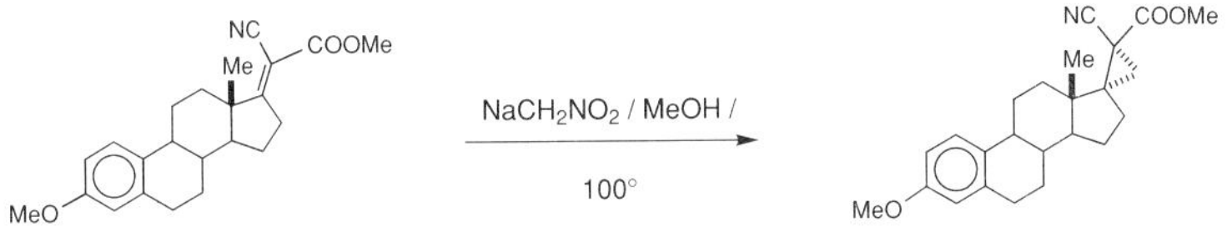

505 **506**

Equation 47

Dimethyloxosulfonium methylide **508**,[160] a long-established reagent for the cyclopropanation of activated alkenes *via* consecutive Michael addition–displacement of dimethylsulfoxide, is employed in the asymmetric cyclopropanation of optically pure methyl α-(*p*-tolylsulfinyl) acrylate **507** (Equation 48).[161] Purification of the crude product yields **510** (82% with de = 98%) and a small amount of minor product **511** (14%). The formation of cyclopropane **510** probably takes place *via* a stereospecific attack of methylide **508** on sulfinyl acrylate **507** to give zwitterionic intermediate **509** as shown, with subsequent elimination of dimethylsulfoxide occurring on the lone-pair-electron side of the *p*-toluenesulfinyl moiety.

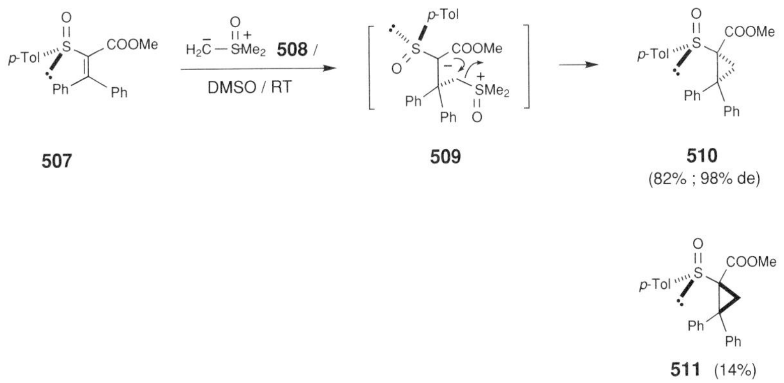

Equation 48

On the other hand, the use of optically active lactone **512** as the chiral Michael acceptor with dimethyloxosulfonium methylide as the cyclopropanating agent does not give the expected result. That is, the reaction produces as the major product cyclopropane **515** from a sterically unfavorable *β*-attack of ylide (**513** in Scheme 47) instead of the isomer **516** from *α*-attack, in an approximate 2:1 to 3:1 ratio. This stereoselectivity is further enhanced with the use of [(diethylamino)phenyl]oxosulfonium methylide.[162] It is quite plausible that the reaction proceeds mostly *via* transition state **514a** because of *π* interaction between the phenyl substituent in the substrate and the incoming ylide, which directs Michael addition on to the *β*-side of the molecule as shown. This reasoning is supported by the result from cyclopropanation with diazomethane, in which the ratio of products is reversed, with **516** becoming the major product as shown in the accompanying table in Scheme 47.[163]

Scheme 47

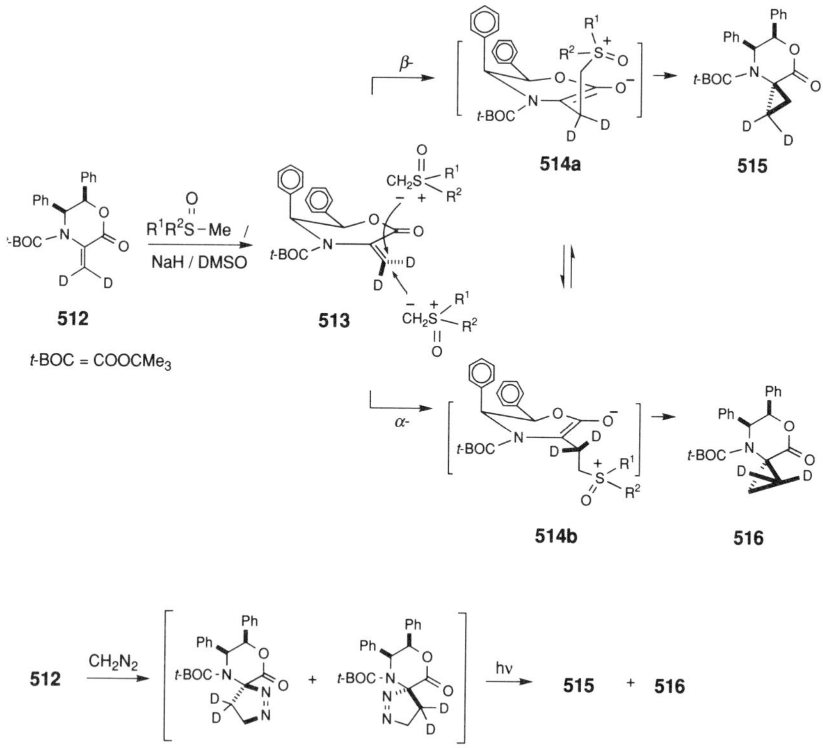

517 **518**

cyclopropanation agent	ratio of **515** : **516**	yield %
CH_2N_2	1 : 2–3	100
$\bar{C}H_2-\overset{+}{S}(O)Me_2$ (from $CH_3-\overset{+}{S}(O)Me_2$ I^-)	2–3 : 1	98
$\bar{C}H_2-\overset{+}{S}(O)(NEt_2)-Ph$ (from $CH_3-\overset{+}{S}(O)(NEt_2)-Ph$ BF_4^-)	11 : 1	97

V. CYCLIZATION OF ALLYLSILANES AND ALLYLSTANNANES

The cyclization of allylsilanes and related reactions are well documented and reports up to the year 1990 have been reviewed.[3–5,9] In general, the cyclization of allylsilanes can be summarized as presented in Scheme 48. Most frequently the reaction is catalyzed by a Lewis acid, and the cyclization center may involve either the α- or γ-carbon of the silane moiety, with the most commonly employed electrophile being the carbonyl or the protected carbonyl. It should be pointed out that the intermediate allyl anion **520** in Scheme 48 is drawn for the purposes of illustration only, and is not intended to imply a true reaction intermediate. Allylsilanes also undergo conjugate addition reactions, therefore in a situation such as **521** it is possible for several products to be formed. This point is well illustrated in the reactions shown in Equations 49 and 50, reported since 1985,[164,165] which display total dependence of products upon the catalyst employed.

The strange failure of allylsilane **534**, derived from allylic alcohol **531**, to cyclize with Lewis acids such as $TiCl_4$ and $EtAlCl_2$, or with the use of Bu_4NF, was recently reported.[166] All attempts failed to deliver the cyclized product **536**, producing only the simple desilylation product **535**. Yet treatment of the parent alcohol **531** itself with *p*-toluenesulfonic acid in benzene smoothly effected cyclization to yield the spirocompound **533** as a mixture of diastereoisomers (Scheme 49).

Scheme 48

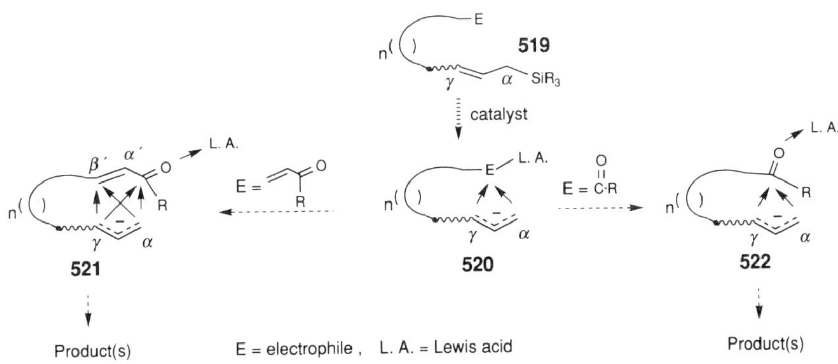

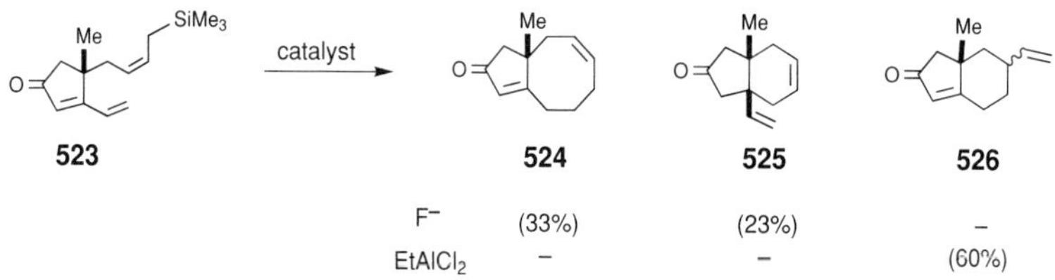

Equation 49

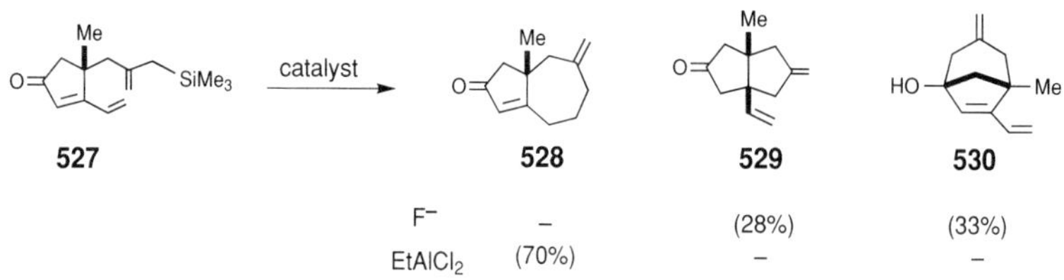

Equation 50

Scheme 49

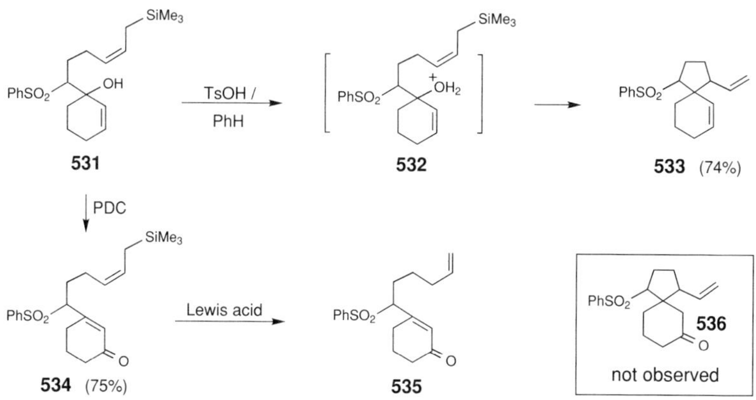

Exploitation of allylsilane cyclization in natural product synthesis is not uncommon. For example, the cyclization of silane **537** is used to construct the spirostructure **538** (obtained along with **539** and **540**) which can be converted easily to α-acoradiene **541**.[167] Similarly, the group of Majetich *et al.*[168] have applied their cumulative experience in allylsilane cyclization to the regio- and stereospecific synthesis of tricyclic enone **543**, which is obtained in excellent yield from starting silane **542** using ethylaluminum dichloride as the catalyst (Equation 52). This is in fact an application of the reaction shown in Equation 50. The product **543** has been converted to the natural product 14-deoxyisoamijiol **544**.

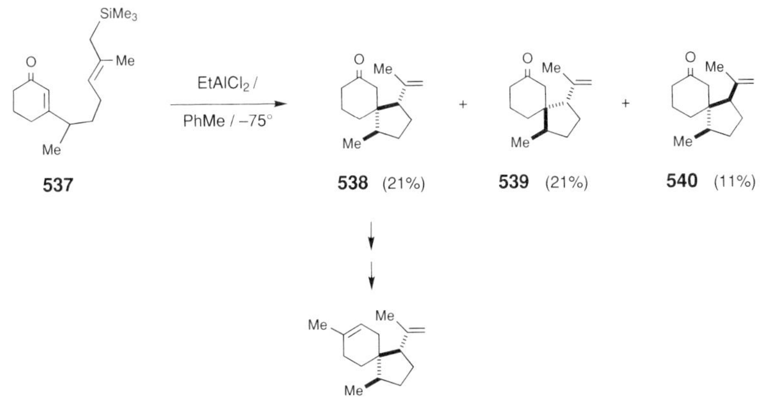

537 **538** (21%) **539** (21%) **540** (11%) **541**

Equation 51

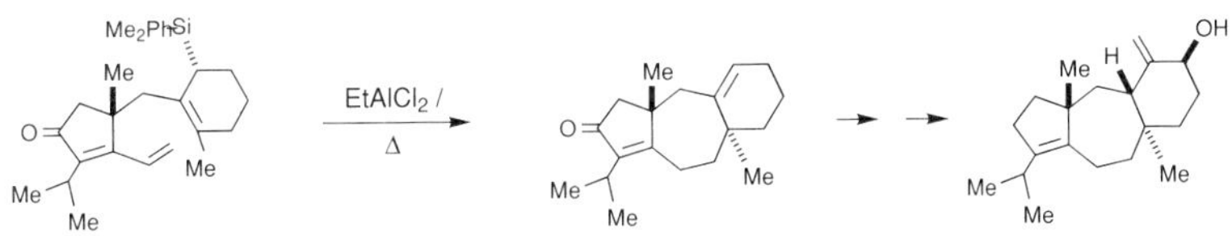

542 **543** (95%) **544**

Equation 52

Allylsilane incorporated with a protected aldehyde group (**545**) constitutes an excellent bifunctional annulating agent.[169,170] Although longer chains in **545** (*e.g.*, n ≥ 3) are vulnerable to intramolecular cyclization to **546**,[171] shorter chains (n = 1, 2) are much less likely to "self destruct" in this manner, and are amenable to Lewis acid-catalyzed reaction with silylnol ether **547** to give **548**, which subsequently cyclizes *via* allylsilane–aldol condensation to yield the bicyclic alcohol **549**.[169] This concept has been applied to the synthesis of eight- and nine-membered ring compounds **554**[172] starting from **550** and **551**, as summarized in Scheme 50.

Scheme 50

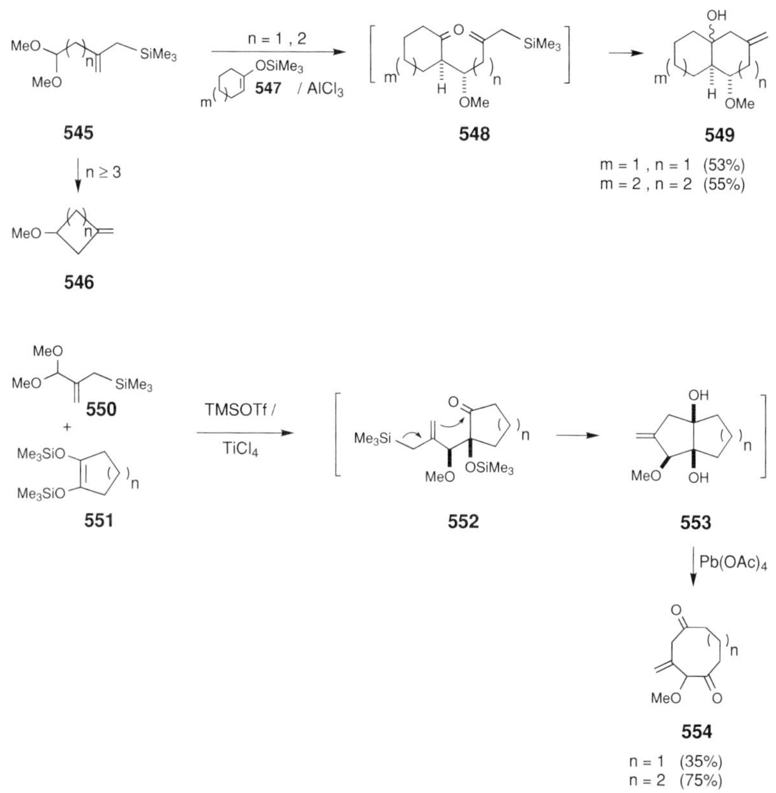

Intramolecular cyclization between the allylsilane and aldehyde functionalities in **555** yields *cis*-hydroxy ester **556** when n = 1, and a mixture of *cis*- (major product) and *trans*-esters when n = 2. Lactonization of **556** can be accomplished by heating with *p*-toluenesulfonic acid in benzene to give lactone **557**. Experiments using a chiral auxiliary R^* on the ester group in substrate **555** show good to excellent chiral induction when R^* = (–)-8-phenylmenthyl, but no induction when R^* = (–)-menthyl or (–)-bornyl. The overall yields and ee values obtained for α-methylene lactone **557** are given in Scheme 51.[173]

Scheme 51

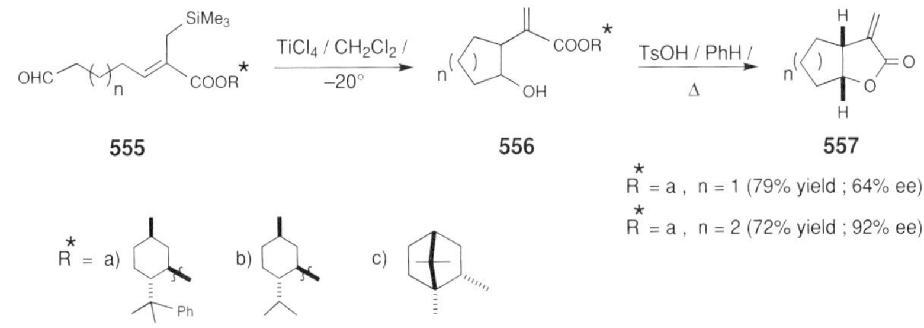

Allylsilane cyclization also provides a good approach to the synthesis of macrocyclic ethers. An example is the Lewis acid-catalyzed cyclization of **558** to an eight- or nine-membered ring ether **560** or **561**, respectively (Equation 53).[166] Despite the modest reaction yields and unclear reason for the observed stereochemical control, the simple procedures involved make the reaction an attractive candidate in the consideration of such a total synthesis.

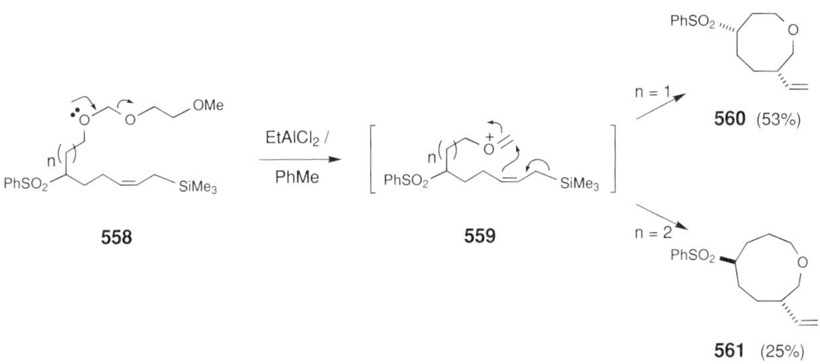

Equation 53

The concentrated efforts to synthesize cyclic ethers are attributable to the discovery of several marine toxins whose structures are composed of fused polyethers of various ring sizes.[174] In this endeavor the allylsilane and allylstannane cyclizations play an important part. One such study is illustrated in Scheme 52. The cyclization of cyclic acetal **562** using $TiCl_3(Oi\text{-}Pr)$ as catalyst yields a mixture of *cis*-(**564**) and *trans*-(**563**) cyclized products when n = 1, but only the *trans*-product when n = 2, and a bare 8% yield of the expected eight-membered ring products when n = 3. However, incorporation of a cyclohexane unit into the chain results in an improved yield of eight-membered ring ether, cyclic ether **566** being obtained in 36% from substrate **565**. The more rigid cyclic hexane structure probably reduces chain flexibility, thereby facilitating cyclization (Scheme 52).[175]

Scheme 52

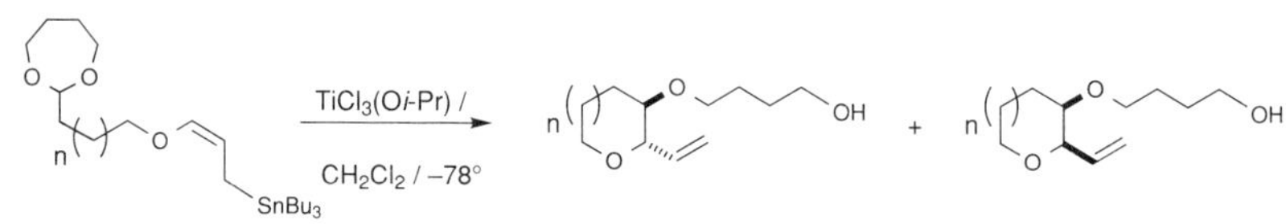

562 **563** **564**

	563		564
n = 1	(65%)		(21%)
n = 2	(59%)		–
n = 3	(trace)	(≈ 8%)	(trace)

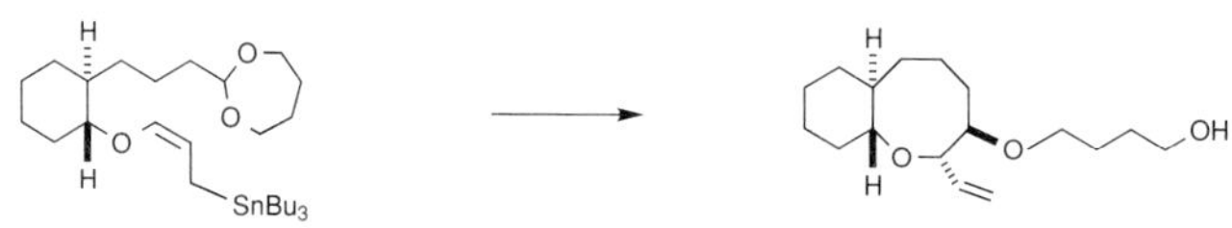

565 **566** (36%)

It turns out that intramolecular cyclization of allylstannanes proceeds better with the unprotected aldehyde function than with its corresponding acetal. For instance, the boron trifluoride etherate-catalyzed cyclization of allylstannanes with an intramolecular aldehyde function to form seven-membered ring ethers (*e.g.*, **567** → **568**) is almost quantitative and completely stereospecific in affording only the *trans*-isomer.[176] This finding has led to the total synthesis of polycyclic ether **573**, as depicted in Scheme 53.[177] An exceptionally beautiful application of the reaction is displayed in this synthesis: stereospecific cyclization to a seven-membered ring ether in near-quantitative yield is exercised twice, first in the cyclization of monocycle **567** to bicycle **568**, and later in the cyclization of bicycle **569** to tricycle **570**, while cyclization to a six-membered ring ether is executed in the final ring closure of tricycle **571** to yield a mixture of tetracycles **572** and **573** in a 1:1 ratio (92%).

The conspicuous nonstereoselectivity of the six-membered ring cyclization in the above synthesis, in complete contrast to the stereospecific seven-membered ring formation, obviously needs an explanation. Using a simple model **574** (Scheme 54), the reaction was investigated in terms of varying catalysts: boron trifluoride– and zinc chloride etherates both yield a mixture of *cis*- and *trans*-products, **575** and **576**, in a 2:1 ratio, but heating the substrate **574** at 120°C without catalyst results in a high *cis*-selectivity. These results have led to the proposal of transition state **577** for the heat-induced six-membered ring cyclization, while the corresponding seven-membered cyclization is thought to enjoy a more flexible geometry which allows the selective formation of the thermodynamically more stable *trans*-isomer.[177] However, more experimental results will be needed before anything more definite can be said concerning the stereochemistry of the products from allylstannane cyclizations.

Scheme 53

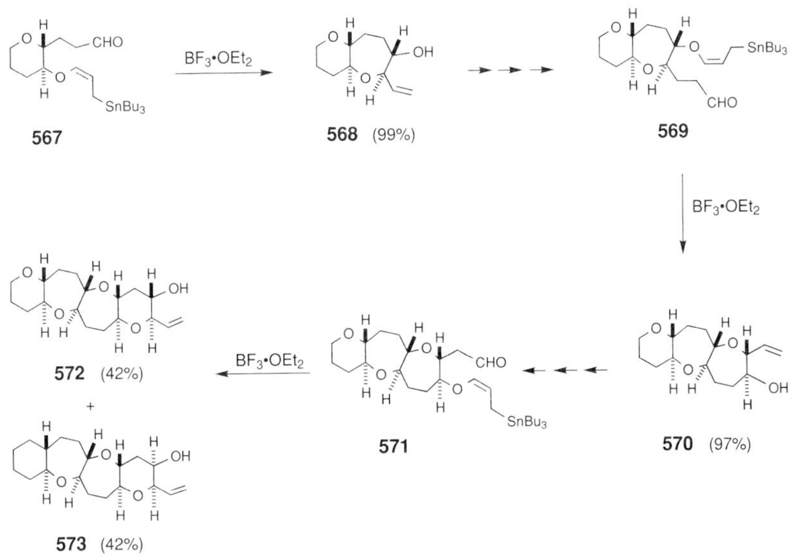

Scheme 54

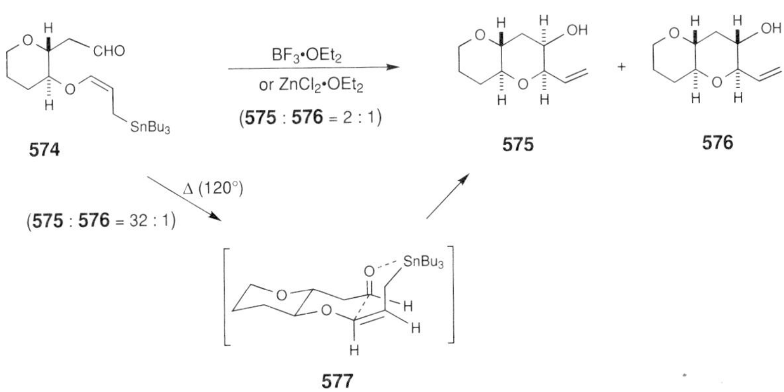

The last reaction to be discussed here is the intramolecular cyclization of allenylstannane with an aldehyde moiety as demonstrated by compound **578**, which yields cyclic homopropargylic alcohol **579** as a mixture of diastereoisomers (Scheme 55).[178] This mode of cyclization complements the earlier method of condensation of acetylide anion with aldehyde, which gives cyclic propargylic alcohol[179] as previously shown in Equation 11.[54] The homopropargylic alcohol **579** can be oxidized with Dess–Martin periodinane **580**[180] to yield a 1:3 mixture of acetylenic ketone **581** and allenone **582**, which mixture smoothly cyclizes to furan **583** upon exposure to $AgNO_3$ / $CaCO_3$. The furan structure **583** forms the main skeleton of a new type of diterpenoids, kallolides A and B, from marine sources.[181]

Scheme 55

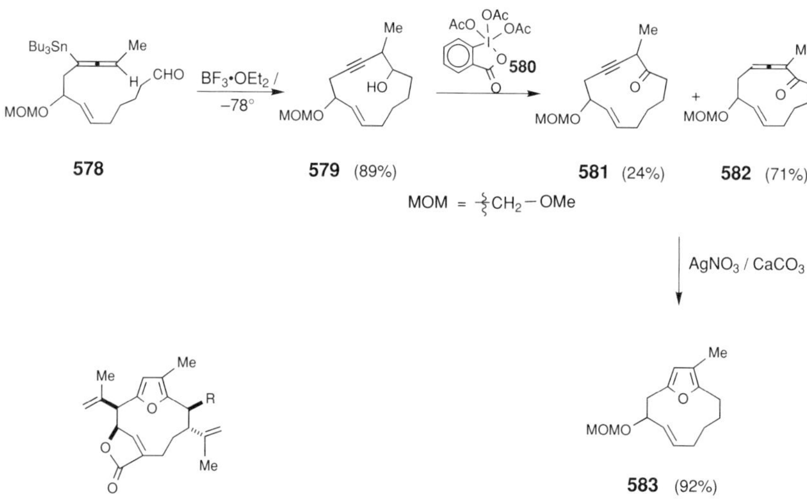

584 R = OH (kallolide A)
585 R = H (kallolide B)

VI. CYCLIZATION INVOLVING ADDITION TO UNACTIVATED OLEFINIC BONDS

Unlike its fully established cation and radical counterparts, the anionic cyclization to unactivated olefinic bonds is still in an early stage of development. Although a fair number of examples of such cyclizations can be found in the literature, they are scattered and unrelated. It will probably take some time for the reaction to reach the status enjoyed by the radical and cationic cyclizations, to be of use in the organic synthesis of complex molecules. Meanwhile, results from recent work certainly look promising for future application of the reaction.

The general reaction can be most simply represented as shown in Scheme 56, which depicts the generated carbanionic center in **586** reacting intramolecularly with the unactivated olefinic moiety in an exclusively *exo* fashion to give **587**. Electronically, the alternative *endo* cyclization to give **588** is very unfavorable, which explains why these reactions are all *exo* cyclizations. Obviously, the success of the reaction depends on the relative reactivity of the starting anionic center in **586** as compared with the resultant anion **587**. This, coupled with the fact that the olefinic bond in **586** is essentially unactivated (resulting in a reactive cyclized anion **587**), demands that the starting anion for this type of reaction be, quite compulsorily, highly reactive.

Scheme 56

588 **586** **587**

Several methods have been developed for the generation of various reactive anions, both alkyl[182–187] and aryl,[188–191] but these have rarely been applied to intramolecular cyclization with unactivated olefins. In fact the use of alkyl anions in such cyclizations was not reported until the second half of the last decade (since 1985), after it was discovered that 6-iodo-1-hexene **589** efficiently undergoes the metal–halogen exchange reaction with *tert*-butyllithium at –78°C,[192–194] in contrast to the slow reaction of the corresponding bromide and the almost inert response of the chloride. The primary alkenyllithium **590** thus formed is stable at low temperature and can be trapped with various electrophiles E^+, but on warming to room temperature it cyclizes to (cyclopentyl)methyllithium **591**, also identified by trapping with electrophiles, as shown in Scheme 57.[195]

Scheme 57

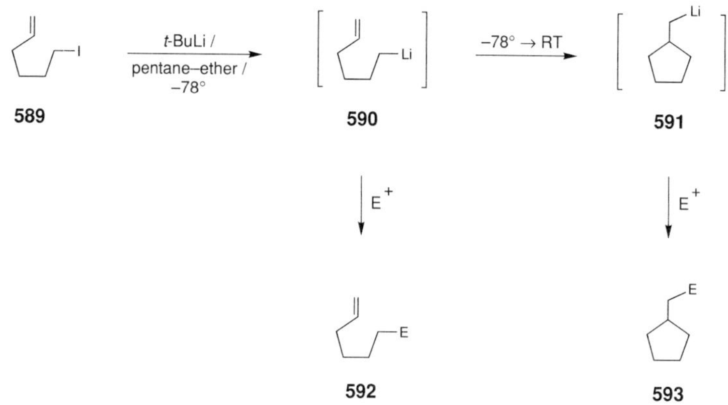

E+	E	592 (%)	593 (%)
MeOH	H	98	89
MeOD	D	87	67
CO_2	COOH	87	54
O_2	OH	78	78

As mentioned earlier, there are various points to be considered in the cyclization of hexenyllithium derivatives. Regarding the comparative stabilities of the starting vs. the cyclized anions, experimental results have borne out the requirement of a more reactive starting anion. For example, 5-hepten-1-yllithium **595**, generated from the corresponding iodide **594**, fails to give any cyclized product under standard cyclizing conditions as shown in Equation 54,[2] confirming that the primary carbanion **595** will not cyclize to the more reactive secondary anion **596**.

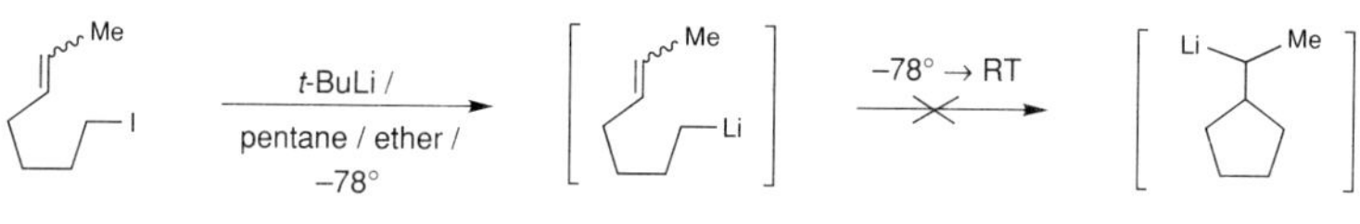

594 595 596

Equation 54

As regards the transition state of the reaction, the ratio of *cis-* and *trans-* products from the cyclization of hexenyl iodides **598**, **600**, and **601** provides experimental evidence for a chair-like conformation for the reaction, as depicted in **597** (Scheme 58),[2] similar to the transition state of the corresponding radical cyclization.[196]

Scheme 58

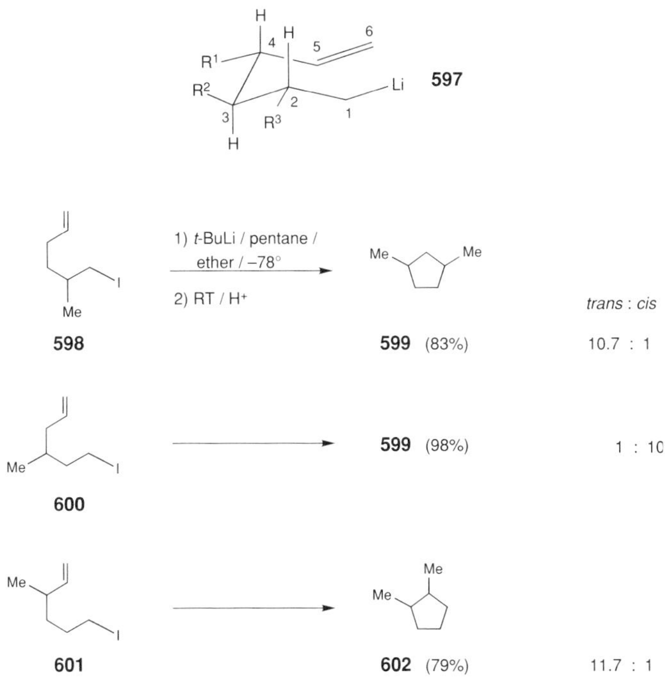

On the preference of the *exo*, over the *endo* mode of cyclization, several examples clearly demonstrate this partiality.[197,198] The synthesis of the naturally occurring hydrocarbon cuparene **606**, for one, nicely illustrates this cyclization behavior (Scheme 59).

Treatment of iodide **603** with 1.2 equivalents of *tert*-butyllithium in *n*-pentane–diethyl ether at –78°C followed by the addition of TMEDA and raising the reaction to room temperature directly affords **606** in 82% yield after quenching the reaction mixture with dry deoxygenated methanol. Besides furnishing the required natural product **606**, the reaction is very informative in many ways. First, it shows the cyclization of the anionic center to the olefinic bond to occur preferentially in the *exo* and not the *endo* mode, since **604** yields only **605** even though an *endo* cyclization would mean attack at the sterically less congested sp^2 carbon to yield the electronically favorable benzylic anion **607** (although, of course, anion **607** is tertiary while **605** is primary). Second, it was found that the reaction must be performed with only 1.2 equivalents of *tert*-butyllithium, and not the usual excess, as is the protocol with tandem metal–halogen exchange cyclizations.[195,197] Such an excess, in this case, resulted in products derived from dianion **608**. Therefore, unlike the intramolecular anionic addition, intermolecular attack of *tert*-butyllithium does occur at the primary carbon to give the tertiary benzylic anion. Last, these results contrast with that of the radical cyclization of starting material **603**, which yields **609** as the sole product from *endo* cyclization of the radical intermediate. Thus, the reaction provides an excellent example of how radical and anionic cyclizations can complement each other very conveniently.[198]

Scheme 59

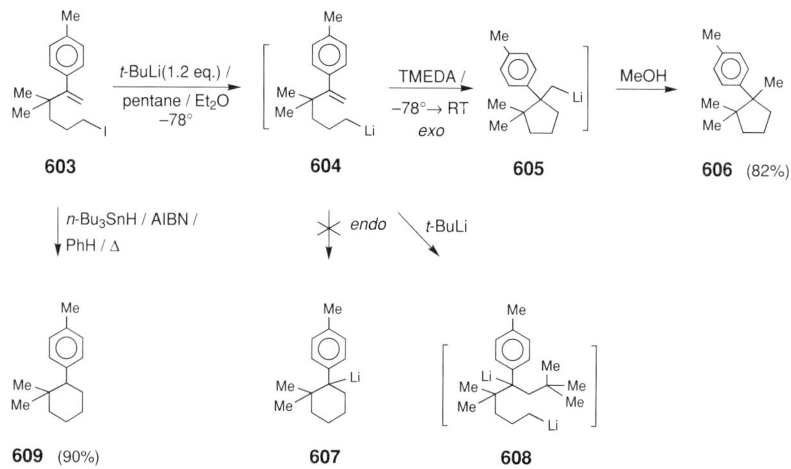

Is the cyclization of hexenyllithium a reversible reaction? It would appear sensible to presume that it is. However, experimental results indicate that this may not be the case. Take, for example, the tandem cyclization of iodide **610** which spectacularly delivers the strained *trans*-fused bicyclic compound **614** in 87% yield.[199] This reaction, rationalized as shown in Scheme 60, entails a first step of selective 5-*exo* cyclization (other possibilities, *e.g.*, 6-*exo* cyclization, are not observed) *via* transition state **611a**, as discussed earlier in Scheme 58, to give **612** which proceeds through a second cyclization to yield the observed product **614**. If it were reversible the reaction would be expected to reverse and recyclize to the thermodynamically much more favorable *cis*-fused product **617** *via* transition state **611b** and intermediates **615** and **616**. No trace of product **617** has been detected.

Scheme 60

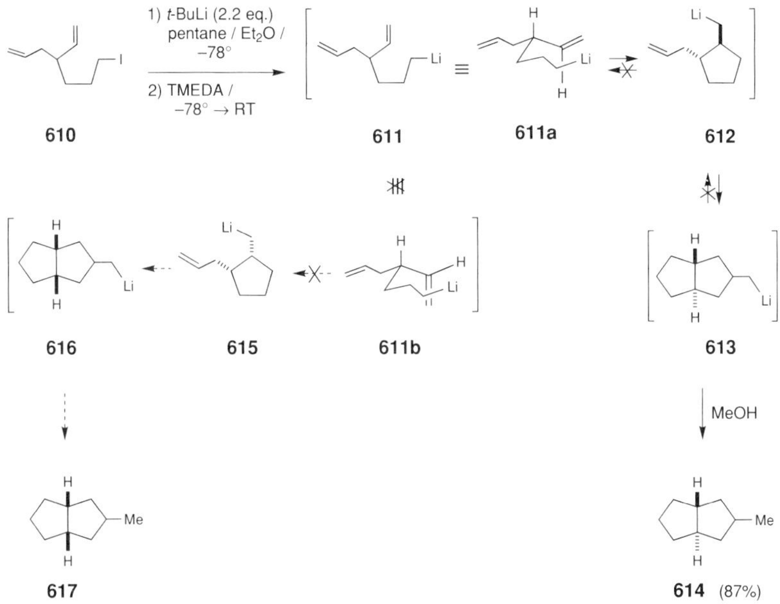

Further examples of tandem cyclization of the anionic center to unactivated olefinic bonds are seen in the cyclizations of **619** and **624**, generated from the corresponding iodides **618** and **623**, which give, respectively, spiro[4.4]nonane **622** and [4.3.3]propellane **627** in fairly good yields. The propellane anion can be trapped with various electrophiles, as shown in Scheme 61.[200]

Scheme 61

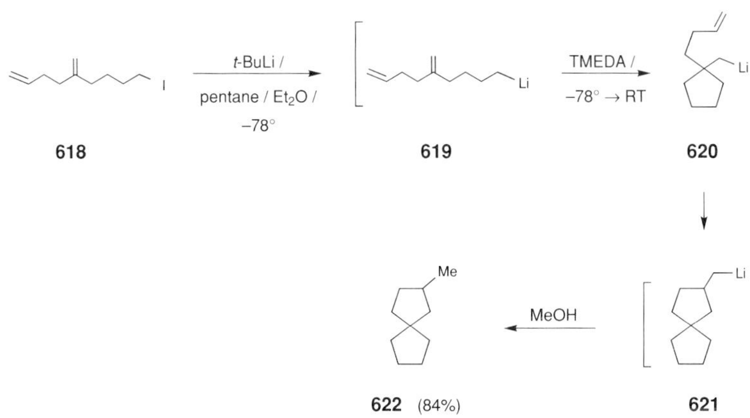

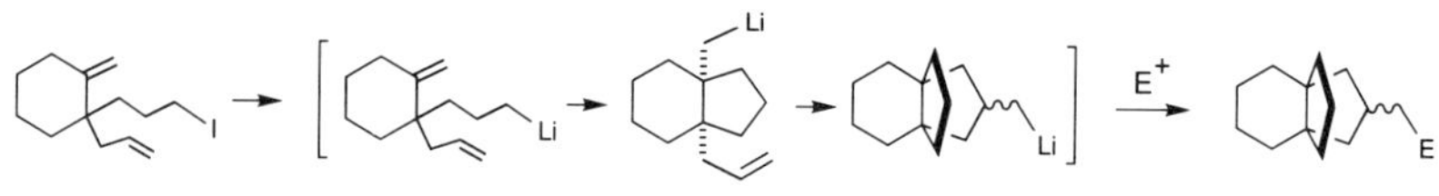

623 624 625 626 627

E^+	E	% yield
MeOH	H	81
MeOD	D	75
CO_2	COOH	60
PhCHO	CH(OH)Ph	80
MeCOMe	$C(OH)Me_2$	80
		82
Me_3SiCl	Me_3Si	73

Although the lithium–halogen exchange reaction appears to be the method of choice for the generation of sp^3 carbanions as aforesaid, other methods, for example the exchange reactions between lithium–selenium (Equation 55),[201] lithium–tin (Equation 56),[202] and lithium–sulfur (Equation 57),[203] are also employed to produce anions for subsequent cyclizations to unactivated olefinic bonds.

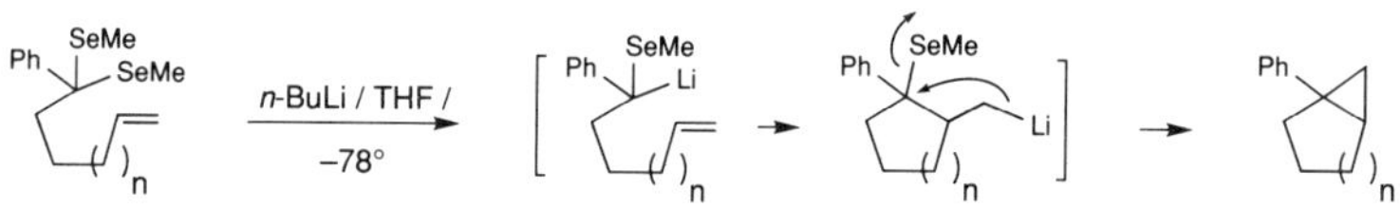

628 629 630 631

n = 1 (82%)
n = 2 (81%)

Equation 55

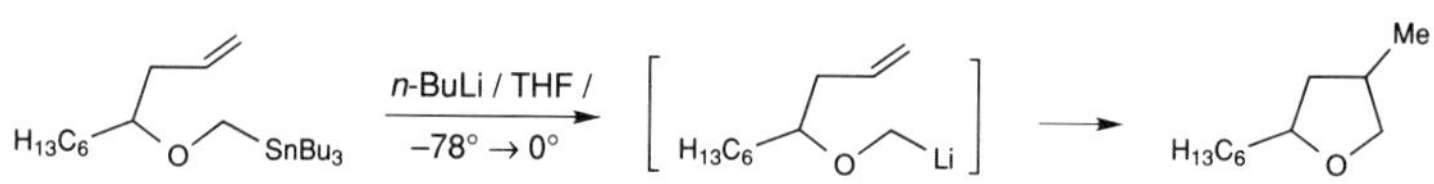

632 633 634

(54% ; *cis* : *trans* = 11 : 1)

Equation 56

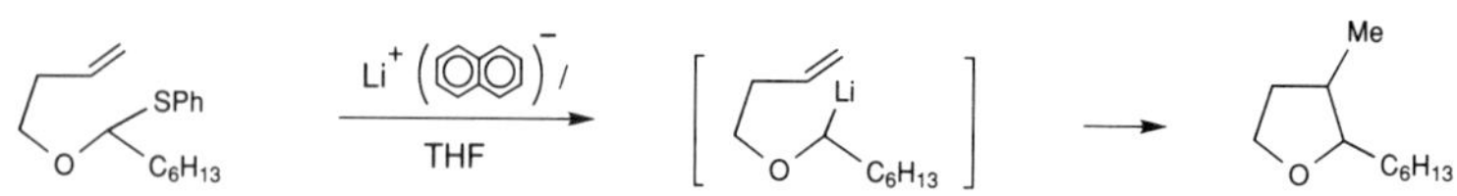

635 636 637

(52% ; *cis* : *trans* = 1 : 7)

Equation 57

With respect to the reactions of anions **633** and **636** in the above equations, an interesting turn of events is observed upon shortening the chain length by one carbon atom. That is, anion **639**, generated from **638**, instead of cyclizing, smoothly undergoes [2,3]-sigmatropic rearrangement to **640**, thence product **641** upon acidification (Equation 58).[204] Furthermore, transposing the oxygen atom along the chain to the vinyl ether position as in **643** also results in an open chain alcohol product, pentenol **645** (Equation 59).[205] Although anion **643** initially cyclizes in the usual 5-*exo* fashion, the cyclized intermediate subsequently ring opens to the alcohol. It should be noted that the reaction shown in Equation 58 is actually the well-known [2,3]-Wittig rearrangement, while the overall process in Equation 59 represents a [1,4]-Wittig rearrangement.

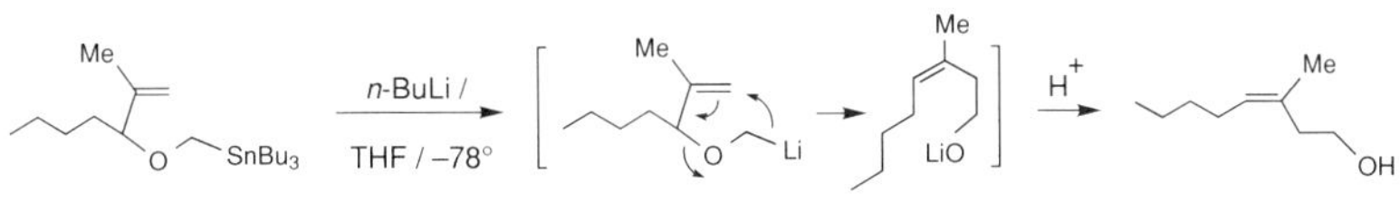

638 **639** **640** **641** (95%)

Equation 58

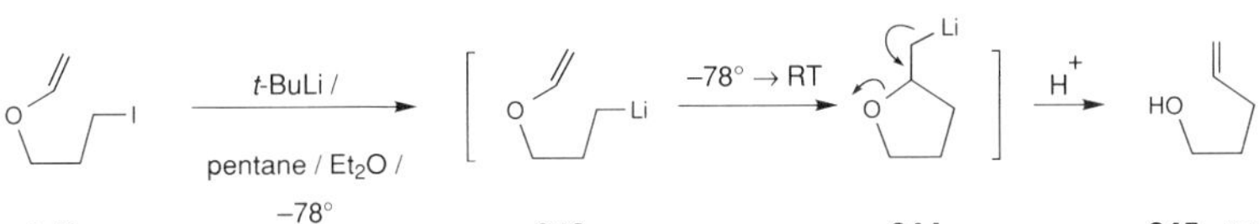

642 **643** **644** **645** (93%)

Equation 59

Cyclization to unactivated olefins is also observed with the vinyllithium reagent. For example, vinyl anion **650**, generated from the corresponding hydrazone **647** by the modified Shapiro reaction,[206] cyclizes to finally give the fused product **651**.[207] Significant features are that the reaction seems favorable only when n = 1 (yields being generally poor when n ≥ 2) and appears to operate quite well with open chain substrates *e.g.*, **653** in Scheme 62.

Scheme 62

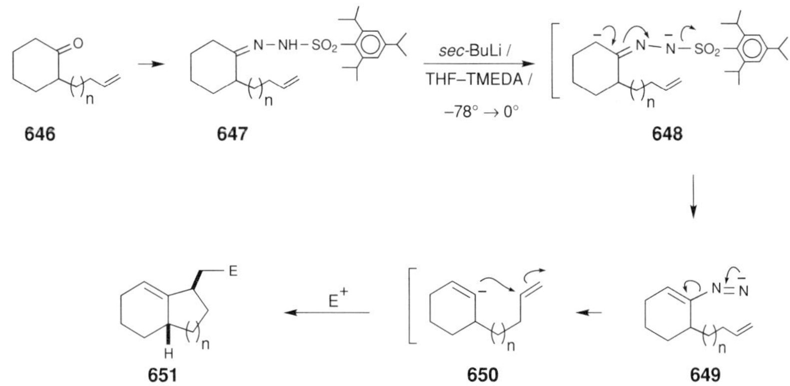

n = 1 , E = D (87%) , E = Br (81%) , E = COOH (50%)
n = 2 , E = H (< 20%)

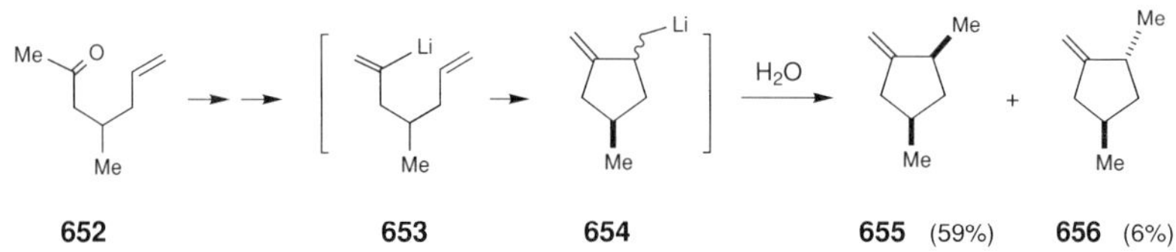

An unusual cyclization that makes interesting mention is the reaction of 6-chloro-5-hexen-1-yllithium **658**. Generated as an isomeric mixture from iodide **657** by the standard method at –78°C and then allowed to warm to room temperature, chlorohexenyllithium **658** curiously yields bicyclo[3.1.0]hexane **661** and methyleneycyclopentane **662** as products (Equation 60).[208] These compounds apparently arise from the carbenoid intermediate **660** formed upon cyclization of **658**.

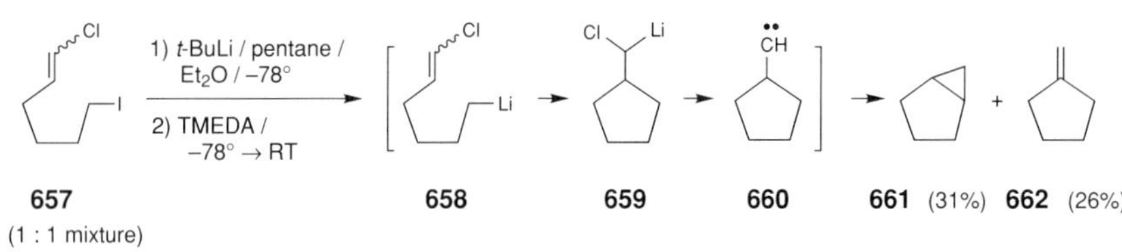

Equation 60

VII. CONCLUSION

Not withstanding the newly developed radical cyclizations discussed in the previous chapter, it would seem that the conventional anionic cyclization is still, so to speak, the bread and butter of synthetic chemists in the business of ring construction. Development in this area is also fast taking place, particularly strong quarters to watch out for being tandem cyclization in the construction of polycyclic compounds and the useful ortho-metalation of arene rings and their subsequent reactions.

Advances frequently being reported encompass such topics as the annulation reaction involving anionic species derived from 1,3-dithiane or ketenedithioacetal derivatives,[209,210] the anionic cyclization reaction leading to furan-3(2*H*)-ones,[211,212] improved procedures for the preparation of substituted tetrahydrofurans,[213,214] lactols,[215,216] and lactones,[217] the application of anionic cyclization as key steps in the synthesis of natural molecules, *e.g.*, iridomyrmecin,[218] and the model synthesis of analogs of taxol from pinnene,[220] *etc*. Lastly, the introduction of various bases and stabilizing groups has enabled the generation of hitherto inaccessible anions and thus opened up new possibilities in the design of anionic cyclization for the construction of ever more challenging molecular frameworks.

REFERENCES

1. Posner, G. H., Multicomponent one-pot annulations forming three to six bonds, *Chem. Rev.*, 86, 831, 1986.
2. Bailey, W. F., Khanolkar, A. D., Gavaskar K., Ovaska, T. V., Rossi, K., Thiel, Y., and Wiberg, K. B., Stereoselectivity of cyclization of substituted 5-hexen-1-yllithiums: regiospecific and highly stereoselective insertion of unactivated alkene into a C–Li bond, *J. Am. Chem. Soc.*, 113, 5720, 1991.
3. Organosilicon chemistry in organic synthesis, Fleming, I., Ed., Tetrahedron Symposia-in-Print no. 32, *Tetrahedron*, 44, 3761, 1988.
4. Fleming, I., Dunoguès, J., and Smithers, R., The electrophilic substitution of allylsilanes and vinylsilanes, *Org. React.*, 37, 57, 1989.
5. Schinzer, D., Intramolecular addition reactions of allylic and propargylic silanes, *Synthesis*, 263, 1988.
6. Snieckus, V., Directed *ortho* metalation. Tertiary amide and *O*-carbamate directors in synthetic strategies for polysubstituted aromatics, *Chem. Rev.*, 90, 879, 1990.
7. Gschwend, H. W., and Rodriguez, H. R., Heteroatom-facilitated lithiations, *Org. React.*, 26, 1, 1979.
8. Thompson, C. M., and Green, D. L. C., Recent advances in dianion chemistry, *Tetrahedron*, 47, 4223, 1991.
9. Thebtaranonth, C., and Thebtaranonth, Y., Developments in cyclisation reactions, *Tetrahedron*, 46, 1385, 1990.
10. Hudlicky, T., and Price, J. D., Anionic approaches to the construction of cyclopentanoids, *Chem. Rev.*, 89, 1467, 1989.
11. Swindell, C. S., Taxane diterpene synthesis strategies. A review, *Org. Prep. Proc. Int.*, 23, 467, 1991.
12. Tius, M. A., and Reddy, N. K., Cembrane synthesis. An advanced intermediate for crassin acetate, *Tetrahedron Lett.*, 32, 3605, 1991.
13. Dauben, W. G., Wang, T., and Stephens, R. W., Total synthesis of (±)-crassin acetate methyl ether, *Tetrahedron Lett.*, 31, 2393, 1990.
14. McMurry, J. E., and Dushin, R. G., Total synthesis of (±)-isolobophytolide and (±)-crassin by titanium-induced carbonyl coupling, *J. Am. Chem. Soc.*, 112, 6942, 1990.
15. Danishefsky, S., and Kahn, M., Regiospecific Michael reactions to an enedione, *Tetrahedron Lett.*, 22, 485, 1981.
16. Siwapinyoyos, T., and Thebtaranonth, Y., Regiospecific Michael additions to α-methylene cyclopentenones, *Tetrahedron Lett.*, 25, 353, 1984.
17. Piers, E., and Renaud, J., Annulations *via* bifunctional reagents. Total syntheses of (±)-methyl cantabrenonate and (±)-methyl epoxycantabronate, *Synthesis*, 74, 1992.
18. Plamondon, J., and Canonne, P., Synthesis of (±)-α-chamigrene, *Tetrahedron Lett.*, 32, 589, 1991.
19. Kim, D., and Lee, Y. K., Stereoselective synthesis of a steroidal *trans*-hydrindanone synthon from (*R*)-limonene, *Tetrahedron Lett.*, 32, 6885, 1991.
20. Kim, D., Kim, S., Lee, J. J., and Kim, H. S., Stereoselective construction of steroidal *trans*-hydrindanes *via* intramolecular ester enolate alkylation, *Tetrahedron Lett.*, 31, 4027, 1990.
21. Piers, E., and Roberge, J. Y., Use of the Me_3Sn function as a readily removable 'anchoring' group in the synthesis of enantiomerically pure bicyclic alcohols and ketones, *Tetrahedron Lett.*, 32, 5219, 1991.
22. Piers, E., and Gavai, A. V., A (*Z*)-ethylidenecyclopentane annulation method. Total syntheses of (±)-anhydrooplopanone, (±)-oplopanone, and (±)-8-*epi*-oplopanone, *J. Org. Chem.*, 55, 2380, 1990.

23. Duhamel, L., Peschard, O., and Plé, G., Cyclization of methyl α-bromocrotonate to a pentasubstituted cyclopropane, *Tetrahedron Lett.*, 32, 4695, 1991.
24. Dauzonne, D., Josien, H., and Demerseman, P., A general synthesis of 1-nitro-2-phenyl-4-oxospiro[2.5]octanes, *Synthesis*, 309, 1992.
25. Nishiyama, H., Isaka, K., Itoh, K., Ohno, K., Nagase, H., Matsumoto, K., and Yoshiwara, H., Metal–halogen exchange between polybromoanisoles and aliphatic Grignard reagents: a synthesis of cyclopenta[*b*]benzofurans, *J. Org. Chem.*, 57, 407, 1992.
26. Trost, B. M., and Verhoeven, T. R., Organopalladium compounds in organic synthesis and in catalysis, In *Comprehensive Organometallic Chemistry*, Vol. 8, Wilkinson, G., Ed., Pergamon Press, Oxford, 1982, 799.
27. Bäckvall, J.-E., Nyström, J.-E., and Nordberg, R. E., Stereo- and regioselective palladium-catalyzed, 1,4-acetoxychlorination of 1,3-dienes. 1-Acetoxy-4-chloro-2-alkenes as versatile synthons in organic transformations, *J. Am. Chem. Soc.*, 107, 3676, 1985.
28. Bäckvall, J.-E., Vågberg, J.-O., and Granberg, K. L., Palladium-catalyzed *cis*- and *trans*-annulations to 1,3-cyclohexadiene and 1,3-cycloheptadiene, *Tetrahedron Lett.*, 30, 617, 1989.
29. Kim, D., Kim, H. S., and Yoo, J. Y., A total synthesis of spirocyclic alkaloids *via* an intramolecular S_N2' lactam enolate alkylation, *Tetrahedron Lett.*, 32, 1577, 1991.
30. Birch, A. J., and Wright, J. J., The brevianamides: a new class of fungal alkaloid, *J. Chem. Soc., Chem. Commun.*, 644, 1969.
31. Williams, R. M., Glinka, T., Kwast, E., Coffman, H., and Stille, J. K., Asymmetric, stereocontrolled total synthesis of (–)-brevianamide B, *J. Am. Chem. Soc.*, 112, 808, 1990.
32. Nicolaou, K. C., Duggan, M. E., and Hwang, C.-K., Synthesis of the FG ring system of brevetoxin B, *J. Am. Chem. Soc.*, 111, 6676, 1989.
33. Stork, G., Kobayashi, Y., Suzuki, T., and Zhao, K., The allylic epoxide cyclization. A method for the control of regiochemistry and stereochemistry in cyclohexane systems, *J. Am. Chem. Soc.*, 112, 1661, 1990.
34. Stork, G., and Zhao, K., Total syntheses of (–)-histrionicotoxin and (–)-histrionicotoxin 235A, *J. Am. Chem. Soc.*, 112, 5875, 1990.
35. Hobbs-Mallyon, D., and Whiting, D. A., Spirocyclisation modes in the formation of tricyclospirodienones, *J. Chem. Soc., Perkin Trans. 1*, 2277, 1991.
36. Lei, B., and Fallis, A. G., A direct total synthesis of (+)-longifolene *via* an intramolecular Diels–Alder strategy, *J. Am. Chem. Soc.*, 112, 4609, 1990.
37. Kimpe, N. D., Sulmon, P., and Boeykens, M., Synthesis of 2,2-dialkylcyclopropylamines from β-chloroimines and application towards the synthesis of 1-amino-2,2-dialkylcyclopropanecarboxylic acids, *Tetrahedron*, 47, 3389, 1991.
38. Kimpe, N. D., Sulmon, P., and Stevens, C., Synthesis of 1-amino-2,2-dialkylcyclopropanecarboxylic acids from β-chloroaldimines, *Tetrahedron*, 47, 4723, 1991.
39. Padwa, A., Murphree, S. S., and Yeske, P. E., Use of 2,3-dibromo-1-(phenylsulfonyl)-1-propene as a reagent for the synthesis of annulated furans, *J. Org. Chem.*, 55, 4241, 1990.
40. Murphree, S. S., Muller, C. L., and Padwa, A., A new route to disubstituted cyclopentenones using 2,3-dibromo-1-(phenylsulfonyl)-1-propene as a pivotal reagent, *Tetrahedron Lett.*, 31, 6145, 1990.
41. Padwa, A., and Ishida, M., Silver tetrafluoroborate induced reaction of trimethylsilyl enol ethers with 2,3-diiodo-1-(phenylsulfonyl)-1-propene as a method for preparing substituted furans, *Tetrahedron Lett.*, 32, 5673, 1991.
42. Magar, S. S., and Fuchs, P. L., Bis-alkylation of dimetallated phenylsulfonylmethyl triflone. A n+1 annulation strategy for synthesis of cyclic vinyl sulfones, *Tetrahedron Lett.*, 33, 745, 1992.
43. Canas, M., Poch, M., Verdaguer, X., Moyano, A., Pericàs, M. A., and Riera, A., Regioselective ring opening of chiral epoxyalcohols by primary amines, *Tetrahedron Lett.*, 32, 6931, 1991.

44. Poch, M., Verdaguer, X., Moyano, A., Pericàs, M. A., and Riera, A., A versatile enantiospecific approach to 3-azetidinols and aziridines, *Tetrahedron Lett.*, 32, 6935, 1991.
45. Wilson, S. R., and Sawicki, R. A., Transannular cyclizations of 1-aza-4-cyclooctene, *J. Org. Chem.*, 44, 287, 1979.
46. Edstrom, E. D., An unexpected reversal in the stereochemistry of transannular cyclizations. A stereoselective synthesis of (±)-epilupinine, *Tetrahedron Lett.*, 32, 5709, 1991.
47. Edstrom, E. D., New methodology for the synthesis of functionalized indolizidine and quinolizidine ring systems, *J. Am. Chem. Soc.*, 113, 6690, 1991.
48. Bennett, R. B., III, Choi, J.-R., Montgomery, W. D., and Cha, J. K., A short, enantioselective synthesis of (–)-swainsonine, *J. Am. Chem. Soc.*, 111, 2580, 1989.
49. Pearson, W. H., Bergmeier, S. C., Degan, S., Lin, K.-C., Poon, Y.-F., Schkeryantz, J. M., and Williams, J. P., Synthesis of pyrrolizidines and indolizidines by the intramolecular cycloaddition of azides with electron-rich 1,3-dienes. A synthetic equivalent of a nitrene–diene cycloaddition, *J. Org. Chem.*, 55, 5719, 1990.
50. Pearson, W. H., and Lin, K.-C., The intramolecular cycloaddition of azides with ω-chloroalkenes. A facile route to (–)-swainsonine and other indolizidine alkaloids, *Tetrahedron Lett.*, 31, 7571, 1990.
51. Pearson, W. H., and Bergmeier, S. C., A synthesis of (–)-slaframine and (–)-1,8a-diepislaframine, *J. Org. Chem.*, 56, 1976, 1991.
52. Pearson, W. H., and Hines, J. V., A synthesis of (+)-7-epiaustraline and (–)-7-epialexine, *Tetrahedron Lett.*, 32, 5513, 1991.
53. Shibata, I., Matsuo, F., Baba, A., and Matsuda, H., Synthesis of hydroxy lactones by bis(tributyltin) oxide promoted ring expansion of halo lactones, *J. Org. Chem.*, 56, 475, 1991.
54. Kende, A. S., and Smith, C. A., Synthesis of a calicheamicin deoxyaglycone model by an intramolecular acetylide cyclization, *Tetrahedron Lett.*, 29, 4217, 1988.
55. Nicolaou, K. C., and Dai, W.-M., Chemistry and biology of the enediyne anticancer antibiotics, *Angew. Chem. Int. Ed. Engl.*, 30, 1387, 1991.
56. Heathcock, C. H., Ruggeri, R. B., and McClure, K. F., *Daphniphyllum* alkaloids. XV. Total syntheses of (±)-methyl homodaphniphyllate and (±)-daphnilactone A, *J. Org. Chem.*, 57, 2585, 1992.
57. Comins, D. L., and Dehghani, A., *N*-Acyldihydropyridones as synthetic intermediates. A short synthesis of (±)-pumiliotoxin C, *Tetrahedron Lett.*, 32, 5697, 1991.
58. Taylor, R. J. K., Turner, S. M., Horwell, D. C., Howarth, O. W., Mahon, M. F., and Molloy, K. C., The synthesis of novel benzomorphan analogues: a new intramolecular acid-catalysed aldol route to benzannelated bicyclo[3.3.1]nonane derivatives. X-ray molecular structure of 8-hydroxy-1-methoxytricyclo[7.3.1.$0^{2,7}$]trideca-2,4,6-trien-10-one, *J. Chem. Soc., Perkin Trans. 1*, 2145, 1990.
59. Heathcock, C. H., The aldol addition reaction, in *Asymmetric Synthesis*, Vol. 3, Morrison, J. D., Ed., Academic Press, New York, 1984, 111.
60. Kuwajima, I., and Nakamura, E., Reactive enolates from enol silyl ethers, *Acc. Chem. Res.*, 18, 181, 1985.
61. Myers, A. G., and Widdowson, K. L., A silicon-directed aldol condensation. Synthesis of enantiomerically pure *anti* aldols, *J. Am. Chem. Soc.*, 112, 9672, 1990.
62. Yasui, K., Fugami, K., Tanaka, S., Tamaru, Y., Ii, A., Yoshida, Z., and Saidi, M. R., Synthesis of 1-allyloxy-1-siloxycyclopropanes and 1-propargyloxy-1-siloxycyclopropanes, *Tetrahedron Lett.*, 33, 785, 1992.
63. Mandai, T., Murakami, T., Kawada, M., and Tsuji, J., A new stereoselective synthetic method for bicyclo[3.3.0]octane framework, *Tetrahedron Lett.*, 32, 3399, 1991.

64. Saito, S., Yamamoto, T., Matsuoka, M., and Moriwake, T., Tandem dianionic dioxy-Cope rearrangement and aldol condensation: a single-pot enantioselective route to *trans*-4,5-disubstituted 1-cyclopentenecarbaldehyde from acyclic bisallylic compounds, *Synlett*, 239, 1992.
65. Altenbach, H.-J., and Holzapfel, W., Synthesis of (+)-terrein from L-tartaric acid, *Angew. Chem. Int. Ed. Engl.*, 29, 67, 1990.
66. Klunder, A. J. H., Bos, W., and Zwanenburg, B., An efficient stereospecific total synthesis of (±) terrein, *Tetrahedron Lett.*, 22, 4557, 1981.
67. Bestmann, H. J., and Moenius, T., Optically active cyclopentanediol synthetic building blocks from tartaric acid, *Angew. Chem. Int. Ed. Engl.*, 25, 994, 1986.
68. Bestmann, H. J., and Roth, D., Synthesis of (–)-neplanocin A from (*R*,*R*)-tartaric acid, *Angew. Chem. Int. Ed. Engl.*, 29, 99, 1990.
69. Neyer, G., and Ugi, I., A novel Dieckmann-type cyclization, the final step of the synthesis of a carbacephem derivative, *Synthesis*, 743, 1991.
70. Cooke, M. P., Jr., and Houpis, I. N., Metal–halogen exchange–initiated cyclization of iodo carbonyl compounds, *Tetrahedron Lett.*, 26, 4987, 1985.
71. Boardman, L. D., Bagheri, V., Sawada, H., and Negishi, E., Mechanistic duality in cyclialkylation of alkenylmetal derivatives, *J. Am. Chem. Soc.*, 106, 6105, 1984.
72. Boatman, R. J., Whitlock, B. J., and Whitlock, H. W., Jr., An anionic equivalent of the Friedel–Crafts cycloacylation, *J. Am. Chem. Soc.*, 99, 4822, 1977.
73. McCarthy, C., and Jones, K., Indole alkaloid synthesis; a stereospecific preparation of functionalised *cis*-hexahydrocarbazoles, *J. Chem. Soc., Chem. Commun.*, 1717, 1989.
74. Storey, J. M. D., McCarthy, C., and Jones, K., Intramolecular alkyllithium additions to lactams; a synthesis of 2,3,9,9a-tetrahydro-1*H*-pyrrolo[1,2-*a*]indoles (pyrrolo[1,2-*a*]indolenines) related to mitomycins, *J. Chem. Soc., Chem. Commun.*, 892, 1991.
75. Lamas, C., Castedo, L., and Domínguez, D., Conversion of dibenzoxepinones to aristocularine alkaloids, *Tetrahedron Lett.*, 31, 6247, 1990.
76. Mori, M., Kaneta, N., Isono, N., and Shibasaki, M., Reaction of metal free Bu_3Sn^- generated from $Bu_3SnSiMe_3$–R_4NX with an aryl or vinyl halide, *Tetrahedron Lett.*, 32, 6139, 1991.
77. Deslongchamps, P., in *Stereoelectronic Effects in Organic Chemistry*, Baldwin, J. E., Ed., Pergamon Press, Oxford, 1984.
78. Dolle, R. E., Osifo, K. I., and Li, C.-S., Enantioselective synthesis of (+)-pinidine, *Tetrahedron Lett.*, 32, 5029, 1991.
79. Dolle, R. E., Li, C.-S., and Shaw, A. N., Concomitant [2,3]-sigmatropic rearrangement of allylic sulfilimines and intramolecular *N*-alkylation. Synthesis of 2-vinyl substituted cyclic amines, *Tetrahedron Lett.*, 30, 4723, 1989.
80. Solé, D., and Bonjoch, J., *cis*-3a-(*o*-Nitrophenyl)octahydroindol-4-ones. Building blocks for the synthesis of indole alkaloids, *Tetrahedron Lett.*, 32, 5183, 1991.
81. Overman, L. E., and Shim, J., First total synthesis of *Amaryllidaceae* alkaloids of the 5,11-methanomorphanthridine type. An efficient total synthesis of (±)-pancracine, *J. Org. Chem.*, 56, 5005, 1991.
82. Vavrecka, M., Janowitz, A., and Hesse, M., Transformation of 4-nitroalkane-1,7-diones into pyrrolizidines, *Tetrahedron Lett.*, 32, 5543, 1991.
83. Janowitz, A., Vavrecka, M., and Hesse, M., Stereoselektive synthese von pyrrolizidin-alkaloiden aus nitroalkanonen, *Helv. Chim. Acta*, 74, 1352, 1991.
84. Texier-Boullet, F., Klein, B., and Hamelin, J., Pyrrole and pyrazole ring closure in heterogeneous media, *Synthesis*, 409, 1986.
85. Pilard, J.-F., Klein, B., Texier-Boullet, F., and Hamelin, J., Fast synthesis of heterobicycles containing a bridgehead nitrogen atom in dry media under microwave irradiation, *Synlett*, 219, 1992.

86. Thompson, C. M., "Remote" dianions. I. Utility of 4-phenylsulfonylbutanoic acid in the mild conversion of aldehydes and ketones to lactones, *Tetrahedron Lett.*, 28, 4243, 1987.
87. Thompson, C. M., Green, D. L. C., and Kubas, R., Remote dianions. III. Novel synthesis of substituted 2-piperidones from imines, *J. Org. Chem.*, 53, 5389, 1988.
88. Green, D. L. C., and Thompson, C. M., "Remote" dianions. V. Synthesis of δ-coniceine, *Tetrahedron Lett.*, 32, 5051, 1991.
89. Woodward, R. B., Cava, M. P., Ollis, W. D., Hunger, A., Daeniker, H. U., and Schenker, K., The total synthesis of strychnine, *Tetrahedron*, 19, 247, 1963.
90. Weisbach, J. A., Macko, E., De Sanctis, N. J., Cava, M. P., and Douglas, B., Synthesis and pharmacology of some β-spiroindolenines and indolines, *J. Med. Chem.*, 7, 735, 1964.
91. Freund, R., Mahboobi, S., Noack, K., Schönholzer, P., and Bernauer, K., Diastereoselective spirocyclization of *C*-(alkyloxycarbonyl)formimines of 2-substituted 1*H*-indole-3-ethanamines (= tryptamines): basic studies, *Helv. Chim. Acta*, 73, 439, 1990.
92. Freund, R., Martinovic´, S., and Bernauer, K., Diastereoselective spirocyclization of imines of 2-substituted 1*H*-indole-3-ethylamines (= tryptamines): variation of the electrophile and the substituent at C(2) and its influence on the steric outcome, *Helv. Chim. Acta*, 75, 282, 1992.
93. Wang, Y., Mukherjee, D., Birney, D., and Houk, K. N., Synthesis and reactions of ester-substituted fulvenes. A new route to $\Delta^{9(12)}$-capnellene, *J. Org. Chem.*, 55, 4504, 1990.
94. Little, R. D., Carroll, G. L., and Petersen, J. L., Total synthesis of the marine natural product $\Delta^{9(12)}$-capnellene. Reversal of regiochemistry in the intramolecular 1,3-diyl trapping reaction, *J. Am. Chem. Soc.*, 105, 928, 1983.
95. Linderman, R. J., and Godfrey, A., A novel synthesis of tetrahydrofurans *via* a synthetic equivalent to a carbonyl ylide, *J. Am. Chem. Soc.*, 110, 6249, 1988.
96. Linderman, R. J., Graves, D. M., Kwochka, W. R., Ghannam, A. F., and Anklekar, T. V., Furanone synthesis *via* an electrophilic capped carbonyl ylide, *J. Am. Chem. Soc.*, 112, 7438, 1990.
97. Kametani, T., Yukawa, H., and Honda, T., Synthesis of pyrrolizidine alkaloids, (±)-trachelanthamidine, (±)-isoretronecanol, and (±)-supinidine, by means of an intramolecular carbenoid displacement (ICD) reaction, *J. Chem. Soc., Chem. Commun.*, 651, 1986.
98. Kametani, T., Chu, S.-D., Itoh, A., Wang, T.-C., Nakayama, A., and Honda, T., Novel construction of penems and carbapenems. Further evidence of a carbene intermediate for the oxalimide cyclization reaction, *J. Chem. Soc., Chem. Commun.*, 544, 1988.
99. Kametani, T., Yukawa, H., and Honda, T., Stereoselective synthesis of pyrrolizidine alkaloids, (+)-heliotridine and (+)-retronecine, by means of an intermolecular carbenoid displacement reaction, *J. Chem. Soc., Chem. Commun.*, 685, 1988.
100. Cooke, M. P., Jr., and Widener, R. K., Lithium–halogen exchange-initiated cyclization reactions. III. Intramolecular conjugate addition reactions of unsaturated acylphosphoranes, *J. Org. Chem.*, 52, 1381, 1987.
101. Cooke, M. P., Jr., Stereochemistry of metal–halogen exchange-initiated intramolecular conjugate addition reactions in the construction of 5-membered carbocycles, *J. Org. Chem.* 57, 1495, 1992.
102. Liu, H.-J., and Han, Y., *S,S´*-Diethyl dithiomalonate as ethanol carbanion equivalent in annelation reactions, *J. Chem. Soc., Chem. Commun.*, 1484, 1991.
103. Stork, G., and Saccomano, N. A., Enantioselective synthesis of an intermediate for the intramolecular Diels–Alder construction of 11-ketosteroids, *Tetrahedron Lett.*, 28, 2087, 1987.
104. Hirai, Y., Terada, T., and Yamazaki, T., An asymmetric intramolecular Michael reaction. Construction of chiral building blocks for the synthesis of several alkaloids, *J. Am. Chem. Soc.*, 110, 958, 1988.

105. Hirai, Y., Terada, T., Katoh, H., Sonohara, S., and Momose, T., An enantioselective construction of pyrrolidones bearing functionalized appendages *via* an asymmetric intramolecular Michael reaction, *Heterocycles*, 32, 7, 1991.
106. Bunce, R. A., Peeples, C. J., and Jones, P. B., Tandem S_N2–Michael reactions for the preparation of simple five- and six-membered-ring nitrogen and sulfur heterocycles, *J. Org. Chem.*, 57, 1727, 1992.
107. Little, R. D., and Dawson, J. R., MIRC (Michael initiated ring closure) reactions. Formation of three, five, six, and seven membered rings, *Tetrahedron Lett.*, 21, 2609, 1980.
108. Prempree, P., Radviroongit, S., and Thebtaranonth, Y., Reaction of methyl 4-bromocrotonate with lithium ester enolates: direct S_N2 displacement *vs* Michael-initiated ring closure, *J. Org. Chem.*, 48, 3553, 1983.
109. Cooke, M. P., Jr., and Jaw, J. Y., Cyclopropane derivatives through charge-directed conjugate addition reactions of unsaturated acylphosphoranes, *J. Org. Chem.*, 51, 758, 1986.
110. Enders, D., Scherer, H. J., and Raabe, G., Diastereo- and enantioselective Michael addition initiated cyclizations to *trans*-substituted cyclopentanecarboxylates, *Angew. Chem. Int. Ed. Engl.*, 30, 1664, 1991.
111. Harrowven, D. C., A concise synthesis of taiwanin E using a Michael initiated ring closure protocol, *Tetrahedron Lett.*, 32, 3735, 1991.
112. Reux, D., and Pochat, F., Single step synthesis of 2,4-dicyanonaphthylamines from synthetic equivalents of α-cyanoalkynes, *J. Chem. Soc., Chem. Commun.*, 1419, 1991.
113. Hendrickson, J. B., Rees, R., and Templeton, J. F., A new general heterocycle synthesis; use of acetylenedicarboxylic esters, *J. Am. Chem. Soc.*, 86, 107, 1964.
114. Mann, J., and Holland, H. J., Cycloaddition reactions of polysubstituted furans with oxyallyl carbocations, *Tetrahedron*, 43, 2533, 1987.
115. Jauch, J., and Schurig, V., An improved procedure for the preparation of substituted 4-hydroxy-4,5-dihydrofurans, *Tetrahedron Lett.*, 32, 4687, 1991.
116. Crimmins, M. T., and Nantermet, P. G., Addition of zinc homoenolates to acetylenic esters: a formal [3+2] cycloaddition, *J. Org. Chem.*, 55, 4235, 1990.
117. Aurell, M. J., Gaviña, P., Gil, S., Parra, M., Tortajada, A., and Mestres, R., Dienediolates of unsaturated carboxylic acids in synthesis. Tandem Michael Dieckmann synthesis of substituted 2-cyclohexenones, *Synth. Commun.*, 21, 1825, 1991.
118. Kodpinid, M., Siwapinyoyos, T., and Thebtaranonth, Y., Simple routes to sarkomycin, *J. Am. Chem. Soc.*, 106, 4862, 1984.
119. Mahidol, C., Thebtaranonth, C., Thebtaranonth, Y., and Yenjai, C., Synthesis of deepoxy-4,5-didehydromethylenomycin A and methylenomycin A methyl esters, *Tetrahedron Lett.*, 30, 3857, 1989.
120. Tarnchompoo, B., Thebtaranonth, C., and Thebtaranonth, Y., Synthesis of 8-methoxy-1-tetralones, *Synthesis*, 785, 1986.
121. Mahidol, C., Tarnchompoo, B., Thebtaranonth, C., and Thebtaranonth, Y., Total synthesis of diospyrol: an anthelmintic drug from *Diospyros mollis* Griff, *Tetrahedron Lett.*, 30, 3861, 1989.
122. Arcamone, F., Cassinelli, G., DiMatteo, F., Forenza, S., Ripamonti, M. C., Rivola, G., Vigevani, A., Clardy, J., and McCabe, T., Structures of novel anthracycline antitumor antibiotics from *Micromonospora peucetica*, *J. Am. Chem. Soc.*, 102, 1462, 1980.
123. Hauser, F. M., and Prasanna, S., A general route for total synthesis of 9-alkyl-9-hydroxy aklavinones and pyrromycinones, *Tetrahedron*, 40, 4711, 1984.
124. Hauser, F. M., and Tommasi, R. A., Total synthesis of (–)-(9*R*)-7,11-dideoxy-13-deoxodaunomycinone, *J. Org. Chem.*, 56, 5758, 1991.

125. Asao, N., Uyehara, T., and Yamamoto, Y., Lithium *N*-benzyltrimethylsilylamide (LSA): a new reagent for conjugate addition–enolate trapping reactions, *Tetrahedron*, 44, 4173, 1988.
126. Uyehara, T., Shida, N., and Yamamoto, Y., New type of cyclization of α,β,χ,ψ-unsaturated dioic acid esters through tandem conjugate additions by using lithium *N*-benzyl-*N*-(trimethylsilyl)amide as a nitrogen nucleophile, *J. Org. Chem.*, 57, 3139, 1992.
127. Uyehara, T., Shida, N., and Yamamoto, Y., Cyclisation of α,β,χ,ψ-unsaturated dioic acid esters *via* tandem conjugate additions by using lithium *N*-benzyltrimcthylsilylamide (LSA) as a nitrogen nucleophile and its application to a total synthesis of (±)-dihydronepetalactone and (±)-isodihydronepetalactone, *J. Chem. Soc., Chem. Commun.*, 113, 1989.
128. Klimko, P. G., and Singleton, D. A., Silyl group-transfer-mediated serial Michael additions, *J. Org. Chem.*, 57, 1733, 1992.
129. Saito, S., Hara, T., Naka, K., Hayashi, T., and Moriwake, T., Enantioselective synthesis of functionalized cyclopentenone and alkylidenecyclopentane derivatives from an acyclic bisallylic diol framework, *Synlett*, 241, 1992.
130. Bunce, R. A., Wamsley, E. J., Pierce, J. D., Shellhammer, A. J., Jr., and Drumright, R. E., Tandem Michael reactions for the construction of highly functionalized five-membered rings, *J. Org. Chem.*, 52, 464, 1987.
131. Barco, A., Benetti, S., Casolari, A., Pollini, G. P., and Spalluto, G., Tandem Michael reactions for the construction of pyrrolidine and piperidine ring systems, *Tetrahedron Lett.*, 31, 3039, 1990.
132. White, K. B., and Reusch, W., The synthesis of bicyclo[2.2.2]octan-2-ones by sequential Michael reactions, *Tetrahedron*, 34, 2439, 1978.
133. Lavallée, J.-F., and Deslongchamps, P., Synthesis of *cis*-decalin *via* Diels–Alder and double Michael cycloaddition with substituted Nazarov reagent, *Tetrahedron Lett.*, 29, 5117, 1988.
134. Spino, C., and Deslongchamps, P., Highly stereoselective double Michael cyclization of 1-phenylsulfinyl and sulfonyl analogues of the Nazarov reagent, *Tetrahedron Lett.*, 31, 3969, 1990.
135. Ruel, R., and Deslongchamps, P., Synthesis of an optically active 13β-methyl 14β-hydroxy steroid *via* base-catalyzed reactions, *Tetrahedron Lett.*, 31, 3961, 1990.
136. Spitzner, D., Wagner, P., Simon, A., and Peters, K., Diastereoselective synthesis of bicyclo[2.2.2]octanes by double Michael addition, *Tetrahedron Lett.*, 30, 547, 1989.
137. Zhao, R.-B., Zhao, Y.-F., Song, G.-Q., and Wu, Y.-L., Double Michael reaction of carvone and its derivatives, *Tetrahedron Lett.*, 31, 3559, 1990.
138. Schinzer, D., and Kalesse, M., Diastereoselective double Michael additions of vinylogous esters and thexyldimethylsilyl triflate-induced cyclizations, *Tetrahedron Lett.*, 32, 4691, 1991.
139. Jaafar, A., Alilou, E. H., Réglier, M., and Waegell, B., Unexpected [2+2+2] MIMIRC annulation between a lithium dienolate and methyl acrylate, *Tetrahedron Lett.*, 32, 5531, 1991.
140. Ihara, M., Katogi, M., Fukumoto, K., and Kametani, T., Total synthesis of (±)-pentalenic acid *via* intramolecular double Michael reaction, *J. Chem. Soc., Chem. Commun.*, 721, 1987.
141. Ihara, M., Takahashi, T., Shimizu, N., Ishida, Y., Sudow, I., Fukumoto, K., and Kametani, T., A novel approach to estranes by an intramolecular double Michael reaction, *J. Chem. Soc., Chem. Commun.*, 1467, 1987.
142. Ihara, M., Suzuki, S., Taniguchi, N., Fukumoto, K., and Kabuto, C., A novel construction of polyfunctionalised *trans*-hydroindanes *via* sulphur-mediated intramolecular double Michael type reaction, *J. Chem. Soc., Chem. Commun.*, 1168, 1991.

143. Ihara, M., Suzuki, M., Fukumoto, K., Kametani, T., and Kabuto, C., Stereoselective total synthesis of (±)-atisine *via* intramolecular double Michael reaction, *J. Am. Chem. Soc.*, 110, 1963, 1988.
144. Ihara, M., Suzuki, M., Fukumoto, K., and Kabuto, C., Asymmetric total synthesis of atisine *via* intramolecular double Michael reaction, *J. Am. Chem. Soc.*, 112, 1164, 1990.
145. Ihara, M., Takino, Y., Fukumoto, K., and Kametani, T., Asymmetric total syntheses of (–)-tylophorine *via* the highly enantioselective intramolecular double Michael reaction, *Tetrahedron Lett.*, 29, 4135, 1988.
146. Hagiwara, H., Abe, F., and Uda, H., Bicycloannulation leading to the tricyclo[4.4.0.0^{1,5}]decane framework and its congeners, *J. Chem. Soc., Chem. Commun.*, 1070, 1991.
147. Cory, R. M., Anderson, P. C., McLaren, F. R., and Yamamoto, B. R., Bicycloannulation with nitroethene and 1-nitropropene. A one-step synthesis of tricyclenone, *J. Chem. Soc., Chem. Commun.*, 73, 1981.
148. Ono, N., The nitro function as a leaving group in organic synthesis, in *Nitro Compounds: Recent Advances in Synthesis and Chemistry*, Feuer, H., and Nielsen, A. T., Eds., VCH Publishers, New York, 1990, 1.
149. Tamura, R., Kamimura, A., and Ono, N., Displacement of aliphatic nitro groups by carbon and heteroatom nucleophiles, *Synthesis*, 423, 1991.
150. Hagiwara, H., Okano, A., and Uda, H., One-step synthesis of tricyclo[5.3.1.0^{3,8}]undecane-4,11-diones by three consecutive Michael reactions. A formal synthesis of (±)-seychellene, *J. Chem. Soc., Chem. Commun.*, 1047, 1985.
151. Hagiwara, H., Okano, A., and Uda, H., One-pot annulation to tricyclo[5.3.1.0^{3,8}]undecane-2,6-diones by sequential three-fold Michael reactions. A formal synthesis of (±)-seychellene, *J. Chem. Soc., Perkin Trans. 1*, 2109, 1990.
152. Thanupran, C., Thebtaranonth, C., and Thebtaranonth, Y., Stereospecific triple Michael addition, *Tetrahedron Lett.*, 27, 2295, 1986.
153. Dean, F. M., Houghton, L. E., Nayyir-Mazhir, R., and Thebtaranonth, C., Cage formation: a new reaction of quinones with carbanions, *J. Chem. Soc., Chem. Commun.*, 159, 1979.
154. Kurihara, T., Santo, K., Harusawa, S., and Yoneda, R., Application of cyanophosphates in organic synthesis. Reactivity of α-cyano-α-diethylphosphonooxy anions, *Chem. Pharm. Bull.*, 35, 4777, 1987.
155. Chen, C., Huang, Y.-Z., and Shen, Y., A new method for the synthesis of cyclopropane with three electron-withdrawing groups, *Tetrahedron Lett.*, 29, 1033, 1988.
156. Nowick, J. S., and Danheiser, R. L., Chemistry of cyclopropylacylsilanes I. α-Functionalized acylsilane reagents for the cyclopropanation of electrophilic alkenes, *Tetrahedron*, 44, 4113, 1988.
157. Hudlicky, T., Radesca-Kwart, L., Li, L., and Bryant, T., Short, enantioselective synthesis of (–)-retigeranic acid *via* [2+3] annulation, *Tetrahedron Lett.*, 29, 3283, 1988.
158. Ono, N., Yanai, T., Hamamoto, I., Kamimura, A., and Kaji, A., Nitro compounds as nucleophilic alkylidene transfer reagents, *J. Org. Chem.*, 50, 2806, 1985.
159. Annen, K., Hofmeister, H., Laurent, H., Seeger, A., and Wiechert, R., Eine neuartige einfache dreiring-synthese, *Chem. Ber.*, 111, 3094, 1978.
160. Corey, E. J., and Chaykovsky, M., Dimethyloxosulfonium methylide ($(CH_3)_2SOCH_2$) and dimethylsulfonium methylide ($(CH_3)_2SCH_2$). Formation and application to organic synthesis, *J. Am. Chem. Soc.*, 87, 1353, 1965.
161. Hamdouchi, C., A new approach to the asymmetric cyclopropanation *via* a chiral sulfinyl group, *Tetrahedron Lett.*, 33, 1701, 1992.

162. Johnson, C. R., and Rogers, P. E., Preparation and applications of (dialkylamino)-methyloxosulfonium methylides. Synthesis of cyclopropanes and oxiranes, *J. Org. Chem.*, 38, 1793, 1973.
163. Williams, R. M., and Fegley, G. J., Asymmetric syntheses of 1-aminocyclopropane-1-carboxylic acid derivatives, *J. Am. Chem. Soc.*, 113, 8796, 1991.
164. Majetich, G., Hull, K., and Desmond, R., A new approach for the formation of fused cyclooctane or cyclohexane rings, *Tetrahedron Lett.*, 26, 2751, 1985.
165. Majetich, G., Hull, K., Defauw, J., and Desmond, R., Regioselectivity in intramolecular addition of allylsilanes to conjugated dienones, *Tetrahedron Lett.*, 26, 2747, 1985.
166. Chakraborty, R., and Simpkins, N. S., Cyclisation chemistry of some functionalised allylsilanes, *Tetrahedron*, 47, 7689, 1991.
167. Yamamoto, Y., and Furuta, T., A multistep rearrangement from 2,2-disubstituted 1,3-cyclohexanediones to 3-substituted 2-cyclohexenones *via* phosphonate anions and its application to a formal synthesis of (±)-α-acoradiene, *J. Org. Chem.*, 55, 3971, 1990.
168. Majetich, G., Song, J. S., Ringold, C., and Nemeth, G. A., A direct stereoselective synthesis of (±)-14-deoxyisoamijiol, *Tetrahedron Lett.*, 31, 2239, 1990.
169. Lee, T. V., Boucher, R. J., Porter, J. R., and Rockell, C. J. M., Reactions of *O*-silylated enolates with allylsilane bifunctional [4+2] and [5+2] annulating reagents, *Tetrahedron*, 45, 5887, 1989.
170. Lee, T. V., and Roden, F. S., Chemoselective reaction of bifunctional aldehydo allylsilanes, *Tetrahedron Lett.*, 31, 2067, 1990.
171. Lee, T. V., Roden, F. S., and Yeoh, H. T.-L., Selectivity of bifunctional annulating reagents: additional rules for ring closure, *Tetrahedron Lett.*, 31, 2063, 1990.
172. Lee, T. V., Porter, J. R., and Roden, F. S., The use of a combined annulation-ring cleavage strategy for the novel synthesis of seven, eight and nine-membered rings, *Tetrahedron*, 47, 139, 1991.
173. Nishitani, K., and Yamakawa, K., Asymmetric synthesis of α-methylene-γ-butyrolactones fused to carbocyclic rings *via* the Hosomi reaction of chiral allylsilanes, *Tetrahedron Lett.*, 32, 387, 1991.
174. Nicolaou, K. C., Prasad, C. V. C., and Ogilvie, W. W., Novel stereocontrolled synthesis of the nonacene ring system of brevetoxin A. Conformational–reactivity effects in nine-membered rings, *J. Am. Chem. Soc.*, 112, 4988, 1990.
175. Yamada, J., Asano, T., Kadota, I., and Yamamoto, Y., A new approach to the construction of β-alkoxy-substituted cyclic ethers *via* the intramolecular cyclization of ω-trialkylplumbyl and ω-trialkylstannyl ether acetals, *J. Org. Chem.*, 55, 6066, 1990.
176. Suzuki, T., Sato, O., Hirama, M., Yamamoto, Y., Murata, M., Yasumoto, T., and Harada, N., Enantioselective synthesis of the AB ring fragment of gambiertoxin 4B. Implication for the absolute configuration of gambiertoxin 4B and ciguatoxin, *Tetrahedron Lett.*, 32, 4505, 1991.
177. Yamamoto, Y., Yamada, J., and Kadota, I., Stereocontrolled intramolecular cyclization of ω-tributylstannyl ether aldehydes. Synthesis of the 6·7·7·6 ring system of polycyclic ethers, *Tetrahedron Lett.*, 32, 7069, 1991.
178. Marshall, J. A., and Wang, X., Synthesis of macrocyclic homopropargylic alcohols through intramolecular $S_{E'}$ addition of allenylstannanes and their subsequent conversion to 2,5-furanocycles, *J. Org. Chem.*, 56, 6264, 1991.
179. Tius, M. A., and Cullingham, J. M., Marine cembranoid synthesis, *Tetrahedron Lett.*, 30, 3749, 1989.
180. Dess, D. B., and Martin, J. C., Readily accessible 12-I-5 oxidant for the conversion of primary and secondary alcohols to aldehydes and ketones, *J. Org. Chem.*, 48, 4155, 1983.

181. Look, S. A., Burch, M. T., Fenical, W., Qi-tai, Z., and Clardy, J., Kallolide A, a new antiinflammatory diterpenoid, and related lactones from the Caribbean octocoral *Pseudopterogorgia kallos* (Bielschowsky), *J. Org. Chem.*, 50, 5741, 1985.
182. Shiner, C. S., Berks, A. H., and Fisher, A. M., Metalation of nonenolizable ketones and aldehydes by lithium dialkylamide bases, *J. Am. Chem. Soc.*, 110, 957, 1988.
183. Ramón, D. J., and Yus, M., 2-(3-Lithiopropyl)- and 2-(3-lithiopropyl)-2-methyl-1,3-dioxolane: new masked lithium bishomoenolates in the synthesis of bifunctionalized compounds, *Tetrahedron Lett.*, 31, 3763, 1990.
184. Ramón, D. J., and Yus, M., Direct synthesis of δ-lactones from 2-(3-lithiopropyl)-1,3-dioxolane and carbonyl compounds, *Tetrahedron Lett.*, 31, 3767, 1990.
185. Yus, M., and Ramón, D. J., One-step synthesis of substituted 6,8-dioxabicyclo[3.2.1]octanes: easy preparation of racemic frontalin, brevicomins, and related systems, *J. Org. Chem.*, 57, 750, 1992.
186. Marumo, K., Inoue, S., Sato, Y., and Kato, H., Lithiation of (dialkylaminomethyl)trimethylsilanes, *J. Chem. Soc., Perkin Trans.* 1, 2275, 1991.
187. Barluenga, J., González, R., and Fañanás, F. J., An easy entry to trilithiated diallyl amine by a regio- and stereo-selective lithiation of heterosubstituted diallyl amines, *Tetrahedron Lett.*, 33, 831, 1992.
188. Ashe, A. J., III, Kampf, J. W., and Savla, P. M., Selective functionalization in the bay region of polycyclic aromatic hydrocarbons *via* dilithiation, *J. Org. Chem.*, 55, 5558, 1990.
189. Fu, J., Zhao, B., Sharp, M. J., and Snieckus, V., Remote aromatic metalation. An anionic Friedel–Crafts equivalent for the regioselective synthesis of condensed fluorenones from biaryl and *m*-teraryl 2-amides, *J. Org. Chem.*, 56, 1683, 1991.
190. Wang, W., and Snieckus, V., Remote directed metalation of biaryl *o*-carbamates. Ring to ring carbamoyl transfer route to biaryls, dibenzo[*b*,*d*]pyranones, and the natural fluorenone dengibsin, *J. Org. Chem.*, 57, 424, 1992.
191. Fisher, L. E., Muchowski, J. M., and Clark, R. D., Heteroatom-directed metalation. Lithiation of *N*-propenylbenzamides and *N*-propenyl-*o*-toluamides. Novel routes to *ortho*-substituted primary benzamide derivatives and N-unsubstituted isoquinolin-1(2*H*)-ones, *J. Org. Chem.*, 57, 2700, 1992.
192. Bailey, W. F., Patricia, J. J., DelGobbo, V. C., Jarret, R. M., and Okarma, P. J., Cyclization of 5-hexenyllithium to (cyclopentylmethyl)lithium, *J. Org. Chem.*, 50, 1999, 1985.
193. Bailey, W. F., Patricia, J. J., Nurmi, T. T., and Wang, W., Metal–halogen interchange between *t*-butyllithium and 1-iodo-5-hexenes provides no evidence for single-electron transfer, *Tetrahedron Lett.*, 27, 1861, 1986.
194. Bailey, W. F., Patricia, J. J., and Nurmi, T. T., Reactions of 1-halo-5-hexenes with alkyllithiums. Evidence for a pronounced halogen effect on the mechanism of the metal–halogen interchange reaction of primary alkyl halides, *Tetrahedron Lett.*, 27, 1865, 1986.
195. Bailey, W. F., Nurmi, T. T., Patricia, J. J., and Wang, W., Preparation and regiospecific cyclization of alkenyllithiums, *J. Am. Chem. Soc.*, 109, 2442, 1987.
196. Beckwith, A. L. J., and Schiesser, C. H., A force-field study of alkenyl radical ring closure, *Tetrahedron Lett.*, 26, 373, 1985.
197. Bailey, W. F., and Khanolkar, A. D., Preparation of 1-substituted bicyclo[2.2.1]heptanes by anionic cyclization of a 5-hexen-1-yllithium, *J. Org. Chem.*, 55, 6058, 1990.
198. Bailey, W. F., and Khanolkar, A. D., Construction of a sterically congested carbon framework *via* 5-hexenyllithium cyclization. Synthesis of (±)-cuparene, *Tetrahedron*, 47, 7727, 1991.

199. Bailey, W. F., and Khanolkar, A. D., Formation of a *trans*-bicyclo[3.3.0]octane by tandem anionic cyclization: evidence for the highly stereoselective irreversible nature of 5-hexen-1-yllithium cyclizations, *Tetrahedron Lett.*, 31, 5993, 1990.
200. Bailey, W. F., and Rossi, K., Tandem anionic cyclization approach to polycarbocyclic products, *J. Am. Chem. Soc.*, 111, 765, 1989.
201. Krief, A., and Barbeaux, P., Regioselective syntheses of polycylic compounds by carbanion-mediated polycyclisation of olefins, *Tetrahedron Lett.*, 32, 417, 1991.
202. Broka, C. A., Lee, W. J., and Shen, T., Anionic cyclizations of α-alkoxy lithiums. New, stereoselective routes to substituted tetrahydrofurans, *J. Org. Chem.*, 53, 1336, 1988.
203. Broka, C. A., and Shen, T., Reductive lithiation mediated anionic cyclizations and [2,3]-sigmatropic rearrangements, *J. Am. Chem. Soc.*, 111, 2981, 1989.
204. Still, W. C., and Mitra, A., A highly stereoselective synthesis of *Z*-trisubstituted olefins *via* [2,3]-sigmatropic rearrangement. Preference for a pseudoaxially substituted transition state, *J. Am. Chem. Soc.*, 100, 1927, 1978.
205. Bailey, W. F., and Zarcone, L. M. J., The conversion of 4-oxa-5-hexenyllithiums to 4-alken-1-ols: a novel [1,4]-Wittig rearrangement, *Tetrahedron Lett.*, 32, 4425, 1991.
206. Jiricny, J., Orere, D. M., and Reese, C. B., The conversion of aldehydes and ketones *via* their 2,4,6-tri-isopropylbenzenesulphonyl hydrazones into nitriles containing one additional carbon atom, *J. Chem. Soc., Perkin Trans. 1*, 1487, 1980.
207. Chamberlin, A. R., Bloom, S. H., Cervini, L. A., and Fotsch, C. H., Vinyllithium cyclizations with unactivated alkenes as internal electrophiles: stereoselective formation of substituted methylenecyclopentanes, *J. Am. Chem. Soc.*, 110, 4788, 1988.
208. Dolbier, W. R., Jr., and Chen, Y., Generation of a carbenoid by cyclization of 6-chloro-5-hexen-1-yllithium, *J. Org. Chem.*, 57, 1947, 1992.
209. Wadi, A., López-Calahorra, F., Alvarez, M., and Joule, J. A., An improved annelation method with methyl 2-(1,3-dithian-2-yl)benzoate as a bidentate synthon, *Tetrahedron Lett.*, 33, 3679, 1992.
210. Harrowven, D. C., The addition of organolithiums to ketenethioacetals: a new cyclopentannulation sequence leading to indanes, *Tetrahedron Lett.*, 33, 2879, 1992.
211. Shao, X., and Tamm, C., Cyclization of α-hydroxy-β-diketones to furan-3(2*H*)-ones under mild basic conditions, *Tetrahedron Lett.*, 32, 2891, 1991.
212. Linderman, R. J., Viviani, F. G., and Kwochka, W. R., Novel methods for the construction of 3-(2*H*) furanone spiroketals, *Tetrahedron Lett.*, 33, 3571, 1992.
213. Dehmlow, H., Mulzer, J., Seilz, C., Strecker, A. R., and Kohlmann, A., Stereocontrolled formation of polysubstituted tetrahydrofurans by debenzylating cycloetherifiation, *Tetrahedron Lett.*, 33, 3607, 1992.
214. Craig, D., and Smith, A. M., Stereoselective synthesis of 3-(phenylsulphonyl)-2,5 disubstituted tetrahydrofurans *via* 5-*endo-trig* ring-closure reactions, *Tetrahedron Lett.*, 33, 695, 1992.
215. Duhamel, L., Guillemont, J., and Poirier, J.-M., A new prenylation method using the lithium enolate of prenal. Reaction with aldehydes and α,β-unsaturated aldehydes, *Tetrahedron Lett.*, 32, 4495, 1991.
216. Tanaka, K., Matsuura, H., Funaki, I., and Suzuki, H., Stereo- and regio-selective synthesis of polyfunctionalized cyclopropanes, *J. Chem. Soc., Chem. Commun.*, 1145, 1991.
217. Ward, R. A., and Procter, G., Allylsilanes in organic synthesis; double asymmetric induction in the dihydroxylation of a chiral allylsilane, *Tetrahedron Lett.*, 33, 3363, 1992.
218. Yokoyama, Y., and Tsuchikura, K., Doubly allylic strain-controlled diastereoselective intramolecular Michael addition and a synthesis of (±)-iridomyrmecin, *Tetrahedron Lett.*, 33, 2823, 1992.

219. Ramig, K., Kuzemko, M. A., McNamara, K., and Cohen, T., Use of dilithiomethane equivalent in a novel one-flask [2+1+2] cyclopentannulation reaction: a highly efficient total synthesis of (±)-hirsutene, *J. Org. Chem.*, 57, 1968, 1992.
220. Wender, P. A., and Mucciaro, T. P., A new and practical approach to the synthesis of taxol and taxol analogues: the pinene path, *J. Am. Chem. Soc.*, 114, 5878, 1992.

Chapter 5

METAL-CATALYZED CYCLIZATION

I. INTRODUCTION

In the chemist's soul-searching article entitled "Organic Synthesis—Where Now?"[1] it was expressed that "... New synthetic *methods* are most likely to be encountered in the fields of biological and organometallic chemistry...". Truly, despite the long years of usage and the host of known important and useful reactions of organometallic reagents, the chemist is still comparatively naive in this enormously vast and fertile field. Doubtless to say, discoveries of new reactions and inventions of new reagents will continue to multiply far into the future.

An active area in the above field is the metal-mediated cyclization process. Reactions that otherwise do not occur have been reported possible with metal catalysis. This has triggered a burst of activity which has produced a flood of research articles on new and efficient reactions, followed by books and reviews compiling and organizing the materials. Such works range from palladium-catalyzed cyclizations and cycloadditions,[2,3] to metal carbene complexes in organic synthesis;[4–6] from cobalt-mediated cyclizations[7] and carbonylative cyclizations,[8,9] to the use of samarium salts in organic synthesis;[10,11] and the use of rhodium-catalyzed reaction of diazo-carbonyl compounds in the formation of carbocycles *via* carbene intermediates,[12] as well as metallo–ene cyclization[13] and the general use of the metal catalyst in organic synthesis.[14–16] Consequently, this chapter will discuss metal-catalyzed cyclizations mostly in view of the more recent reports, yet it is hoped that, by the end, it will have given the reader a good glimpse of the expansive terrain of metal-promoted cyclizations.

II. PALLADIUM–CATALYZED CYCLIZATION

The most well known and widely used metal-catalyzed C–C bond-forming reaction is that employing the palladium catalyst either in the form of Pd(0) or as various palladium salts. The reaction has been rigorously developed and its intramolecular version allows the construction of endless types of carbo- and heterocycles. The most well-known palladium-catalyzed cyclizations are the Heck reaction,[17] which involves the vinylation of organic halides, and the Stille reaction,[18] which is a palladium-catalyzed coupling reaction of organotin reagents. More recently, the coupling reaction of π-allyl palladium has also been extensively developed, particularly by Trost and his co-workers.[2]

A. THE HECK AND RELATED REACTIONS

The Heck reaction can be represented most simply as shown in Scheme 1. The dependence of the reaction pathway on the choice of metal catalyst, substrate structure, and reaction conditions permits manipulations to yield products **4** and/or **5** *via exo* and/or *endo* cyclizations, respectively.

Scheme 1

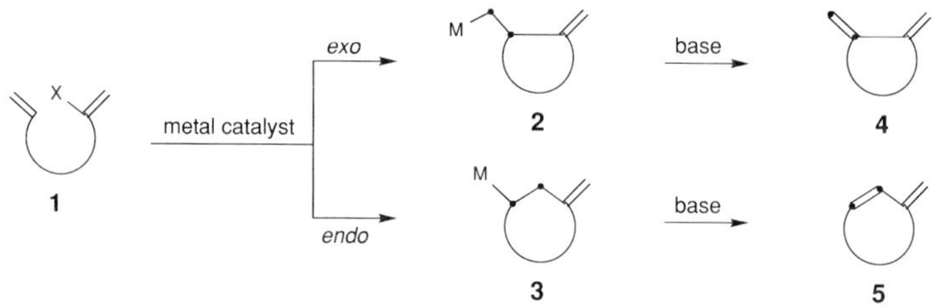

Similar to that observed with the radical and ionic cyclizations, the 5-*exo* mode seems, in general, to be the preferred pathway unless otherwise maneuvered. A good example is shown in Scheme 2, where the palladium acetate-catalyzed reaction of **6a** yields a mixture of **8** and **10a** in a 4:1 ratio (80%) while the modified substrate **6b** gives only **10b** (57%).[19] The rationale for the latter behavior is that the organopalladium intermediate **7b** from a 5-*exo* cyclization of **6b** lacks the necessary β-hydrogen to eliminate to product, hence the cyclization is forced to the take the 6-*endo* path *via* intermediate **9b** to the observed product **10b**.

Scheme 2

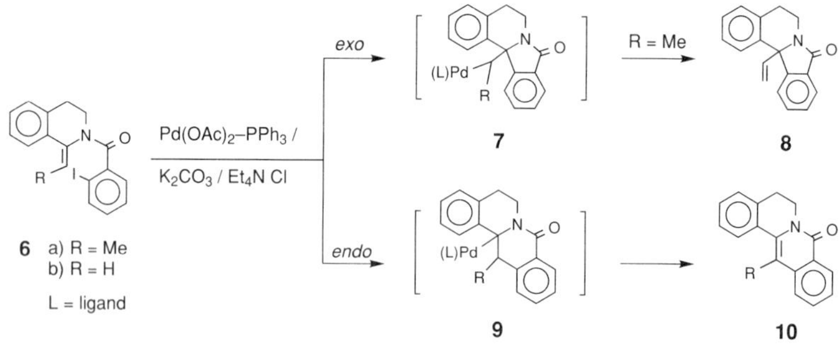

The above type of situation is frequently encountered, as further exemplified in the reactions of allylic alcohols shown in Scheme 3. Both the open chain and cyclic allylic alcohols **11** and **15** undergo palladium-catalyzed cyclizations in *exo* fashion to yield aldehyde **14**[20] and bicyclic ketone **18**,[21] respectively. A similar treatment of substrate **19**, in contrast, yields *endo*-cyclized products **22** and **23**. Here, again, the 5-*exo*-cyclized intermediate **20** with its quarternary β-

carbon cannot eliminate palladium, and cyclization is compelled to take the 6-*endo* route to **21** which regiospecifically eliminates hydrogen from the ring carbon to yield **22**, accompanied by **23**, obviously from isomerization of **22**.[21]

Scheme 3

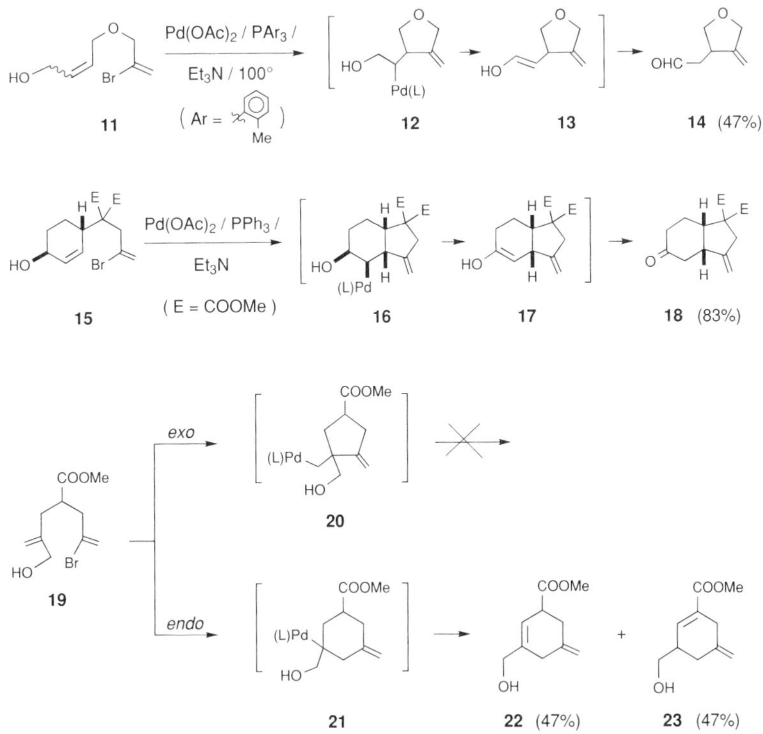

Applications of the Heck reaction to more complex molecules include the synthesis of annelated indoles **25** (Equation 1)[22] and 2-azabicyclo[3.3.1]nonanes **27**,[23] which latter forms a subunit of the *Strychnos* alkaloids (Equation 2).

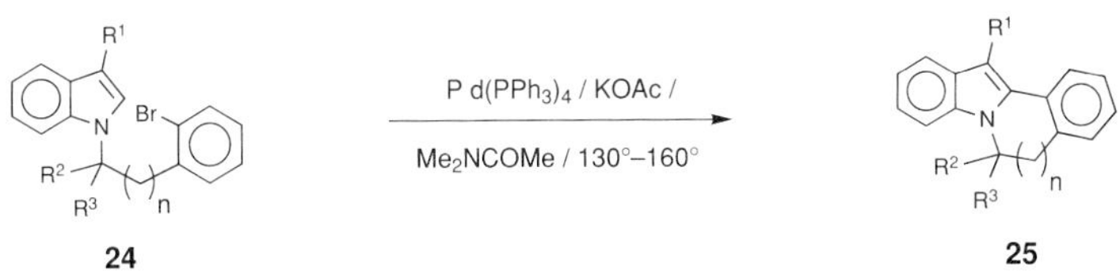

$R^1 = R^2 = R^3 = H$, n = 0	(86%)
$R^1 = H$, R^2 , $R^3 = O$, n = 0	(72%)
$R^1 = (n\text{-Pr})_2N\text{–CO–}CH_2\text{–}$, $R^2 = R^3 = H$, n = 0	(74%)
$R^1 = (n\text{-Pr})_2N\text{–CO–}CH_2\text{–}$, $R^2 = R^3 = H$, n = 2	(81%)

Equation 1

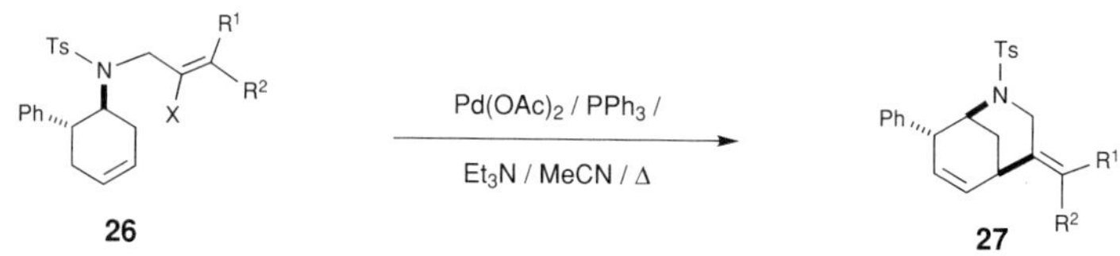

X = Br , R^1 = R^2 = H (85%)
X = I , R^1 = H , R^2 = Me (90%)
X = I , R^1 = Me , R^2 = H (90%)

Equation 2

Catalytic asymmetric Heck reaction is also possible. This was first realized in the cyclization of vinyl iodide **28** using 2,2´-bis(diphenylphosphino)-1,1´-binaphthyl (*R*)-BINAP,[24] as the chiral ligand, and silver carbonate as trapping agent for the hydriodic acid formed in the reaction, whereupon *cis*-decalin derivative **29** was obtained with an enantiomeric excess of 46% (Equation 3).[25] Higher asymmetric induction has been observed in the Heck-type cyclization of vinyl triflate **30** employing (*S*)-BINAP as the catalyst. In this case the ee value for product **31** is as high as 80% (Equation 4).[26] It should be pointed out here that, according to the proposed mechanism in Equation 5,[27] it is totally feasible for the intermediate **33** (see also **2** in Scheme 1) in the Heck reaction to be captured by a hydride or other nucleophiles, before elimination can take place. The mechanism accounts for the formation of bicyclic acetate **31** (Equation 4), a key intermediate in the synthesis of triquinane derivatives such as **32**.[26]

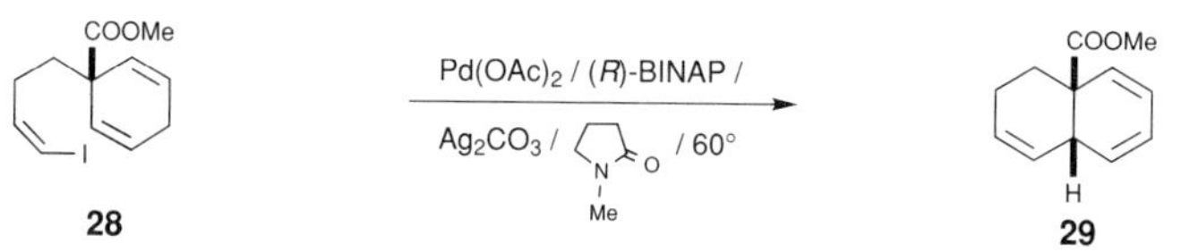

BINAP = 2,2´-bis(diphenylphosphino)-1,1´-binaphthyl

(74% ; 46% ee)

Equation 3

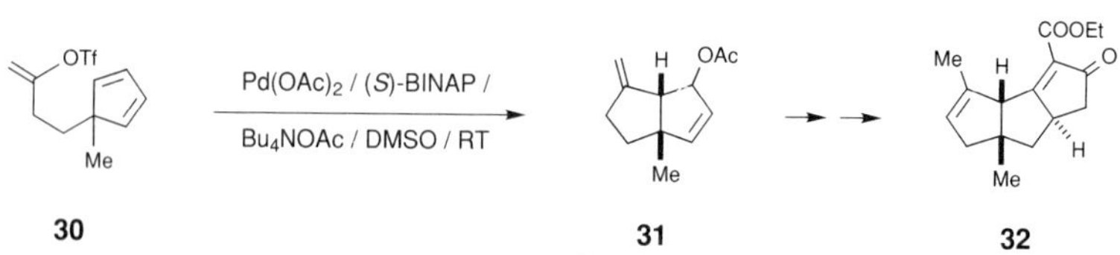

(89% ; 80% ee)

Equation 4

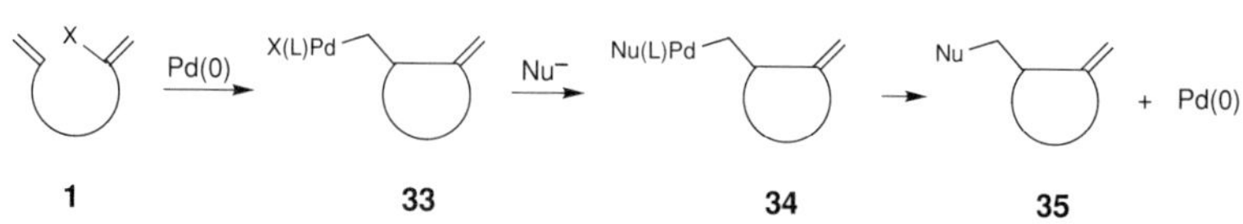

Equation 5

It was almost predictable, with the arrival of the Heck reaction, that its application to polycyclization, or tandem cyclization, would soon follow. A tandem Heck cyclization would involve successive cyclizations in one pot that would allow the simultaneous construction of multiple rings. The concept is illustrated in Equations 6 and 7. It should be reminded that although the cyclization steps depicted are all in the *exo* mode, the unfavorable *endo* cyclization is also possible, provided that the substrate possesses the appropriate structural features.

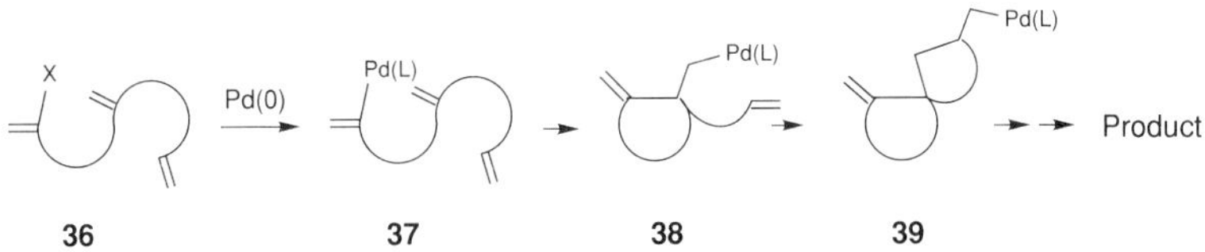

Equation 6

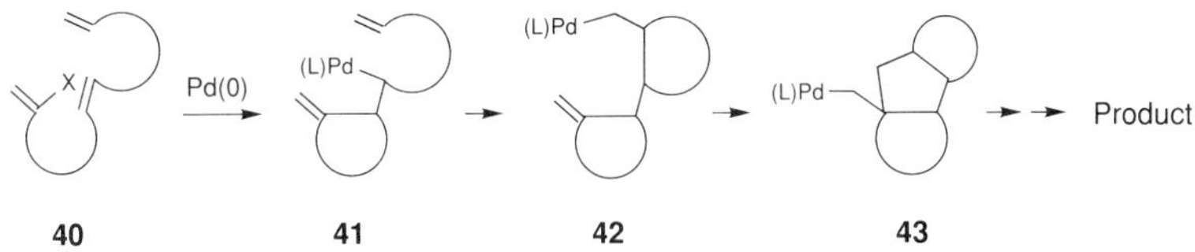

Equation 7

A good example of the tandem Heck reaction is the cyclization of iodo compound **44** which includes two consecutive 5-*exo* cyclizations *via* the organopalladium intermediates **45** and **46** to afford spiroproducts **47** in good yields (Equation 8).[28] Likewise, a tandem construction of the spirotricyclic hydrindenone **51** can be accomplished with trienyl triflate **48** as substrate (Equation 9).[29]

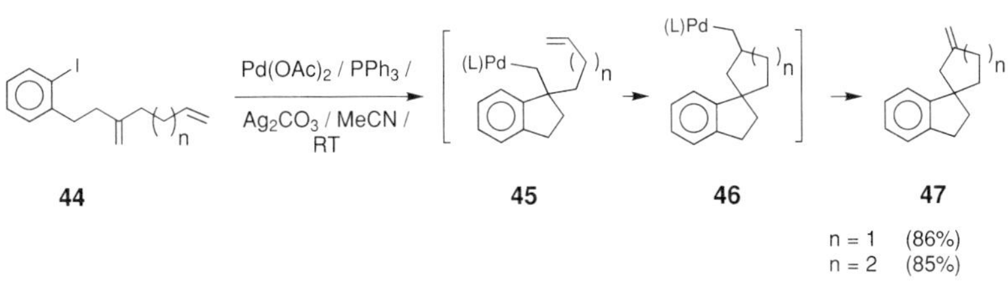

Equation 8

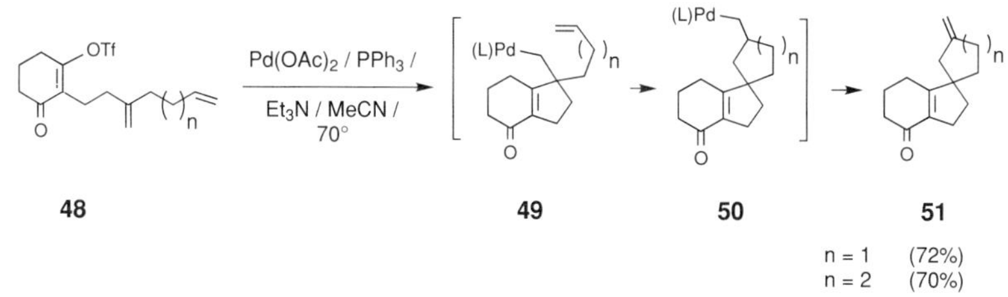

Equation 9

Even more dramatic tandem Heck cyclizations are demonstrated in the reactions of bromodienynes **52** and **59**, as shown in Schemes 4 and 5, respectively.[30] The double Heck cyclization of dienyne **52** follows the normal reaction course through intermediates **53** and **54**, and subsequent *β*-hydride elimination to give triene **55**, which, however, does not survive the prevailing thermal reaction conditions and spontaneously cyclizes in an electrocyclic reaction to yield tricycle **56**, thus completing the construction of three rings in one single operation. The reaction works equally well with or without protection of the hydroxy substituent (R = OH or OMe). Likewise, the tetracyclic carbocycle **63** can be assembled in a super short, and efficient, two-step synthesis: first aldehyde **57** is reacted with bromoacetylene **58**, to provide precursor **59**, which is then subjected to the second, one-pot, tandem double Heck cyclization–electrocyclic reaction, as illustrated in Scheme 5.

Scheme 4

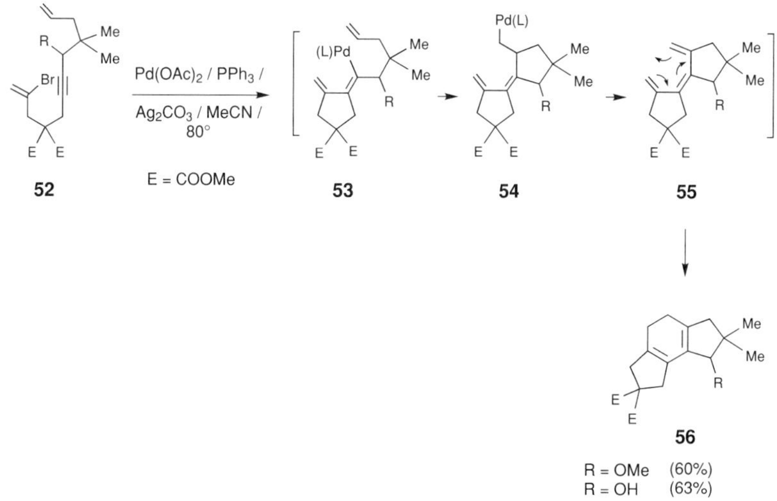

Scheme 5

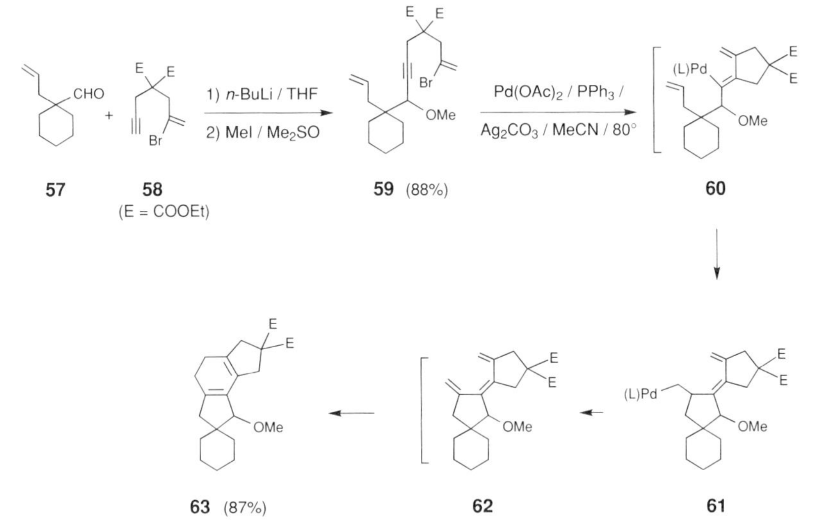

An interesting case for study is the cyclization of system **64**, whose molecular design prompts a triple-Heck cyclization sequence. Following the double cyclization to form intermediate **65**, the third ring closure has, in principle, two options for *exo* mode cyclization, *i.e.*, the 5-*exo* mode to yield **66** or the 3-*exo* mode to give the three-membered ring intermediate **69** whose strain is offset by the stable π-allylpalladium complex thus formed. Experimental results show the cyclization of substrate **64** to yield exclusively **70** *via* 3-*exo* cyclization (Scheme 6).[31]

Scheme 6

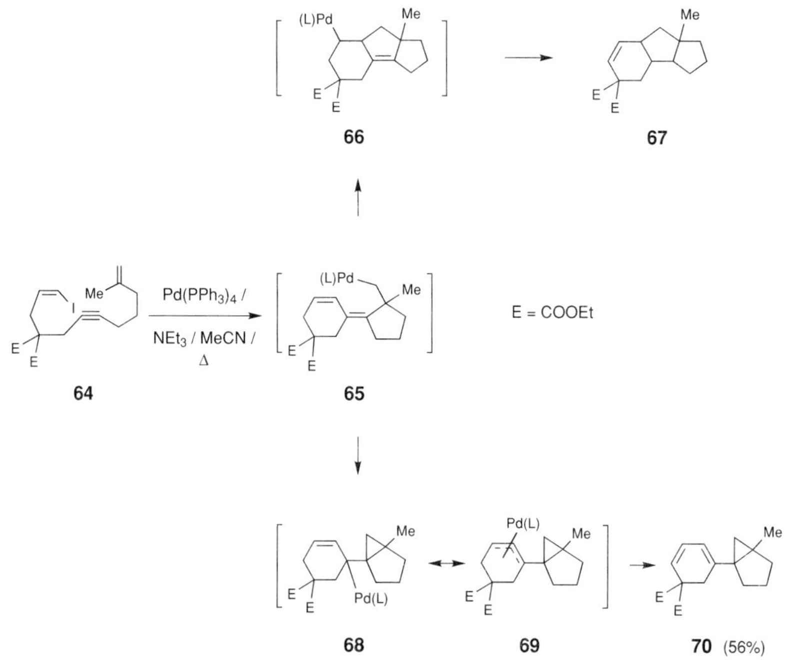

Nevertheless, with suitable modifications of the open chain substrate, the above reaction can be maneuvered to yield the alternative triquinane skeleton **67**. Accordingly, iodotrienyne **71** undergoes triple cyclization to afford **73** in very high yields (90 to 95%), as depicted in Equation 10.[32] Meanwhile, increasingly higher orders of tandem Heck cyclizations are being applied to the construction of increasingly more complex molecules. The reaction shown in Equation 11[32] is one version of a quadruple cyclization involving three consecutive 6-*exo* and a final 5-*exo* cyclization. Another, the cyclization of bromodienyne **77** shown in Scheme 7, closely resembles the reactions discussed in Schemes 4 and 5. In this case, however, the organopalladium intermediate **79** resulting from the double cyclization of **78** is unable to eliminate to the triene (see **54** and **61** in Schemes 4 and 5) since its β-position is completely blocked, and therefore reverts to further tandem cyclization in 5-*exo* and 3-*exo* fashions, successively, to yield ultimately tetracyclic **81** as the product.[33]

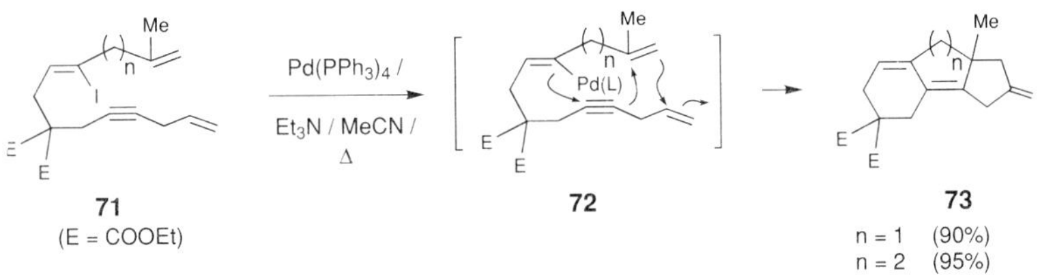

Equation 10

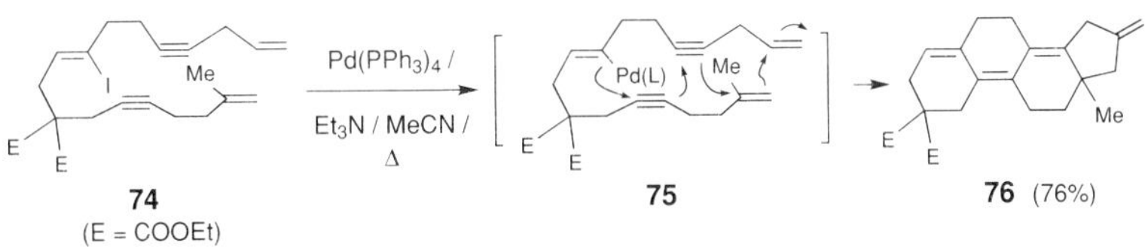

Equation 11

Scheme 7

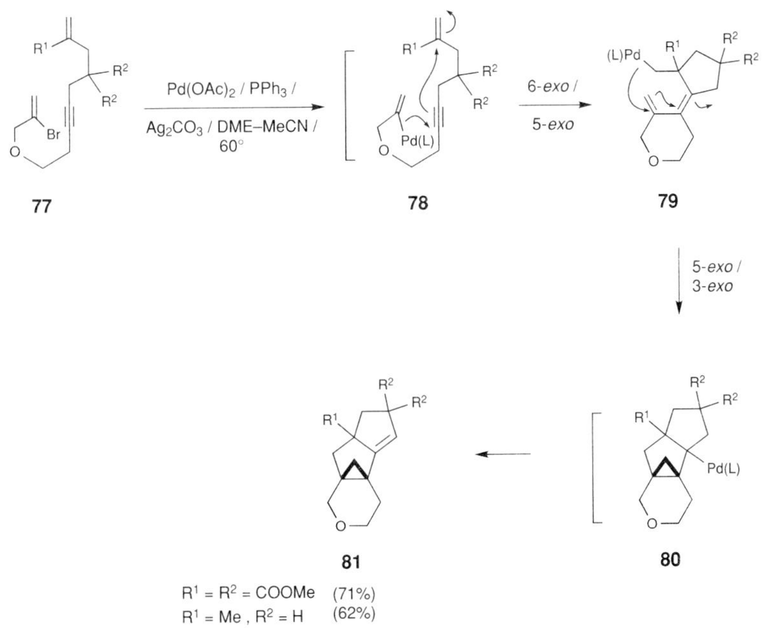

The only limitation of polycyclization, it appears, is the imagination of the chemist in the design of substrate for cyclization. Hence, with time, we can probably expect the reaction to be developed to yield more and more sophisticated structures. In addition, modifications of reaction conditions are offering increasing variations, *e.g.*,the double cyclization of substrates **82** and **84** under anion capture conditions (as in Equation 5) leads to tricyclic heterocycles **83** and **85**, respectively. In these examples (Equations 12 and 13),[34] sodium formate is used as the hydride ion source.

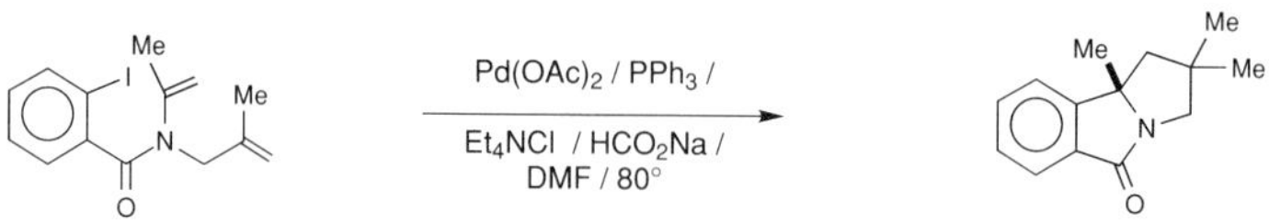

82 **83** (70%)

Equation 12

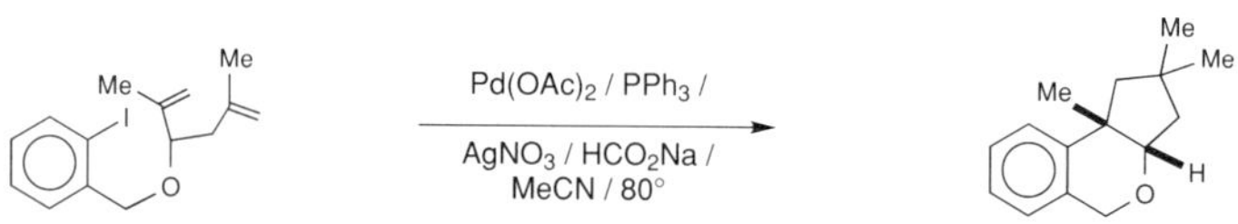

84 **85** (70%)

Equation 13

The complementary nature of the Heck and radical cyclizations is beautifully demonstrated in the reaction of substrate **86**, which yields different stereoisomeric products under the different reaction conditions, as shown in Scheme 8. Radical cyclization of substrates such as **86** had formerly been demonstrated in the synthesis of prostaglandins by Stork *et al.*,[35] and trapping of the resulting intermediate radical with β-stannyl enone was also reported.[36] Accordingly, reacting iodide **86** with stannyl enone **87** under radical cyclization conditions yields product **89**. On the other hand, subjecting substrate **86** to intramolecular Heck cyclization followed by intermolecular Heck reaction of the resulting complex **92** with enone **90** leads to product **94**,[37] an epimer of compound **89**.

Scheme 8

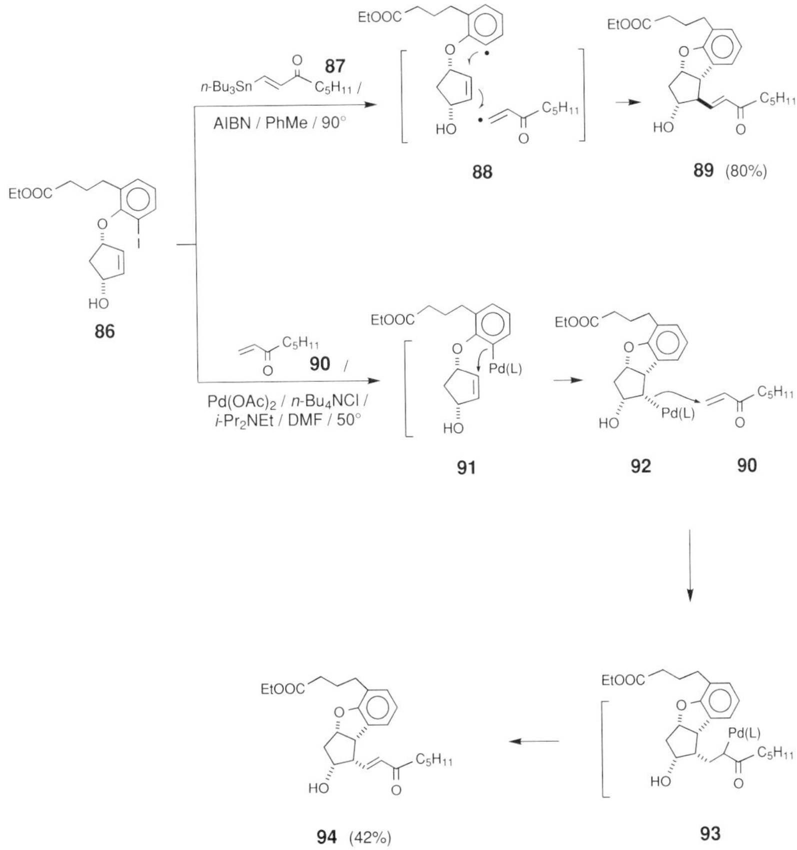

Metal-catalyzed cyclizations of enynes (*e.g.*, **95**), diynes (**96**), and enallenes (**97**) have been widely studied (Scheme 9). Numerous examples employing an assortment of metal complexes have been reported and several reviews have appeared.[2,3,14,38] A general working mechanism for these reactions is outlined in Scheme 10. The initial addition of palladium complex Pd(H)LX (L = ligand, X = leaving group, *e.g.*, acetate) to the alkyne bond of **95** forms complex **101**, which then inserts into the olefinic bond to give the cyclized intermediate **102** followed by β-elimination to yield the diene product **99**. However, if the reacting system contains a hydride source such as polymethylhydrosiloxane (PMHS, a very convenient and suitable hydride source for palladium-mediated cyclizations[39,40]), the hydride will displace the leaving group in **102** to give **103** and then insert between the carbon–palladium bond to yield **98** as the final product, rendering the overall process a reductive cyclization.

Scheme 9

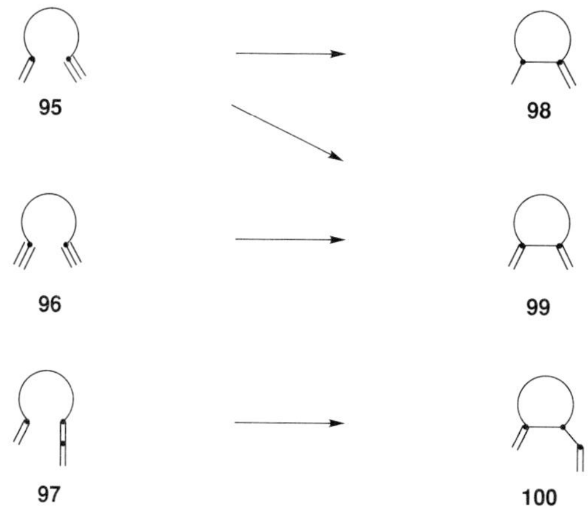

Scheme 10

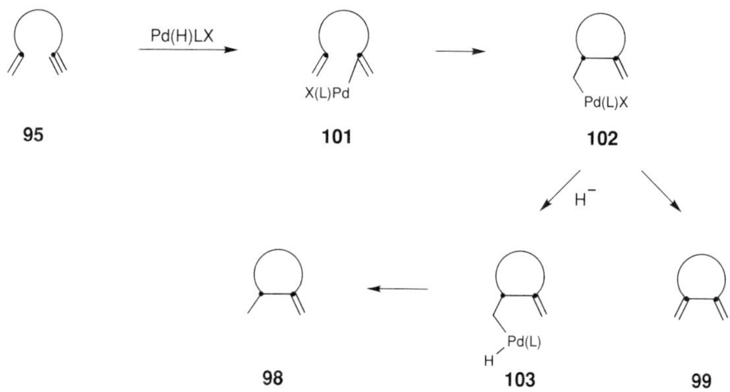

Examples of reactions of the types represented in Scheme 9 are shown in Equations 14 to 21. Equations 14 and 15 demonstrate the use of reductive cyclization of enyne (see **95** → **98** in Scheme 9) in the syntheses of phyllanthocin **106**[41] and the tumor promoter, phorbol **109**,[42] respectively. Also exemplified are the syntheses of cyclic dienes from enynes (Equations 16 and 17)[38] and diynes (Equations 18 and 19),[43] as well as enallenes (Equations 20 and 21),[44] the latter using the polymer support nickel–chromium complex [(Ph_2P-*P*-polystyrene)$NiCl_2$–$CrCl_2$], a novel and very effective catalyst for the cyclization of enynes.[45]

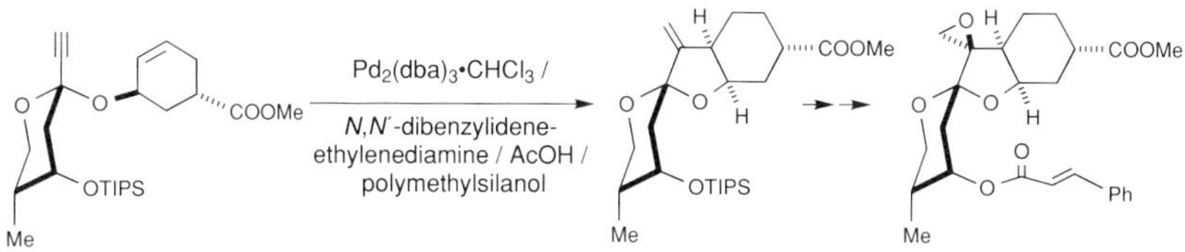

104 **105** (78%) **106**

TIPS = $(i\text{-Pr})_3Si$
dba = dibenzylideneacetone

Equation 14

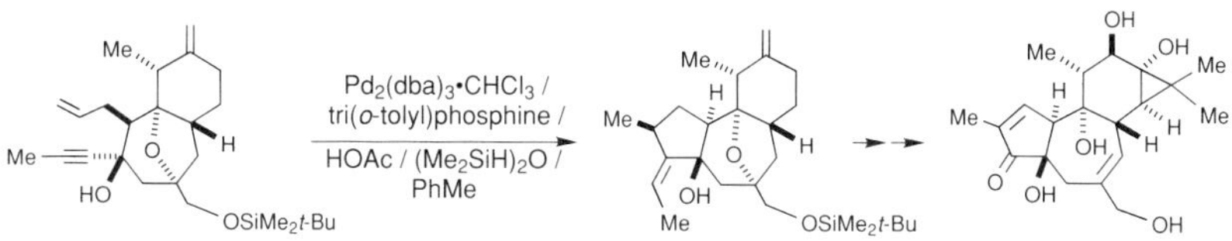

107 **108** (58%) **109**

Equation 15

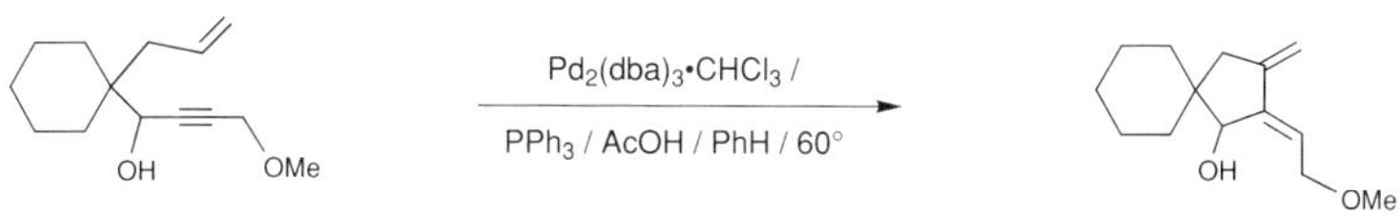

110 **111** (76%)

Equation 16

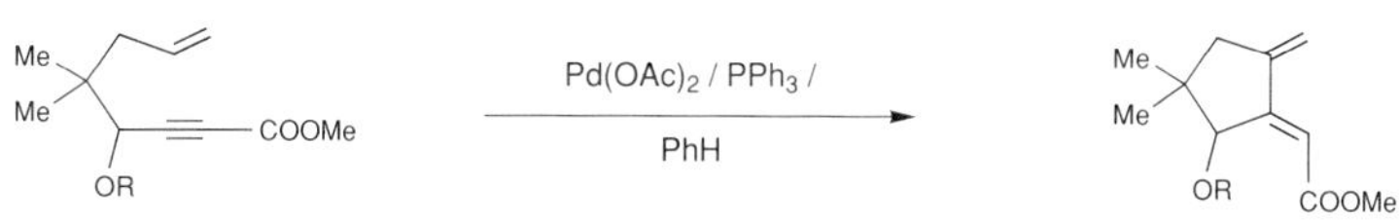

112 **113**

R = H , $SiMe_3$, $SiMe_2t\text{-Bu}$, COMe (50–71%)

Equation 17

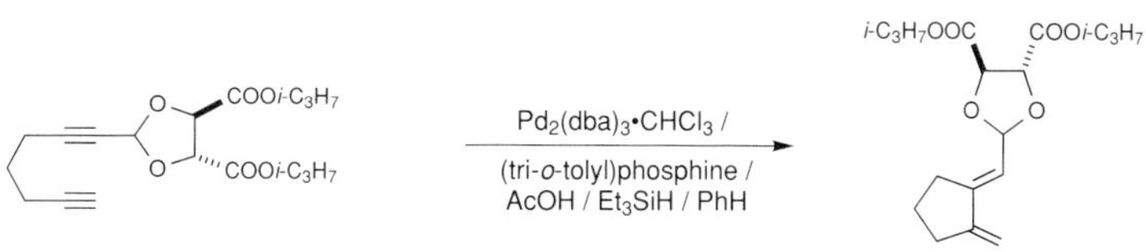

114 **115** (51%)

Equation 18

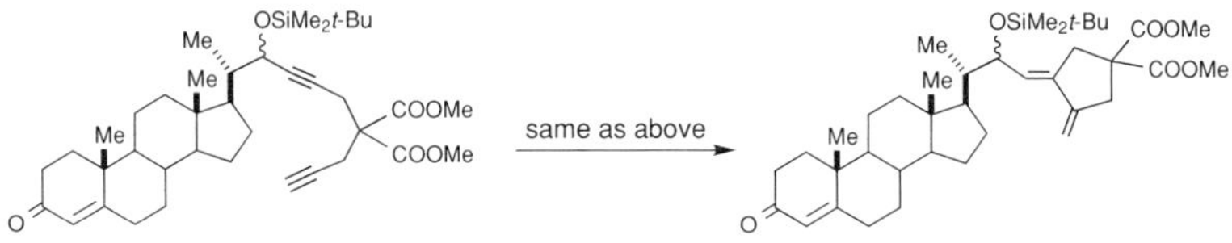

116 **117** (84%)

Equation 19

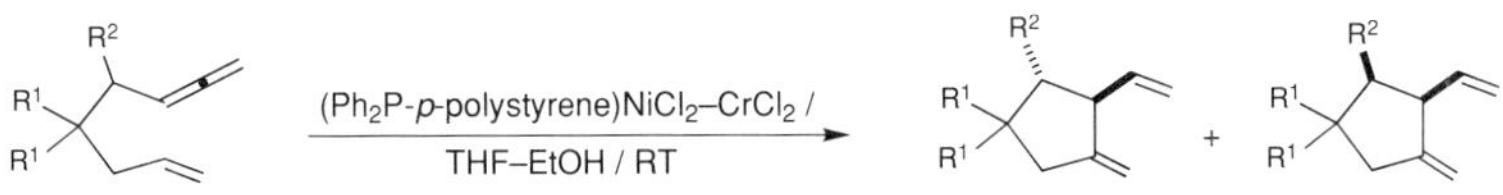

118 **119** **120**

R^1 = COOMe , R^2 = H (80%)

R^1 = Me , R^2 = OH (55%) 3 : 1

Equation 20

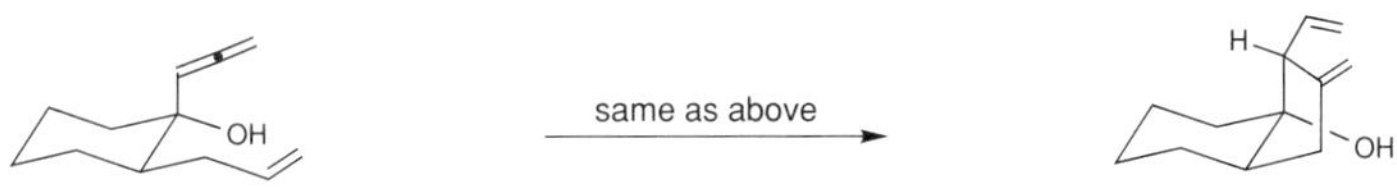

121 **122** (72%)

Equation 21

Extension of the metal-catalyzed cyclization of the enyne system (which forms diene derivatives according to Equations 16 and 17) has led to palladium-catalyzed alkylative cyclization, as shown in Scheme 11. Alkylative palladation of the acetylenic bond in **123** with vinyl bromide **124** leads to intermediate **125** which cyclizes in Heck fashion to yield triene **127** *via* the metal complex **126**. Under the prevailing reaction conditions the triene further undergoes electrocyclic cyclization to form stereospecifically bicyclic compound **128** in 60% overall yield.[46] Such alkylative cyclization has been applied successfully to the synthesis of calcidiol **132**, a vitamin D derivative, employing the reaction between enyne **129** and vinyl bromide **130**, as illustrated in Scheme 12.[47]

Scheme 11

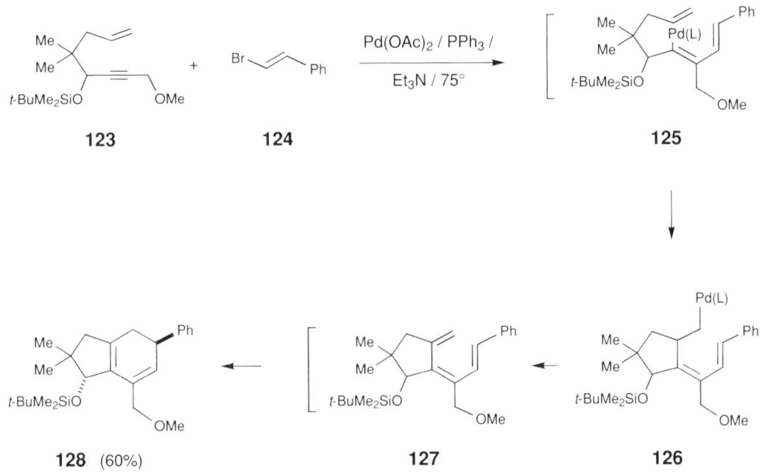

Scheme 12

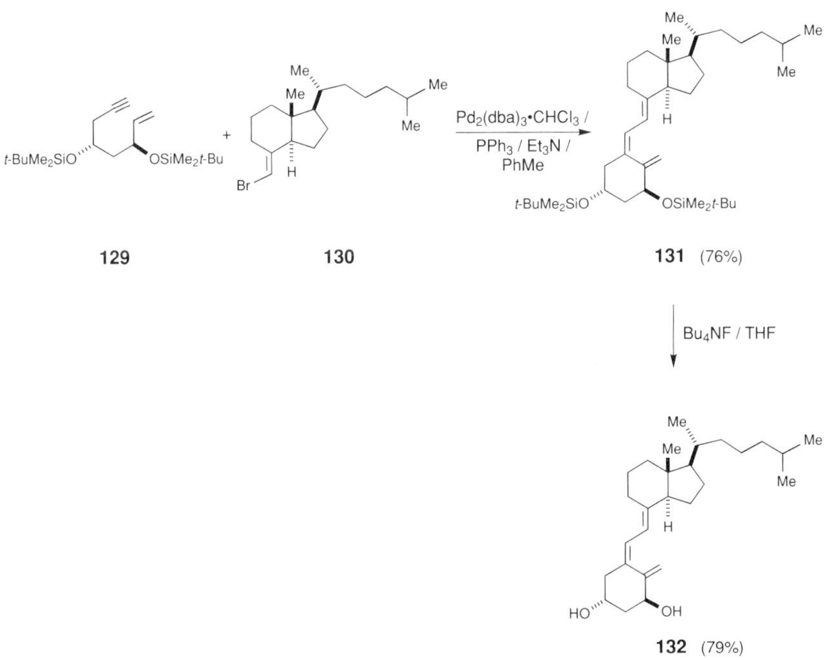

Ring formation *via* palladium-catalyzed displacement of halides or other leaving groups from aromatic substrates, as shown in Scheme 13, is essentially the same type of reaction as discussed above. Here the intermediate **135** is intramolecularly captured by the stabilized enolate moiety to give palladium ring complex **136** which eventually collapses to the carbocycle **137**.[2]

Scheme 13

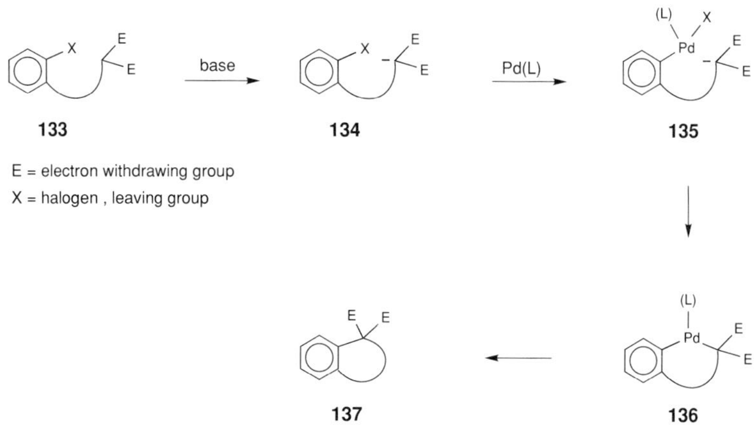

Synthetic applications of the palladium-catalyzed enolate cyclization outlined in Scheme 13 are quite varied. The reaction offers access to diverse structures such as the spirostructure **139** (Equation 22),[48] bicyclic keto alkene **141** (Equation 23),[49] angular triquinane **143** (Equation 24),[49] and the methylenecyclopentane skeleton **145** (Equation 25).[50] The last synthesis exemplifies the less common insertion of palladium at the allylic carbon, using PhO as the leaving group in place of the usual halogen. This reaction has been explored and developed using a variety of substrates under various reaction conditions.[51–54]

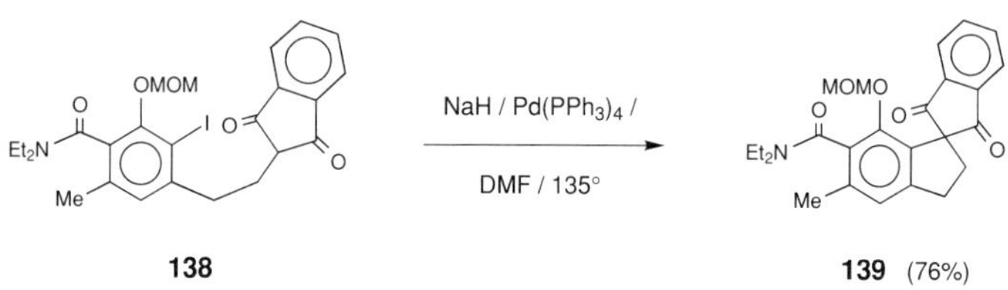

Equation 22

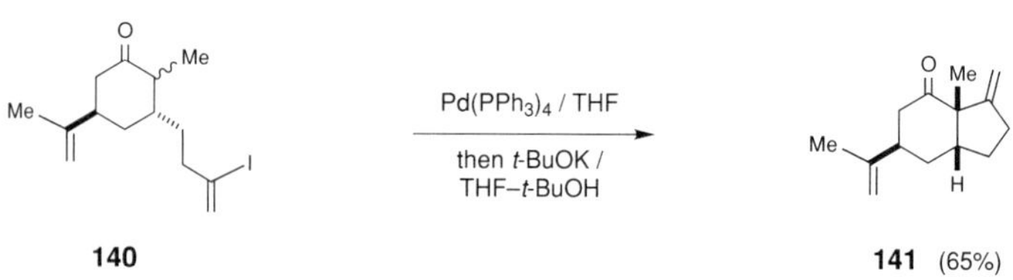

Equation 23

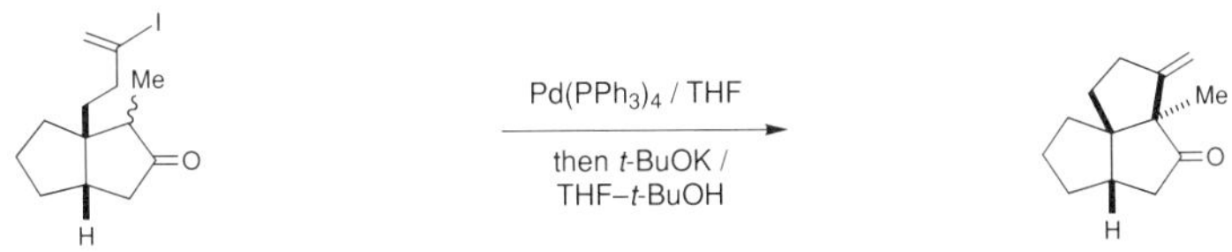

142 **143** (65%)

Equation 24

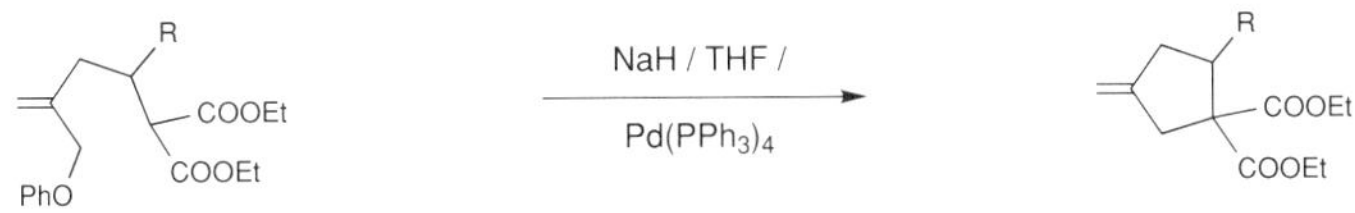

144 **145**

R = Ph (85%)
R = Me_2CH (83%)

Equation 25

B. CYCLIZATION INVOLVING π-ALLYL METAL COMPLEXES

Ring formation *via* π-allyl metal complexes is, strictly speaking, a variation of the Heck cyclization. The process, generally represented as shown in Scheme 14, involves the generation of a π-allyl complex **147** from the reaction of substrate **146** with a metal catalyst, followed by cyclization by insertion into the olefinic bond in proximity to yield **148**, which may then eliminate to product **100** or undergo further reaction with reactive groups present in the molecule. However, because the β-elimination reaction (**148** $\rightarrow$ **100**) is fast, further reactions of **148** are not often encountered. It is interesting to note that the strategic transformation of **147** $\rightarrow$ **148** is, in actual fact, the formal (and well-known) metallo–ene reaction.[13,55] Here the most popularly used metal is palladium or nickel, and the most commonly employed group X the acetate (OAc) or carbonate (OCOOMe) group. An alternative mode of generation of the π-allyl complex **147** is from enallene **97**, therefore, in retrospect, the polymer support nickel–chromium-catalyzed cyclizations of enallenes depicted in Equations 20 and 21 are also π-allyl metal complex reactions.[44]

Scheme 14

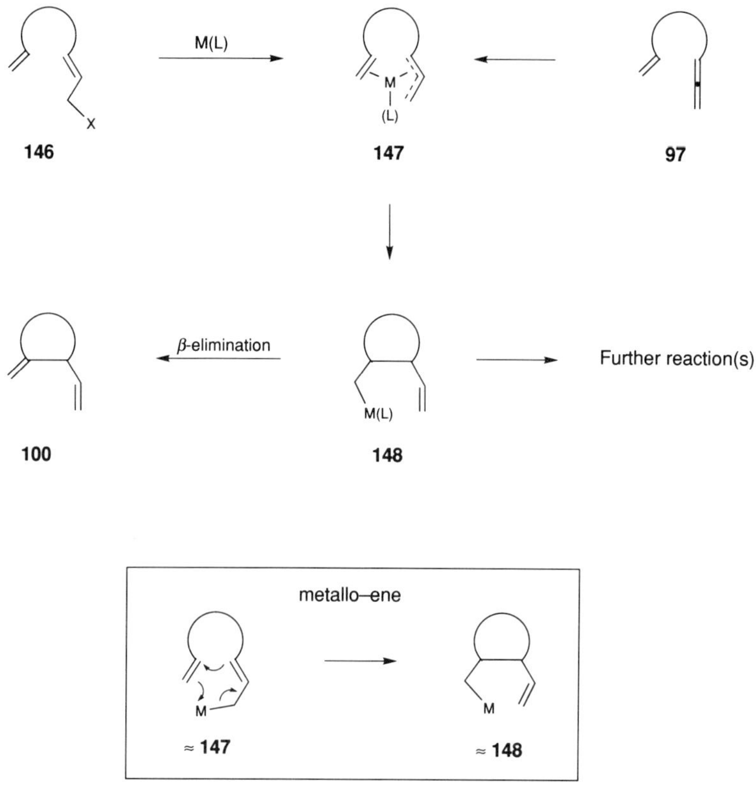

Heating allyl acetate **149** at 118°C in acetic acid with $Pd(dba)_2$–PPh_3 yields the "palladium–ene"–*β*-elimination product **153**, which is thought to arise *via* intermediates **150**, **151**, and **152** (Scheme 15). According to the mechanism, removing the steric interaction inherent in transition state **151** (between the sulfone and palladium moieties) should enable the reaction to proceed more efficiently under less severe conditions. True enough, substrate **154** was subsequently found to undergo a regio- and stereocontrolled cyclization at 80°C to give product **155** in a very much higher yield.[56]

Scheme 15

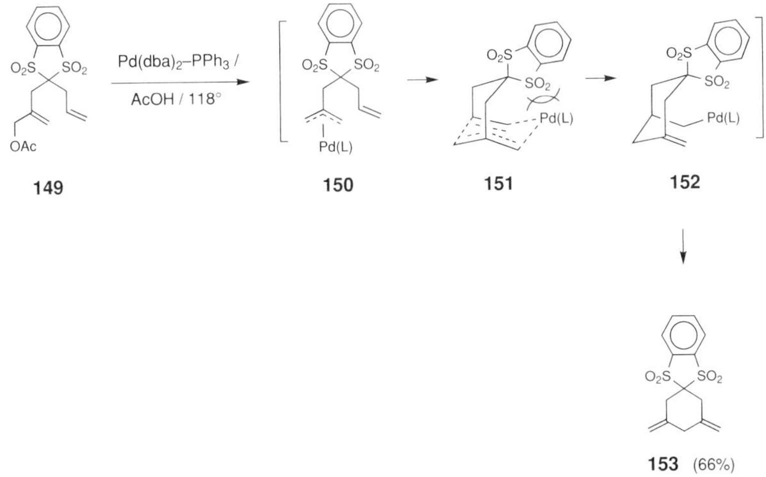

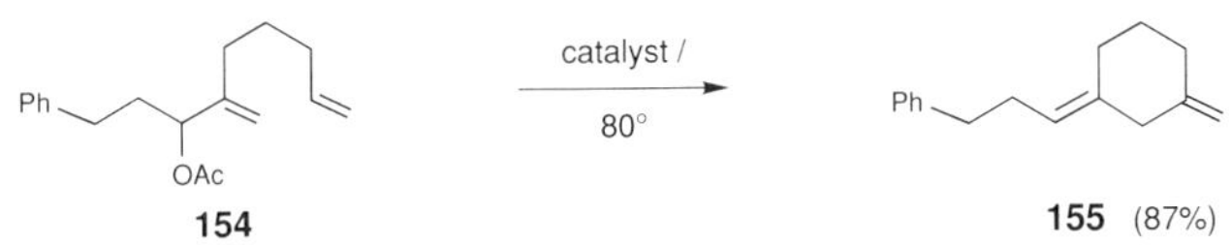

A similar reaction involving the π-allylpalladium intermediate is the cyclization of isomeric allyl acetates **156** and **157** to give the cyclopentane derivative **158**. The overall transformation is ring formation with migration of the acetate group, probably *via* intermediates **159** and **160**, as depicted in Scheme 16. The geometry of the double bond in the substrate apparently has no influence on either the yield (62 to 63%) or the stereochemistry of product (*trans*:*cis* = 1.4:1).[57]

Scheme 16

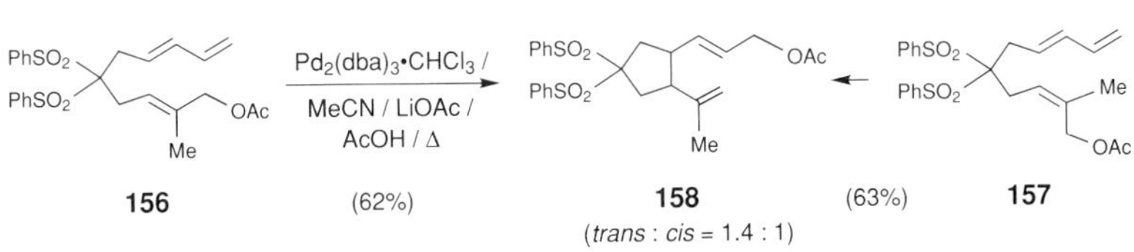

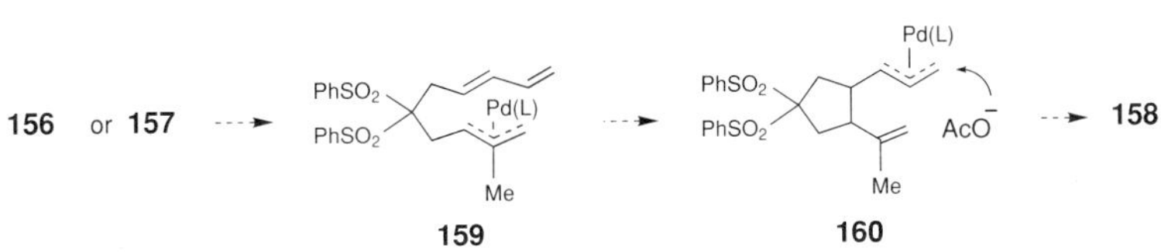

An interesting observation regarding the generation of π-allylpalladium complexes was made in the study of dienol **161** (Scheme 17),[58] whose treatment with palladium catalyst and acetic acid yielded bicyclic pentadiene **162** accompanied by triene **163**. This latter was found to convert readily to **162** upon retreatment with the catalyst under the former reaction conditions. This conversion reveals that the resourceful palladium catalyst is able to form a π-allylpalladium complex with diene **163**.[58] Other examples of such generation of π-allylpalladium from dienes include the reaction of iodoarene **167** with diene **168** and palladium acetate in the presence of base (Na_2CO_3), which yields cyclized product **171**. In this case the mechanism probably entails the generation of Pd(0) from palladium acetate and insertion into the carbon–iodine bond to form **169**, which then reacts intermolecularly with diene **168** to produce the π-allyl complex **170**. Once generated, the intermediate complex is intramolecularly captured by the ester enolate, the whole sequence thus culminating in the formation of product **171**, as summarized in Scheme 18.[59]

Scheme 17

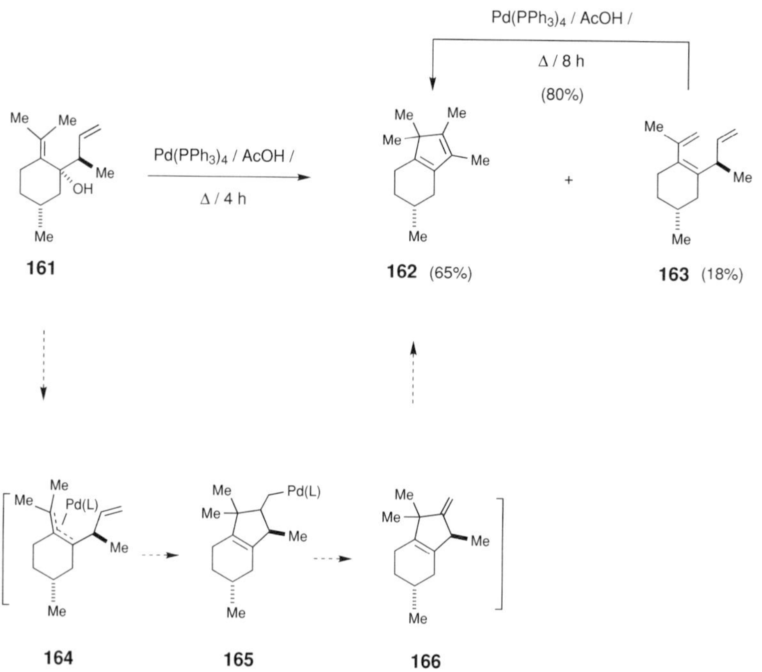

Scheme 18

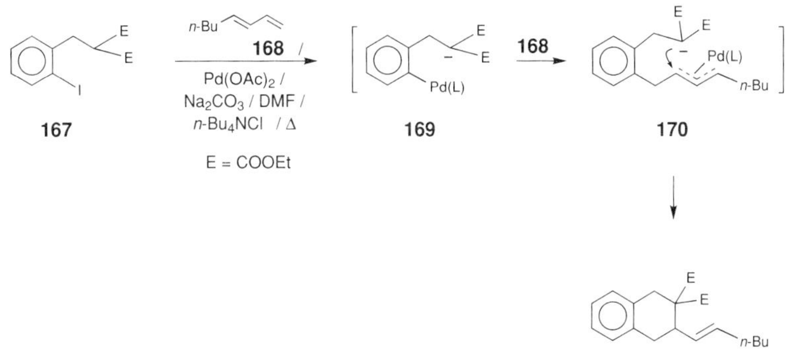

167 **169** **170** **171** (87%)

Incidentally, one of the most interesting reactions to be reported concerns the regiospecific generation of π-allylpalladium complexes from diene intermediates. In brief, identical treatment of bromotosylate **172** and its isomer **177** with palladium catalyst in the presence of sodium carbonate leads to the 5,5-fused bicyclic nitrogen heterocycle **176** in the former, but to the 5,6-fused heterocycle **181** in the latter, case. The rationale for this disparity is illustrated in Scheme 19. Both substrates **172** and **177** first undergo the Heck cyclization to yield diene complexes **174** and **179**, respectively. Reorientation of palladium then follows in a completely regiospecific manner to generate distinct π-allylpalladium complexes **175** and **180** which lead, respectively, to products **176** and **181**. Since no cross-over products are detected in the cyclization of the two substrates it is rationalized that either the π-complexes **174** and **179** do not interconvert, or that the rate of their interconversion is so slow as to be negligible when compared to the regiospecific formation of the π-allyl complexes **175** and **180**.[60]

Scheme 19

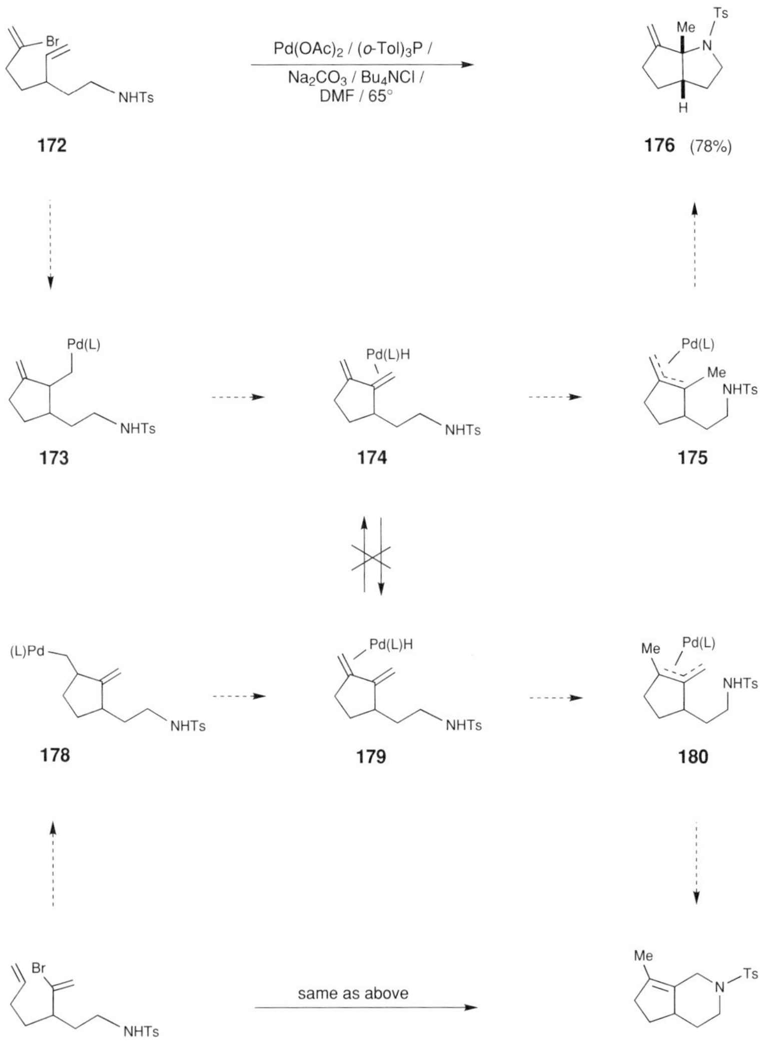

The use of 2-(trimethylsilylmethyl)allyl acetate **182** as a dipole synthon **182a** in palladium-catalyzed cyclization has been investigated and the reaction mechanism considerably well studied.[61] For example, methylenecyclopentane **184** is easily prepared in high yield from the reaction between **182** and crotonate ester **183** with palladium catalyst and base. Extension of the reaction has already found much use in organic synthesis.[62–64] For instance, the methylene bicyclic compound **190** can be synthesized in one pot from substrate **186**, which is a derivative of **182**. The course of reaction is proposed to progress successively *via* π-allylpalladium complexes **187** and **188**, then Michael addition ring closure followed by capture of the π-allyl complex by the resultant anion to yield the bicyclic product, as shown in Scheme 20.[64]

Scheme 20

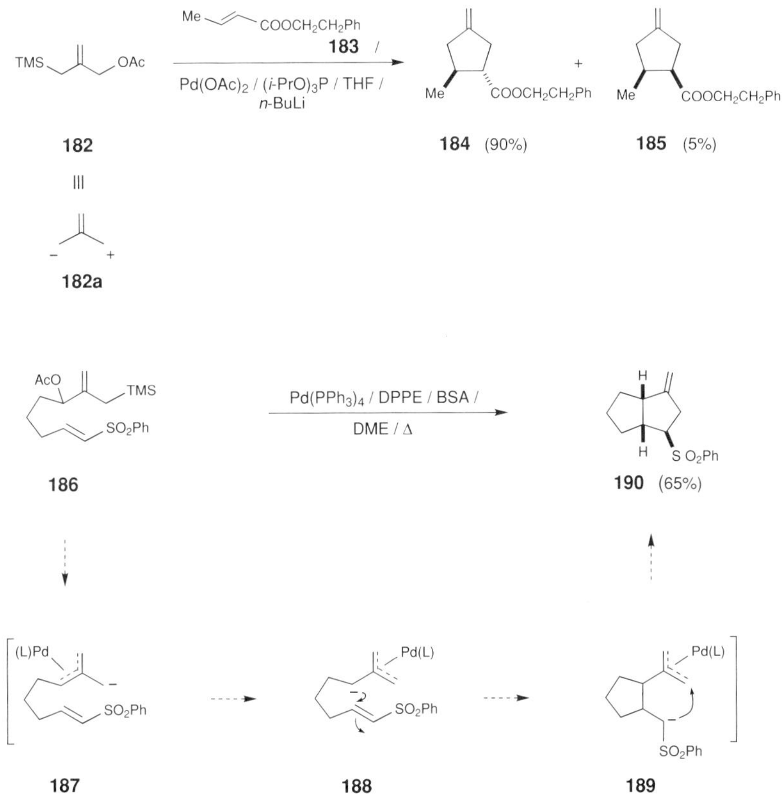

DPPE = 1,2-bis(diphenylphosphino)ethane
BSA = bis(trimethylsilyl)acetamide

Recalling that the cyclized intermediate (**148**) from the reaction of the π-allylmetal complex (generalized in Scheme 14) may either suffer β-elimination or undergo further reactions under favorable circumstances, examples of the latter type of situation are provided in Scheme 21. Palladium-catalyzed cyclization of allylic acetate **191** proceeds *via* π-allylpalladium **192** to the cyclized intermediate **193**, which, lacking hydrogen at the β-position for elimination, can consequently be decomposed with formic acid[65] or sodium tetraphenylborate[66] to afford products **194** and **195**, respectively. However, if performed under the established β-elimination conditions (R–A = triethylamine) the reaction would resort to an alternative course *via* intermediate **196** to yield the strained 6,5,4-fused system **197** (61%).[65]

Scheme 21

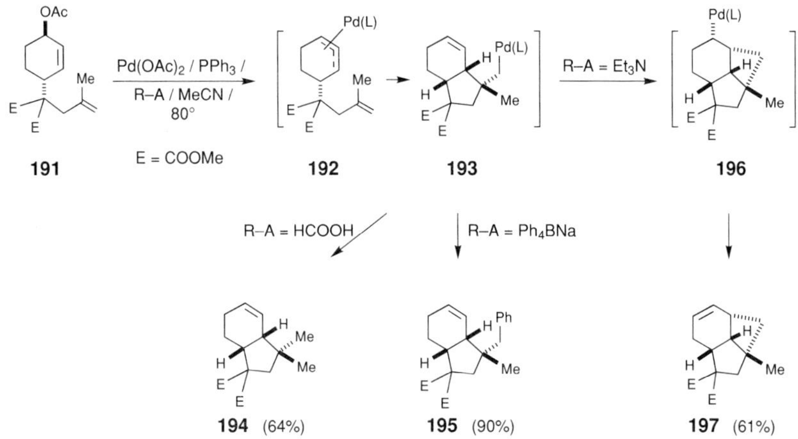

Metal-catalyzed tandem cyclizations that involve π-allylpalladium intermediates are also known and are exemplified in Schemes 22 and 23. The π-allylpalladium complex **199** formed from allyl acetate **198** first cyclizes to the bicyclic complex **200** with a *cis* stereochemistry ring junction,[67] and then endures a second cyclization to yield tricyclic products **202** and **203**,[68] the latter arising from double bond isomerization of the former. On the other hand, starting with the homologous seven-membered ring substrate **204**, the initial cyclization results in a *trans*-fused bicyclic complex **205**, a stereochemical variance attributed to the higher flexibility of the larger seven-membered ring. A sequence of successive cyclizations finally terminate in the tetracyclic product **208** as shown.[68]

Scheme 22

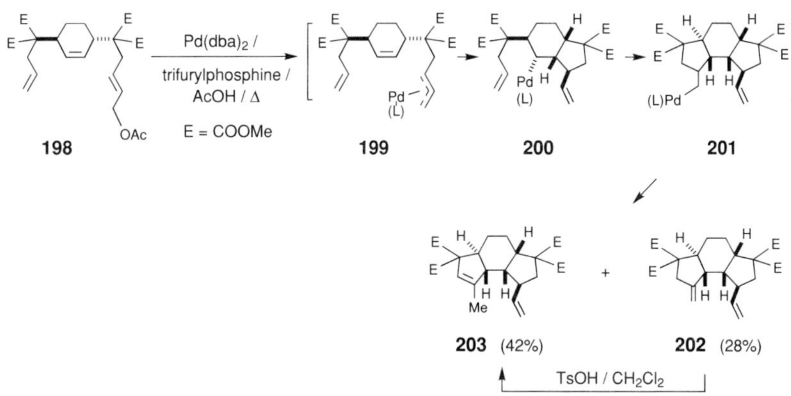

Scheme 23

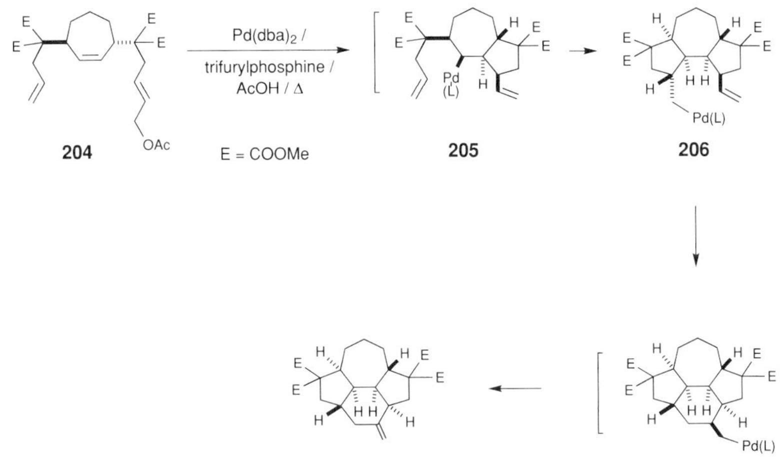

204 **205** **206** **207** **208** (50%)

C. THE STILLE REACTION

The intramolecular Stille coupling reaction is a palladium-catalyzed cross-coupling reaction between an organotin and an organic electrophilic function within the same molecule. The reaction has been explored extensively and widely utilized.[18] Generally represented as shown in Equation 26, the reaction can be viewed as the capture of palladium complex by an internal nucleophile derived from a C–Sn function (**210** → **211** → **99**). The intermolecular version of this anion capture process is also quite widely employed.[66]

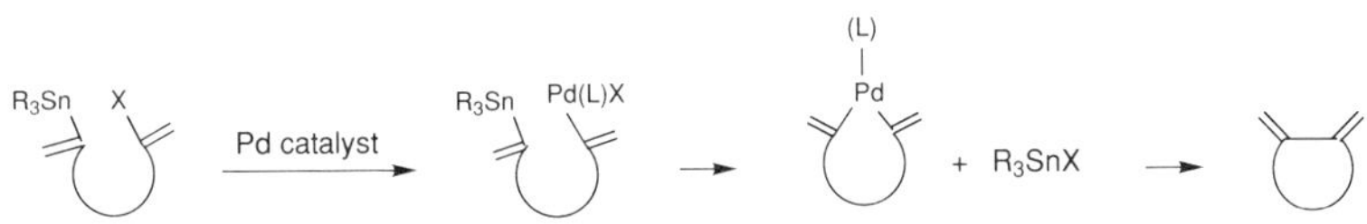

209 **210** **211** **99**

Equation 26

A good example is the palladium-catalyzed cyclization of the geometrical isomers bromostannanes **214** and **217** which gives products **215** and **218**, respectively, *via* the Stille cyclization (Equations 27 and 28).[69] The starting bromostannanes can be regio- and stereospecifically prepared by deprotonation and alkylation of the corresponding vinylstannanes **212** and **216**. The reaction provides a very good general method for the preparation of 1,2-dialkylidenecyclobutanes and has been extended to the synthesis of 2-benzylidenebenzocyclobutenones **221** and **222** from bromoaryl acetylenic ketone **219** (Equation 29). Presumably, under the reaction conditions the intermediate vinylstannane **220** is produced, which, upon heating, couples to give the products.[70]

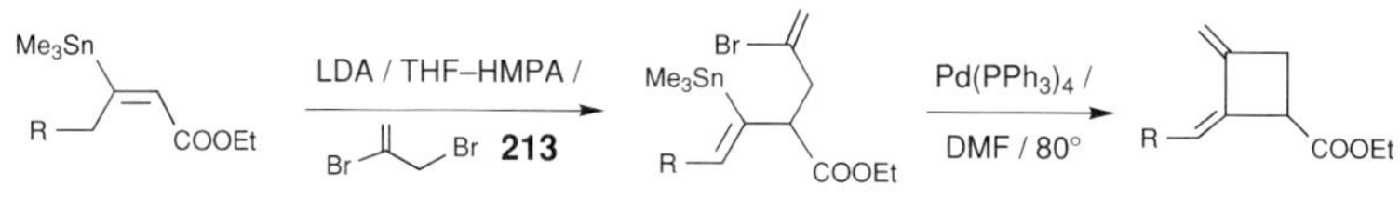

212 **214** **215** (70–95%)

Equation 27

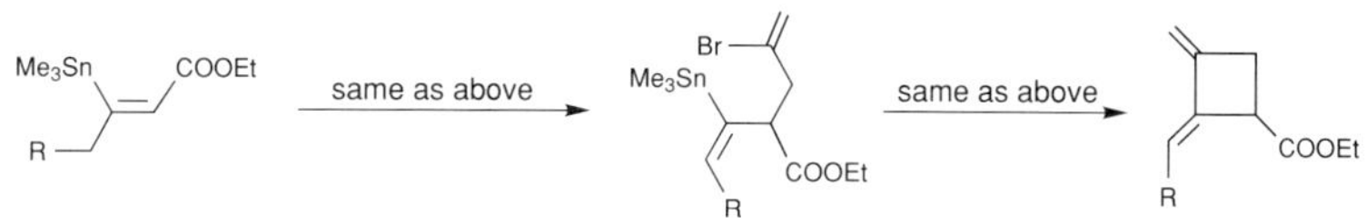

216 **217** **218**

R = Me , -$(CH_2)_2$–O–CH_2–OMe , -$(CH_2)_3$–OSiMe$_2$*t*-Bu (70–95%)

Equation 28

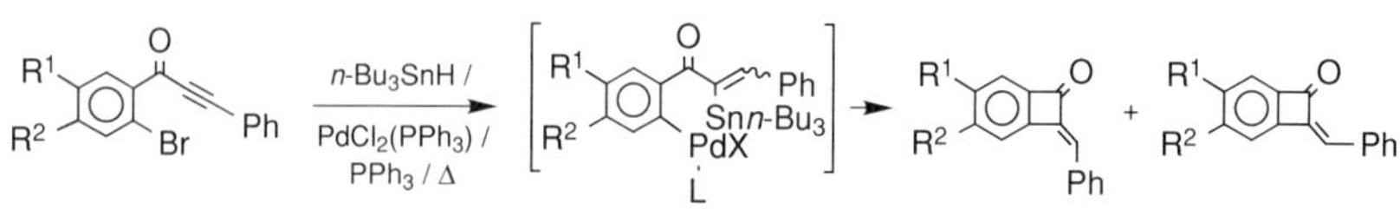

219 **220** **221** **222**

R^1, R^2 = O–CH_2–O	(58%)	1	:	1.1
R^2 = R^2 = H	(70%)	1	:	1.5
R^1 = OMe , R^2 = H	(49%)	1	:	1.3

Equation 29

Quite unexpectedly, the treatment of bromo alcohol **223** under free radical reaction conditions (*n*-Bu$_3$SnH / AIBN / toluene / reflux) yields, not products from the radical intermediate **224**, but vinylstannane **225**, which can undergo the Stille coupling reaction to give 2-benzylidenebenzocyclobutenol **226**, as depicted in Scheme 24.[70]

Scheme 24

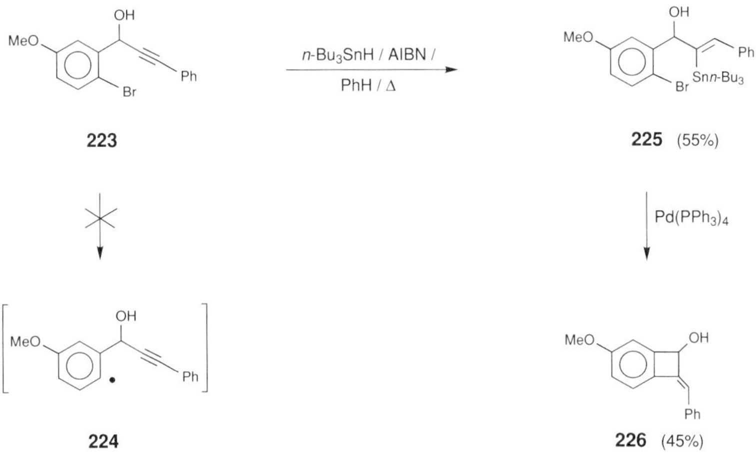

Another reaction thought to involve cross-coupling of the Stille reaction type is the tandem cyclization of **227** with palladium catalyst plus two added equivalents of hexamethylditin ($Me_3Sn–SnMe_3$), which results in spiroheterocycle **229** in 70% yield (Equation 30).[71] The reaction is believed to proceed *via* intermediate **228** which cross couples to give the product, since in the absence of hexamethylditin (*i.e.*, Heck cyclization conditions : $Pd(OAc)_2$ / PPh_3 / K_2CO_3) the reaction, though viable, affords product **229** in a lower yield (60%).

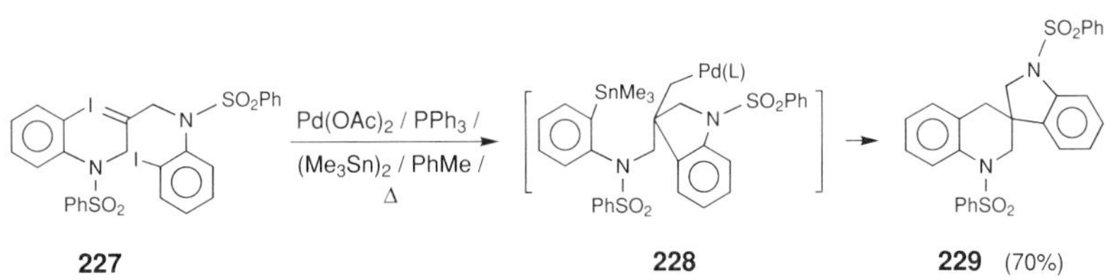

Equation 30

Taking advantage of the above methodology, various bicyclic diene systems with the general structure **234** can be prepared conveniently by the simple method outlined in Equation 31.[72] Alkylation of the cyclic ketone **230** with vinylstannane **231**, followed by conversion of the alkylated ketone **232** to its corresponding enol triflate **233**[18] sets the stage for intramolecular palladium-catalyzed Stille coupling reaction to furnish the desired bicycle **234**.

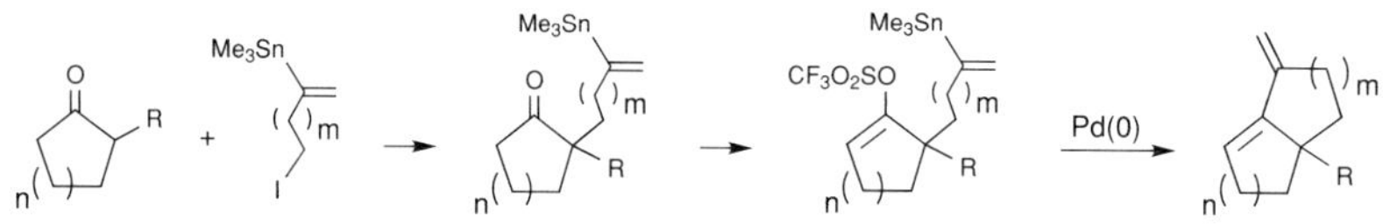

230 **231** **232** **233** **234**

R = H , Me , COOR′ m , n = 1–3

Equation 31

The crucial role of the Stille coupling reaction in the total synthesis of macrolide (*S*)-zearalenone **237** is also illustrative of the synthetic use of the process. Several catalysts and various different reaction conditions were studied for the intramolecular aryl iodide–vinylstannane coupling to form the 14-membered ring lactone **236** from iodostannane **235**, selected examples of which are given in Scheme 25. Although the coupling is afflicted by a fair amount of polymerization, the use of a $Pd(PPh_3)_4$ catalyst on a 20% cross-linked polystyrene support alleviates the problem and delivers 54% yield of the desired product **236** upon refluxing in toluene. Deprotection by removal of the MEM-group readily affords the natural product **237** as shown.[73]

Scheme 25

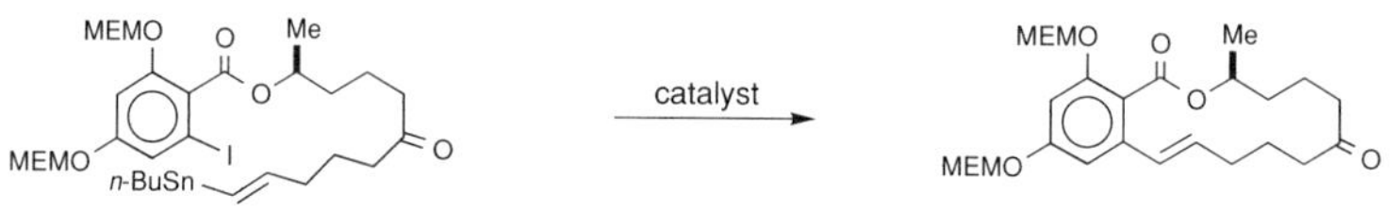

235 **236**

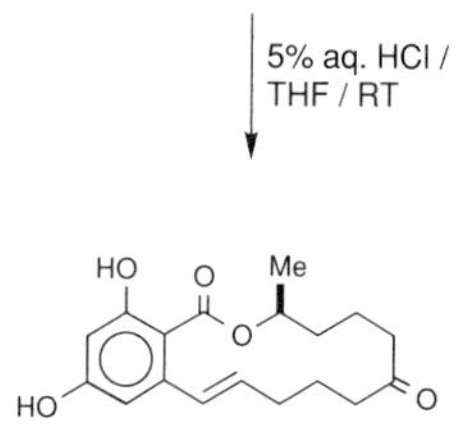

catalyst / conditions	% yield of **236**
$Pd(PPh_3)_4$ / PhMe / 105° / 19 h	30
$PdCl_2(MeCN)_2$ / DMF / 25° / 69 h	–
$Pd(PPh_3)_4$ / PhMe / 105° / 14 h	39
Ⓟ–$Pd(PPh_3)_4$ / PhMe / 105° / 61 h	54

237 (80%)

III. METAL–CATALYZED CARBONYLATIVE CYCLIZATION

The carbonyl group, in the form of carbon monoxide gas or metal ligand, is capable of inserting into the carbon–metal bond prior to cyclization of the latter. In this manner the overall process of metal-catalyzed carbonylative cyclization becomes the linking of two reactive centers to a common intermediary carbon of an added carbonyl unit.[13,55]

A. CARBONYLATIVE CYCLIZATION RELATED TO PALLADIUM-CATALYZED CYCLIZATION

Apart from the foregoing Stille coupling reaction, an intermolecular version of a carbonylative coupling reaction has also been demonstrated,[74] a long time ago, in the total synthesis of (±)-$\Delta^{9,12}$-capnellene **244**. That is, conducting the palladium-catalyzed Stille coupling reaction under an atmosphere of carbon monoxide leads to the corresponding ketone, and reaction with enol triflate **238** and vinylstannane **239** yields divinyl ketone **240**. After successive conversions to **241** and **242** followed by a second palladium-catalyzed carbonylative coupling with **239**, the product **243** is obtained and can be converted easily to capnellene **244** (Scheme 26).[74] The mechanism of the reaction is envisaged as similar to the direct palladium coupling reaction, only in this case the sequence has an added step of carbon monoxide insertion into the carbon–palladium bond (*e.g.*, **246** → **247**) prior to coupling, resulting in the eventual formation of the ketone product **251**, as depicted in Scheme 27.[18]

Scheme 26

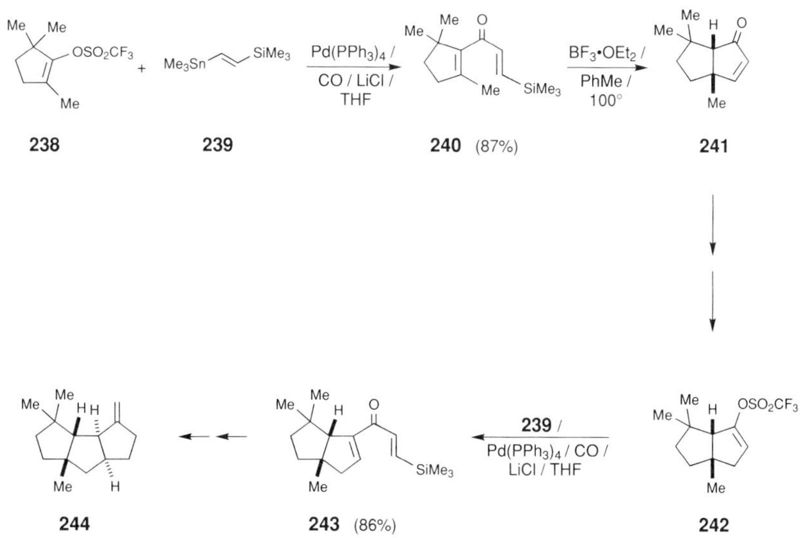

Scheme 27

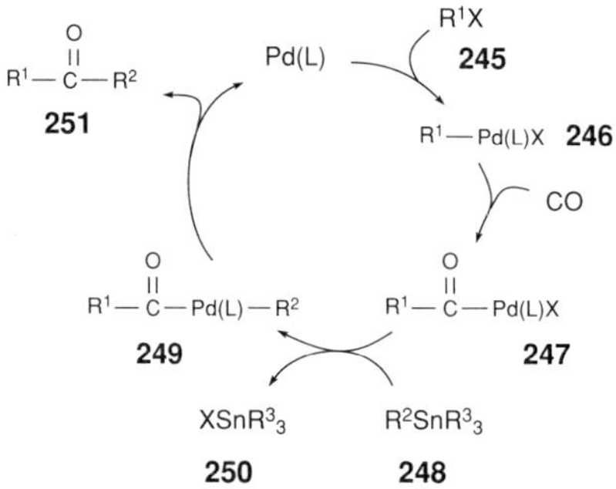

An elegant demonstration of the intramolecular version of the above reaction is seen in the total synthesis of jatrophone **256**, shown in Scheme 28.[75] Important steps of the synthesis include the crucial TASF-catalyzed furanone formation, *i.e.*, from **252** to **253** [TASF = $(Me_2N)_3SMe_3SiF_2$],[76] as other methods lead to a loss of the vinylic tin moiety. The spirofuranone alcohol **253** is then converted to the *Z*-triflate **255** for the carbonylative Stille coupling reaction with palladium catalyst under a carbon monoxide atmosphere at 50 psi, which yields jatrophone **256**.

Scheme 28

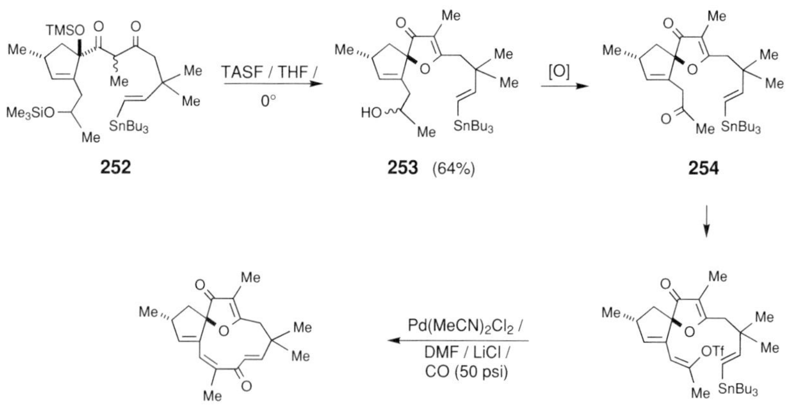

TASF = $(Me_2N)_3SMe_3SiF_2$ [tris(dimethylamino)sulfur (trimethylsilyl)difluoride]
Tf = SO_2CF_3 (triflate)

As regards the carbonylative Heck reaction, there are many reports of both inter- and intramolecular processes which have been expediently exploited in organic synthesis. For example, the carbonylative coupling reaction between aryl iodide **257** and arylacetylene **258** catalyzed by palladium in the presence of 1,1´-bis(diphenylphosphino)ferrocene (DPPF) and 1,8-diazabicyclo[5.4.0]undec-7-ene (DBU) gives the cross-coupling intermediate, acetylenic ketone **259**, which, under the employed reaction conditions, yields a mixture of flavone **260** and aurone **261** from *endo* and *exo* modes of cyclization, respectively.[77] Many examples of enol acting as the nucleophile in cyclization have been reported, *e.g.*, the intramolecular carbonylative Heck reaction shown in Equation 33.[78] In this case the intermediate palladium complex **263** formed undergoes exclusively *O*-cyclization to afford lactone **264**, with no trace whatever of the strained *C*-cyclized product **265**. Meanwhile, Equations 34[78] and 35[79] show extension of the reaction to the preparation of tricyclic lactones which are formed *via* successive carbonylative Heck cyclizations.

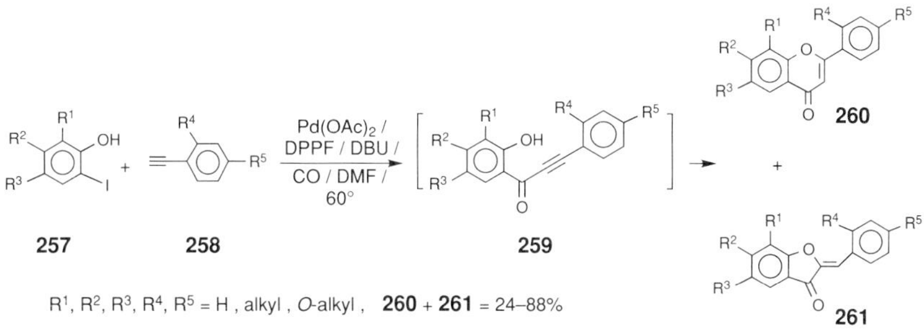

R¹, R², R³, R⁴, R⁵ = H , alkyl , *O*-alkyl , **260** + **261** = 24–88%

DPPF = 1,1´-bis(diphenylphosphino)ferrocene

DBU = 1,8-diazabicyclo[5.4.0]undec-7-ene

Equation 32

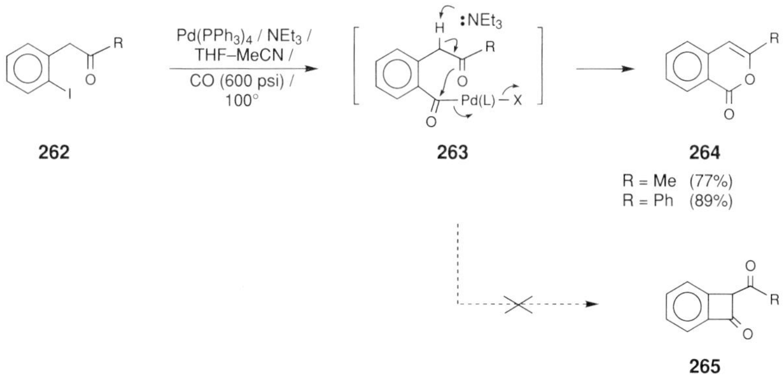

R = Me (77%)
R = Ph (89%)

Equation 33

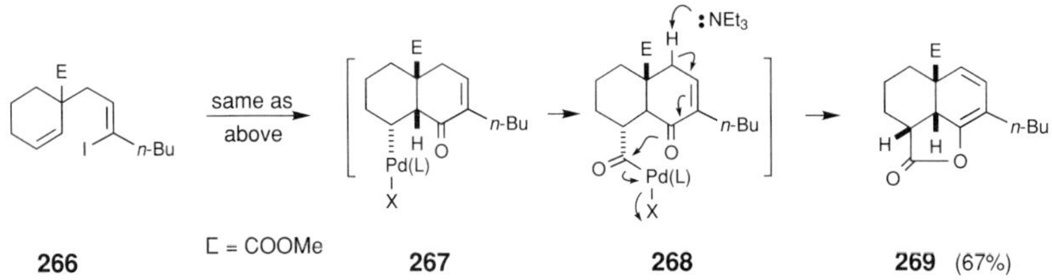

266 **267** **268** **269** (67%)

Equation 34

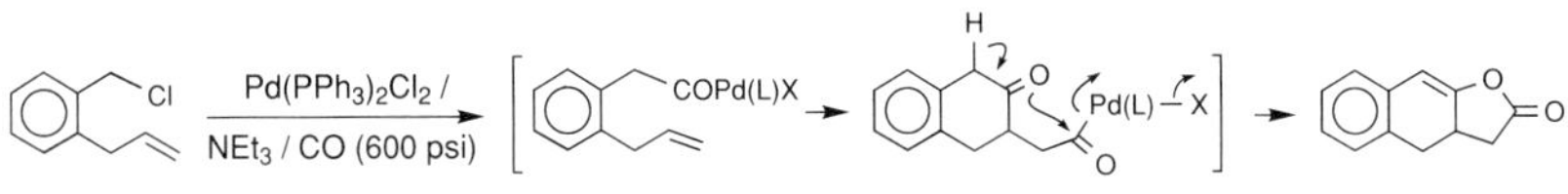

270 **271** **272** **273** (67%)

Equation 35

With respect to carbonylative cyclization that involves π-allylmetal intermediates, cyclization precedes carbonylation. The process has been extensively studied by the group of Oppolzer.[80] Scheme 29 outlines the general reaction path, with more detailed examples provided in the accompanying table. For example, π-allylmetal intermediate **275** (generated from the corresponding allyl acetate or iodide **274**), cyclizes yielding *cis*- or *trans*-metal complexes **276** or **277**. These may be terminated by esterification, either by trapping directly with methanol, or with water followed by esterifying with diazomethane, to afford the corresponding *cis*- or *trans*-vinyl esters **278** or **279**. As an alternative to termination, however, the *cis*-metal complex **276** can also further cyclize to **280** which has the option of β-elimination to methylenecyclopentanone **281** or carbonylation–esterification to bicyclic keto ester **283** *via* the intermediate complex **282**.

The stereochemical outcome of the cyclization of the allylmetal intermediate **275** has been found to be dependent upon the metal catalyst employed. Thus, with nickel catalyst the *cis*-product is formed as the major product,[81] whereas with palladium catalyst[82] the *trans*-product predominates. This can be seen from the table in Scheme 29, where entries 3 and 4 (palladium catalyst) yield predominantly the *trans*-vinyl ester **279**, while entries 1, 2, 5, and 6 (nickel catalyst) yield mainly products derived from the *cis*-complex **276**, either as its direct termination product **278**, or as the double cyclization product, bicyclic keto ester **283**. It should be noted that in only one case (entry 4) is termination by β-elimination to yield methylenecyclopentanone **281** observed, in a small amount.

Scheme 29

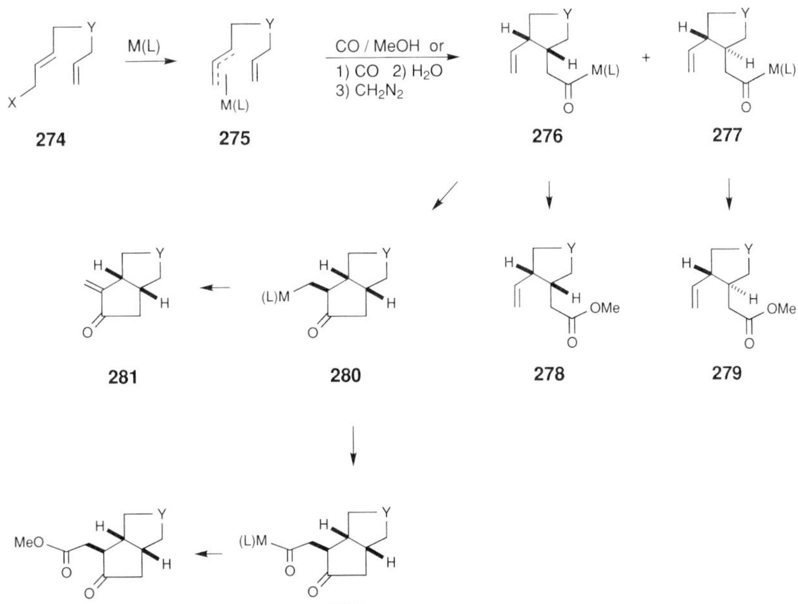

Entry	Y	X	Catalyst	Conditions	%			
					279	278	281	283
1	NTs	I	$Ni(CO)_3PPh_3$	THF : MeOH (4 : 1)/ CO	1	28	–	47
2	NTs	I	$Ni(COD)_2$ / dppb	THF : MeOH (4 : 1)/ CO	–	–	–	80
3	NTs	OAc	$Pd(dba)_2$ / PPh_3	1) CO / HOAc 2) H_2O 3) CH_2N_2	53	1	–	16
4	$C(COOMe)_2$	OAc	$Pd(dba)_2$ / PPh_3	1) CO / HOAc 2) H_2O 3) CH_2N_2	62	1	5	10
5	$C(COOMe)_2$	I	$Ni(CO)_3PPh_3$	THF : MeOH (4 : 1)/ CO	3	15	–	62
6	$C(COOMe)_2$	I	$Ni(COD)_2$ / dppb	THF : MeOH (4 : 1)/ CO	4	4	–	66

The synthesis of [5.5.5.5]fenestranes **290** and **291** from bicyclic allylacetate **284** is an elegant example of application of metal-catalyzed carbonylative cyclization. The process is detailed in Scheme 30.[83]

Scheme 30

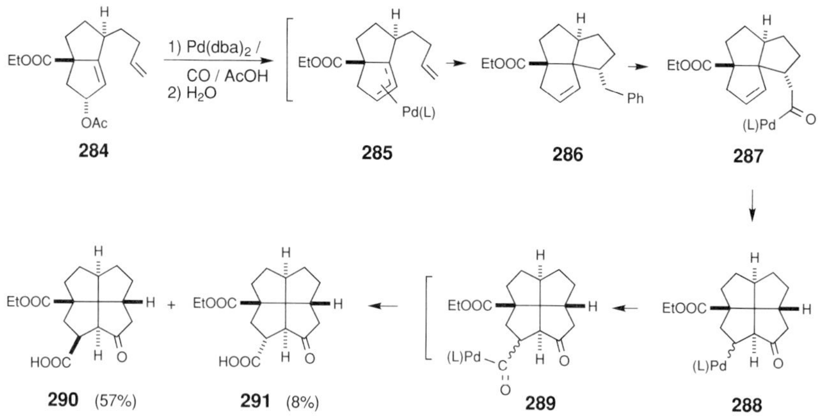

Cyclization of the enyne system *via* the π-allylmetal complex has also been studied, with both nickel and palladium catalysts.[80,82] The reaction has been exploited in the synthesis[84] of bicyclic alcohol **299**, an important precursor in the synthetic route to pentalenolactone E methyl ester **300** obtained from esterification of the extracts of *Streptomyces UC 5319*.[85] Briefly, palladium-catalyzed carbonylative cyclization of acetylenic allyl carbonate **292** under the hydroxylative-esterification conditions discussed earlier (see Scheme 29)[80] furnishes a single stereoisomeric product, bicyclic keto ester **298** (56% yield).[86] This can be converted to **299**, thence the natural product **300** (Scheme 31).

Scheme 31

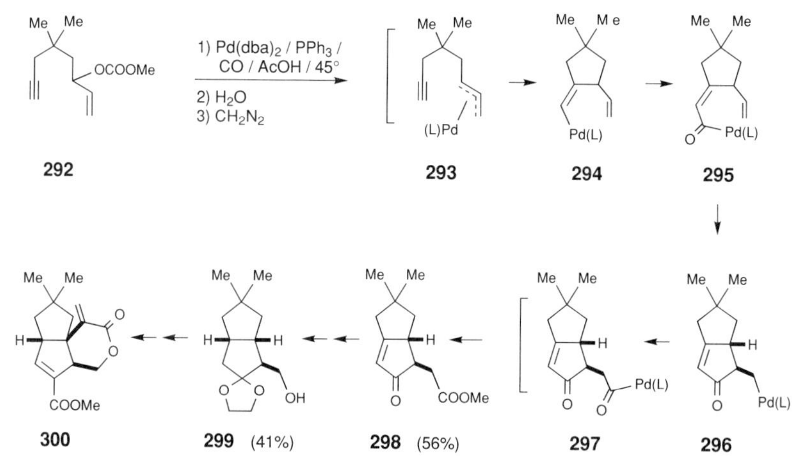

An interesting study is the reaction of the enyne system **301**, an analog of bicyclic allyl acetate **284** (shown in Scheme 30). Under the same reaction conditions as for the cyclization of **284**, the enyne yields tetracyclic enone **305** as the sole product (Equation 36).[83] This termination of reaction by β-elimination is in remarkable contrast to the carbonylative-hydrolysis fate of the final palladium complex earlier observed.

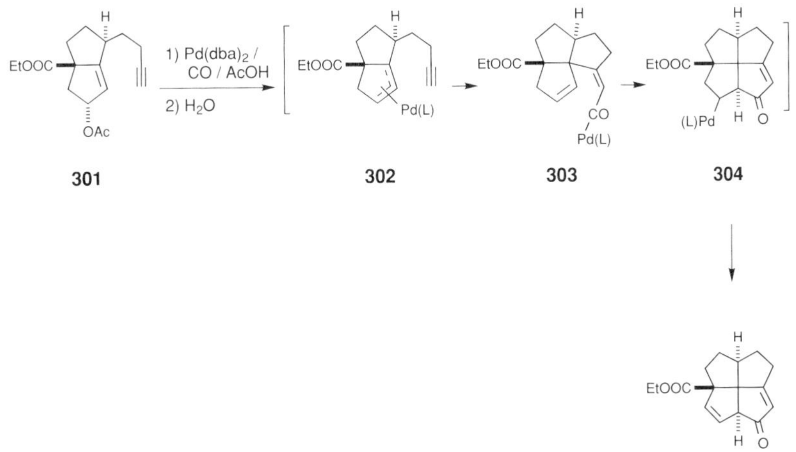

301 **302** **303** **304** **305**

Equation 36

B. THE PAUSON–KHAND AND RELATED REACTIONS

First reported[87] in 1973 and now known by the name of Pauson and Khand, the reaction is essentially a carbonylative cyclization between acetylene, olefin, and carbon monoxide, with dicobalt octacarbonyl as catalyst, to yield cyclopentenone **314**.[8,9] The mechanism of reaction, proposed according to the route depicted in Scheme 32, involves first the generation of stable acetylenehexacarbonyldicobalt complex **309**, into which the olefin and carbon monoxide successively insert, forming intermediates **311** and **312**, respectively. Subsequent decomplexation finally yields cyclopentenone **314**. Recognition of its immense synthetic potential has spurred comprehensive study of the reaction which encompasses wide-ranging aspects such as the regio- and stereoselectivity of reaction,[88] the viability of catalyst recycling in view of an economical industrial process,[89] the catalyst adsorbents for acceleration of reaction,[90] as well as the effects of ultrasound,[91] phosphine oxides[91] and *N*-oxides,[92] on the reaction.

Scheme 32

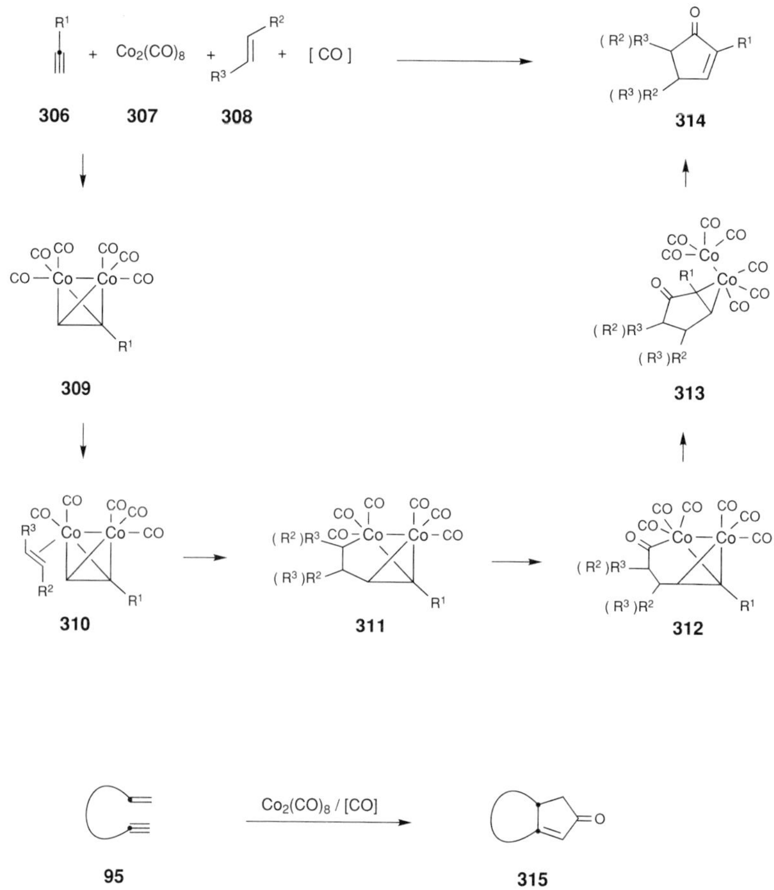

Conventionally, the intermolecular Pauson–Khand reaction is best carried out by separately preparing the acetylenehexacarbonylcobalt complex **309**, then allowing it to react with the olefin, with or without carbon monoxide. This type of complex (**309**) finds other uses too, *e.g.*, as a protecting group for the acetylene function in the synthesis of complex natural products. Once reaction at the required sites have been completed deprotection is then just a matter of simple decomplexation.[93,94] In the case of the intramolecular Pauson–Khand reaction (**95** → **315** in Scheme 32), the net result is the production of bicyclic cyclopentenone **315**. This reaction is exceedingly useful in organic synthesis with its superior yield and minimum problem of regioselectivity commonly associated with the intermolecular process.

However, even in the case of the intermolecular Pauson–Khand reaction it is sometimes possible to achieve regiochemical control in the intermediate **310** (Scheme 32). With substituted acetylenes and alkenes, the appropriate choice of substituents can control the approach of olefin to the acetylene–cobalt complex. For example, the experimental results in Scheme 33 and the accompanying table show the reaction between sulfur-substituted alkene **316** and acetylene-cobalt complex **317** to yield varying ratios of **321** and **322**, but with **321** consistently as the major product.[88] This predominance can be explained in terms of transition state **318**, in which the combined effects of the acetylene methyl group and the heteroatom directing group L (in this case the sulfur atom) dictate the approach of the olefin to the acetylene–cobalt complex. Subsequent reaction leads to 2,5-disubstituted cyclopentenone **321**. Other heteroatoms such as nitrogen or oxygen can also serve to direct the reaction, but not to the same extent as sulfur.[88]

Scheme 33

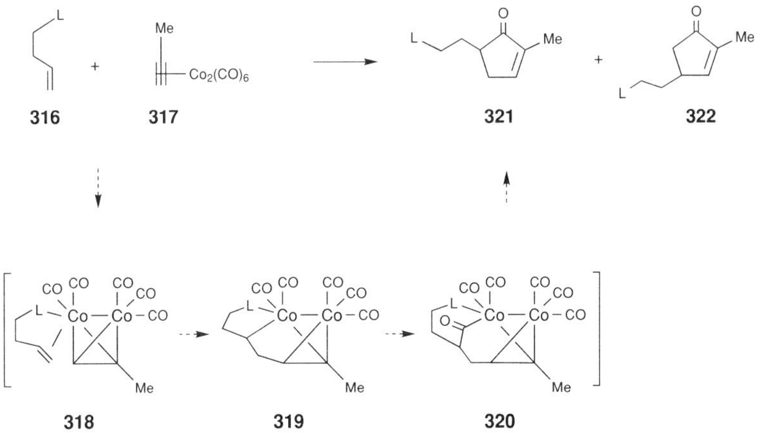

L	% yield (321 + 322)	ratio 321 : 322
SPh	65	3 : 1
S-(4-OMe-C_6H_4)	50	4 : 1
S-*i*-Pr	60	9 : 1
S-C_6H_{13}	65	20 : 1

The heteroatom-directed intermolecular Pauson–Khand reaction outlined above has been applied to the synthesis of cyclopentene acetal **326**, a key intermediate in the synthesis of prostaglandin, PGA_2. The process employs olefinic sulfide **323** and (trimethylsilyl)acetylene complex **324**, as shown in Equation 37.[95]

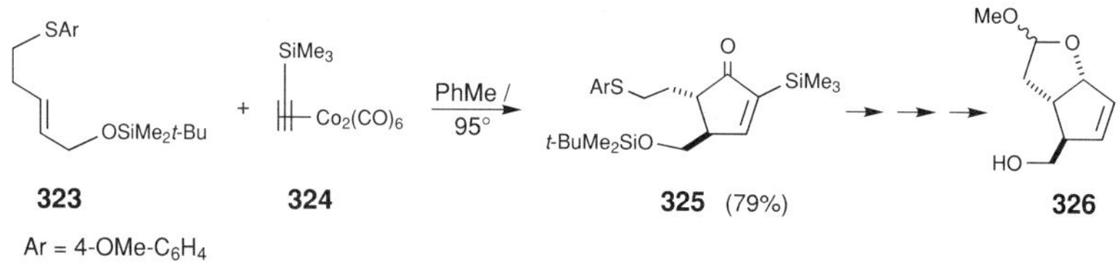

Equation 37

The reaction of complex **327** to yield bicyclic product **328**, shown in Equation 38,[92] represents a well-known and widely used intramolecular Pauson–Khand reaction. However, in contrast to normal reaction conditions which involve heating the starting material in a solvent, here the use of *N*-methylmorpholine-*N*-oxide (NMO) promotes reaction at room temperature and thus effectively avoids damaging functional groups that may be present in the molecule. Comparison of reaction conditions is elaborated in Equation 39, which shows reactions heated at 82°C, or 45°C, and promoted by ultrasound to be all inferior to the *N*-oxide-promoted reaction which proceeds at room temperature, gives a respectable product yield, and, most important of all, results in a much higher selectivity of products (**330**:**331**) over the other conditions.[92] However, the mechanism by which the *N*-oxide functions is still unclear.

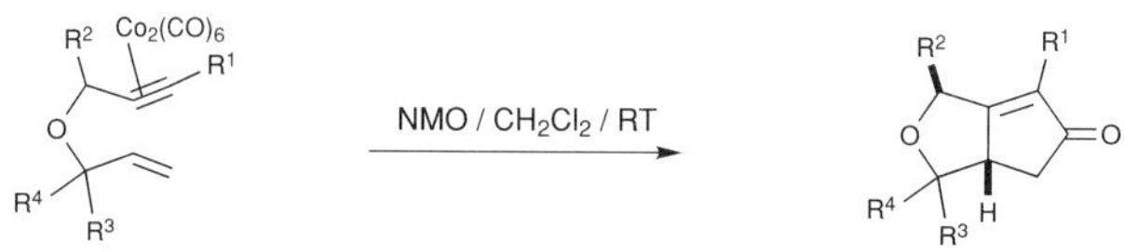

327 **328**

$R^1 = R^2 = R^3 = R^4 = H$ (85%)
$R^1 = R^2 = H$, $R^3 = R^4 = Me$ (92%)
$R^1 = R^3 = R^4 = Me$, $R^2 = H$ (77%)
$R^1 = CH_2-(CH_2)_3$ ∼∼ $SiMe_3$, $R^2 = O$ ∼∼, R^3, $R^4 = H$ (70% ; selectivity = 4 : 1)

NMO = *N*-methylmorpholine-*N*-oxide

Equation 38

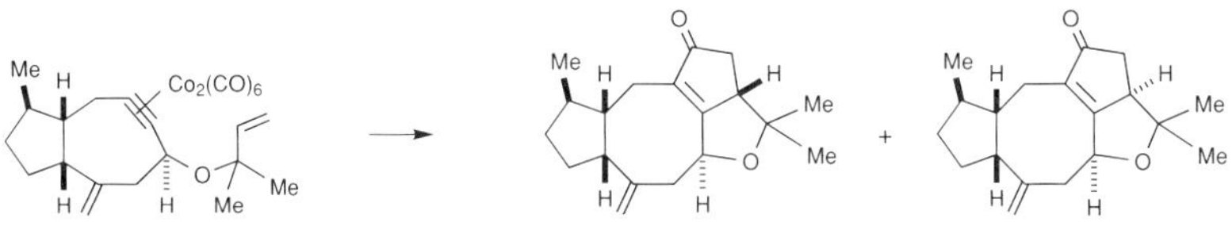

329 **330** **331**

conditions	% yield (330 + 331)	ratio 330 : 331
MeCN, 82°	75	4 : 1
MeCN, ultrasound, 45°	45	3 : 1
CH_2Cl_2, NMO, RT	68	11 : 1

Equation 39

Since a triple bond conjugated with a transition metal carbene moiety becomes highly polarized,[96] it stands to reason that such a system should generally facilitate the intramolecular Pauson–Khand reaction. True enough, upon treatment with dicobalt octacarbonyl in THF solution at room temperature the acetylenic Fischer carbene complex of tungsten and chromium **332** smoothly cyclizes to give the corresponding bicyclic product in respectable yields (Equation 40).[97] In similar manner the tricyclic product **335** can also be obtained from starting material **334**, although in markedly lower yield (Equation 41).[97]

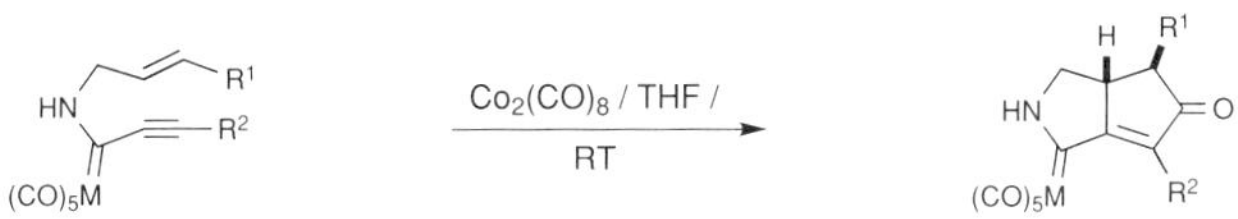

332 **333**

$R^1 = H$, $R^2 = Ph$, $M = W$ (75%)
$R^1 = H$, $R^2 = Ph$, $M = Cr$ (72%)
$R^1 = H$, $R^2 = Et$, $M = W$ (70%)
$R^1 = Me$, $R^2 = Ph$, $M = W$ (72%)

Equation 40

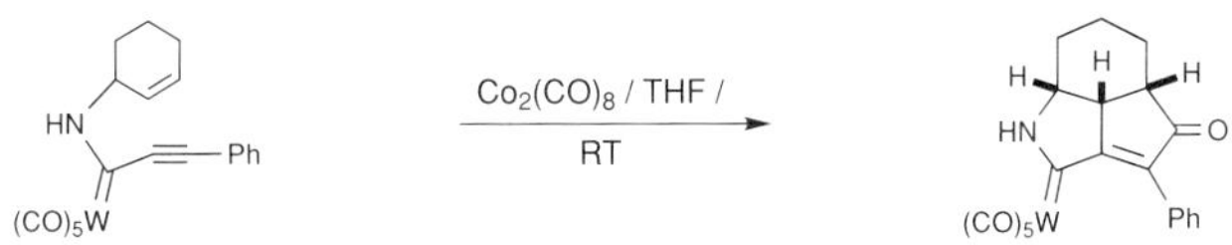

334 **335** (35%)

Equation 41

Formation of quarternary centers is demonstrated in the construction of the angular triquinane skeleton (Scheme 34). Starting from acetylenic cyclopentene

336, products **339** and **340** are obtained in a ratio of >7:1 (51%). Upon inspection of the reaction transition states **337** and **338**, leading to the respective products **339** and **340** (see transition state **311** in Scheme 32), it can be seen that stereoselectivity is governed solely by the stereochemistry of the methyl group. In view of the comparable energies of the two transition states, it is thus amazing that such high selectivity should be observed (Scheme 34).[98]

Scheme 34

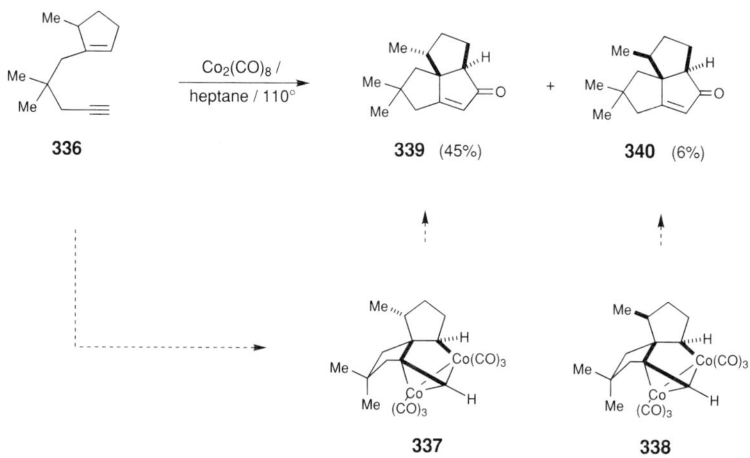

The asymmetric Pauson–Khand reaction has also been studied using a chiral template, as shown in Scheme 35.[99] Treatment of acetylenic *E*-enol ether **344** (obtained from lithium aluminum hydride reduction of the homochiral diyne **343**) with dicobalt octacarbonyl effects bis-cyclization to the bicyclic enone system with an induction level of 6:1, the major product **345** (55%) of which is easily separable. Further manipulations on **345** lead to (+)-hirsutene **347**.

With regard to the [5.5.5.5]fenestrane system, its preparation *via* a tandem Pauson–Khand tetracyclization of enediyne **348** has also been pioneered, albeit in poor yield, as shown in Equation 42.[100] The reaction will doubtlessly be further developed, since the assembly of four rings in one single operational step from **348** makes it a worthwhile competitor to other methods (see Scheme 30).

Scheme 35

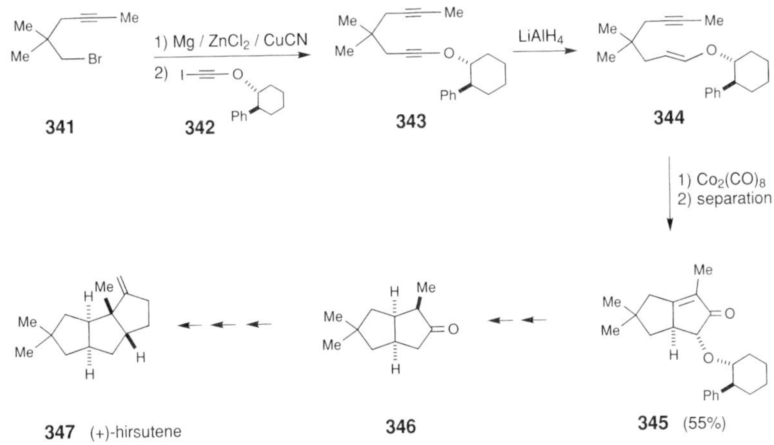

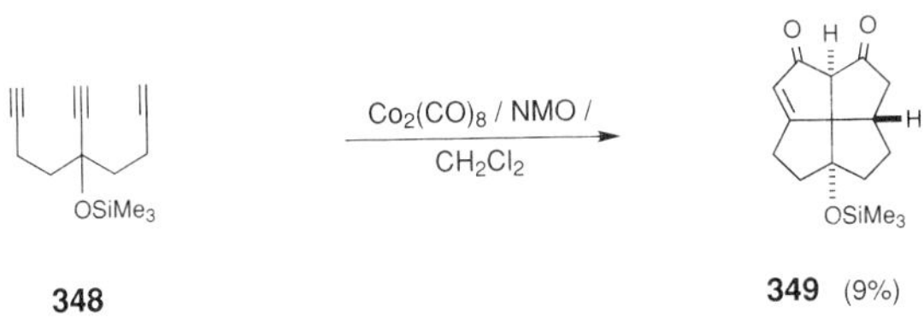

Equation 42

Related to the use of dicobalt octacarbonyl in the Pauson–Khand reaction, the use of zirconocene, $ZrCp_2$ (Cp = η^5-C_5H_5), has also been reported by Negishi *et al.*[101] Generated from the reaction of Cl_2ZrCp_2 with magnesium and mercuric chloride, or *n*-butyllithium, zirconocene reacts with enyne **95** and diyne **96** to form zirconabicycles **350** and **351**, respectively. Carbonylation of **350** yields bicyclic enone **315** while hydrolysis yields **98**, and exocyclic diyne **99** is obtained from the hydrolysis of **351** as shown in Scheme 36. These reactions superbly compliment the palladium-catalyzed cyclization (see Scheme 9) and the Pauson–Khand reaction (see Scheme 32). As a matter of fact, the conversion of diyne **96** into exocyclic diene **99** had earlier been reported using $TiCp_2$ generated from the reaction of Cl_2TiCp_2 with sodium amalgum in the presence of diphenylmethylphosphine (PPh_2Me).[102] However, the reaction remained comparatively undeveloped from the point of view of organic synthesis.

Scheme 36

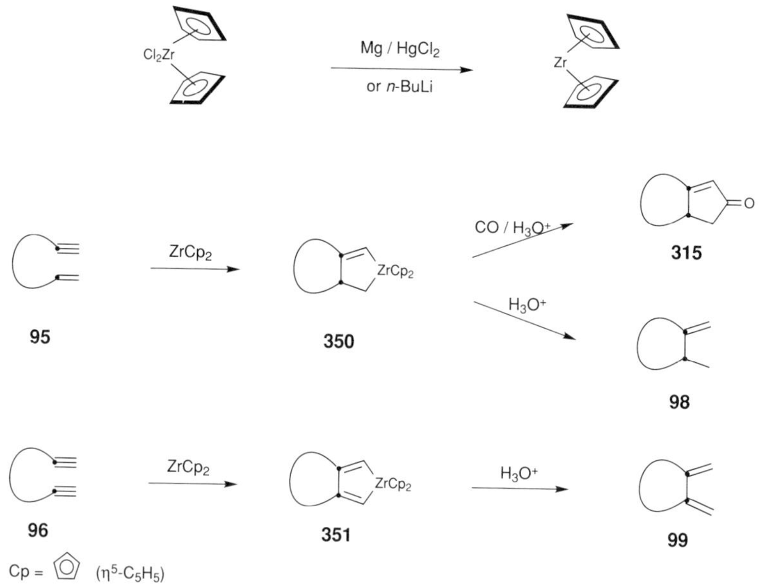

Equation 43[101] exemplifies the preparation of zirconabicycle **353** from enyne **352**, and the subsequent carbonylation of **353** to the bicyclic enone **354** substituted with various functional groups "R". On the whole, the overall reaction yields are fairly reasonable.

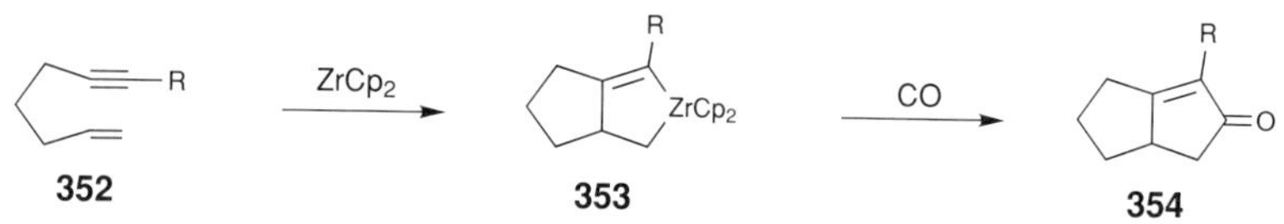

R	353 %	354 %
$SiMe_3$	95	65
$SnMe_3$	92	63
Me	95	65
Ph	90	55
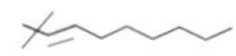	80	50

Equation 43

Zirconocene-promoted bis-cyclization of enynes carrying an allylic hydroxy or methyl substituent can be diastereoselective. For example, (*R*)-hydroxyenyne **355** cyclizes to bicyclic hydroxy enone **356** in substantial yields with extremely high enantiomeric purity (Equation 44).[103] The reaction has been exploited in the synthesis of naturally occurring triquinane, pentalenic acid **361**,[104] wherein zirconocene-promoted carbonylative cyclization of enyne **357**, carrying a protected allylic alcohol, is employed to prepare bicyclic enone **358** which constitutes rings A and B of the target molecule (Scheme 37).[103]

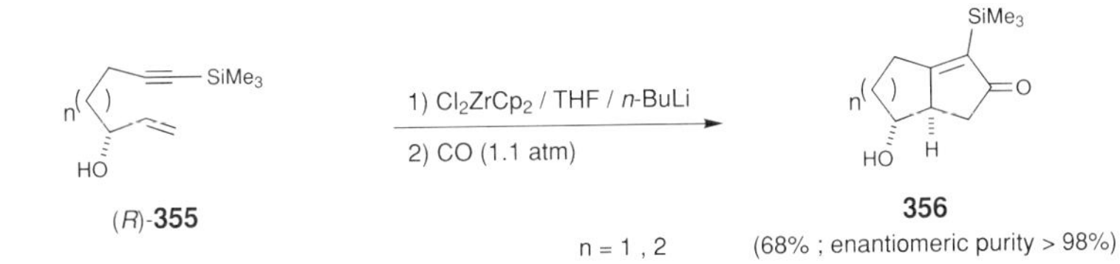

Equation 44

Scheme 37

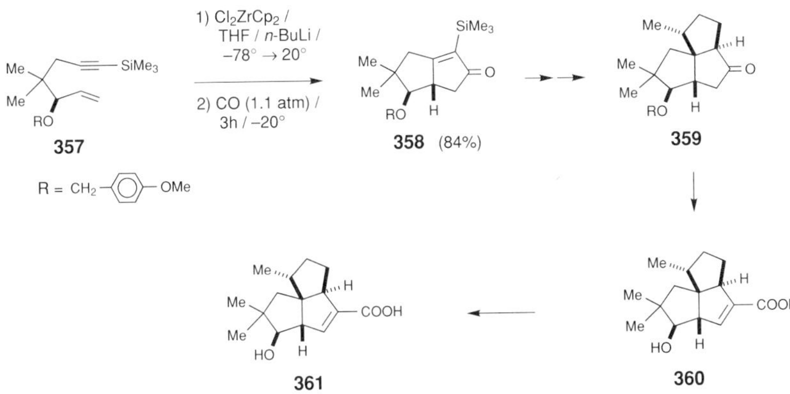

In likewise manner, the cyclization of optically active, allylically methyl-substituted enyne **362** *via* the corresponding zirconabicycle yields (after carbonylation followed by quenching with iodine)[105] enone **363** with a diastereomeric excess (de) of more than 90%. Purification to give a product of de >98% is easily possible. Hydrogenation of **363** and conversion of the hydrogenation product **364** to optically active iridomyrmecin **365** can be carried out according to established methods (Scheme 38).[106]

Scheme 38

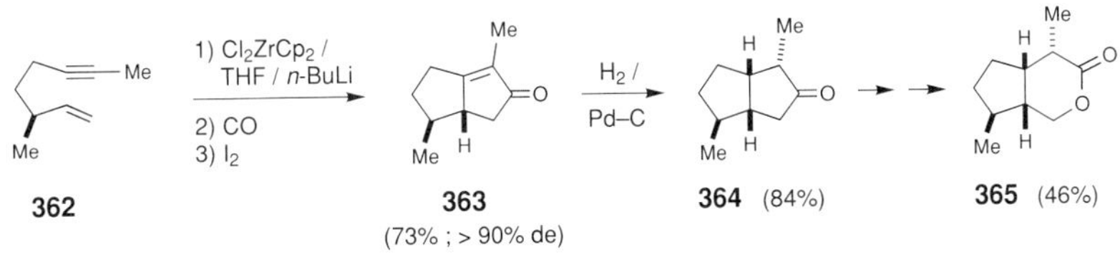

Examples of carbonylative cyclizations are abundant. Apart from the more popularly used metal catalysts derived from palladium, cobalt, or zirconium, other metal catalysts such as the iron-carbonyl,[107–110] nickel carbonyl,[111,112] and chromium carbonyl[113–115] complexes are also effective. Predictably this type of reaction will see much development in the near future.

IV. METAL-CATALYZED CARBON-HETEROATOM BOND FORMATION

Palladium, silver, and mercury complexes enjoy popular use as metal catalysts for cyclizations involving carbon-heteroatom bond formation, particularly for the construction of five- and six-membered oxygen and nitrogen heterocycles. Other metal salts such as those of copper and rhodium, although also used, are not as widely appreciated.

A. PALLADIUM–CATALYZED CARBON–HETEROATOM BOND FORMATION

In retrospect, the earlier shown five- and six-membered ring lactone formation *via* carbonylative cyclization of precursors **262**, **266**, and **270** (Equations 33, 34,[78] and 35,[79] respectively) are, in actual fact, ring closures *via* carbon-heteroatom bond formation catalyzed by palladium.

Another palladium-catalyzed carbon–oxygen bond formation reaction, the cyclization of acetylenic alcohol **366** to yield furan **369** is best carried out in tetrahydrofuran solution, with either $PdCl_2(PhCN)_2$ or $PdCl_2(MeCN)_2$ as catalyst. The reaction entails an *endo* attack of the hydroxy group on the palladium-activated acetylenic bond to first yield complex **367** which then loses water or methanol to give the furan intermediate **368**. Decomplexation of the latter by trapping with a proton or an allyl group leads, respectively, to products **370** or **369** in moderate to excellent yields (Scheme 39).[116] Attempts to convert the palladium-free furan **369** to the allyl-substituted product **370** were, as to be expected, unsuccessful.

Acetylenic ketones, *e.g.*, **371**, are also amenable to cyclization *via* oxypalladation, as discussed above, yielding furan **376**. The reaction is feasible under both anhydrous and hydrated conditions, and is thought to involve either the enol **372** or the hydrate **373**, depending upon the prevailing circumstances. The table in Scheme 40[117] compares the reaction yields obtained using various solvent systems for products with different "R" substituents.

Scheme 39

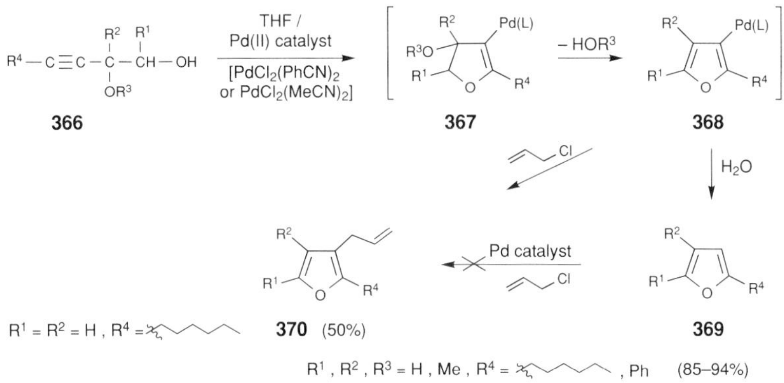

Scheme 40

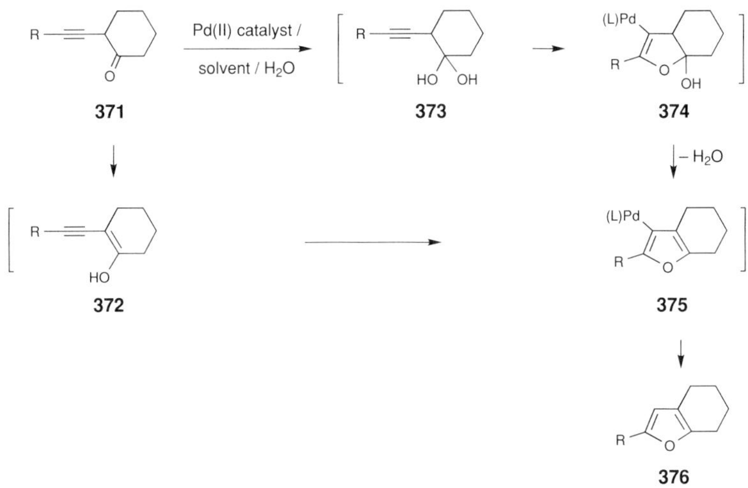

Solvent system	% 376	
	R = n-C_6H_{13}	R = Et
MeCN	73	65
MeCN–H_2O	90	75
THF	76	74
THF–H_2O	26	8

Given a longer connecting chain between the reactive centers, the *exo* rather the *endo* mode of cyclization (see Schemes 39 and 40) is observed. Accordingly the stereodefined 2-alkylidenetetrahydrofurans **378** and **380** can be prepared by treatment of the corresponding acetylenic alcohols **377** and **379** with *n*-butyllithium in THF at 0°, followed by addition of palladium acetate or chloride and triphenyl phosphine, and finally iodobenzene. The mechanism of reaction is as illustrated in Scheme 41.[118] It is noteworthy that, when n = 1, no *trans*-bicyclic furan **380** is formed at all, due to ring strain. Other features to note are that palladium acetate is by far the best catalyst for this reaction, and the use of solvents such as benzene or toluene inhibits the reaction, causing an almost complete loss of product.

Scheme 41

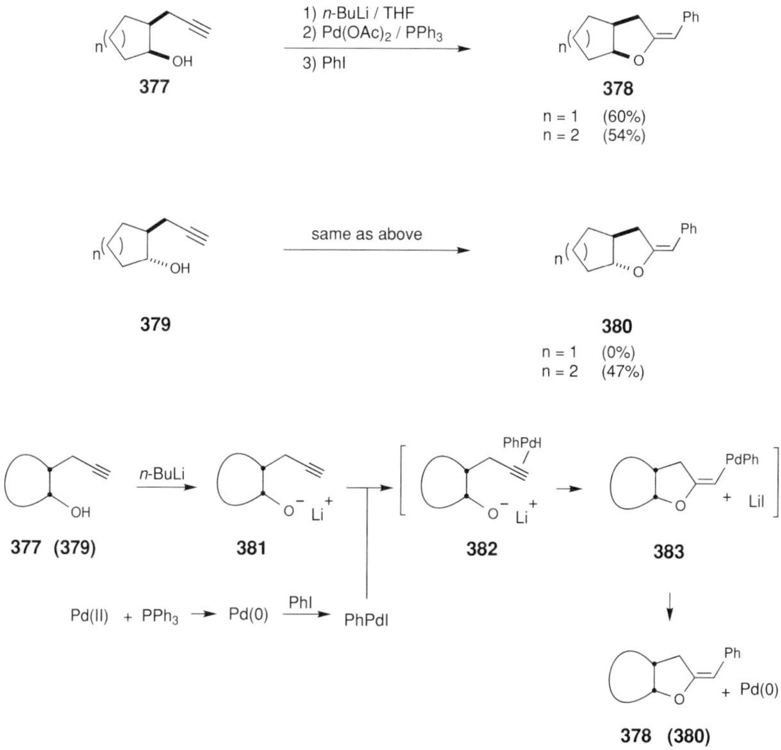

According to literature reports, stereoselective oxypalladation relies upon the use of chiral palladium complexes[119] or the intrinsic asymmetric nature of the starting molecule.[120] This is still the current practice but, perhaps due to improved understanding, techniques, and new reagents, later reports are claiming higher selectivity and yields, not to mention increased complexity of target molecules.

The optically active diol and amino alcohol **384** both cyclize with high enantiomeric purity (86% and 97% ee, respectively) to the corresponding dihydrofuran or pyrrole **385**,[121] as shown in Equation 45. In this *endo* cyclization the chirality of the newly created center in the product is obviously induced by the allylic hydroxy substituent in the starting olefin, an induction already discussed in Equation 44[103] and Scheme 37.[104] To gain further insight into this stereoselective process the reaction was carried out on the diastereomerically pure enediol **386**, whereupon a disubstituted dihydrofuran **387** with zero optical rotation ($[\alpha]_D$) was obtained, indicating it to be the *meso* steroisomer.[121] Again, the result can be attributed to chiral induction by the allylic hydroxy group as in the previous example (Equation 46).

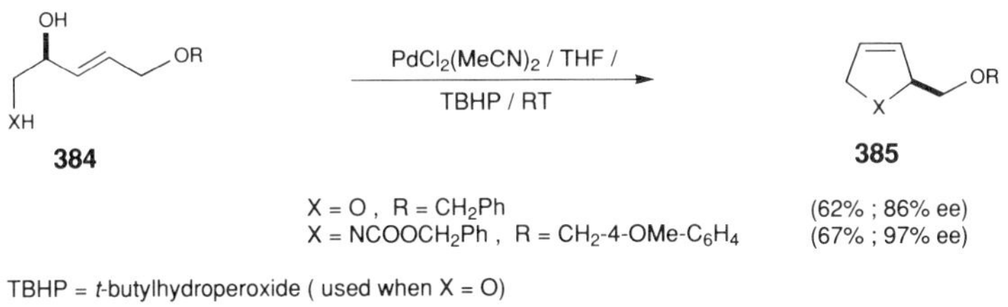

Equation 45

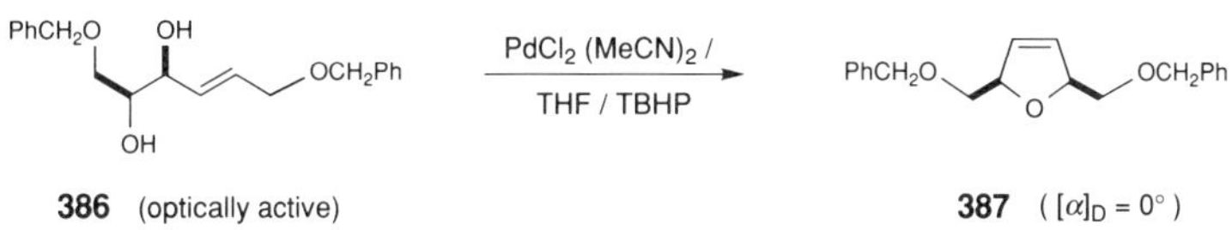

Equation 46

Another example of stereochemical induction of palladium-catalyzed carbon-heteroatom cyclization is the carbonylative cyclization of prochiral vinyliodide **391** (prepared from cyclopentanedione **388**) to hydroxy α-methylene-γ-butyrolactone **392** (Scheme 42).[122] As for Scheme 43, the first reaction is similar to that already discussed in Schemes 11 and 12, *i.e.*, alkylative cyclization of enyne using the catalyst system $Pd(OAc)_2$ / PPh_3 / Ag_2CO_3. Enyne **393** reacts at the acetylenic function to give complex **394**, which undergoes carbocyclization to product **395**. However, upon changing the catalyst system to $Pd_2(dba)_3$•$CHCl_3$ the reaction takes the totally different course of carbon–oxygen cyclization, which proceeds *via* alkylative addition at the olefinic bond to give complex **396** followed by *O*-alkylation to yield tetrahydrofuran **397**.[46]

Scheme 42

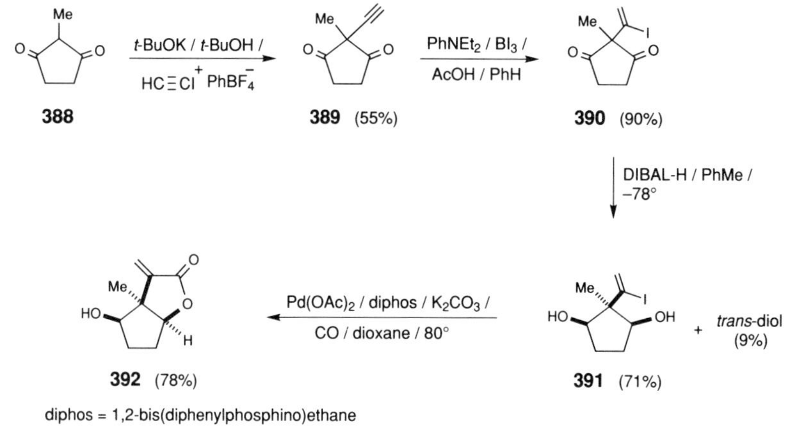

diphos = 1,2-bis(diphenylphosphino)ethane

Scheme 43

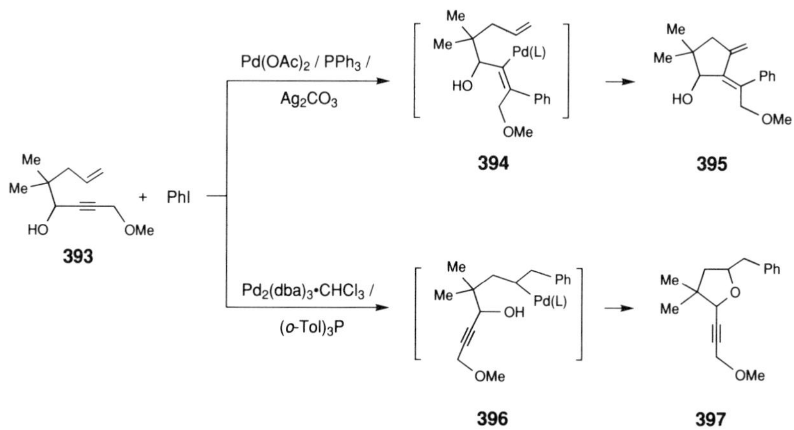

Also providing a versatile and useful methodology for organic synthesis is the cyclization of oxygen onto π-allylpalladium complex to yield vinyl cyclic ethers. As shown in the table in Scheme 44, the smaller ring is formed preferentially in all cases except in entry 1, where the strained oxetane ring formation is unfavorable. The reaction yields are eminently high, as also is the stereoselectivity observed with the five-membered ring products (entries 1 to 3).[123]

Scheme 44

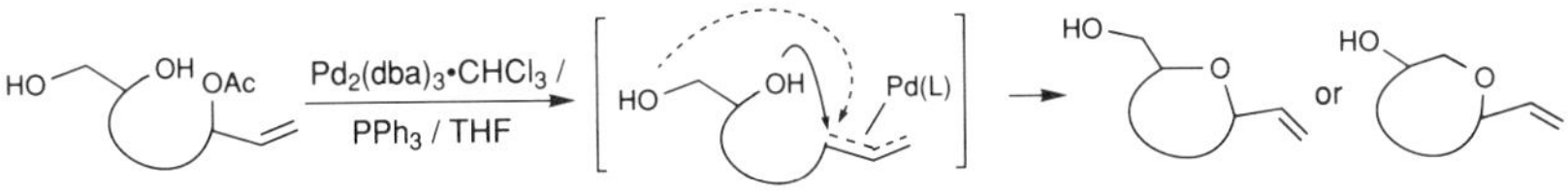

398 **399** **400** **401**

Entry	Starting material	Product	% yield	Diastereomeric ratio
1	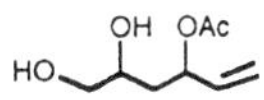	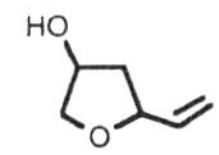	95	91 : 9
2	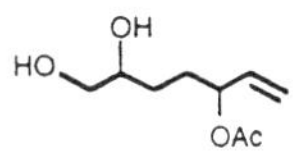	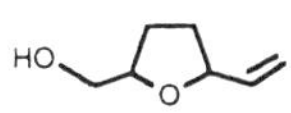	86	100 : 0
3	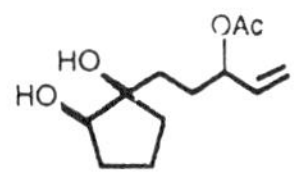	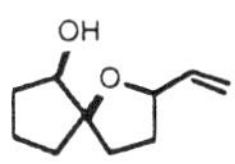	93	100 : 0
4	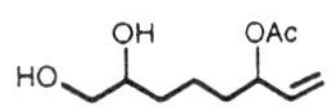	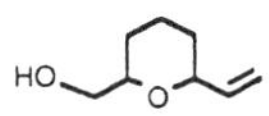	95	1 : 1

B. SILVER AND MERCURY SALT-CATALYZED CARBON–HETEROATOM BOND FORMATION

Although frequently used, the reactions with Ag(I) and Hg(II) have the disadvantage of being unpredictably irregular with respect to regio- and stereochemistry, which are influenced by several factors such as the catalyst itself, the nature of the substrate, and reaction conditions (exerting kinetic or thermodynamic control). For example, the amidomercuration of **402**, reported over a decade ago,[124] is both regio- and stereospecific, giving **405** as the sole product in immensely high yield (Equation 47), and the analogous amidomercuration of substrate **406**, even though not completely stereospecific, still yields predominantly the *trans*-product, with the ratio becoming extremely high with increasingly larger R groups (Equation 48).[125] In contrast, the almost identical substrate **409** undergoes amidomercuration in both endocyclic and exocyclic fashions to give, respectively, products **410** and **411** in a 1:4 ratio (60%, both as mixtures of *cis*- and *trans*-isomers).[126] The explanation put forward is that the oxygen atom presence in the cyclizing chain destabilizes the carbocation character at the position *β*- to itself, forcing ring closure to occur at the alternative γ-position in an *endo* fashion to yield **411** (Equation 49). Support for this rationale is provided by the observations shown in Equation 50. When substrate **412**, which has balanced oxygen presence at both termini of the double bond, is subjected to cyclization followed by reduction with borohydride, only oxazolidine **413** from *exo* cyclization is obtained.[126]

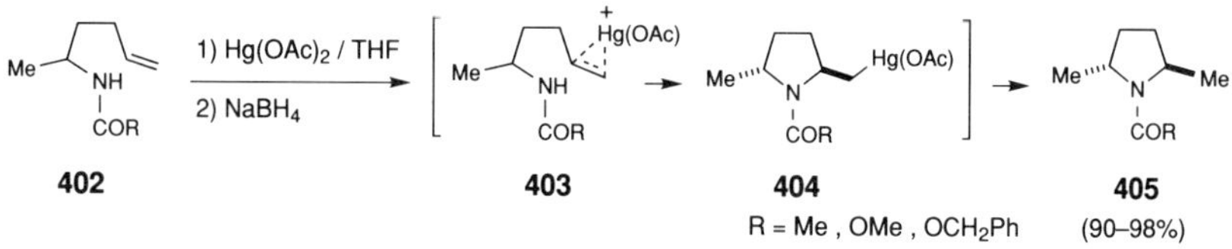

Equation 47

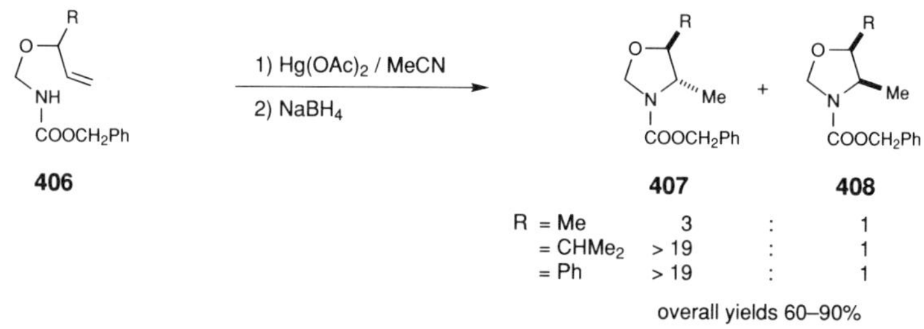

Equation 48

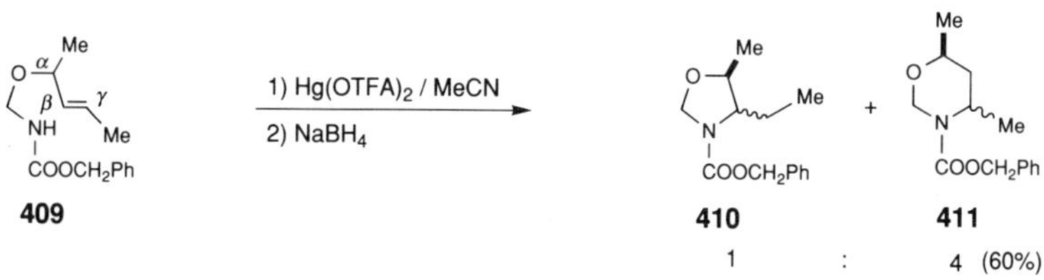

Equation 49

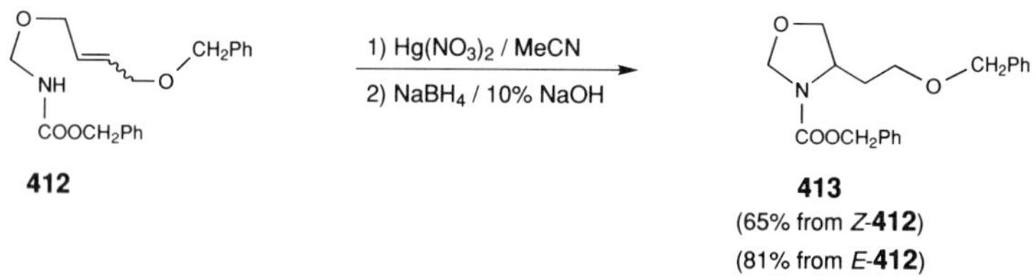

Equation 50

However, the above observation is by no means the only influence affecting *exo* and *endo* regiospecificity of the amidomercuration reaction, as shown by the reaction of **414** in Scheme 45. Possessing the combined characteristics of **409** (with α-substituent) and **412**, substrate **414** undergoes cyclization catalyzed by mercuric trifluoroacetate in ethyl acetate in the presence of sodium bicarbonate (to neutralize acid formed in the reaction) to stereospecifically yield *trans*-oxazolidine **415** from *exo* cyclization, as well as a stereomeric mixture of tetrahydrooxazine **416** from *endo* cyclization, in a ratio of 3.8:1 (82% overall yield). Compound **415** can be converted easily to the racemic lactone **418**, which, except for being a racemate, is identical with the optically active lactone obtained from acid-catalyzed lactonization of naturally occurring *threo*-γ-hydroxy-β-lysine **419**, as shown in Scheme 45.[127]

Scheme 45

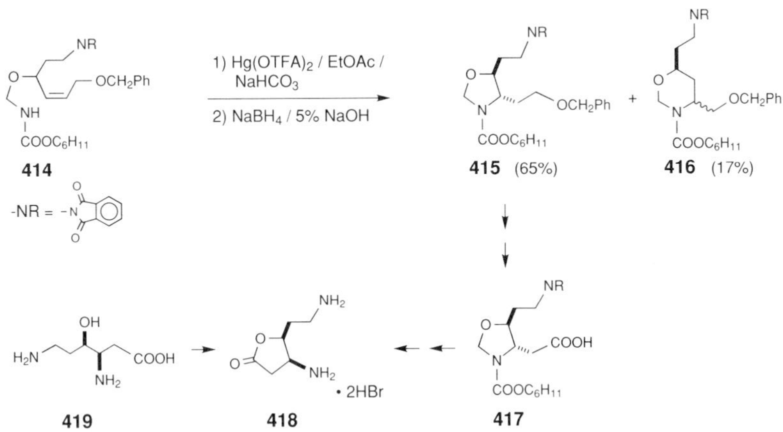

Furthermore, the influence of kinetically and thermodynamically controlled reaction conditions on stereoselectivity in amidomercuration has also been studied[128] and the information obtained has led to the successful manipulation of amino acid synthesis.[129]

In view of reaction regiospecificity, mercury salt-catalyzed cyclizations are (as mentioned earlier) admittedly inferior to other methods of ring formation, *e.g.*, radical cyclization. This is quite clearly demonstrated by the reaction of diene hydroperoxide under various cyclizing conditions, as shown in Scheme 46. Thus, peroxymercuration of **420** using mercuric nitrate followed by trapping with bromide yields a 2:1 mixture of *trans*-dioxolane **421** and dioxane **422**, from 5-*exo* and 6-*exo* cyclizations, respectively. Radical cyclization of **420**, on the other hand, yields exclusively the product from 5-*exo* cyclization: **423** from *N*-iodosuccinimide-initiated, and **424** from *t*-butoxyl-initiated reactions. In neither case is the 6-*exo*-cyclized product observed.[130]

Scheme 46

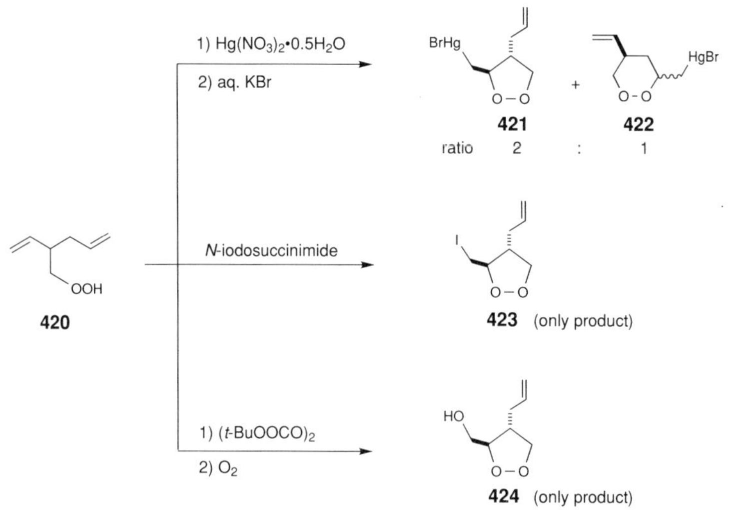

Besides catalyzing the reaction of hydroperoxide with the olefinic bond, mercury(II) salts also catalyze the intramolecular cyclization of hydroperoxide to cyclopropane (Equation 51).[131] Subsequent reduction of the carbon–mercury bond with borohydride proceeds smoothly without affecting the peroxide linkage.[132]

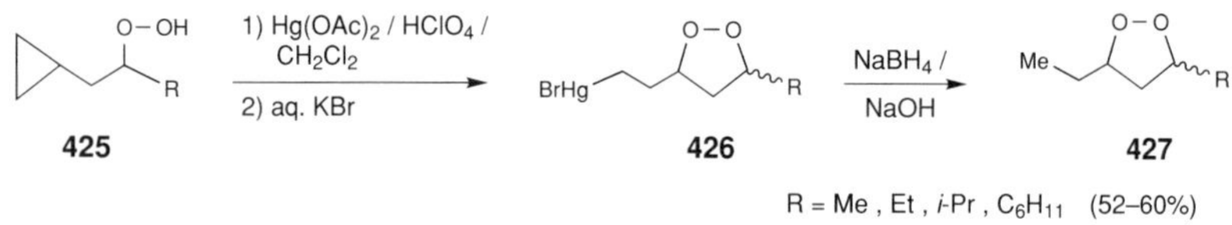

Equation 51

Polycyclization is also possible with mercury(II) catalysis. For example, the mercuric triflate-*N,N*-dimethylaniline complex catalyzes the polycyclization of farnesol derivative **428** to tricyclic mercury salt **429**, which delivers product **430** upon treatment with borohydride, as shown in Equation 52.[133] Although the yields are not impressive, the cyclization, which also involves carbon–oxygen bond formation, is very useful. In truth, the mercuric triflate-*N,N*-dimethylaniline complex was introduced specifically for the purpose of polyene cyclization,[134,135] but proved effective in catalyzing carbon–oxygen bond formation, as in the cyclization of **431** to bicyclic ether **432** shown in Equation 53.[134]

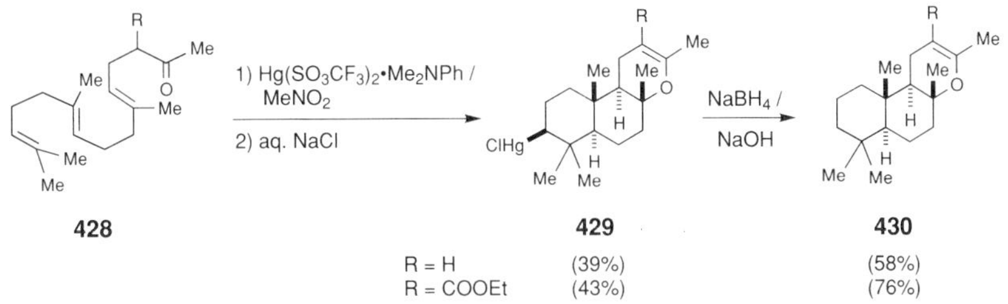

Equation 52

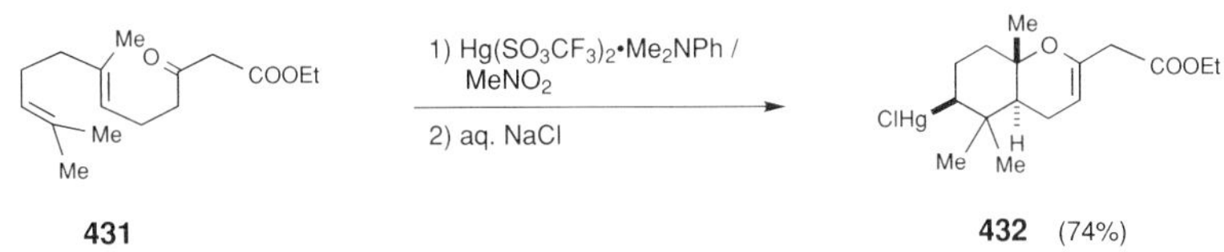

Equation 53

A popular use of silver(I) salts is in the cyclization of heteroatom to the allene system. Depending upon the size of the ring being formed, this type of cyclization can proceed either in the *exo* or *endo* mode.[136,137] Equation 54 shows the cyclization of allenic amine **433**, using silver tetrafluoroborate, which results in *cis*- and *trans*-pyrrolidines **434** and **435**, respectively.[138] A similar cyclization of the substituted carbamate **436**, as expected, gives the corresponding oxazolidinones **437** and **438**, as shown in Equation 55.[139] In the former case, where the product is 1,3-disubstituted, the *cis*-isomer **435** is the major product, whereas with the 1,2-disubstituted product in Equation 55, the *trans*-isomer **437** predominates.

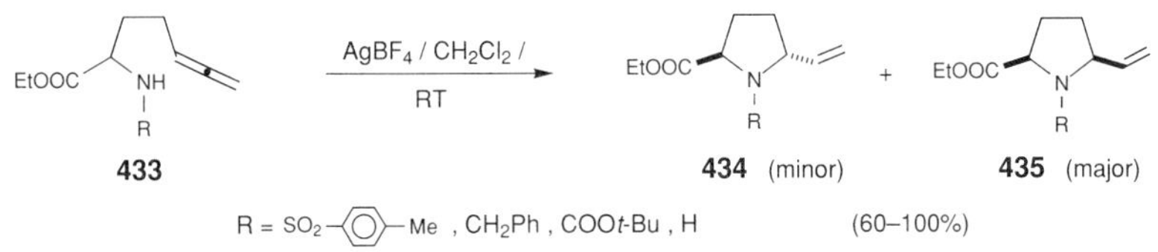

Equation 54

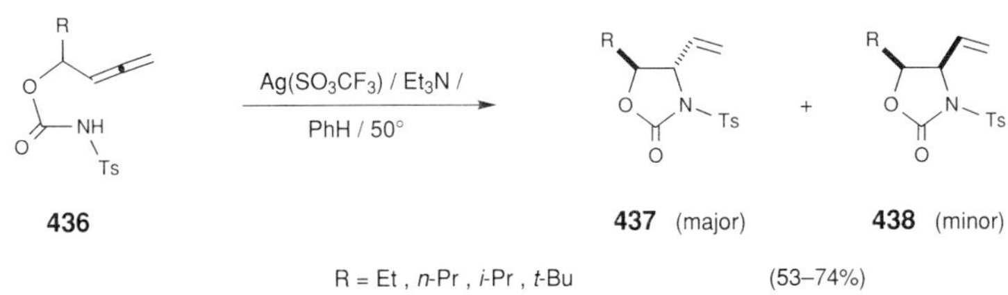

Equation 55

Asymmetric induction of the silver triflate induced cyclization of amino allene **439**, as shown in Equation 56,[140] is found to be influenced not simply by the chiral carbon adjacent to the amino group, but to an even greater extent by the functional group attached to the benzylic residue (group X). For example, with X as CH_2PPh_2 or CH_2SePh, moderate yields of pyrrolidine **440** are obtained with modest diastereomeric excess, but if the groups sulfide or sulfoxide are used (X = CH_2SPh, $CH_2{-}\overset{+}{S}(O^-){-}Ph$, or $CH_2{-}\overset{+}{S}(O^-){-}Ph$) the yields are dramatically improved and near 100% de are obtained. Interestingly, chemical correlations and X-ray crystallography have revealed that the new chiral carbon in the products from the latter cyclizations always possesses the stereochemistry shown in **441**, whether group X is sulfide or sulfoxide, and whatever the stereochemistry of sulfoxide.[140]

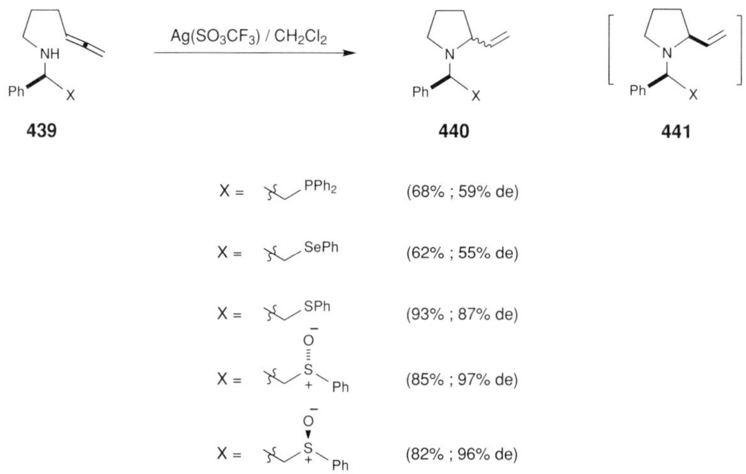

Equation 56

As stated above, *endo* cyclization of the heteroatom to allene proceeds well and is not uncommon, occurring when *exo* cyclization is unfavorable. For example, the synthesis of carbapenems **444** utilizes silver-mediated cyclization of 4-allenylazetidinones **442** which involves an *endo* instead of an *exo* mode of cyclization, as the latter would result in the unfavorably strained four-membered ring. The reaction proceeds very smoothly with terminal allenes **442a** and **442b**, requiring as little as 0.1 equivalent of catalyst. However, with nonterminal allenes **442c** to **442e**, the formation of the nitrogen–carbon bond is slower, and a minimum of 0.5 to 1 equivalent of silver tetrafluoroborate is necessary (Equation 57).[141] It should be noted that the reaction is also catalyzed by palladium. This latter catalyst offers the added advantage of further functionalization, that is, capture of the palladium intermediate with electrophiles, *e.g.*, an activated alkene, yields functionalized carbapenems **447** (Equation 58).[142]

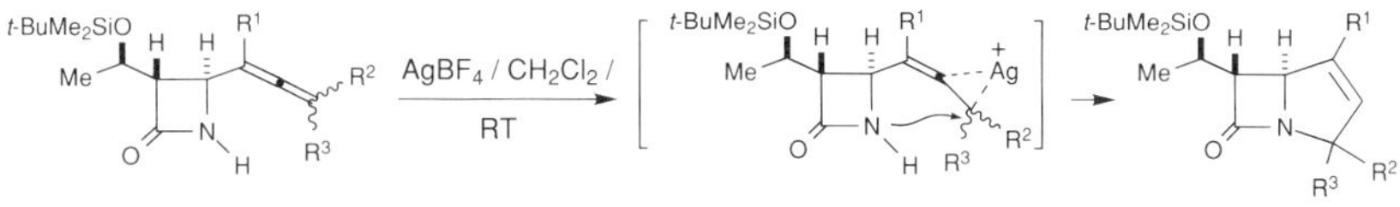

442 **443** **444**

	Diastereomeric ratio	%
a) $R^1 = R^2 = R^3 = H$	–	75
b) $R^1 = Me$, $R^2 = R^3 = H$	–	70
c) $R^1 = H$, $R^2, R^3 = H, Me$	1 : 2	65
d) $R^1 = Me$, $R^2, R^3 = H, Me$	1 : 2.8	62
e) $R^1 = H$, $R^2, R^3 = Me$, *n*-Bu	1 : 2.2	58

Equation 57

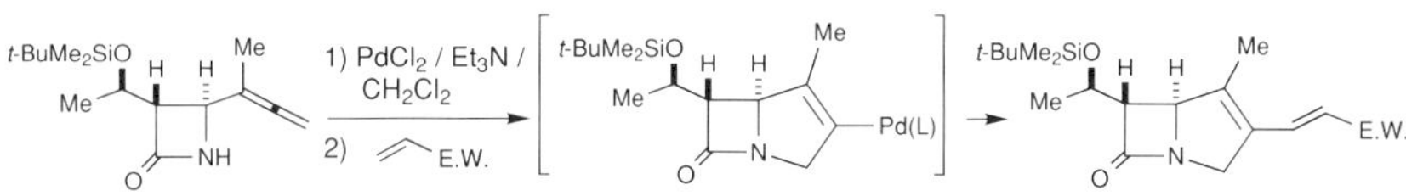

445 **446** **447**

E.W. = COOEt, COMe, CHO, CN (40–53%)

Equation 58

Endo cyclization of allenes extends to the use of carbonyl oxygen as the heteroatom nucleophile. Thus, allenal (**448**, $R^1 = H$) and allenone (**448**, $R^1 = Me, C_7H_{15}$), upon treatment with silver nitrate or tetrafluoroborate, in acetonitrile, cyclize to give furan **450** in good to excellent yields.[143,144] The reaction has been successfully applied to the synthesis of model compounds for bridged cembranoids. As shown in Scheme 47, the starting macrocyclic acetylene **451** can be oxidized by the Dess–Martin[145] method to provide allenone **452**, a substrate upon which silver-catalyzed furan formation can be readily carried out to furnish **453**. Alternatively, the starting macrocycle **451** may be converted to **454** by the Wittig rearrangement prior to oxidation which subsequently yields allenone **455**, a precursor for cyclization to furano-bridged macrocycle **456** *via* treatment with silver nitrate in acetonitrile.[143] Apparently, attempts to effect the Wittig rearrangement of **453** to **456** by treatment with butyllithium were unsuccessful.

Scheme 47

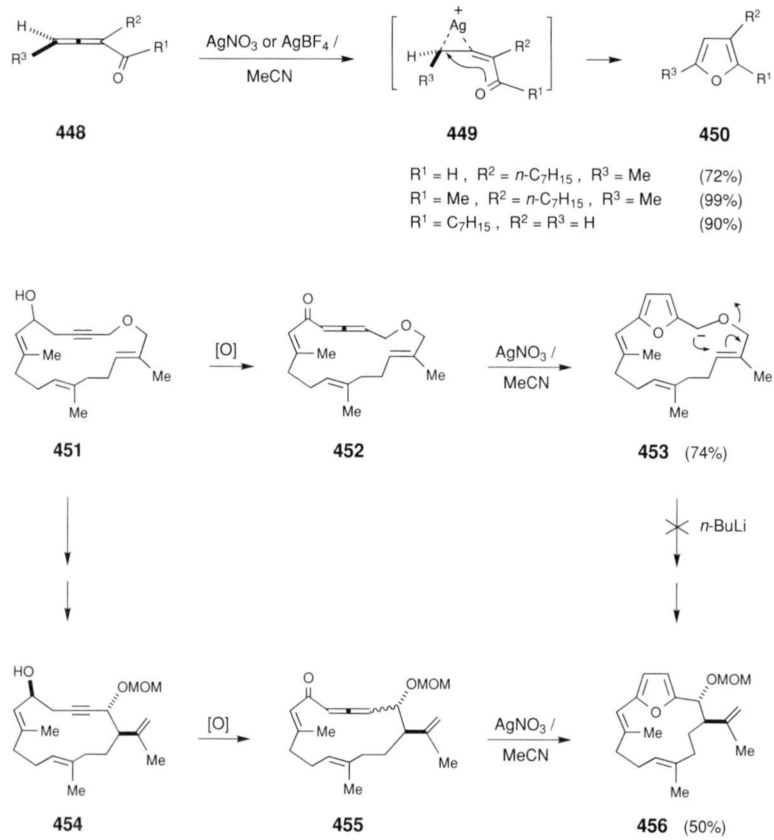

Application of hydroxy–allene cyclization to the synthesis of chiral dihydrofurans has been reported. Optically active propargylic acid **457**, available in ~90% ee, undergoes stereospecific Wittig rearrangement when treated with lithium diisopropylamide (LDA) in tetrahydrofuran to give, after esterification, the allenic ester **458** which readily cyclizes *via* silver-catalyzed endocyclic ring closure to give the *trans*-dihydrofuran **460**. Inversion of configuration of the alcoholic carbon in **458** was also carried out, using the method of Mitsunobu,[146] to yield the stereoisomer **461** for a similar cyclization to the *cis*-dihydrofuran **462**, thereby confirming the stereochemistry of the products (Scheme 48).[147]

Scheme 48

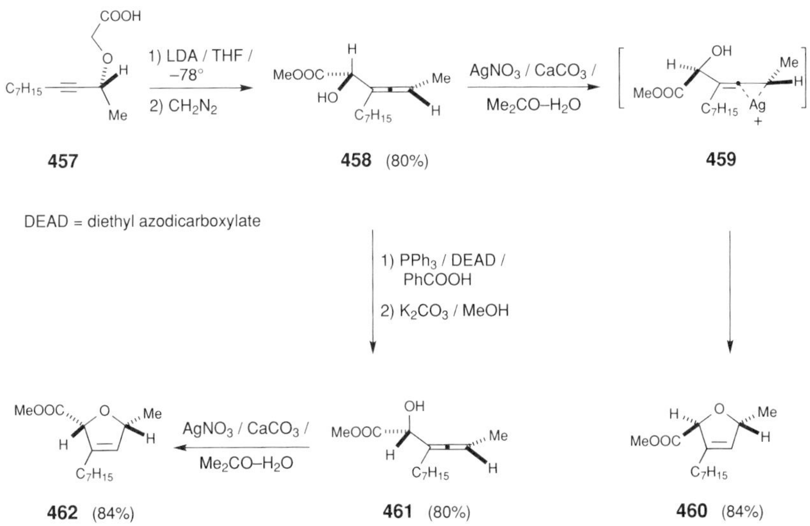

In relation to the above cyclization, an interesting rearrangement–cyclization process mediated by silver salt has been observed with optically active acetylenic acetate **463**, as illustrated in Scheme 49. Treatment of acetylenic ester **463** with silver tetrafluoroborate in benzene at 60°C in the dark yields dihydrofuran **468** with a high percentage of enantiomeric excess. The mechanism of reaction is probably as detailed in the scheme.[148] The process has provided the methodology for a formal enantioselective synthesis of (–)-(*S*)-ascofuranone **469**, an antibiotic from *Ascochyta viciae*.[149]

Scheme 49

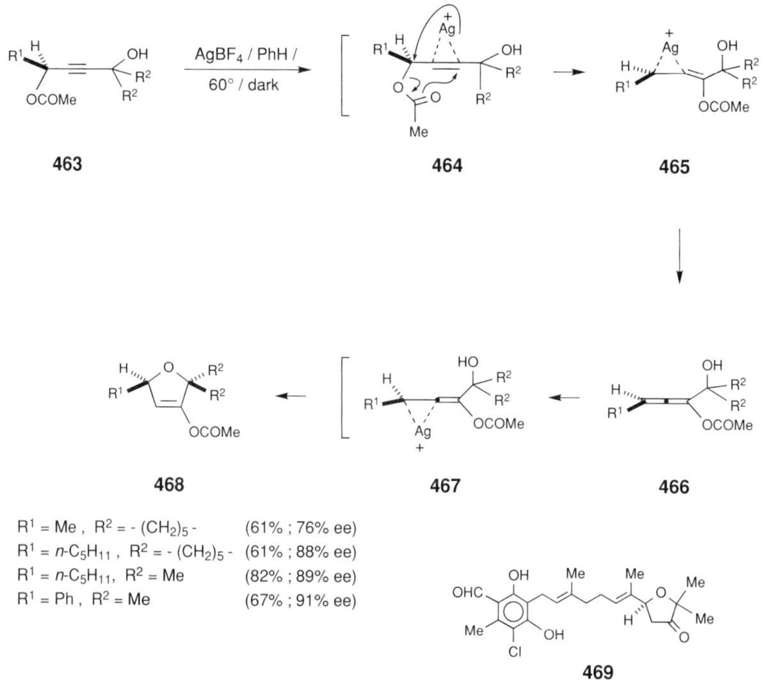

V. SAMARIUM (II) IODIDE–CATALYZED CYCLIZATION

Since its introduction to the chemical community in 1980[150] samarium(II) iodide has developed into an extremely versatile tool in organic synthesis. Despite the very recent reviews on SmI_2 in synthesis,[10,11,151] the reagent's wide usage (often being the "reagent of choice") warrants here a section on samariun iodide-mediated cyclization, although discussion will be limited mostly to later work that has not yet been reviewed.

Due to its powerful action as a reducing agent samarium iodide can initiate various kinds of chemical transformations, which could be either radical or carbanionic (organosamarium) in nature. Several examples of the advantageous use of SmI_2 in the construction of complex molecules have appeared lately,[152] one such application being the preparation of key intermediate **473**, *via* samarium iodide-mediated tandem cyclization of aldehyde **470**,[153] in the synthesis of hypnophilin **474**[153] and coriolin **475**[154] as formerly discussed in Scheme 44 of Chapter 3 and shown here again in Scheme 50.

Scheme 50

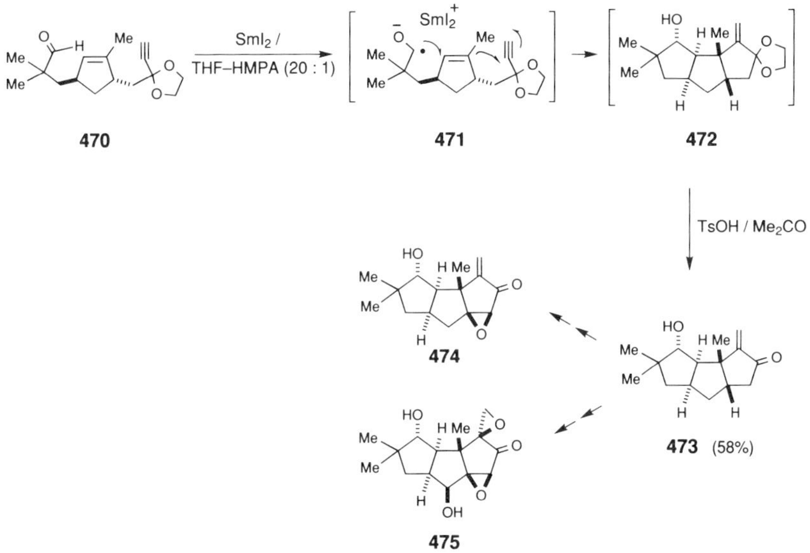

One of the most important, useful, and widely employed samarium iodide-mediated cyclization is the Barbier reaction,[150] which involves cyclization of a halo ketone as depicted simply in Equation 59.[155] Actual applications encompass substrates of diverse structures. The syntheses in Equations 60[156] and 61[157] demonstrate the versatility of the reaction. An interesting feature of the cyclization in Equation 61 is the stereospecific formation of *cis*-6,6-fused ring product **481** (65% yield) under the particular conditions employing both samarium iodide and zinc. Under normal conditions without zinc, however, a mixture of *cis*- and *trans*-ring junctions is produced.[155]

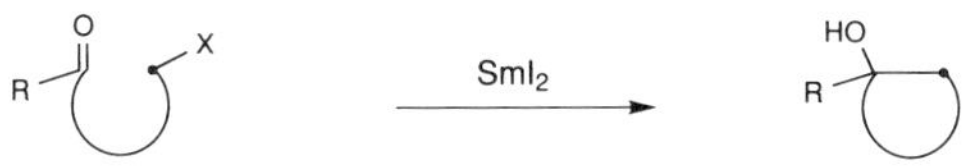

476 **477**

Equation 59

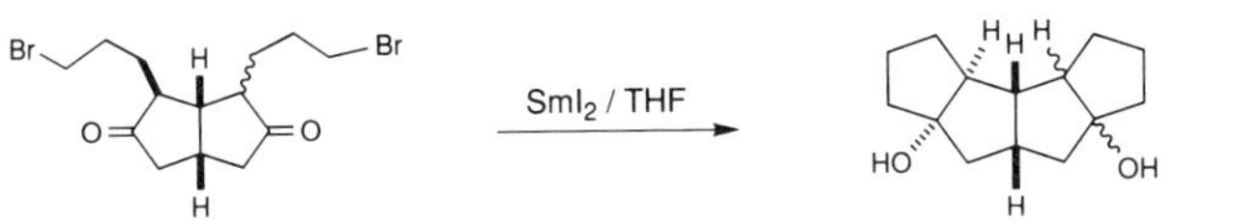

478
($\alpha : \beta$ = 1 : 9)

479 (68%)
H , OH (*cis* , α) : (*cis* , β) = 15 : 2

Equation 60

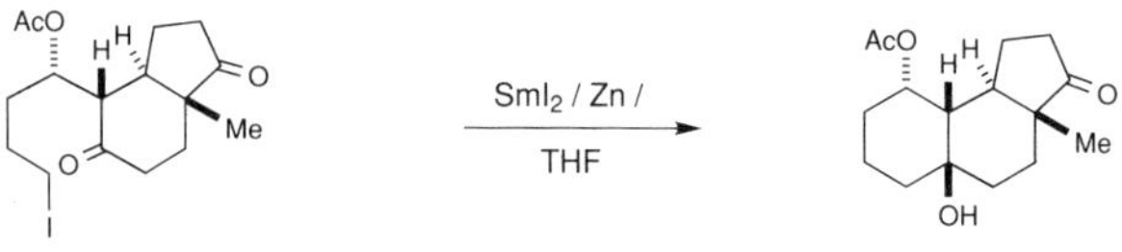

480 **481** (65%)

Equation 61

The intramolecular samarium iodide-mediated Barbier reaction has been applied successfully to the synthesis of bicyclo[m.n.1]alkan-1-ol **483** from the corresponding iodoketone **482** as shown in Equation 62.[158] The reaction proceeds under very mild reaction conditions in the presence of ferrous or ferric salts such as $FeCl_2$, $FeCl_3$, $Fe(acac)_3$ and $Fe(DBM)_3$. Illustrated in Equation 62 is the use of the catalyst $Fe(DBM)_3$ which is air-stable, THF-soluble, and nonhygroscopic. The facile preparation of the strained bicyclo[2.1.1]hexan-1-ol (**483**, n = 2, m = 1) is quite remarkable. For as yet unknown reasons, however, a similar preparation of bicyclo[2.2.2]octan-1-ol **485** fails (Equation 63).

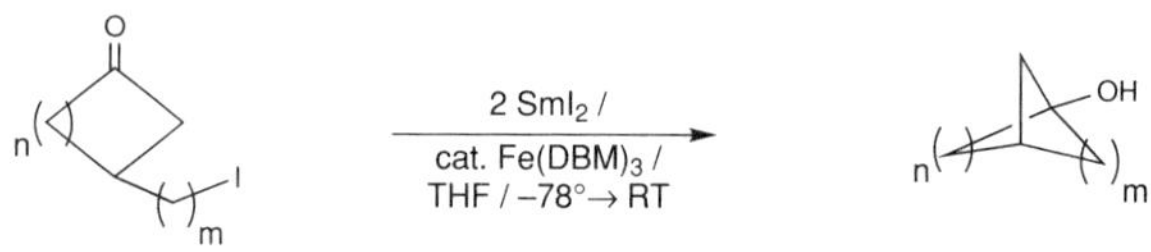

482 **483**

DBM = dibenzoylmethido

n = 2 ,	m = 1–3	(54–71%)
	m = 4	(22%)
n = 3 ,	m = 1–3	(69–77%)
	m = 4	(15%)
n = 4 ,	m = 2–3	(73–76%)
n = 5 ,	m = 2–3	(86–87%)

Equation 62

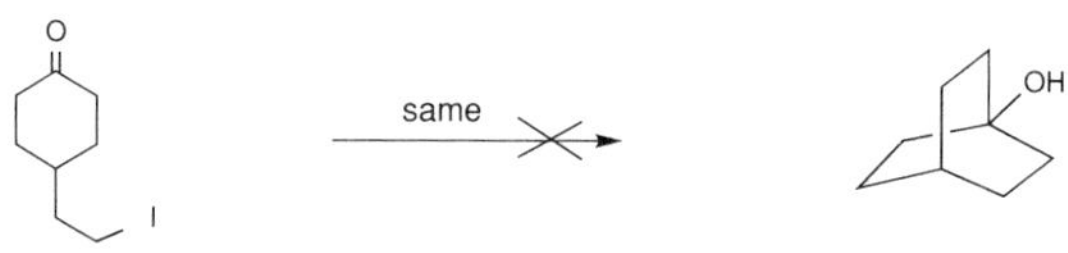

484 **485**

Equation 63

Regarding the mechanism of the intramolecular Babier-type reaction, two possibilities exist: one with anionic character *via* organosamarium **486**[158] (Scheme 51) and the other with radical characteristics involving coupling between alkyl and ketyl radicals, *e.g.*, intermediate **487**.[159] Apparently there is experimental evidence in support of both types of mechanism, and care must be exercised in weighing between the two, bearing in mind that factors such as reaction conditions, catalyst, as well as the nature of substrate all have the potential to tip the balance quite dramatically from one mechanism to the other.[158]

Scheme 51

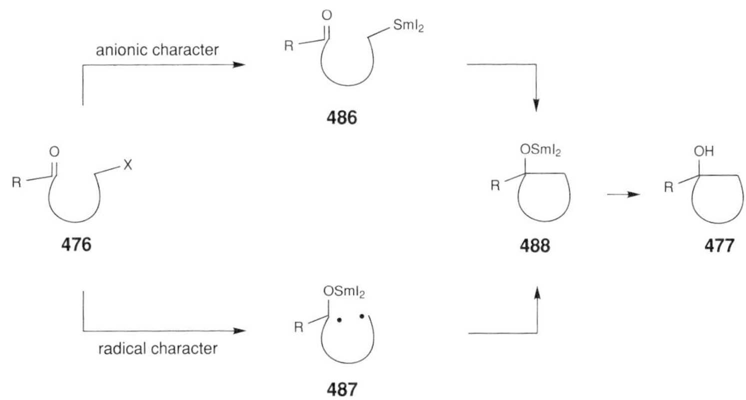

Another important reaction is the samarium iodide-mediated reductive cyclization of unactivated olefinic ketones. Already mentioned in Scheme 50, a complete outline of the sequence, including examples of applications, is now given in Scheme 52. After ketyl-olefin cyclization the resulting radical intermediate **491** may be further reduced by SmI_2 to organosamarium **492** which can be trapped by various electrophiles to yield the final product **493**.[160] The presence of HMPA, so it has been found, is absolutely essential for efficient cyclization,[161] conceivably[160] because it increases the reducing ability of SmI_2 solutions (Scheme 52).

Scheme 52

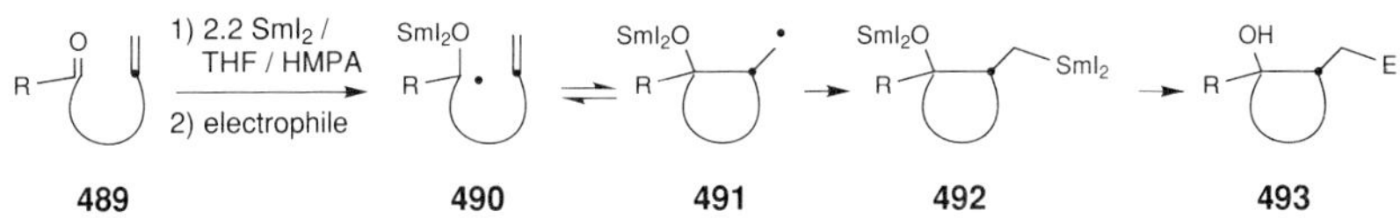

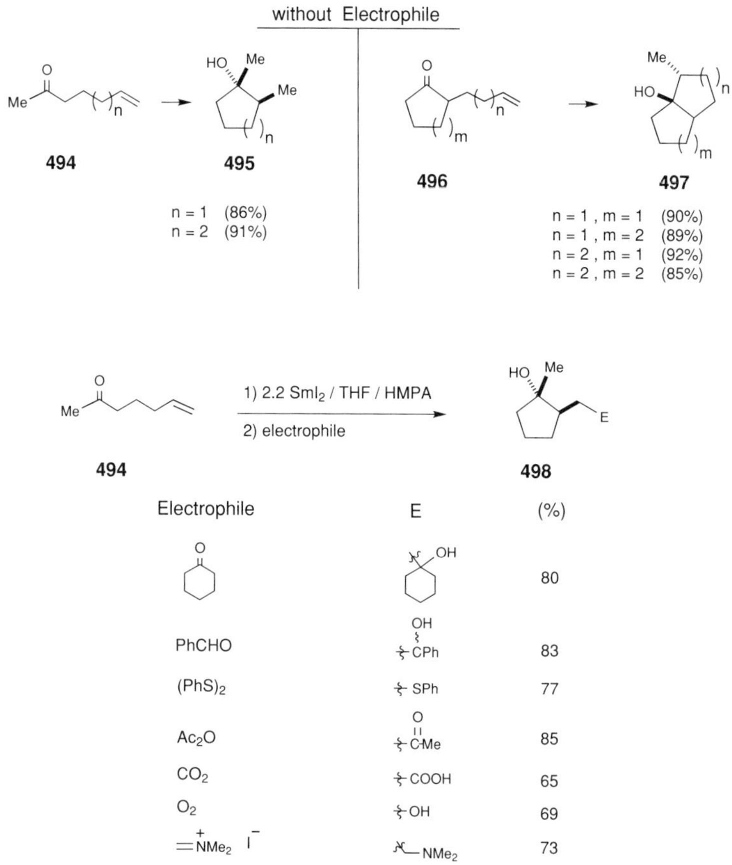

Organosamarium can also add in conjugate fashion to α,β-unsaturated ketones. For example, the organosamarium **501**, generated either by the cyclization of iodo olefin **499** or by the reduction of **500**, undergoes conjugate addition with ketone **503**. The reaction can be enhanced by transmetalating the organosamarium **501** with a range of copper salts to form **502** prior to addition, CuI•P(OEt)$_3$ having been found to give the best results. However, final yields are still poor to moderate but the reaction will probably continue to be improved upon and further developed for future use (Scheme 53).[162]

Scheme 53

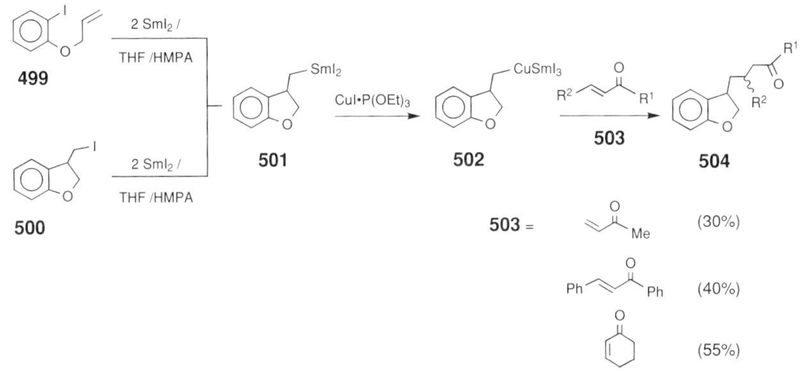

As might be anticipated, samarium iodide promotes the intramolecular reductive cyclization of aldehydes and ketones to alkynes. Nevertheless, as shown by the accompanying data in Scheme 54, without an activating group on the acetylenic moiety ($R^2 = H$, entries 1 and 2), the yield of cyclized product **506** is low or even negligible. With R^2 as an electron-withdrawing group, however, the situation changes and product **506** is obtained in substantial yields. The source of protons is here provided by *t*-BuOH (Schene 54).[163]

Scheme 54

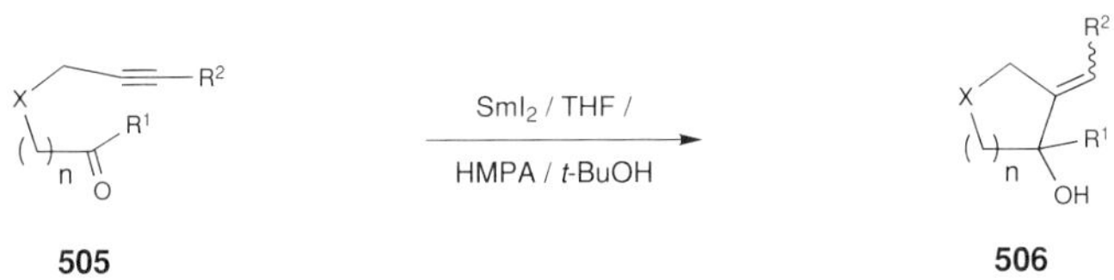

Entry		
1)	X = CH_2 , n = 1 , R^1 = H , Me , R^2 = H	(30–55%)
2)	X = CH_2 , n = 2 , R^1 = H , Me , R^2 = H	(0%)
3)	X = CH_2 , n = 1, 2 , R^1 = H , Me , R^2 = COOEt	(60–75%)
4)	X = O , n = 1, 2 , R^1 = H , Me , R^2 = COOEt	(60–70%)
5)	X = *N*-Boc , n = 1, 2 , R^1 = H , Me , R^2 = COOMe	(53–62%)

Samarium-catalyzed pinacol coupling of dialdehydes to yield cyclic vicinal *cis*-diols as shown in Equations 64[164] and 65[165] is quite a useful reaction in synthesis, particularly when considering the ease with which the *cis*-diol **511** may be formed from its *trans*-isomer **509** (Equation 65). Even more significant, however, is the stereochemical conversion of diols **512** and **515** to the major products **514** and **517** (Equations 66 and 67, respectively), which demonstrates how an α-alkoxy group influences the stereochemistry of reaction to yield mainly products with the *cis*-diol oriented *trans* to the α-alkoxy group.[165] This stereochemical outcome can be interpreted in terms of the transition states shown in Equation 68.[165]

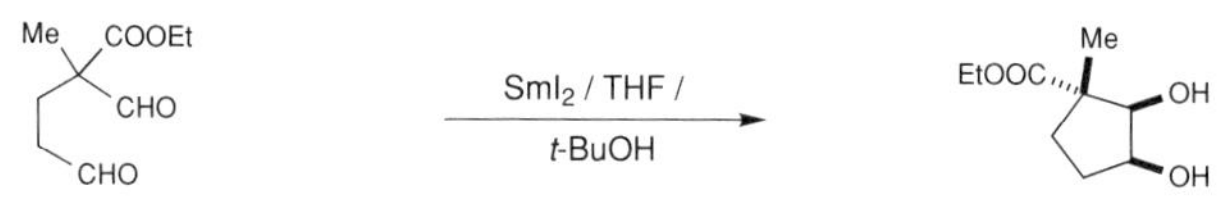

507 **508** (46%)
(*cis* : *trans* > 200 : 1)

Equation 64

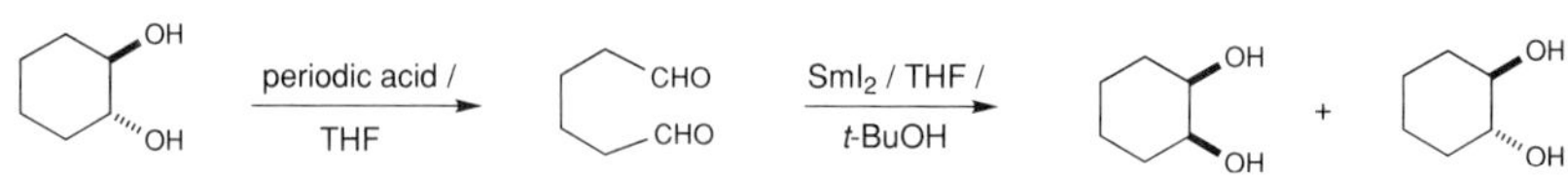

509 **510** **511** (75%) **509** (6%)

Equation 65

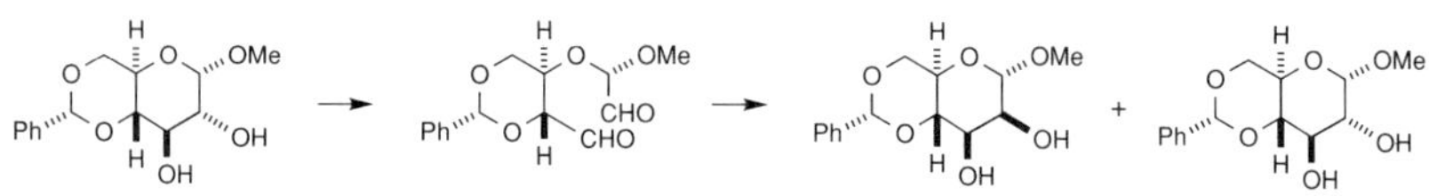

512 **513** **514** (44%) **512** (9%)

Equation 66

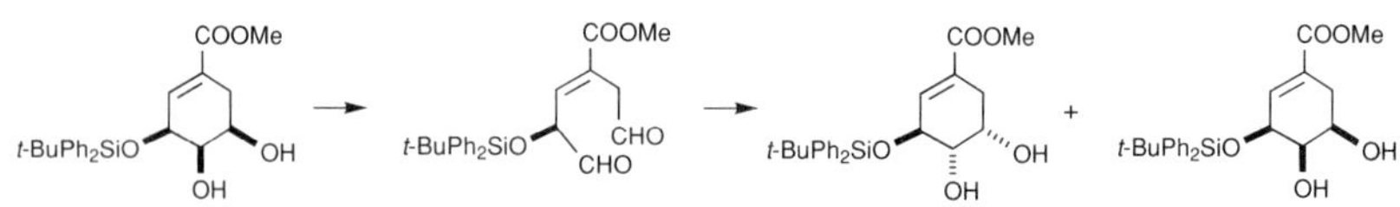

515 **516** **517** (69%) **515** (16%)

Equation 67

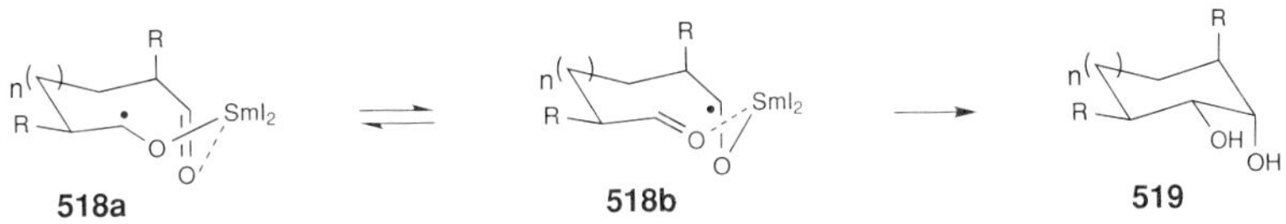

Equation 68

Medium to large carbocycles are suitable targets for the application of SmI_2-promoted intramolecular Reformatsky reaction. Ring sizes in the range 8- to 15-membered can be synthesized in fairly good yields. The mechanism of reaction could well involve a two-electron reduction of the α-bromo ester to form the samarium enolate **521** which would then proceed to cyclize as depicted in Equation 69.[166]

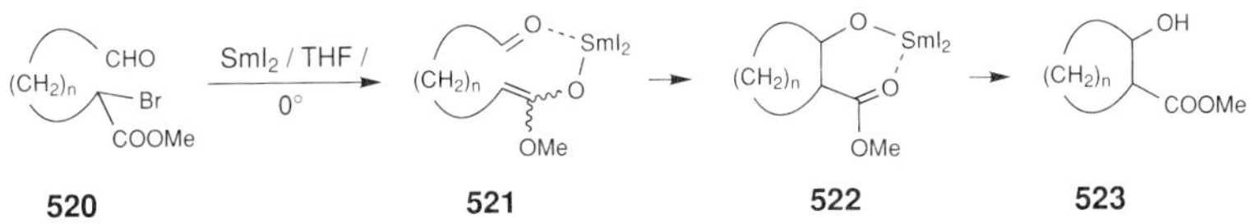

n = 6	(68%)
n = 7	(70%)
n = 9	(74%)
n = 12	(82%)
n = 13	(82%)

Equation 69

A new method reported recently for the synthesis of spiro-γ-butyrolactone makes use of the samarium iodide-induced cyclization of cyclopentenone **525**. The reaction most likely progresses successively through ketyl **526**, allyl radical **527**, and allyl anion **528** whose anionic cyclization leads to lactol **529**. The latter can be oxidized readily by pyridinium dichromate to furnish spirolactone **530**. The presence of the allyl anion has been proved by trapping with D_2O to obtain deuterated products **531** and **532**, as illustrated in Scheme 55.[167] The process is applicable to substrates of various ring sizes.

It would seem that, thus far, organic synthesis is only at the dawning of the age of samarium chemistry. The gathering momentum of reports on advances in samarium iodide chemistry (including improvements of known reactions)[168] seems to be signalling the busy role that the reagent will most certainly play in future organic synthesis.

Scheme 55

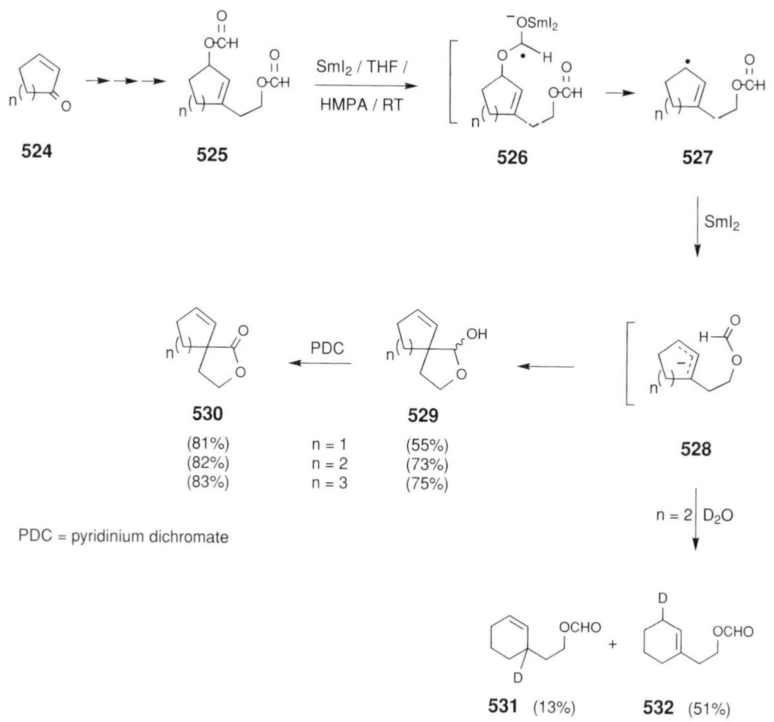

VI. CONCLUSION

In brief, the foregoing chapter represents but a glimpse of the domain of metal-catalyzed cyclization reactions. Regrettably it has not been expedient to include many other interesting topics, namely, the use of low-valent titanium in carbonyl coupling reactions;[169–172] the titanocene chloride catalyst in regioselective cyclopentane ring formation;[173] the role of transition metal carbene complexes in cycloaddition and annulation reactions,[4–6] in particular the Fischer carbene complexes of chromium (briefly mentioned in Equations 40 and 41), molybdenum, and tungsten;[174–185] and the rhodium-catalyzed diazo-carbonyl compound decomposition to form Rh(II) carbene complexes (Fischer-type carbenoid)[12,186] for use in further reactions such as insertion, ylide formation, cyclopropanation, and annulations.[187–194] As for other organometallic reagents apart from those mentioned in the text, organomagnesium[195–197] is one that is becoming important, while the newly introduced organocerium[198,199] is already starting to have an impact on organic synthesis.[199] Truly, we are moving inevitably towards a world of organometallics!

REFERENCES

1. Seebach, D., Organic synthesis—where now?, *Angew. Chem. Int. Ed. Engl.*, 29, 1320, 1990.
2. Trost, B. M., Cyclizations *via* palladium-catalyzed allylic alkylations, *Angew. Chem. Int. Ed. Engl.*, 28, 1173, 1989.
3. Trost, B. M., Transition metal templates as guides for cycloadditions, *Pure & Appl. Chem.*, 60, 1615, 1988.
4. Wulff, W. D., Tang, P.-C., Chan, K.-S., McCallum, J. S., Yang, D. C., and Gilbertson, S. R., Cycloadditions and annulations of transition metal carbene complexes, *Tetrahedron*, 41, 5813, 1985.
5. Dötz, K. H., Carbene complexes in organic synthesis, *Angew. Chem. Int. Ed. Engl.*, 23, 587, 1984.
6. Dötz, K. H., in *Organometallics in Organic Synthesis: Aspects of a Modern Interdisciplinary Field*, TomDieck, H., and de Meijere, A., Eds., Springer–Verlag, Berlin, 1988, 85.
7. Vollhardt, K. P. C., Cobalt-mediated [2+2+2]-cycloadditions: a maturing synthetic strategy, *Angew. Chem. Int. Ed. Engl.*, 23, 539, 1984.
8. Pauson, P. L., The Khand reaction. A convenient and general route to a wide range of cyclopentenone derivatives, *Tetrahedron*, 41, 5855, 1985.
9. Schore, N. E., Transition–metal-mediated cycloaddition reactions of alkynes in organic synthesis, *Chem. Rev.*, 88, 1081, 1988.
10. Soderquist, J. A., Samarium(II) iodide in organic synthesis, *Aldrichimica Acta*, 24, 15, 1991.
11. Molander, G. A., Application of lanthanide reagents in organic synthesis, *Chem. Rev.*, 92, 29, 1992.
12. Adams, J., and Spero, D. M., Rhodium(II) catalyzed reactions of diazo-carbonyl compounds, *Tetrahedron*, 47, 1765, 1991.
13. Oppolzer, W., Intramolecular, stoichiometric (Li, Mg, Zn) and catalytic (Ni, Pd, Pt) metallo-ene reactions in organic synthesis, *Angew. Chem. Int. Ed. Engl.*, 28, 38, 1989.
14. Negishi, E., Controlled carbometalation as a new tool for carbon–carbon bond formation and its application to cyclization, *Acc. Chem. Res.*, 20, 65, 1987.
15. Yasuda, H., and Nakamura, A., Diene, alkyne, alkene, and alkyl complexes of early transition metals: structures and synthetic applications in organic and polymer chemistry, *Angew. Chem. Int. Ed. Engl.*, 26, 723, 1987.
16. Abramovitch, R. A, Barton, D. H. R., and Finet, J.-P., Newer methods of arylation, *Tetrahedron*, 44, 3039, 1988.
17. Heck, R. F., Palladium-catalyzed vinylation of organic halides, *Org. React.*, 27, 345, 1982.
18. Stille, J. K., The palladium-catalyzed cross-coupling reactions of organotin reagents with organic electrophiles, *Angew. Chem. Int. Ed. Engl.*, 25, 508, 1986.
19. Grigg, R., Sridharan, V., Stevenson, P., and Worakun, T., Palladium(II) catalysed construction of tetrasubstituted carbon centres, and spiro- and bridged-ring compounds from enamides of 2-iodobenzoic acids, *J. Chem. Soc., Chem. Commun.*, 1697, 1986.
20. Shi, L., Narula, C. K., Mak, K. T., Kao, L., Xu, Y., and Heck, R. F., Palladium-catalyzed cyclizations of bromodialkenyl ethers and amines, *J. Org. Chem.*, 48, 3894, 1983.
21. Gaudin, J.-M., Intramolecular Heck reaction with substrates possessing an allylic alcohol moiety, *Tetrahedron Lett.*, 32, 6113, 1991.
22. Kozikowski, A. P., and Ma, D., Palladium catalyzed synthesis of annelated indoles, *Tetrahedron Lett.*, 32, 3317, 1991.
23. Rawal, V. H., and Michoud, C., A general solution to the synthesis of 2-azabicyclo[3.3.1]nonane unit of *Strychnos* alkaloids, *Tetrahedron Lett.*, 32, 1695, 1991.
24. Noyori, R., and Takaya, H., BINAP: an efficient chiral element for asymmetric catalysis, *Acc. Chem. Res.*, 23, 345, 1990.

25. Sato, Y., Sodeoka, M., and Shibasaki, M., Catalytic asymmetric C–C bond formation: asymmetric synthesis of *cis*-decalin derivatives by palladium-catalyzed cyclization of prochiral alkenyl iodides, *J. Org. Chem.*, 54, 4738, 1989.
26. Kagechika, K., and Shibasaki, M., Asymmetric Heck reaction: a catalytic asymmetric synthesis of the key intermediate for $\Delta^{9(12)}$-capnellene-3β,8β,10α-triol and $\Delta^{9(12)}$-capnellene-3β,8β,10α,14-tetrol, *J. Org. Chem.*, 56, 4093, 1991.
27. Burns, B., Grigg, R., Sridharan, V., and Worakun, T., Palladium catalysed tandem cyclisation–anion capture processes. Hydride ion capture by vinylpalladium species, *Tetrahedron Lett.*, 29, 4325, 1988.
28. Abelman, M. M., and Overman, L. E., Palladium-catalyzed polyene cyclizations of dienyl aryl iodides, *J. Am. Chem. Soc.*, 110, 2328, 1988.
29. Carpenter, N. E., Kucera, D. J., and Overman, L. E., Palladium-catalyzed polyene cyclizations of trienyl triflates, *J. Org. Chem.*, 54, 5846, 1989.
30. Meyer, F. E., Brandenburg, J., Parsons, P. J., and de Meijere, A., Palladium-catalysed polycyclization of dienynes to give carbocyclic angularly bisanellated cyclohexadiene derivatives, *J. Chem. Soc., Chem. Commun.*, 390, 1992.
31. Zhang, Y., and Negishi, E., Palladium-catalyzed cascade carbometalation of alkynes and alkenes as an efficient route to cyclic and polycyclic structures, *J. Am. Chem. Soc.*, 111, 3454, 1989.
32. Zhang, Y., Wu, G., Agnel, G., and Negishi, E., One-step construction of fused tricyclic and tetracyclic structures from acyclic precursors *via* cyclic carbopalladation, *J. Am. Chem. Soc.*, 112, 8590, 1990.
33. Meyer, F. E., Parsons, P. J., and de Meijere, A., Palladium-catalyzed polycyclization of dienynes: surprisingly facile formation of tetracyclic systems containing a three-membered ring, *J. Org. Chem.*, 56, 6487, 1991.
34. Grigg, R., Dorrity, M. J., Malone, J. F., Sridharan, V., and Sukirthalingam, S., Palladium-catalysed polycyclisation–anion capture processes, *Tetrahedron Lett.*, 31, 1343, 1990.
35. Stork, G., Sher, P. M., and Chen, H.-L., Radical cyclization–trapping in the synthesis of natural products. A simple, stereocontrolled route to prostaglandin $F_{2\alpha}$, *J. Am. Chem. Soc.*, 108, 6384, 1986.
36. Keck, G. E., and Burnett, D. A., β-Stannyl enones as radical traps: a very direct route to $PGF_{2\alpha}$, *J. Org. Chem.*, 52, 2958, 1987.
37. Larock, R. C., and Lee, N. H., Efficient free-radical and palladium-catalyzed tandem alkene insertions: a new approach to benzoprostacyclins, *J. Org. Chem.*, 56, 6253, 1991.
38. Trost, B. M., Palladium-catalyzed cycloisomerizations of enynes and related reactions, *Acc. Chem. Res.*, 23, 34, 1990.
39. Keinan, E., and Greenspoon, N., Highly chemoselective reductions with PMHS and palladium(0) catalyst, *J. Org. Chem.*, 48, 3545, 1983.
40. Trost, B. M., and Rise, F., A reductive cyclization of 1,6- and 1,7-enynes, *J. Am. Chem. Soc.*, 109, 3161, 1987.
41. Trost, B. M., and Edstrom, E. D., Synthesis of (±)-phyllanthocin *via* a metal-catalyzed cycloreduction, *Angew. Chem. Int. Ed. Engl.*, 29, 520, 1990.
42. Wender, P. A., and McDonald, F. E., Studies on tumor promoters. IX. A second-generation synthesis of phorbol, *J. Am. Chem. Soc.*, 112, 4956, 1990.
43. Trost, B. M., and Lee, D. C., A catalytic reductive cyclization of 1,6-diynes, *J. Am. Chem. Soc.*, 110, 7255, 1988.
44. Trost, B. M., and Tour, J. M., A metal-catalyzed cyclization of enallenes, *J. Am. Chem. Soc.*, 110, 5231, 1988.
45. Trost, B. M., and Tour, J. M., A novel nickel–chromium catalyst system for cyclization *via* intramolecular carbametalation, *J. Am. Chem. Soc.*, 109, 5268, 1987.
46. Trost, B. M., Pfrengle, W., Urabe, H., and Dumas, J., Palladium-catalyzed alkylative cyclization of 1,6- and 1,7-enynes, *J. Am. Chem. Soc.*, 114, 1923, 1992.
47. Trost, B. M., and Dumas, J., New strategy for the total synthesis of 1α-hydroxyvitamin D derivatives, *J. Am. Chem. Soc.*, 114, 1924, 1992.

48. Ciufolini, M. A., Qi, H.-B., and Browne, M. E., Intramolecular arylations of soft enolates catalyzed by zerovalent palladium, *J. Org. Chem.*, 53, 4149, 1988.
49. Piers, E., and Marais, P. C., A new five-membered ring annulation method based on Pd(0)-catalyzed intramolecular coupling of vinyl iodide and enolate anion functions, *J. Org. Chem.*, 55, 3454, 1990.
50. van der Heide, T. A. J., van der Baan, J. L., Bickelhaupt, F., and Klumpp, G. W., Konjugierte addition von 2-(bromzinkmethyl)-3-phenoxy-1-propen an methylenmalonsäurederivate; weitere umsetzung zu methylencyclopentanen, *Tetrahedron Lett.*, 33, 475, 1992.
51. Fournet, G., Balme, G., Van Hemelryck, B., and Gore, J., Stereospecific synthesis of arylidene and allylidene cyclopentanes by a palladium-catalyzed cyclisation, *Tetrahedron Lett.*, 31, 5147, 1990.
52. Bouyssi, D., Balme, G., Fournet, G., Monteiro, N., and Gore, J., Palladium-mediated cyclopentanation of alkenes bearing a nucleophilic substituent. Stereochemistry and mechanism, *Tetrahedron Lett.*, 32, 1641, 1991.
53. Monteiro, N., Balme, G., and Gore, J., Cyclisation of ω-unsaturated β-dicarbonylcompounds catalyzed by a palladium hydride species, *Tetrahedron Lett.*, 32, 1645, 1991.
54. Bouyssi, D., Balme, G., and Gore, J., Synthesis of ethynylidene cyclopentanes by a cyclisation catalyzed by the simultaneous presence of palladium(0) and copper(I), *Tetrahedron Lett.*, 32, 6541, 1991.
55. Oppolzer, W., Regio- and stereocontrolled catalytic palladium- and nickel 'ene-type' cyclizations, *Pure & Appl. Chem.*, 62, 1941, 1990.
56. Oppolzer, W., Swenson, R. E., and Gaudin, J.-M., Catalytic intramolecular 2-(4-alkenyl)allylpalladium insertions, *Tetrahedron Lett.*, 29, 5529, 1988.
57. Trost, B. M., and Luengo, J. I., Metal-catalyzed cyclization *via* isomerization of α-dienyl-ω-allyl acetates, *J. Am. Chem. Soc.*, 110, 8239, 1988.
58. Zair, T., Santelli-Rouvier, C., and Santelli, M., Palladium-mediated cyclization of 1,5-hexadien-3-ols to 1-methyl-1,3-cyclopentadienes, *Tetrahedron Lett.*, 32, 4501, 1991.
59. Larock, R. C., and Fried, C. A., Palladium-catalyzed carboannulation of 1,3-dienes by aryl halides, *J. Am. Chem. Soc.*, 112, 5882, 1990.
60. Harris, G. D., Jr., Herr, R. J., and Weinreb, S. M., A palladium-mediated approach to construction of nitrogen heterocycles, *J. Org. Chem.*, 57, 2528, 1992.
61. Trost, B. M., and Miller, M. L., Criteria for concertedness in cycloadditions, *J. Am. Chem. Soc.*, 110, 3687, 1988.
62. Trost, B. M., and Grese, T. A., Intramolecular palladium-catalyzed cycloadditions with a cleavable tether, *J. Am. Chem. Soc.*, 113, 7363, 1991.
63. Trost, B. M., and Grese, T. A., The acyl effect on intramolecular palladium-catalyzed trimethylenemethane cycloadditions, *J. Org. Chem.*, 57, 686, 1992.
64. Trost, B. M., Grese, T. A., and Chan, D. M. T., Intramolecular palladium-catalyzed trimethylenemethane cycloadditions: initial studies, *J. Am. Chem. Soc.*, 113, 7350, 1991.
65. Grigg, R., Sridharan, V., and Sukirthalingam, S., Alkylpalladium(II) species. Reactive intermediates in a bis-cyclisation route to strained polyfused ring systems, *Tetrahedron Lett.*, 32, 3855, 1991.
66. Grigg, R., Sukirthalingam, S., and Sridharan, V., Palladium catalysed tandem cyclisation–anion capture processes initiated by alkyl- and π-allyl-palladium species, *Tetrahedron Lett.*, 32, 2545, 1991.
67. Oppolzer, W., Keller, T. H., Kuo, D. L., and Pachinger, W., Palladium(0)- and nickel(0) catalyzed "metallo-ene-type" cyclizations: stereodirecting resident chirality, *Tetrahedron Lett.*, 31, 1265, 1990.
68. Oppolzer, W., and DeVita, R. J., Direct syntheses of polyfused ring systems by intramolecular tandem palladium–ene / Heck insertion reactions, *J. Org. Chem.*, 56, 6256, 1991.
69. Piers, E., and Lu, Y.-F., Deconjugation–alkylation of ethyl 3-(trimethylstannyl)-2-alkenoates. Stereocontrolled synthesis of ethyl 2-alkylidene-3-methylenecyclobutanecarboxylates, *J. Org. Chem.*, 53, 926, 1988.

70. Bradley, J. C., and Durst, T., Synthesis of 2-benzylidenebenzocyclobutenones *via* an intramolecular Stille coupling reaction, *J. Org. Chem.*, 56, 5459, 1991.
71. Grigg, R., Teasdale, A., and Sridharan, V., Palladium catalysed intramolecular coupling of aryl and benzylic halides and related tandem cyclisations. A simple synthesis of hippadine, *Tetrahedron Lett.*, 32, 3859, 1991.
72. Piers, E., Friesen, R. W., and Keay, B. A., Intramolecular palladium(0)-catalyzed coupling of enol triflate and vinylstannane functions. New annulation sequences leading to bicyclic diene systems, *Tetrahedron*, 47, 4555, 1991.
73. Kalivretenos, A., Stille, J. K., and Hegedus, L. S., Synthesis of β-resorcyclic macrolides *via* organopalladium chemistry. Application of the total synthesis of (*S*)-zearalenone, *J. Org. Chem.*, 56, 2883, 1991.
74. Crisp, G. T., Scott, W. J., and Stille, J. K., Palladium-catalyzed carbonylative coupling of vinyl triflates with organostannanes. A total synthesis of $(\pm)\Delta^{9(12)}$-capnellene, *J. Am. Chem. Soc.*, 106, 7500, 1984.
75. Gyorkos, A. C., Stille, J. K., and Hegedus, L. S., Total synthesis of (±)-*epi*-jatrophone and (±)-jatrophone using palladium-catalyzed carbonylative coupling of vinyl triflates with vinylstannanes as the macrocycle-forming step, *J. Am. Chem. Soc.*, 112, 8465, 1990.
76. Noyori, R., Nishida, I., and Sakata, J., Tris(dialkylamino)sulfonium enolates. Synthesis, structure, and reactions, *J. Am. Chem. Soc.*, 105, 1598, 1983.
77. Ciattini, P. G., Morera, E., Ortar, G., and Rossi, S. S., Preparative and regiochemical aspects of the palladium-catalyzed carbonylative coupling of 2-hydroxyaryl iodides with ethynylarenes, *Tetrahedron*, 47, 6449, 1991.
78. Shimoyama, I., Zhang, Y., Wu, G., and Negishi, E., Synthesis of enol esters and enol lactones *via* palladium-catalyzed carbonylation of aryl and alkenyl halides, *Tetrahedron Lett.*, 31, 2841, 1990.
79. Wu, G., Shimoyama, I., and Negishi, E., Palladium-catalyzed carbonylative cyclization of *o*-allylbenzyl halides to produce benzo-annulated enol lactones and/or bicyclo[3.3.0]hept-3-en-6-ones. An efficient route to U-68,215, *J. Org. Chem.*, 56, 6506, 1991.
80. Oppolzer, W., Keller, T. H., Bedoya-Zurita, M., and Stone, C., Diastereocontrolled nickel(0)- and palladium(0)-catalyzed "metallo-ene-type" cyclizations/carbonylations, *Tetrahedron Lett.*, 30, 5883, 1989.
81. Oppolzer, W., Bedoya-Zurita, M., and Switzer, C. Y., Nickel-catalyzed intramolecular allylation of alkenes and alkynes coupled with β-elimination or carbonylation, *Tetrahedron Lett.*, 29, 6433, 1988.
82. Oppolzer, W., Gaudin, J.-M., Bedoya-Zurita, M., Hueso-Rodriguez, J., Raynham, T. M., and Robyr, C., Synthesis of nitrogen- or oxygen-containing ring systems by palladium-catalyzed intramolecular olefin allylations, *Tetrahedron Lett.*, 29, 4709, 1988.
83. Keese, R., Guidetti-Grept, R., and Herzog, B., Synthesis of [5.5.5.5]fenestranes by Pd-catalyzed carbonylation–cyclisation, *Tetrahedron Lett.*, 33, 1207, 1992.
84. Cane, D. E., and Thomas, P. J., Synthesis of *dl*-pentalenolactones E and F, *J. Am. Chem. Soc.*, 106, 5295, 1984.
85. Cane, D. E., and Rossi, T., The isolation and structural elucidation of pentalenolactone E, *Tetrahedron Lett.*, 2973, 1979.
86. Oppolzer, W., Xu, J.-Z., and Stone, C., Preparation of a key intermediate for the synthesis of the (±)-pentalenolactone E methyl ester by catalytic 'palladium–ene cyclization'/ methoxycarbonylation, *Helv. Chim. Acta*, 74, 465, 1991.
87. Khand, I. U., Knox, G. R., Pauson, P. L., Watts, W. E., and Foreman, M. I., Organocobalt complexes. II. Reaction of acetylenehexacarbonyldicobalt complexes, $(R^1C_2R^2)Co_2(CO)_6$, with norbornene and its derivatives, *J. Chem. Soc., Perkin Trans. 1*, 977, 1973.
88. Krafft, M. E., Juliano, C. A., Scott, I. L., Wright, C., and McEachin, M. D., The directed Pauson–Khand reaction, *J. Am. Chem. Soc.*, 113, 1693, 1991.
89. Rautenstrauch, V., Mégard, P., Conesa, J., and Küster, W., 2-Pentylcyclopent-2-en-1-one by catalytic Pauson–Khand reaction, *Angew. Chem. Int. Ed. Engl.*, 29, 1413, 1990.

90. Smit, W. A., Gybin, A. S., Shashkov, A. S., Strychkov, Y. T., Kyz'mina, L. G., Mikaelian, G. S., Caple, R., and Swanson, E. D., New route to the synthesis of polycyclic compounds based on a stepwise Ad_E-reaction of dicobalt hexacarbonyl complexes of conjugated enynes with a subsequent intramolecular Khand–Pauson type reaction, *Tetrahedron Lett.*, 27, 1241, 1986.
91. Billington, D. C., Helps, I. M., Pauson, P. L., Thomson, W., and Willison, D., The effect of ultrasound and of phosphine and phosphine oxides on the Khand reaction, *J. Organomet. Chem.*, 354, 233, 1988.
92. Shambayati, S., Crowe, W. E., and Schreiber, S. L., *N*-Oxide promoted Pauson–Khand cyclizations at room temperature, *Tetrahedron Lett.*, 31, 5289, 1990.
93. Magnus, P., Annoura, H., and Harling, J., Studies on the synthesis of the antitumor agents esperamicin A_1/calicheamicin γ_1: stereospecific synthesis of the 12β-hydroxybicyclo[7.3.1] diynene core structure *via* a $Co_2(CO)_6$–acetylene-mediated synclinal aldol reaction, *J. Org. Chem.*, 55, 1709, 1990.
94. Magnus, P., Carter, P., Elliott, J., Lewis, R., Harling, J., Pitterna, T., Bauta, W. E., and Fortt, S., Synthetic and mechanistic studies on the antitumor antibiotics esperamicin A_1 and calicheamicin γ_1: synthesis of 2-ketobicyclo[7.3.1] enediyne and 13-ketocyclo[7.3.1] enediyne cores mediated by η^2 dicobalt hexacarbonyl alkyne complexes. Cycloaromatization rate studies, *J. Am. Chem. Soc.*, 114, 2544, 1992.
95. Krafft, M. E., and Wright, C., A formal total synthesis of PGA_2 using the directed Pauson–Khand reaction, *Tetrahedron Lett.*, 33, 151, 1992.
96. Camps, F., Moretó, J. M., Ricart, S., Viñas, J. M., Molins, E., and Miravitlles, C., Reaction of ethyl diethoxyacrylate with alkynylalkoxycarbene metal (Cr, W) complexes: new procedure for preparation of 6-ethoxy-2*H*-pyrones, *J. Chem. Soc., Chem. Commun.*, 1560, 1989.
97. Camps, F., Moretó, J. M., Ricart, S., and Viñas, J. M., Fischer carbene complexes facilitate the intramolecular Pauson–Khand reaction, *Angew. Chem. Int. Ed. Engl.*, 30, 1470, 1991.
98. Schore, N. E., and Rowley, E. G., Diastereofacialselectivity in intramolecular Pauson–Khand cycloaddition: highly stereoselective synthesis of pentalenene, *J. Am. Chem. Soc.*, 110, 5224, 1988.
99. Castro, J., Sörensen, H., Riera, A., Morin, C., Moyano, A., Pericàs, M. A., and Greene, A. E., Asymmetric approach to Pauson–Khand bicyclization. Enantioselective formal synthesis of hirsutene, *J. Am. Chem. Soc.*, 112, 9388, 1990.
100. van der Waals, A., and Keese, R., Synthesis of a [5.5.5.5]fenestrenedione *via* tandem Pauson–Khand tetracyclisation, *J. Chem. Soc., Chem. Commun.*, 570, 1992.
101. Negishi, E., Holmes, S. J., Tour, J. M., Miller, J. A., Cederbaum, F. E., Swanson, D. R., and Takahashi, T., Novel bicyclization of enynes and diynes promoted by zirconocene derivatives and conversion of zirconabicycles into bicyclic enones *via* carbonylation, *J. Am. Chem. Soc.*, 111, 3336, 1989.
102. Nugent, W. A., and Calabrese, J. C., Titanium-mediated synthesis of *E,E*-exocyclic dienes. Application to the preparation of polycyclic compounds, *J. Am. Chem. Soc.*, 106, 6422, 1984.
103. Agnel, G., and Negishi, E., Highly stereo- and regiocontrolled cyclopentannulation *via* allylphosphonate conjugate addition and hydroboration–oxidation–elimination. Synthesis of pentalenic acid with virtually complete stereo- and regiocontrol, *J. Am. Chem. Soc.*, 113, 7424, 1991.
104. Seto, H., Sasaki, T., Uzawa, J., Takeuchi, S., and Yonehara, H., Studies on the biosynthesis of pentalenolactone. II. Isolation of pentalenic acid and pentalenolactone H., *Tetrahedron Lett.*, 4411, 1978.
105. Swanson, D. R., Rousset, C. J., Negishi, E., Takahashi, T., Seki, T., Saburi, M., and Uchida, Y., Regioselective and diastereoselective alkyl–alkene and alkene–alkene coupling promoted by zirconocene and hafnocene, *J. Org. Chem.*, 54, 3521, 1989.

106. Agnel, G., Owczarczyk, Z., and Negishi, E., Diastereoselective zirconocene-promoted bicyclization–carbonylation of allylically methyl-substituted enynes. Synthesis of (+)-iridomyrmecin, *Tetrahedron Lett.*, 33, 1543, 1992.
107. Annis, G. D., and Ley, S. V., Formation of lactones from dienes *via* iron carbonyl complexes, *J. Chem. Soc., Chem. Commun.*, 581, 1977.
108. Annis, G. D., Hebblethwaite, E. M., and Ley, S. V., Formation of β-lactams from tricarbonyliron lactone complexes, *J. Chem. Soc., Chem. Commun.*, 297, 1980.
109. Bates, R. W., Fernández-Moro, R., and Ley, S. V., Synthesis of the β-lactone esterase inhibitor valilactone using π-allyltricarbonyliron lactone complexes, *Tetrahedron Lett.*, 32, 2651, 1991.
110. Bates, R. W., Fernández-Moro, R., and Ley, S. V., The use of π-allyltricarbonyliron lactone complexes in the synthesis of the β-lactone esterase inhibitor (–)-valilactone, *Tetrahedron*, 47, 9929, 1991.
111. Camps, F., Llebaria, A., Moretó, J. M., and Pagès, L., Fused bicyclic systems through a Ni-mediated intermolecular cyclopentenone annulation, *Tetrahedron Lett.*, 33, 109, 1992.
112. Camps, F., Llebaria, A., Moretó, J. M., and Pagès, L., New practical synthesis of spirocyclopentenones, *Tetrahedron Lett.*, 33, 113, 1992.
113. Hegedus, L. S., de Weck, G., and D'Andrea, S., Evidence for the intermediacy of chromium–ketene complexes in the synthesis of β-lactams by the photolytic reaction of chromium–carbene complexes with imines. Use in amino acid synthesis, *J. Am. Chem. Soc.*, 110, 2122, 1988.
114. Sierra, M. A., and Hegedus, L. S., Formation of cyclobutanones by the photolytic reaction of $(CO)_5Cr{=}C(OMe)Me$ with electron-rich olefins, *J. Am. Chem. Soc.*, 111, 2335, 1989.
115. Söderberg, B. C., Hegedus, L. S., and Sierra, M. A., Synthesis of cyclobutanones by the photolytic reaction of chromium carbene complexes with olefins: inter- and intramolecular reactions, *J. Am. Chem. Soc.*, 112, 4364, 1990.
116. Wakabayashi, Y., Fukuda, Y., Shiragami, H., Utimoto, K., and Nozaki, H., Preparation of furans from alkynols utilizing palladium catalyzed intramolecular addition of alcohol to acetylene as a key reaction, *Tetrahedron*, 41, 3655, 1985.
117. Fukuda, Y., Shiragami, H., Utimoto, K., and Nozaki, H., Synthesis of substituted furans by palladium-catalyzed cyclization of acetylenic ketones, *J. Org. Chem.*, 56, 5816, 1991.
118. Luo, F.-T., Schreuder, I., and Wang, R.-T., Intramolecular oxypalladation and cross-coupling of acetylenic alkoxides, *J. Org. Chem.*, 57, 2213, 1992.
119. Hosokawa, T., Uno, T., Inui, S., and Murahashi, S.-I., Palladium(II)-catalyzed asymmetric oxidative cyclization of 2-allylphenols in the presence of copper(II) acetate and molecular oxygen. Study of the catalysis of the Wacker-type oxidation, *J. Am. Chem. Soc.*, 103, 2318, 1981.
120. Semmelhack, M. F., and Zask, A., Synthesis of racemic frenolicin *via* organochromium and organopalladium intermediates, *J. Am. Chem. Soc.*, 105, 2034, 1983.
121. Saito, S., Hara, T., Takahashi, N., Hirai, M., and Moriwake, T., Synthesis of optically active substituted dihydrofurans and dihydropyrroles *via* palladium(II)-promoted diastereoselective dehydroxylative heterocyclization, *Synlett*, 237, 1992.
122. Suzuki, T., Uozumi, Y., and Shibasaki, M., A catalytic asymmetric synthesis of α-methylene lactones by the palladium-catalysed carbonylation of prochiral alkenyl halides, *J. Chem. Soc., Chem. Commun.*, 1593, 1991.
123. Trost, B. M., and Tenaglia, A., Pd catalyzed chemoselective cyclization to cyclic ethers, *Tetrahedron Lett.*, 29, 2927, 1988.
124. Harding, K. E., and Burks, S. R., Synthesis of *trans*-2,5-dimethylpyrrolidine by intramolecular amidomercuration, *J. Org. Chem.*, 46, 3920, 1981.
125. Harding, K. E., Stephens, R., and Hollingsworth, D. R., Stereoselective conversion of allylic alcohols to *N*-acyl-*trans*-4,5-dialkyloxazolidines, *Tetrahedron Lett.*, 25, 4631, 1984.
126. Harding, K. E., and Hollingsworth, D. R., Selectivity in the amination of allylic alcohols *via* intramolecular amidomercuration, *Tetrahedron Lett.*, 29, 3789, 1988.

127. Harding, K. E., and Nam, D., Stereoselective synthesis of (±)-*threo*-γ-hydroxy-β-lysine lactone, *Tetrahedron Lett.*, 29, 3793, 1988.
128. Harding, K. E., and Marman, T. H., Stereoselectivity in intramolecular amidomercuration. Kinetic *vs.* thermodynamic control, *J. Org. Chem.*, 49, 2838, 1984.
129. Harding, K. E., Marman, T. H., and Nam, D., Stereoselective synthesis of (±)-*erythro*- and *threo*-γ-hydroxynorvaline, *Tetrahedron Lett.*, 29, 1627, 1988.
130. Bloodworth, A. J., Curtis, R. J., and Mistry, N., 1,2-Dioxolane *versus* 1,2-dioxane formation in the cyclization of an α,ω-diene hydroperoxide under polar and free radical conditions, *J. Chem. Soc., Chem. Commun.*, 954, 1989.
131. Bloodworth, A. J., and Korkodilos, D., Mercury(II)-mediated cyclisation of hydroperoxyalkylcyclopropanes: a new route to cyclic peroxides, *Tetrahedron Lett.*, 32, 6953, 1991.
132. Porter, N. A., Cudd, M. A., Miller, R. W., and McPhail, A. T., A fixed-geometry study of the S_N2 reaction on the peroxide bond, *J. Am. Chem. Soc.*, 102, 414, 1980.
133. Gopalan, A. S., Prieto, R., Mueller, B., and Peters, D., Polyene cyclizations using mercury(II) triflate-*N,N*-dimethylaniline complex–participation by internal nucleophiles, *Tetrahedron Lett.*, 33, 1679, 1992.
134. Nishizawa, M., Takenaka, H., Nishide, H., and Hayashi, Y., A new olefin cyclization agent, mercury(II) trifluoromethanesulfonate–amine complex, *Tetrahedron Lett.*, 24, 2581, 1983.
135. Nishizawa, M., Nishide, H., and Hayashi, Y., Total synthesis of (±)-*a,g*-onoceradienedione and lansic acid, *Tetrahedron Lett.*, 25, 5071, 1984.
136. Grimaldi, J., and Cormons, A., Cyclisation d'oximes β-alleniques. Synthese de dihydro-4,7-oxazepines-1,2, *Tetrahedron Lett.*, 26, 825, 1985.
137. Grimaldi, J., and Cormons, A., Cyclisation d'amides β-alleniques en dihydro-3,6 1*H*-pyridones-2, *Tetrahedron Lett.*, 27, 5089, 1986.
138. Kinsman, R., Lathbury, D., Vernon, P., and Gallagher, T., Stereoselectivity in the synthesis of 2,5-disubstituted pyrrolidines, *J. Chem. Soc., Chem. Commun.*, 243, 1987.
139. Kimura, M., Fugami, K., Tanaka, S., and Tamaru, Y., Silver(I) catalyzed amino cyclization of *O*-(2,3-butadienyl) carbamates: an efficient and stereoselective synthesis of 4-vinyl-2-oxazolidinones, *Tetrahedron Lett.*, 32, 6359, 1991.
140. Davies, I. W., Gallagher, T., Lamont, R. B., and Scopes, D. I. C., Asymmetric synthesis of functionalised pyrrolidines. Highly diastereoselective cyclisations mediated by sulfide and sulfoxide ligands, *J. Chem. Soc., Chem. Commun.*, 335, 1992.
141. Prasad, J. S., and Liebeskind, L. S., Silver mediated cyclizations of 4-allenyl- and 4-(2-propynyl)azetidinones. A stereoselective synthesis of 3-substituted Δ^1-carbapenems *via* N–C3 closure, *Tetrahedron Lett.*, 29, 4253, 1988.
142. Prasad, J. S., and Liebeskind, L. S., Palladium mediated formation of Δ^1- and Δ^2-carbapenems by cyclofunctionalization of 4-allenylazetidinones and 4-(2-propynyl)azetidinones, *Tetrahedron Lett.*, 29, 4257, 1988.
143. Marshall, J. A., and Robinson, E. D., A mild method for the synthesis of furans. Application to 2,5-bridged furano macrocyclic compounds, *J. Org. Chem.*, 55, 3450, 1990.
144. Marshall, J. A., and Wang, X., Synthesis of furans by Ag(I)-promoted cyclization of allenyl ketones and aldehydes, *J. Org. Chem.*, 56, 960, 1991.
145. Dess, D. B., and Martin, J. C., Readily accessible 12-I-5 oxidant for the conversion of primary and secondary alcohols to aldehydes and ketones, *J. Org. Chem.*, 48, 4155, 1983.
146. Mitsunobu, O., The use of diethyl azodicarboxylate and triphenylphosphine in synthesis and transformation of natural products, *Synthesis*, 1, 1981.
147. Marshall, J. A., and Wang, X., [2,3] Wittig rearrangement of nonracemic propargyloxyacetic acids and esters. Synthesis of optically active 2,5-dihydrofurans, *J. Org. Chem.*, 55, 2995, 1990.
148. Shigemasa, Y., Yasui, M., Ohrai, S., Sasaki, M., Sashiwa, H., and Saimoto, H., Enantioselective transformation of propargyl esters to dihydrofurans, *J. Org. Chem.*, 56, 910, 1991.

149. Sasaki, H., Okutomi, T., Hosokawa, T., Nawata, Y., and Ando, K., Ascofuranone, a new antibiotic from Ascochyta viciae, *Tetrahedron Lett.*, 2541, 1972.
150. Girard, P., Namy, J. L., and Kagan, H. B., Divalent lanthanide derivatives in organic synthesis. I. Mild preparation of SmI_2 and YbI_2 and their use as reducing or coupling agents, *J. Am. Chem. Soc.*, 102, 2693, 1980.
151. Kagan, H. B., Divalent samarium compounds: perspectives for organic chemistry, *N. J. Chem.*, 14, 453, 1990.
152. Molander, G. A., and Harring, L. S., Reductive radical cyclizations of haloalkenes promoted by samarium diiodide. Sequential cyclization / intermolecular carbonyl addition reactions, *J. Org. Chem.*, 55, 6171, 1990.
153. Fevig, T. L., Elliott, R. L., and Curran, D. P., A samarium(II) iodide promoted tandem radical cyclization. The total synthesis of (±)-hypnophilin and the formal synthesis of (±)-coriolin, *J. Am. Chem. Soc.*, 110, 5064, 1988.
154. Danishefsky, S., Zamboni, R., Kahn, M., and Etheredge, S. J., Total synthesis of *dl*-coriolin, *J. Am. Chem. Soc.*, 102, 2097, 1980.
155. Molander, G. A., and Etter, J. B., Lanthanides in organic synthesis. III. A general procedure for five- and six-membered ring annulation, *J. Org. Chem.*, 51, 1778, 1986.
156. Lannoye, G., and Cook, J. M., General approach to the synthesis of polyquinenes *via* the Weiss reaction. V. Progress toward the synthesis of dicyclopentapentalenes, *Tetrahedron Lett.*, 29, 171, 1988.
157. Daniewski, A. R., and Uskoković,M. R., Compactin from Hajós (-)-enedione, *Tetrahedron Lett.*, 31, 5599, 1990.
158. Molander, G. A., and McKie, J. A., A facile synthesis of bicyclo[*m.n.*1]alkan-1-ols. Evidence for organosamarium intermediates in the samarium(II) iodide promoted intramolecular Barbier-type reaction, *J. Org. Chem.*, 56, 4112, 1991.
159. Kagan, H. B., Namy, J. L., and Girard, P., Divalent lanthanide derivatives in organic synthesis. II. Mechanism of SmI_2 reactions in presence of ketones and organic halides, *Tetrahedron*, 37, (Suppl. 1), 175, 1981.
160. Molander, G. A., and McKie, J. A., Samarium(II) iodide-induced reductive cyclization of unactivated olefinic ketones. Sequential radical cyclization/intermolecular nucleophilic addition and substitution reactions, *J. Org. Chem.*, 57, 3132, 1992.
161. Otsubo, K., Inanaga, J., and Yamaguchi, M., SmI_2-induced reductive cross-coupling of carbonyl compounds with α,β-unsaturated esters, *Tetrahedron Lett.*, 27, 5763, 1986.
162. Totleben, M. J., Curran, D. P., and Wipf, P., Conjugate addition reactions of organosamarium species *via in situ* transmetalation to Cu(I) salts, *J. Org. Chem.*, 57, 1740, 1992.
163. Shim, S. C., Hwang, J.-T., Kang, H.-Y., and Chang, M. H., Intramolecular reductive cyclization of aldehydes and ketones with alkynes promoted by samarium(II) iodide, *Tetrahedron Lett.*, 31, 4765, 1990.
164. Molander, G. A., and Kenny, C., Intramolecular reductive coupling reactions promoted by samarium diiodide, *J. Am. Chem. Soc.*, 111, 8236, 1989.
165. Chiara, J. L., Cabri, W., and Hanessian, S., The stereocontrolled formation of cyclic vicinal *cis*-diols *via* a samarium diiodide pinacol coupling of dialdehydes, *Tetrahedron Lett.*, 32, 1125, 1991.
166. Inanaga, J., Yokoyama, Y., Handa, Y., and Yamaguchi, M., Preparation of medium- and large-sized carbocycles by means of SmI_2-promoted intramolecular Reformatsky reaction, *Tetrahedron Lett.*, 32, 6371, 1991.
167. Shibuya, K., Nagaoka, H., and Yamada, Y., A new method for the synthesis of spiro γ-butyrolactone *via* samarium(II) iodide-induced cyclization, *J. Chem. Soc., Chem. Commun.*, 1545, 1991.
168. Ruder, S. M., Improved method for samarium diiodide induced reductive coupling reactions of acid halides, *Tetrahedron Lett.*, 33, 2621, 1992.
169. McMurry, J. E., Carbonyl-coupling reactions using low-valent titanium, *Chem. Rev.*, 89, 1513, 1989.

170. Pons, J.-M., and Santelli, M., Reductions promoted by low valent transition metal complexes in organic synthesis, *Tetrahedron*, 44, 4295, 1988.
171. Clive, D. L. J., Zhang, C., Murthy, K. S. K., Hayward, W. D., and Daigneault, S., New low-valent titanium reagents for dicarbonyl coupling and their use in a general method of annulation, *J. Org. Chem.*, 56, 6447, 1991.
172. Fürstner, A., Jumbam, D. N., and Weidmann, H., Titanium-induced syntheses of furans and indoles, *Tetrahedron Lett.*, 32, 6695, 1991.
173. Rigollier, P., Young, J. R., Fowley, L. A., and Stille, J. R., Regioselective cyclopentane ring formation mediated by titanocene chloride, *J. Am. Chem. Soc.*, 112, 9441, 1990.
174. Korkowski, P. F., Hoye, T. R., and Rydberg, D. B., Fischer carbene mediated conversions of enynes to bi- and tricyclic cyclopropane-containing carbon skeletons, *J. Am. Chem. Soc.*, 110, 2676, 1988.
175. Dragisich, V., Murray, C. K., Warner, B. P., Wulff, W. D., and Yang, D. C., Heteroannulation of chromium carbene complexes. A novel and efficient pyrrole synthesis, *J. Am. Chem. Soc.*, 112, 1251, 1990.
176. Brandvold, T. A., Wulff, W. D., and Rheingold, A. L., Efficient entry to bicyclic lactones *via* van Halban–White cyclizations, *J. Am. Chem. Soc.*, 112, 1645, 1990.
177. Hoye, T. R., and Rehberg, G. M., (1-Oxidoalkylidene)pentacarbonylchromium anion $[R(O^-)C{=}Cr(CO)_5] \leftrightarrow$ acylpentacarbonylchromate $[acyl{-}Cr^-(CO)_5]$ chemistry: *in situ* preparation and reactions with alkynes and enynes, *J. Am. Chem. Soc.*, 112, 2841, 1990.
178. Harvey, D. F., and Brown, M. F., Molybdenum carbene complexes: trapping of *in situ* generated vinylcarbene complexes with electron-poor olefins, *J. Am. Chem. Soc.*, 112, 7806, 1990.
179. Harvey, D. F., and Brown, M. F., Synthesis and reactivity of alkynyloxy molybdenum carbene complexes: formation of cyclopropyl substituted dihydrofurans and dihydropyrans, *Tetrahedron Lett.*, 32, 5223, 1991.
180. Katz, T. J., and Yang, G. X.-Q., Reactions of enynes with adsorbed metal–carbenes, *Tetrahedron Lett.*, 32, 5895, 1991.
181. Harvey, D. F., and Lund, K. P., Cyclization reactions of a molybdenum carbene complex with 1,3-nonadien-8-ynes, *J. Am. Chem. Soc.*, 113, 5066, 1991.
182. Brandvold, T. A., Wulff, W. D., and Rheingold, A. L., Chromium-mediated cyclizations of cross-conjugated ketoketenes in 8- and $10e^-$ processes, *J. Am. Chem. Soc.*, 113, 5459, 1991.
183. Merlic, C. A., and Xu, D., Cyclization reactions of chromium dienylcarbene complexes. Entry to *ortho*-substituted aromatic alcohols *via* a designed photoreaction, *J. Am. Chem. Soc.*, 113, 7418, 1991.
184. Herndon, J. W., Chatterjee, G., Patel, P. P., Matasi, J. J., Tumer, S. U., Harp, J. J., and Reid, M. D., Cyclopropylcarbene–tungsten complexes + alkynes: a [4+2+1] cycloaddition route for the construction of seven-membered rings, *J. Am. Chem. Soc.*, 113, 7808, 1991.
185. Harvey, D. F., and Lund, K. P., Cyclopropanation of unactivated 1,3-dienes by Fischer carbene complexes, *J. Am. Chem. Soc.*, 113, 8916, 1991.
186. Hubert, A. J., Noels, A. F., Anciaux, A. J., and Teyssié, P., Rhodium(II) carboxylates: novel highly efficient catalysts for the cyclopropanation of alkenes with alkyl diazoacetates, *Synthesis*, 600, 1976.
187. Padwa, A., Krumpe, K. E., and Zhi, L., Cycloalkenone formation by the intramolecular addition of an α-diazoketone to an acetylenic *pi*-bond, *Tetrahedron Lett.*, 30, 2633, 1989.
188. Padwa, A., Chiacchio, U., Garreau, Y., Kassir, J. M., Krumpe, K. E., and Schoffstall, A. M., Generation of vinylcarbenes by the intramolecular addition of α-diazo ketones to acetylenes, *J. Org. Chem.*, 55, 414, 1990.
189. Pirrung, M. C., Brown, W. L., Rege, S., and Laughton, P., Total synthesis of (+)-griseofulvin, *J. Am. Chem. Soc.*, 113, 8561, 1991.
190. Hoye, T. R., and Dinsmore, C. J., Double (internal/external) alkyne insertion reactions of α-diazoketones, *Tetrahedron Lett.*, 32, 3755, 1991.

191. Padwa, A., Austin, D. J., and Xu, S. L., Rhodium(II) catalyzed cyclizations of α-diazo substituted alkynes. A new mode of reaction, *Tetrahedron Lett.*, 32, 4103, 1991.
192. Ceccherelli, P., Curini, M., Marcotullio, M. C., and Rosati, O., Regiocontrol by the carbon–carbon double bond in the $Rh_2(OAc)_4$ mediated carbon–hydrogen insertion of α-diazoketones, *Tetrahedron*, 47, 7403, 1991.
193. Davies, H. M. L., and Hu, B., Regioselective [3+2] annulations with rhodium(II)-stabilized vinylcarbenoids, *Tetrahedron Lett.*, 33, 453, 1992.
194. Padwa, A., Austin, D. J., Hornbuckle, S. F., Semones, M. A., Doyle, M. P., and Protopopova, M. N., Control of chemoselectivity in catalytic carbenoid reactions. Dirhodium(II) ligand effects on relative reactivities, *J. Am. Chem. Soc.*, 114, 1874, 1992.
195. Xiong, H., and Rieke, R. D., Facile formation of substituted 2-butene-1,4-diylmagnesium using highly reactive magnesium: a simple approach to complex carbocycles and functionalized ketones, *J. Org. Chem.*, 54, 3247, 1989.
196. Rieke, R. D., and Xiong, H., Chemistry of substituted (2-butene-1,4-diyl)magnesium: a facile approach to complex carbocycles, functionalized ketones and alcohols, and silicon-containing heterocycles, *J. Org. Chem.*, 56, 3109, 1991.
197. Xiong, H., and Rieke, R. D., The magnesium complexes of 1,2-dimethylenecycloalkanes: a new method for a one-step spiroannelation, *Tetrahedron Lett.*, 32, 5269, 1991.
198. Imamoto, T., Takiyama, N., Nakamura, K., Hatajima, T., and Kamiya, Y., Reactions of carbonyl compounds with Grignard reagents in the presence of cerium chloride, *J. Am. Chem. Soc.*, 111, 4392, 1989.
199. Mudryk, B., Shook, C. A., and Cohen, T., Simple one-pot syntheses of spiroketals and oxaspirolactones by addition of γ- and δ-cerioalkoxides to lactones and cyclic anhydrides, *J. Am. Chem. Soc.*, 112, 6389, 1990.

Chapter 6

MISCELLANEOUS CYCLIZATIONS

I. INTRODUCTION

Apart from the preceding categories of cationic, radical, anionic, and metal-mediated ring closures, there are other interesting and important cyclization reactions that call for a special section to themselves, without which a book on cyclization would be quite deficient. These processes lend themselves to organization into small groups which are arbitrarily formed and discussed briefly below.

II. MACROLACTONIZATION

Natural products possessing large ring lactones as their molecular core (commonly known as "macrolides") exhibit an interesting spectrum of biological activity which has attracted attention to their synthesis since the early 1970s. The formation of large ring lactones is not altogether a simple task as both entropy and polymerization factors work against macrolactonization. Nevertheless, several methods have been developed for the undertaking, the more important and frequently used of which are outlined below.

A. THE COREY–NICOLAOU METHOD[1–3]

This method involves simultaneously activating both the carbonyl and hydroxyl groups by the use of 2-pyridinethiol ester **3**, itself obtained from 2,2′-dipyridyl disulfide **1** and hydroxy acid **2**. The driving force for the reaction probably comes from proton transfer between the hydroxyl and pyridine to form zwitterion **4** which induces lactonization, as shown in Scheme 1.

Scheme 1

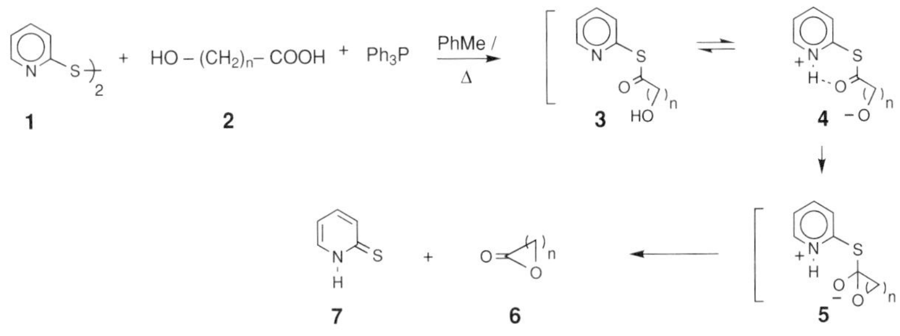

B. THE MASUMUNE METHOD[4–7]

In this method lactonization is effected through the use of *S-t*-butylthioester **9**, with mercuric trifluoroacetate as the activating agent, as shown in Scheme 2. The substrate for cyclization, the thioester **9**, is prepared from the corresponding hydroxy acid **2** *via* the acid chloride **8**.

Scheme 2

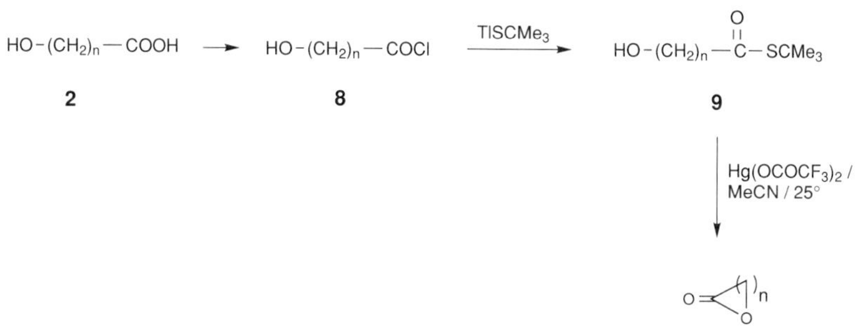

C. THE MUKAIYAMA METHOD[8]

Somewhat similar to the method of Corey,[1–3] this one-step process reacts 1-methyl-2-chloropyridinium iodide with the required hydroxy acid in boiling acetonitrile in the presence of triethylamine, as depicted in Scheme 3.

Scheme 3

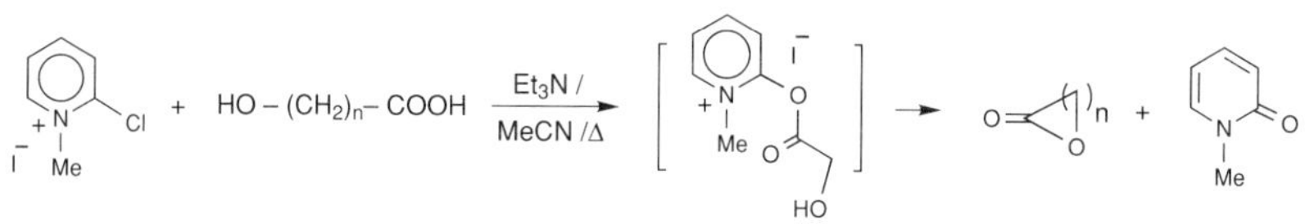

D. THE YAMAGUCHI METHOD[9]

As outlined in Scheme 4, the process involves activating the acid group by conversion to a mixed anhydride which is subsequently reacted with 4-dimethylaminopyridine (DMAP) in benzene solution to effect macrocyclization to the requisite lactone.

Scheme 4

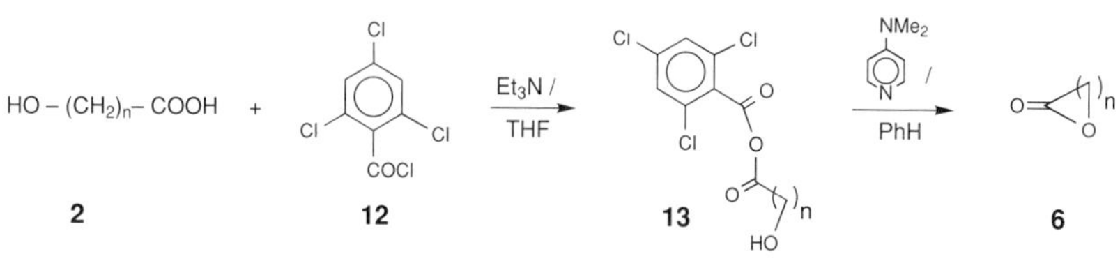

E. THE MITSUNOBU METHOD[10]

By far the most popular macrolactonization method, the process can be viewed as a redox reaction between diethyl azodicarboxylate (DEAD) and triphenylphosphine: DEAD is reduced to diethyl hydrazinedicarboxylate **17** and triphenylphosphine is oxidized to triphenylphosphine oxide at the expense of the water molecule from the hydroxy acid, thus forming the lactone. The process is summarized in Scheme 5. Stereochemical deductions suggest the last step in the reaction sequence to be a simple S_N2 displacement.

Scheme 5

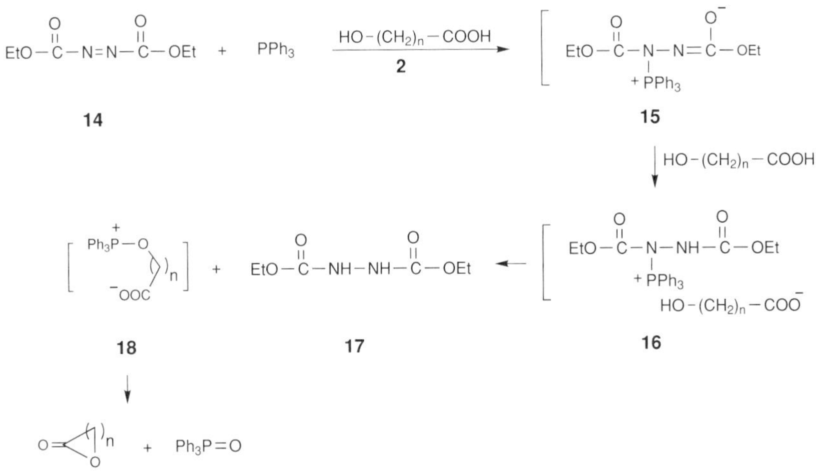

14 15 16 17 18 6

F. THE STEGLICH AND KECK METHOD

Originally developed by Steglich[11] for simple esterification of carboxylic acids using dicyclohexylcarbodiimide (DCC) and 4-dimethylaminopyridine (DMAP) as activating reagents, the process was later modified by Keck[12] by substituting the hydrochloride salt of DMAP for the free base, for solubility reasons. The results were astonishingly impressive, since, although effective for intermolecular esterifications the DCC/DMAP system (Steglich's) fails when applied to macrolactonization. For example, in the cyclization of 15-hydroxypentadecanoic acid **19** it delivers macrolide **20** in only 4% yield while the modified DCC / DMAP • HCl condition (Keck's) acheives a 95% production of macrocyclic lactone **20**![12]

Scheme 6

19 **20**

DCC = dicyclohexylcarbodiimide	reagents	DCC /DMAP	(4%)
DMAP = 4-dimethylaminopyridine		DCC / DMAP•HCl	(95%)

To complicate matters, the main procedures for macrolactonization outlined above are unfortunately not general, that is, one system might work efficiently for a particular substrate, yet fail completely with another. Recently a comparative study of the relative efficiencies of the different processes was carried out,[13,14] but again only on certain hydroxy acids. In spite of this, it would seem that with trial and error the chances of effecting macrolactonization of a given substrate by one or the other of the above methods are quite high. The following are examples of the use of macrolactonization in the key steps in the syntheses of various natural products. Some of the examples also serve to illustrate the unpredictable and highly sensitive nature of these reactions.

Application of the Corey–Nicolaou method to the preparation of a nine-membered ring lactone, leading to the nonacene ring system of brevetoxin A, is shown in Equation 1,[15] while Equations 2[16] and 3[17] demonstrate the lactonization process of Steglich and Keck in the total syntheses of colletol and bryostatin **7**, respectively. In fact the process in Equation 3 is a further modification of the Steglich and Keck process, since the use of the original DMAP, rather than pyridine, gives an intractable mixture.[17] Under the modified conditions shown the acetate and methoxy groups on the "northern" pyran ring also hydrolyze simultaneously. Interestingly, lactonization is regiospecific, occurring exclusively at the specific hydroxyl that yields the smaller ring lactone.

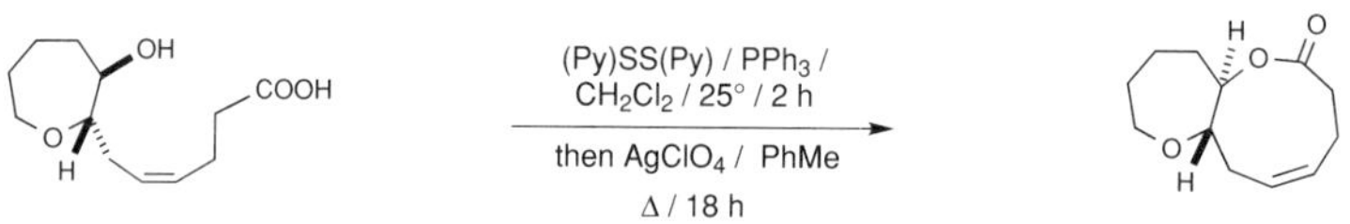

21 **22** (79%)

Equation 1

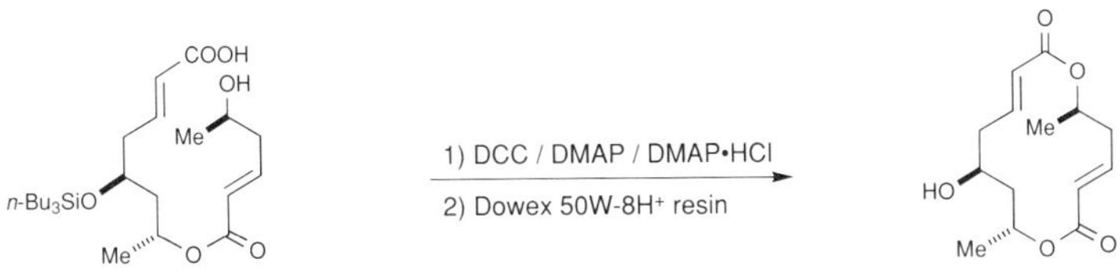

23 **24** (71%)

Equation 2

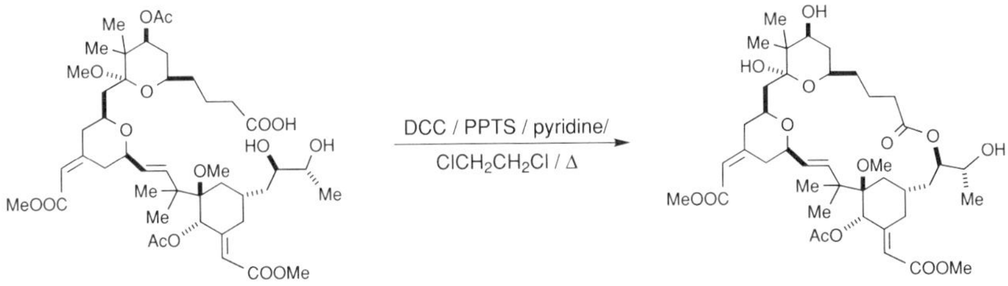

25 **26** (51%)

Equation 3

In the synthesis of the macrolide antibiotic cytovaricin, the Steglich and Keck process[11,12] is again employed for the formation of the lactone ring (**27** → **28**, Equation 4).[18] It should be noted that Evans[18] had failed earlier to effect this conversion by the method of Mukaiyama, although White *et al*[19] was very successful with the same method in conducting ring closure of a 34-membered ring lactone (**29** → **30**),[19] an advance intermediate in the synthesis of aplasmomycin, a metabolite from *Streptomyces griseus* (Equation 5).

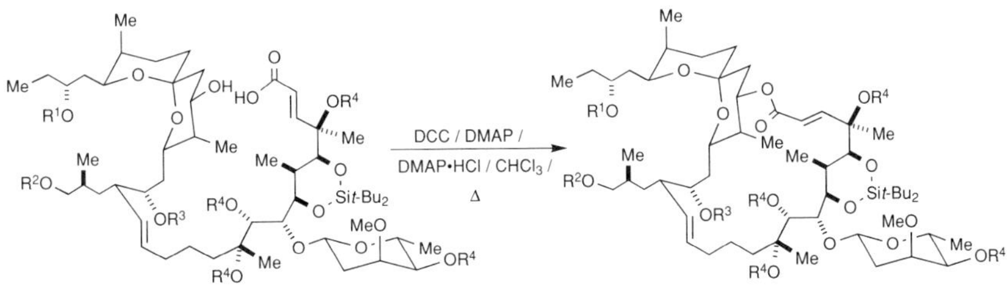

27 **28** (92%)

R^1 = diethylisopropylsilyl (DEIPS) R^2 = *t*-butyldimethylsilyl (TBDMS)

R^3 = *p*-methoxybenzyl (PMB) R^4 = triethylsilyl (TES)

Equation 4

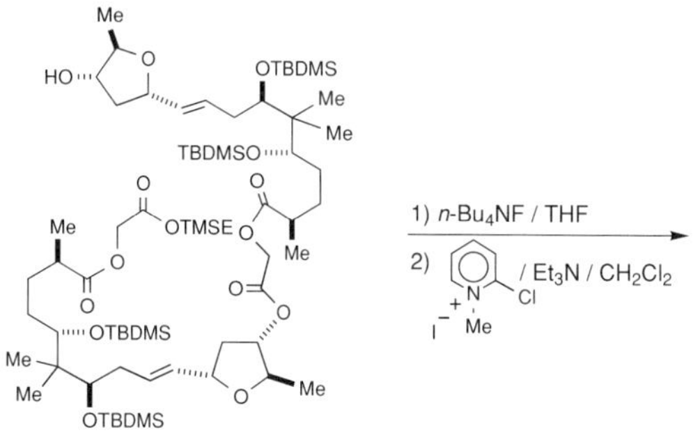

29

TMSE = 2-(trimethylsilyl)ethyl

30 (86%)

Equation 5

Further examples of the critical need for delicate matching of process with substrate are the cyclizations of the closely related systems **31** and **33** to the respective 11-membered ring macrolides **32** and **34**. In the former case (Equation 6),[20] saponification of **31** followed by macrolactonization using the various available techniques gave poorer results than the Mitsunobu method,[10] which, anyhow, managed only a low (25%) yield of the required macrolide **32**. In contrast, the Mukaiyama process reportedly gave satisfactory results in effecting the cyclization of **33** to **34** (Equation 7).[21]

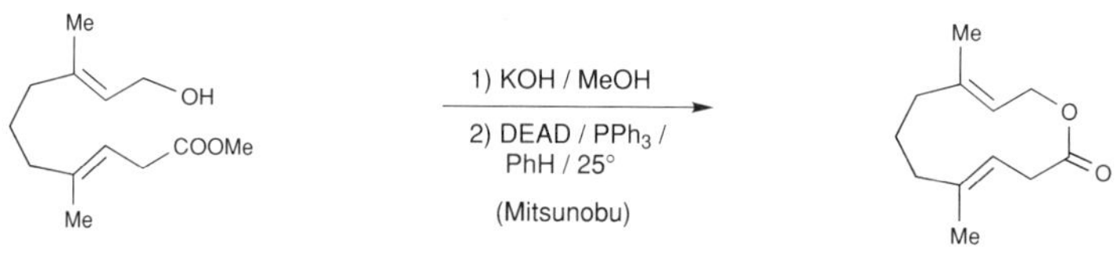

31

32 (~ 25%)

Equation 6

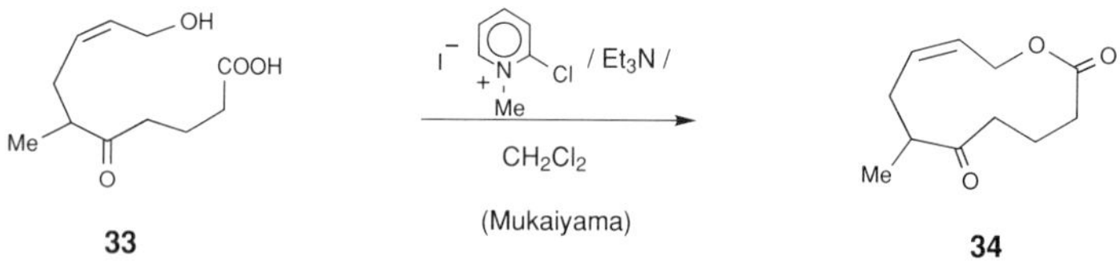

33

34

Equation 7

Generally speaking, the Mitsunobu lactonization method is quite popular and, with its highly regarded reputation, is usually among the first to be considered in the planning of a lactonization. Its applications span a wide range, from small rings, *e.g.*, the simple formation of four-membered ring β-lactam **36** (Equation 8),[22] to macrolides, *e.g.*, the synthesis of 14- and 16-

membered naturally occurring latrunculin B, **39** (Equation 9)[23] and latrunculin A, **42** (Equation 10),[23–25] both toxins from the Red Sea sponge. The stereochemistry of product in every case is consistent with the mechanism put forward by Mitsunobu[10] (outlined in Scheme 5) which proposes S_N2 displacement for the cyclization step (**18** → **6**).

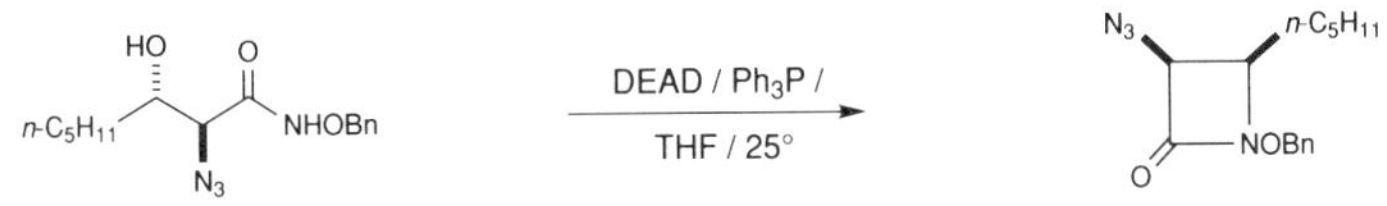

35 **36** (> 53%)

Equation 8

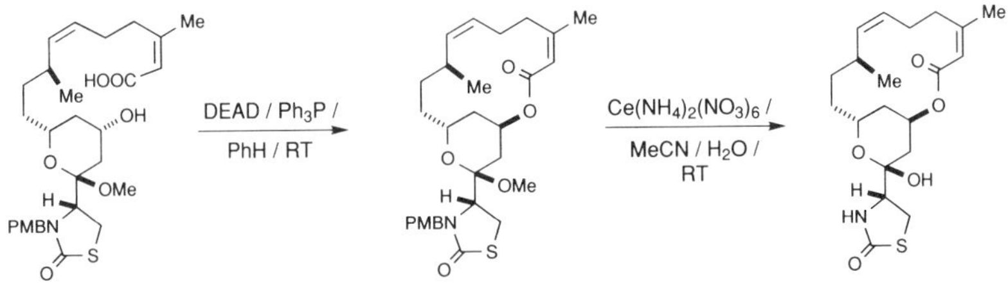

37 **38** (69%) **39** (47%)

PMB = p-methoxybenzyl

Equation 9

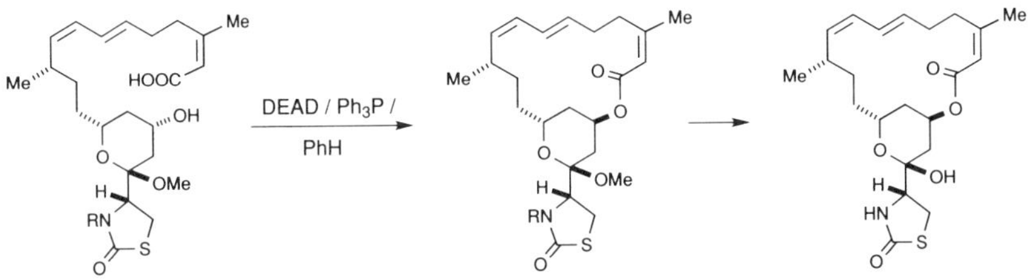

40 **41** **42**

ref 23 ; R = $Me_3SiCH_2CH_2O-\overset{O}{\overset{||}{C}}-$ (28%)

ref 24 ; R = H (67%)

Equation 10

III. VICINAL TRICARBONYL CYCLIZATION

Although the chemistry of vicinal polyketones has long been studied[26] the introduction of the vicinal tricarbonyl system (*e.g.*, **43**) to organic synthesis only seriously began in the last decade. Admittedly the layout of molecule **43** invites a confusion of cyclizing nucleophilic attacks at all three carbonyl sites, not to mention the other functional groups R^1 and R^2. These possibilities have been cleverly explored by Wasserman and his group, and have led to the synthesis of a wide variety of important organic molecules such as the penems, carbacephams, simple heterocycles, and even complex alkaloids, by incredibly short and simple procedures.

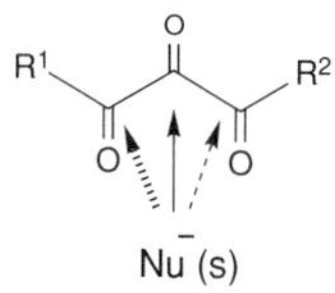

43

To be fair, vicinal tricarbonyl cyclization owes part of its success to the achievement in the preparation of the starting tricarbonyl unit.[27] This is obtained easily by oxidative cleavage of enamine **45**[28] or ylide **48**,[29] in turn prepared from dicarbonyl **44** or acid chloride **46**, respectively, *via* oxidation with singlet oxygen. The sequence is outlined in Scheme 7. In the case of ylide **48**, oxidation can also be accomplished with ozone, or, better still, with the use of potassium peroxymonosulfate,[30] reported in 1990, which converts the ylide to the tricarbonyl without touching other functional groups such as π bonds in the substituents R^1 and R^2.

Scheme 7

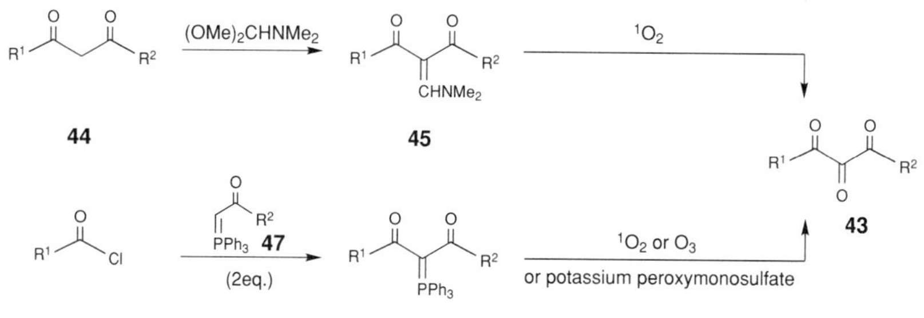

A. CYCLIZATION INVOLVING VICINAL TRICARBONYL SYSTEM

Vicinal tricarbonyl compounds **43** carrying suitable nucleophilic substituents at R^1 or R^2 can readily undergo intramolecular cyclization, as exemplified by the syntheses of carbapenam[31] and carbacepham[28] derivatives **54** and **55**, respectively. As shown in Scheme 8, starting from β-keto ester **49** it is feasible to generate the tricarbonyl derivative **51** *in situ* from enamine **50**. Desilylation of **51** then results in spontaneous cyclization to **53** which is subsequently converted to **54** and **55**.

Scheme 8

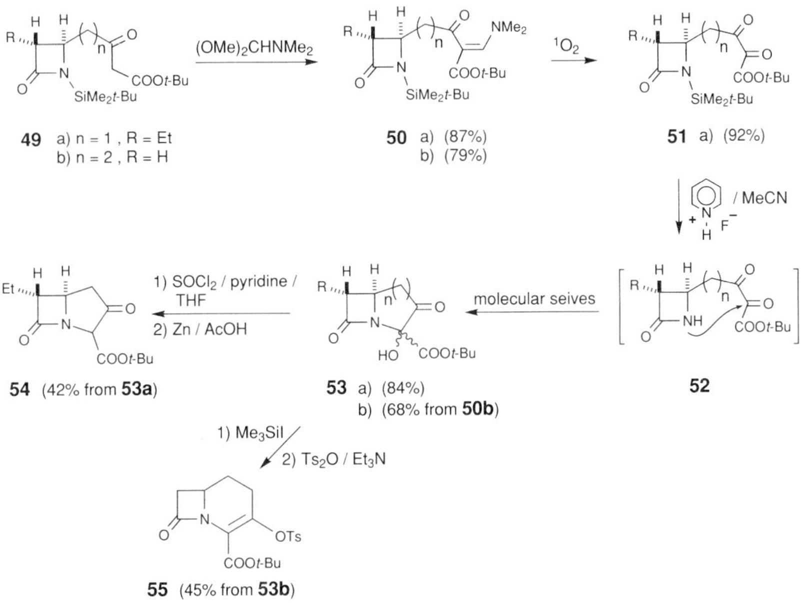

With a slight modification of the starting material the above process is also applicable to the synthesis of penem **61**. Thus, starting from thioester **56** the tricarbonyl derivative **58** can be generated and cyclized to **59**, which, after removal of the hydroxy group *via* conversion to the chloride and reducing with zinc–acetic acid, yields product **60**. Treatment of the latter with diazomethane affords penem **61** (56%) along with its isomer **62** (26%).[32] Compound **61** thus obtained is identical in all respects with the sample prepared *via* the established route.[33]

Scheme 9

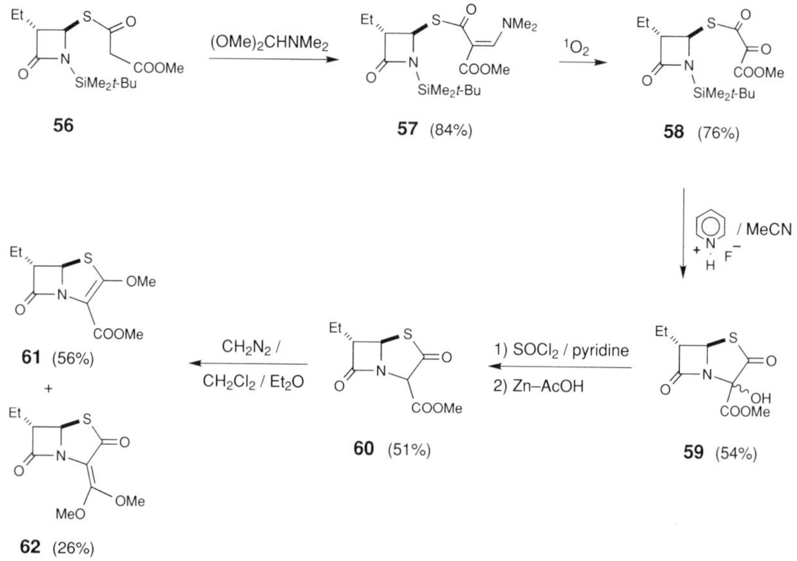

The vicinal tricarbonyl moiety appears well disposed to take part in cyclizations that can lead to various alkaloid skeletons. For example, intermolecular condensation between arylethylamine **63** and tricarbonyl derivative **64** occurs at the middle carbonyl group to give a Schiff base that is predisposed to cyclize, under suitable conditions, in the Pictet–Spengler fashion to yield the isoquinoline alkaloid skeleton, as schematically depicted at the top of Scheme 10. In actual practice, reacting tricarbonyl **64** (generated by the method discussed above)[29] with arylethylamine **63** in benzene solution, followed by treatment of the reaction mixture with trifluoroacetic acid, results in the isolation of a single product, 3,4-dihydropapaveraldine **68**, in 71% yield.[34]

Scheme 10

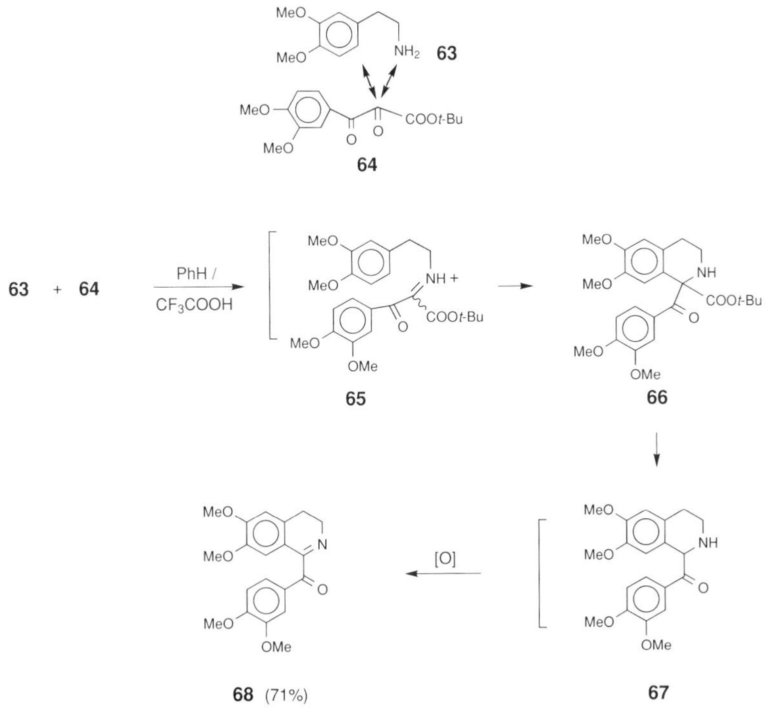

The mechanism of reaction probably involves the intermediacy of iminium ion **65**, formed by attack of the amino group on the most reactive carbonyl of the tricarbonyl unit, followed by a Pictet–Spengler cyclization and rapid decarboxylation of the β-keto ester. Finally, the last step of the sequence appears to be air-oxidation of intermediate **67** under the employed reaction conditions, to yield the observed alkaloid **68**.

Similar results are obtained with the use of tryptamine **69** and tricarbonyl **70**, leading to the natural product eudistomin T **75** after air-oxidation of the primary cyclized product **73** (Scheme 11).[35] By slightly varying the tricarbonyl portion the reaction can easily be extended to the synthesis of nitrogen tetracycle **79** which is readily convertible to the alkaloid eburnamonine **80** by standard methods (Scheme 12).[36] Apart from the given examples, other variations of the donor and acceptor (tricarbonyl) molecules are also feasible, and can lead to other molecular types of products, namely, the carbazole derivatives.[37]

Scheme 11

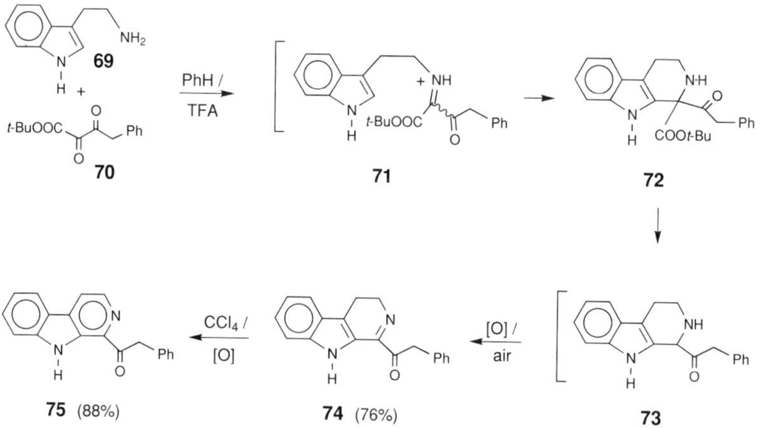

Scheme 12

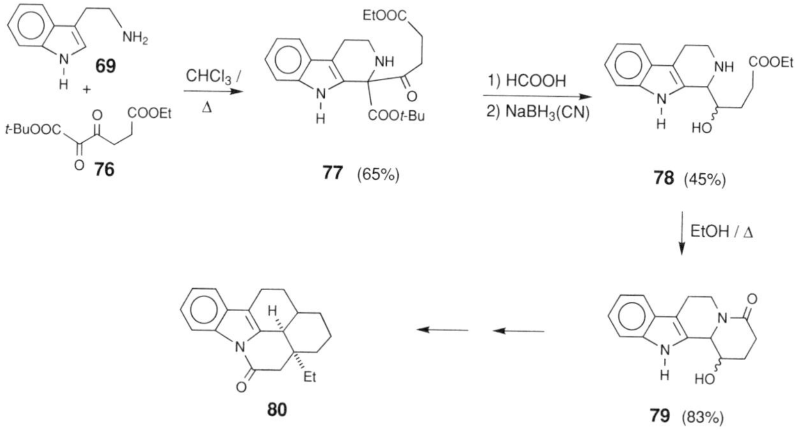

B. CYCLIZATION INVOLVING THE VINYL TRICARBONYL SYSTEM

Vicinal tricarbonyl hydrate with the general structure **81** is a stable compound which can be prepared by oxidative cleavage of the corresponding phosphorus ylide as depicted in Scheme 7.[29,30] With its unique molecular features, the compound provides the opportunity for many kinds of chemical manipulations, possessing two exceedingly good electrophilic sites in the electron-deficient vinyl (undergoes Michael additions) and central reactive carbonyl functions, not to mention the two other carbonyl groups that flank the central unit.

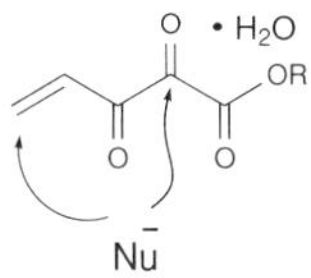

81

Ever since its discovery in 1989 the vinyl tricarbonyl **81** has been put to use by the group of Professor Wasserman, who has reaped innumerable benefits from shrewd exploitation of the molecule. The examples in Scheme 13 represent straightforward syntheses of pyrrole,[38] pyrrolizidine,[39] and indolizine[39] derivatives **87** and **90** (n = 1 and 2), respectively, from vinyl tricarbonyl hydrate **82**. Scheme 14, meanwhile, is an extension of the reaction in Scheme 13, employing tryptamine **91** as the nucleophile to react with **82** to yield **92**, which compound can further cyclize by treatment with boron trifluoride etherate to furnish the tetracyclic system **94** (73% yield from vinyl tricarbonyl **82**).[29] The mechanism of the final cyclization most likely proceeds *via* the iminium ion intermediate **93**.

Scheme 13

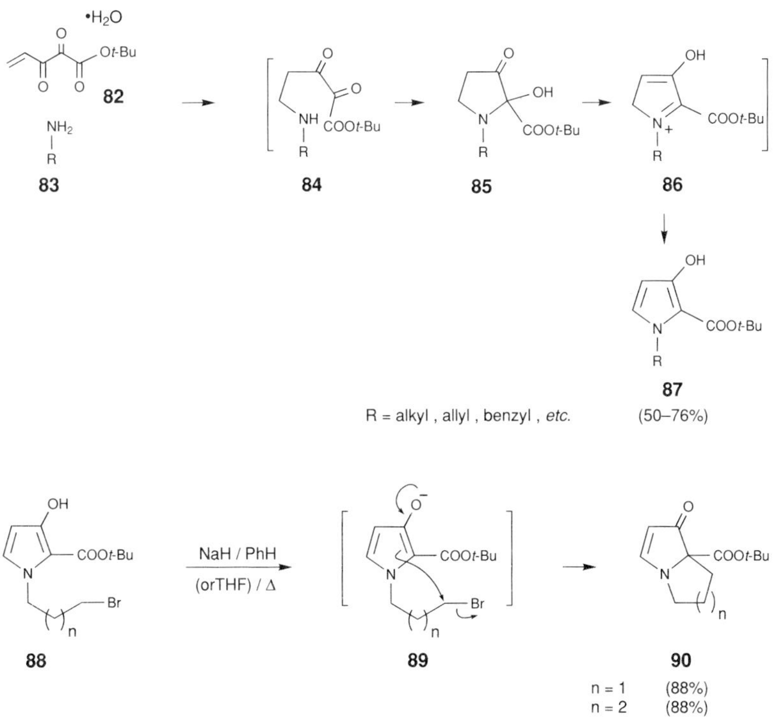

Scheme 14

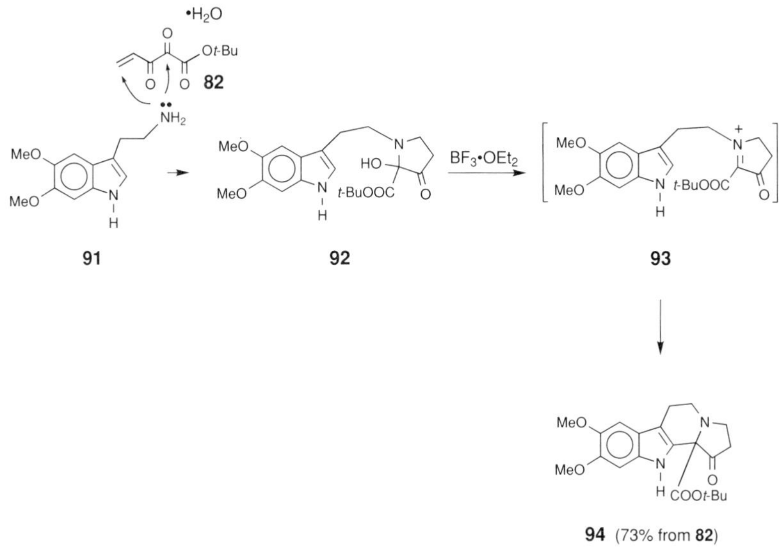

Besides the example in Scheme 14, intramolecular trapping of the iminium ion formed from the reaction of vinyl tricarbonyl with an amine (*e.g.*, **93**) has also been reported with other types of nucleophilic moiety attached to the amine group. As illustrated in Scheme 15,[40,41] these could be the vinylsilane (*e.g.*, **97**), the enol ether (*e.g.*, **99**) or the propargylsilane (*e.g.*, **101**), which all lead to indolizidine derivatives. In the case of the enol ether **99**, the cyclization stereospecifically yields the *cis*-product **100**. It should perhaps be reminded that such iminium ion cyclizations have already been discussed under "Cationic Cyclization" in Chapter 2, the only difference here being the method of generation of iminium ion which employs the reaction of vinyl tricarbonyl with an amine.

Intramolecular trapping of the iminium ion with an amino group comprising the lactam nucleus offers an extremely short and easy route to tricyclic lactams **106** (Equation 11).[42,43] Trapping with an aromatic amine, in likewise manner, provides a very efficient synthesis of tricyclic structures such as **109**, a precursor for conversion to the alkaloid vasicine **110** (Equation 12).[44] This last reaction illustrates how advantage can be taken of the difference in nucleophilicity between the two amino groups on the donor substrate, exploitation of the difference in electrophilicity of the groups on the acceptor, the vinyl tricarbonyl, having already been exemplified above.

Scheme 15

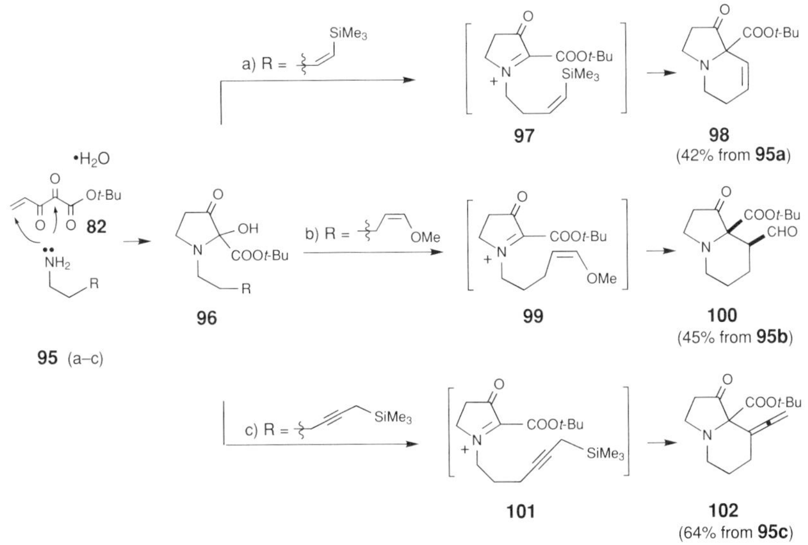

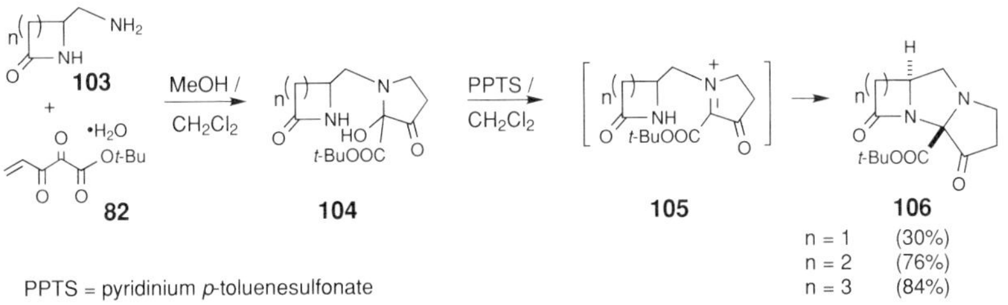

Equation 11

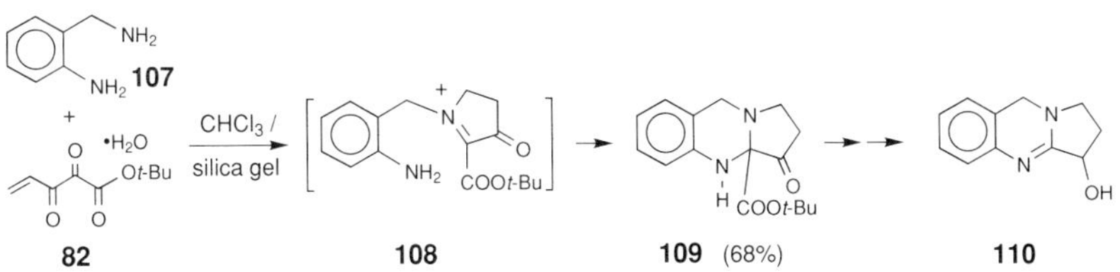

Equation 12

The limits of application of these important building blocks, the vicinal tricarbonyl (*e.g.*, **43**) and vinyl tricarbonyl (*e.g.*, **82**) systems, are, as yet, unknown. Product variation probably rests with the imagination in the design of donor molecules, and possibly in the derivatization of the tricarbonyl system which is likely to follow in the near future.

IV. CYCLIZATION OF ENEDIYNES AND RELATED SYSTEMS

It is probably not over-exaggerating to say that one of the currently most important reactions which has potential clinical use is the cyclization of enediynes and related systems. The reason for this lies in the discovery of important anticancer antibiotics from microorganisms, *i.e.*, the calicheamicins **111**,[45–48] esperamicins **112**,[49,50] dynemicins **113**,[51,52] and neocarzinostatin chromophore **114**.[53,54] These structurally related compounds possess the enediyne unit and display not only high potency against a number of animal tumor models, but also subpicogram potency against Gram-positive bacteria. The hue and cry surrounding these molecules are caused by their unusual yet esthetically engaging molecular structures which offer stimulating challenge to the synthetic chemist, their potent biological activities, and, most important of all, their mode of action. It is not surprising then that 1991 alone saw at least three independent reviews of these compounds[45,53,55] in addition to earlier publications in the scientific news.[56]

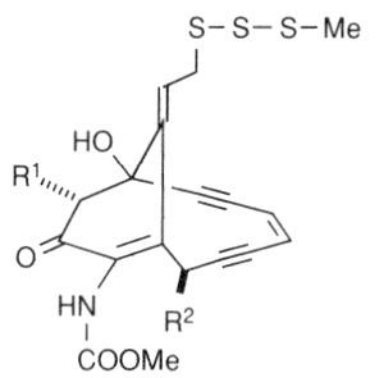

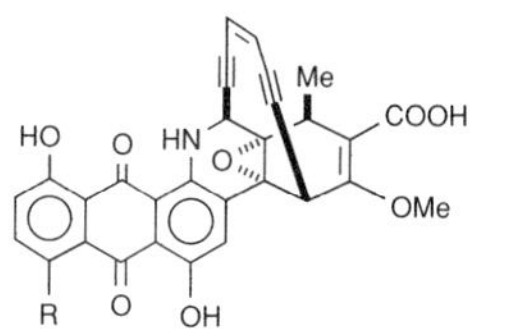

113 (R = H , OH)

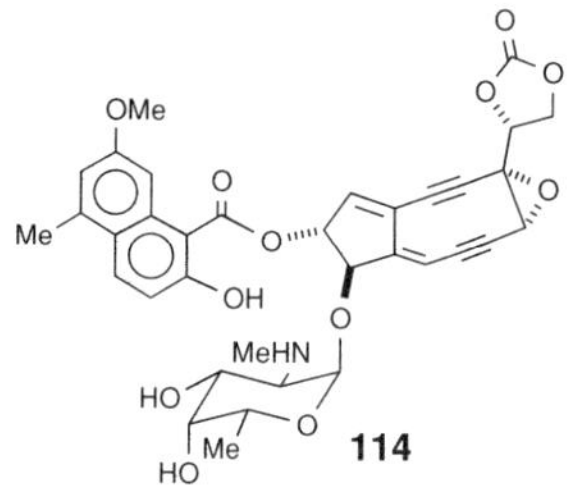

114

111 R¹ = H , R² =

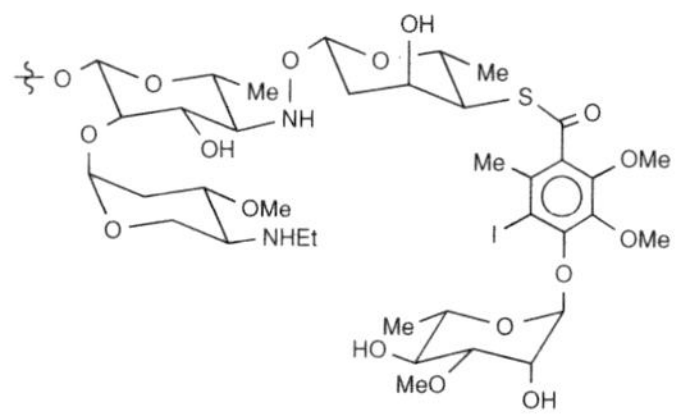

112 R¹ =

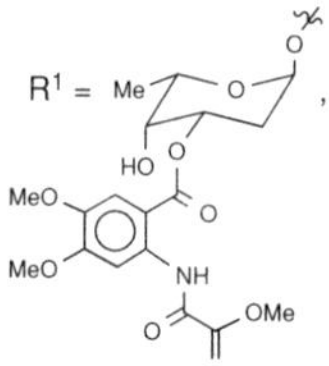

, R² =

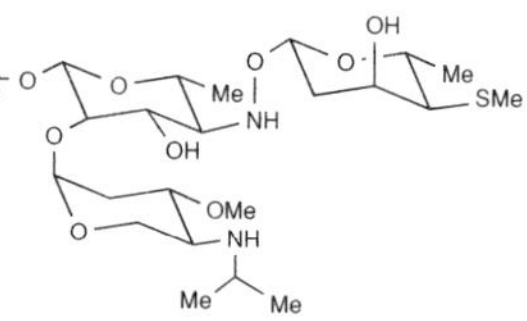

A. THE BERGMAN CYCLIZATION AND RELATED REACTIONS

Before considering the biological action of the natural products **111–114** it is well to take a look at the reactions of enediynes. As far back as two decades ago, Bergman[57] reported the thermal cyclization of systems such as **115** to the 1,4-dehydrobenzene diradical intermediate **116**, which is capable of abstracting radicals, either intermolecularly from the solvent or donor to yield product **117** (a popularly used radical donor being 1,4-cyclohexadiene), or intramolecularly to form new radicals **120** and **121**, to result finally in a mixture of products **119**, **122**, and **123**, as shown in Scheme 16.[58]

Scheme 16

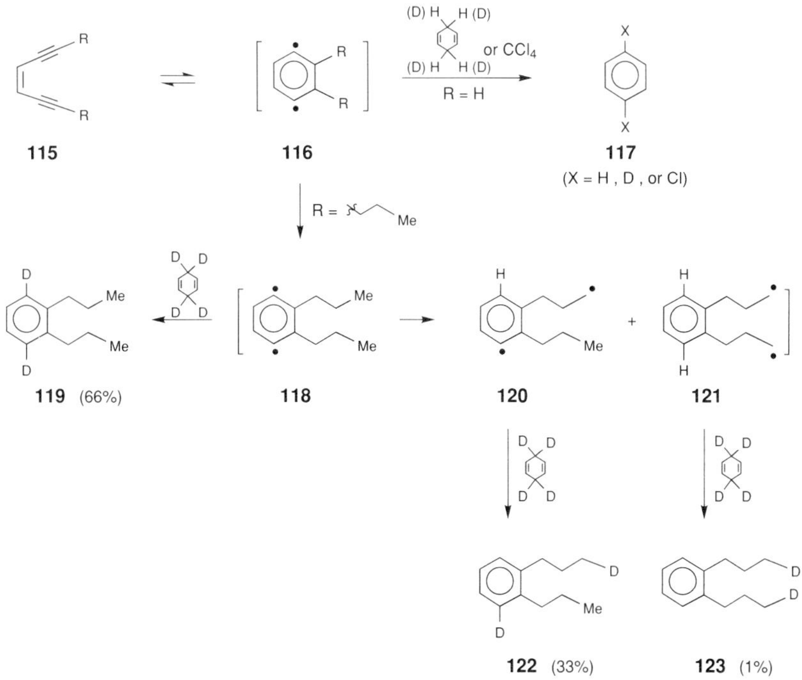

Several studies of the Bergman cyclization have appeared in the literature employing various types of enediynes as model systems.[58–67] It has been concluded from experimental data that the feasibility of the Bergman cyclization of an enediyne system such as **124** (Scheme 17) is dependent upon the distance between C_c and C_d. That is to say, for an inter C_c–C_d distance in the range of 3.31 to 3.20 Å or less, cyclization will be facile or spontaneous at 25°C, but for a C_c–C_d distance of >3.35 Å the molecule will be quite stable and Bergman cyclization will require heating. For example, in the parent open chain unsubstituted enediyne **115** (R = H) the distance between the cyclizing centers (C_c–C_d) is 4.12 Å and the compound is found to cyclize only upon heating to 200°C ($t_{1/2}$ = 30 s).[55,57,58] On the other hand, the cyclic enediyne **130**, prepared from the reaction of **128** with **129**, is unstable and cyclizes spontaneously *via* radical **131**[68] to deliver **132**. In this latter case the distance between C_c and C_d is calculated to be approximately 3.03 Å[60] (Equations 13 and 14).

Scheme 17

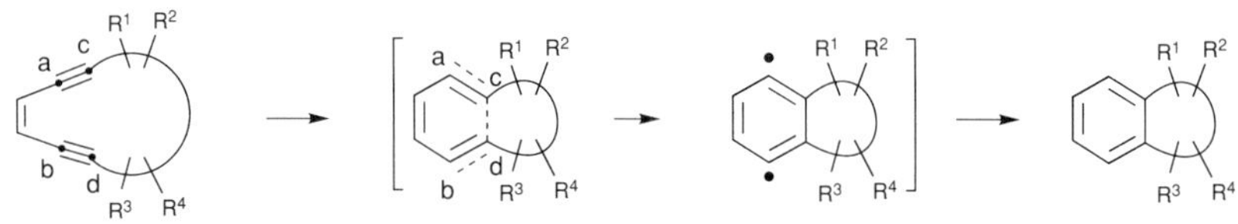

124 **125** **126** **127**

115 **116** **117**

Equation 13

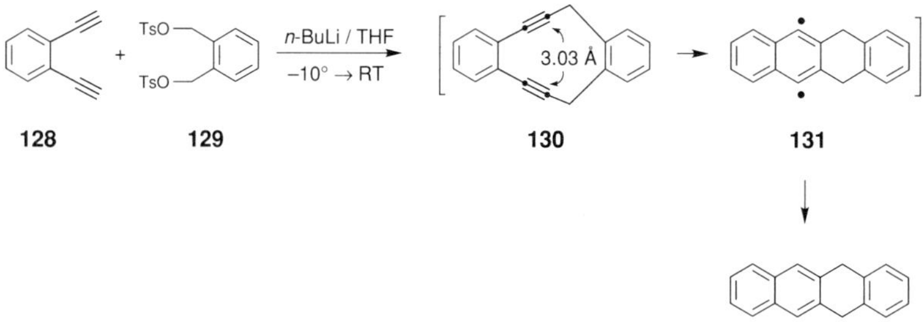

128 **129** **130** **131**

132

Equation 14

However, independent works by the groups of Magnus and Snyder[61–65,69] have revealed that the feasibility of the Bergman cyclization of systems such as **124** is not determined by the distance between the cyclizing centers C_d and C_d alone, but also in part by the steric strain of the cyclization transition state (*i.e.*, **124** → **125** → **126**), which entity is a direct consequence of the nature of substituents R^1–R^4. In other words, the ease of cyclization is dependent upon the difference in strain between the enediyne and the diradical species. Magnus and Snyder have shown this by designing various monocyclic enediynes (*e.g.*, **124**)[65] and bicyclic enediynes (*e.g.*, **124** with R^1 or R^2 forming a bridge with

R^3 or R^4)[61–63] and calculating the C_c–C_d distances as well as measuring the rates of cyclization. For example, in the case of compounds **133** and **136**, with inter-C_c–C_d distances of 3.391 and 3.368 Å respectively (near the predicted < 3.35 Å for spontaneous cyclization),[60] molecule **136** with the shorter C_c–C_d distance is found to be more stable and to cyclize at 124°C at a rate that is 650 times slower than **133**.[63] This result is ascribable to the higher strain involved in the transformation from **136** to diradical **137** when compared with the corresponding change from **133** to **134**. Furthermore, inasmuch as compound **133** is completely converted to **135** at 80°C, compound **139** is recovered unchanged even though the difference in the C_c–C_d distance is small. Again the finding can be attributed to the strain involved in the formation of a cyclized radical **140** whose bicyclic ring is not large enough to accommodate a double bond on the bridge head (Bredt's rule),[61] as shown in Scheme 18.

Scheme 18

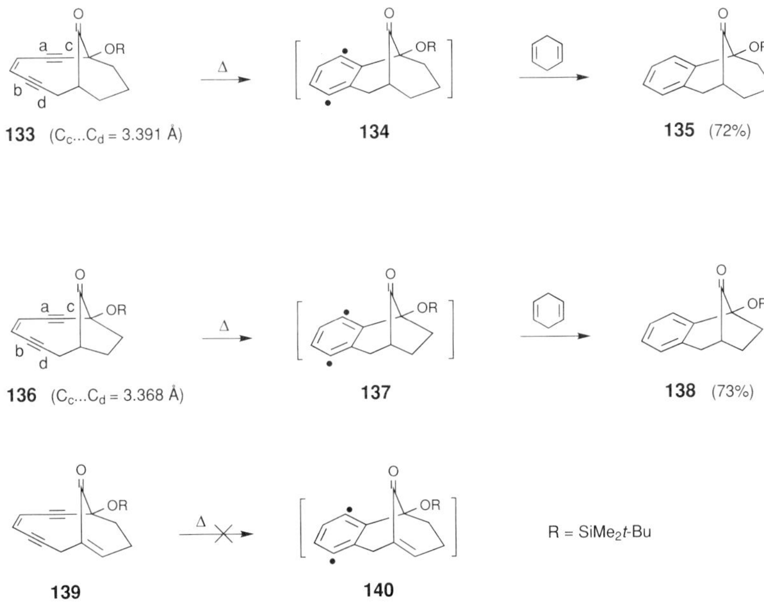

Nevertheless, comparison of the rates of Bergman cyclizations must be conducted with caution, since very recent work[67] has clearly demonstrated that the half-life of compound **141** at 84°C is directly proportional to the concentration of 1,4-cyclohexadiene, the commonly used hydrogen radical source for trapping diradicals from such Bergman cyclizations of enediynes (Equation 15 and the accompanying table).

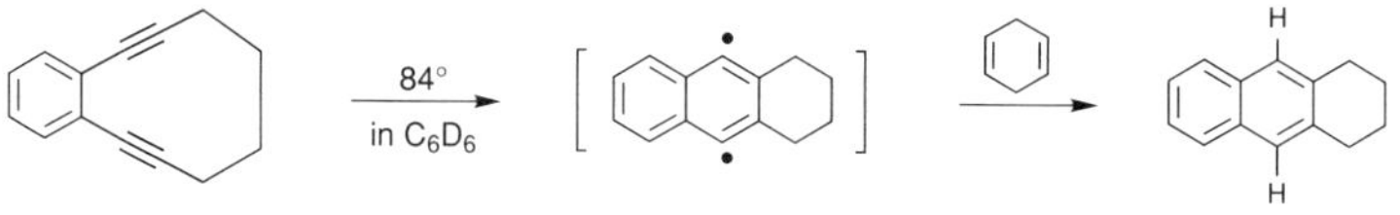

141 142 143

Equation 15

Conversion of 141 to 143 at 84° in C_6D_6

conc. of 1,4-cyclohexadiene	$t_{1/2}$ (based on disappearance of 141)
0.00 M	129 h
0.25 M	39 h
0.50 M	24 h
10.50 M	10.5 h

Scheme 19 and the table below offer further interesting examples. The sulfur analog **144** has a calculated C_c–C_d distance of 3.30 Å , yet it is stable and only cyclizes upon refluxing in benzene (80°C) in the presence of 1,4-cyclohexadiene, to yield **146**[70] (entry 1). Under these conditions sulfone **147**, obtained from the oxidation of sulfide **144**, also behaves similarly (entry 3). Amazingly however, with base as catalyst (entries 4, 5) the sulfone undergoes spontaneous cyclization at 25°C while no such behavior can be observed with sulfide **144** (entry 2).

Acid, on the other hand, has no apparent influence on the reaction, no acceleration of cyclization being observed with **147** (entry 6). From these results it is postulated that base promotes the isomerization of enediyne **147** to enyne–allene **148** which then cyclizes in a fashion similar to the Bergman cyclization (see **148**) to give diradical **149**, thence product **150**. The presence of **149** has been confirmed by trapping with carbon tetrachloride. Accordingly, conducting the reaction of **147** in a carbon tetrachloride solution in the presence of DBU yields the trichloro derivative **151**, as shown in the scheme.[71]

Scheme 19

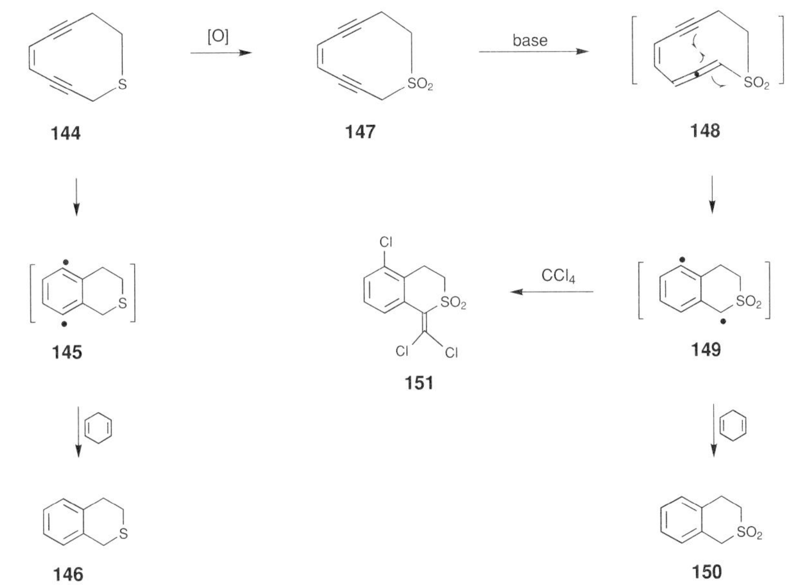

Entry	Compound	Solvent	Catalyst (mol. ratio)	Temp.°C	$t_{1/2}$
1	**144**	PhH	–	80	9 h
2	**144**	PhH	Et_3N (1.0)	25	no reaction
3	**147**	PhH	–	80	9 h
4	**147**	PhH or MeCN or THF	Et_3N (1.0)	25	< 1 min
5	**147**	PhH	*t*-BuOK (0.3) or DBU (0.1)	25	< 1 min
6	**147**	PhH	TFA (1.0)	25	no reaction

In fact, the above type of cyclization of enyne–allene has been reported earlier in 1989 by the groups of Myers[72] and Saito[73] for the simple open chain enyne–allene system **152** (Scheme 20). The reaction is rather slow (requiring some heating) with the unsubstituted substrate (R^1 = H),[72,74] but proceeds well at 37°C when substituents such as $P(O)Ph_2$, S(O)Ph, or Me are present.[73,74] The following Scheme 21 demonstrates the facile cyclization of various enyne–allenes carrying terminally $P(O)Ph_2$-substituted allene moieties. The compounds **160** to **162** are stable enough for isolation but on slight warming cyclize to give **163** to **165** respectively.[75]

Scheme 20

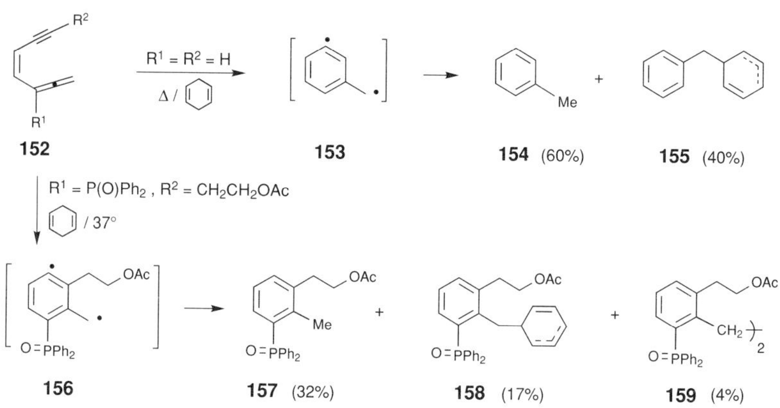

Scheme 21

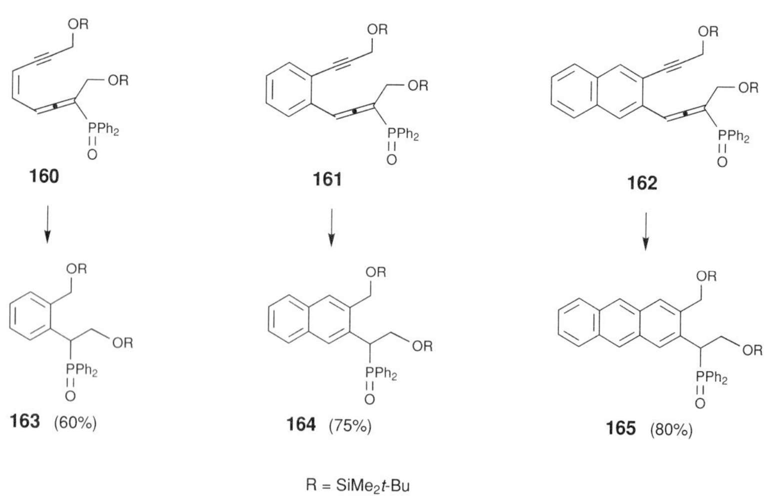

The odds are overwhelming that the arene diradical intermediate in the Bergman cyclization (*e.g.*, **116** in Scheme 16) will be put to use in intramolecular cyclizations with strategically placed radical acceptors. The overall reaction would thus be a tandem radical cyclization initiated by a Bergman cyclization of the enediyne moiety. A report along this line, by the group of Bergman himself, has already appeared,[76] according to which the thermolysis of dilute solutions of **166** in a variety of solvents results in double cycloaromatization to yield naphthalene **169**, but in poor yields (~10%). Conducting the reaction in acetone-d_6 results in naphthalene-d_2 with deuterium located exclusively at the β-position, confirming the intermediacy of diradical **168** which could arise *via* a stepwise (**166** → **167** → **168**) or concerted (**166** → **168**) cyclization as shown in Scheme 22.[76]

Scheme 22

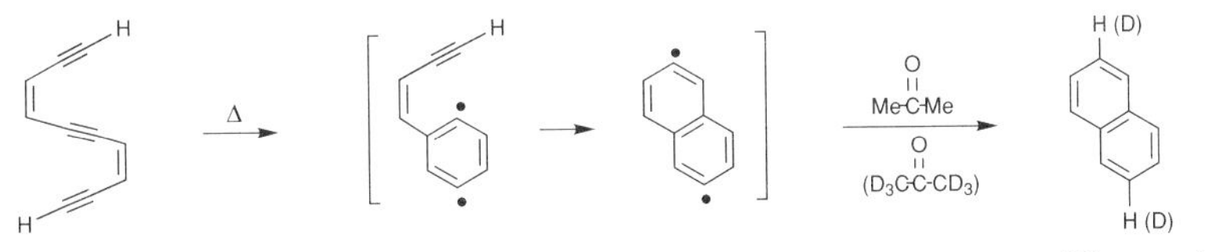

166 **167** **168** **169** (~10%)

The low yield of naphthalene **169** from the tandem cyclization of **166** is not surprising, as the mechanism, stepwise or concerted, necessitates cyclization of the intermediate radical onto a not-so-reactive acetylenic bond in the second ring closure step (**167** → **168**). The situation is much improved upon changing to substrate **170**. Here, besides providing a very reactive radical acceptor for the second cyclization (**171** → **172**) the design of the molecule also allows the ring closure to proceed in a favorable *exo* manner. In the case of n = 1, only the bis-cyclization product **173** (n =1) is obtained (72%), but with n = 2 a substantial amount of the mono cyclization product **174** is also produced, understandably because the 6-*exo* mode of cyclization is slower than the 5-*exo* mode (**171** → **172**), as previously discussed in Chapter 3 (Scheme 23).[77]

It is probably not erroneous to assume as a foregone conclusion that a lot more of the tandem cyclization of the type shown in Schemes 22 and 23 will be appearing in the literature in the near future.

Scheme 23

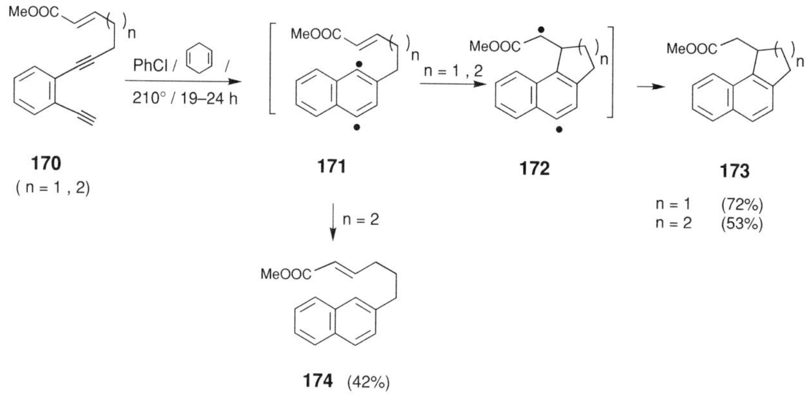

B. CYCLIZATION OF CALICHEAMICINS, ESPERAMICINS, DYNEMICINS, AND NEOCARZINOSTATIN CHROMOPHORE

At this point a correlation can be drawn between the structures of the Bergman cyclization products and the naturally occurring anticancer antibiotics: calicheamicins, esperamicins, dynemicins, and neocarzinostatin. It happens that an important biological action of these compounds involves the Bergman and related cyclizations, which are responsible for inflicting damage upon the DNA strain. Regarding the calicheamicins **111** and esperamicins **112**, which have the most closely related diagrammed architectures, the chemical process that accounts for their biological activity can be schematized as shown in Scheme 24.[45–47,49,50,55] Initially, the enediyne moieties of **111** and **112** are unable to undergo the Bergman cyclization owing to the C_c, C_d cyclizing atoms being too far apart to interact spontaneously,[60] or, to put it another way, because the strain energy of the would-be cyclization intermediate is too high[61–65] (see also **139** with bridge head double bond). However, following nucleophile (*e.g.*, glutathione) attack on the central atom of the trisulfide group the thiolate ion **175** formed is in a position to undergo internal Michael addition to yield **176**, and in so doing, bring the C_c–C_d distance closer (*i.e.*, 3.16 Å).[55] The new molecule is thus set for the Bergman cyclization *via* an intermediate that has comparatively little strain (see also **133** → **134** in Scheme 18).[61] This intermediate diradical **177** abstracts hydrogen atoms from DNA and in the process generates DNA radicals which can further react with molecular oxygen, leading, ultimately, to strand cleavage as indicated in the scheme. The entire phenomenon has been beautifully summed up by Nicolaou and Dai,[55] to the effect that calicheamicins (**111**) and esperamicins (**112**) can be perceived to consist of three components, namely, (1) the sugar fragment (R^1 and R^2 in

111 and **112**) which functions as the "delivery system" to bring the "lethal warhead to its target", (2) the trisulfide which acts as the "triggering device", and (3) the enediyne system or the "warhead" which acts to inflict damage upon the genetic material.

In the case of the dynemicins, enediyne cyclization once again figures as the key step in their biological action, the reaction becoming feasible after removal of the strained epoxide ring. According to one hypothesis, the mechanism, shown in Scheme 25,[55] entails first the bioreduction of the starting quinone to **179**, followed by epoxide ring opening mediated by the nitrogen lone pair electrons, and subsequent nucleophile attack to result in **181**. This molecule is susceptible to cyclization (calculated distance between cyclizing centers ≅ 3.17–3.18 Å) which proceeds *via* diradical **182**, consequently causing DNA cleavage.

Scheme 24

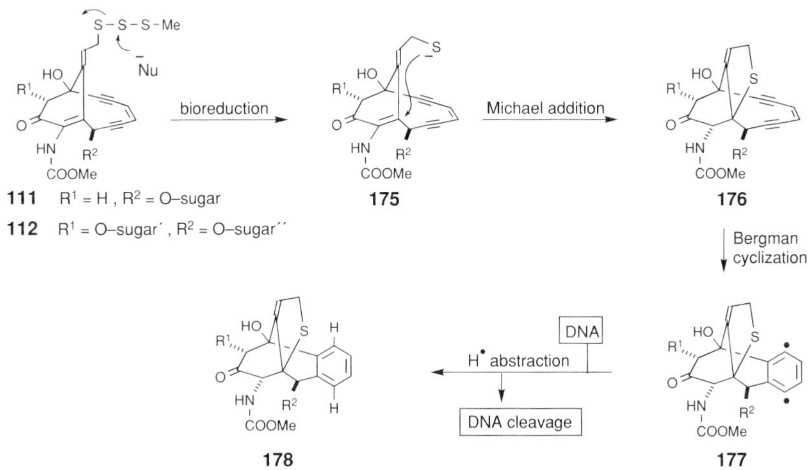

Scheme 25

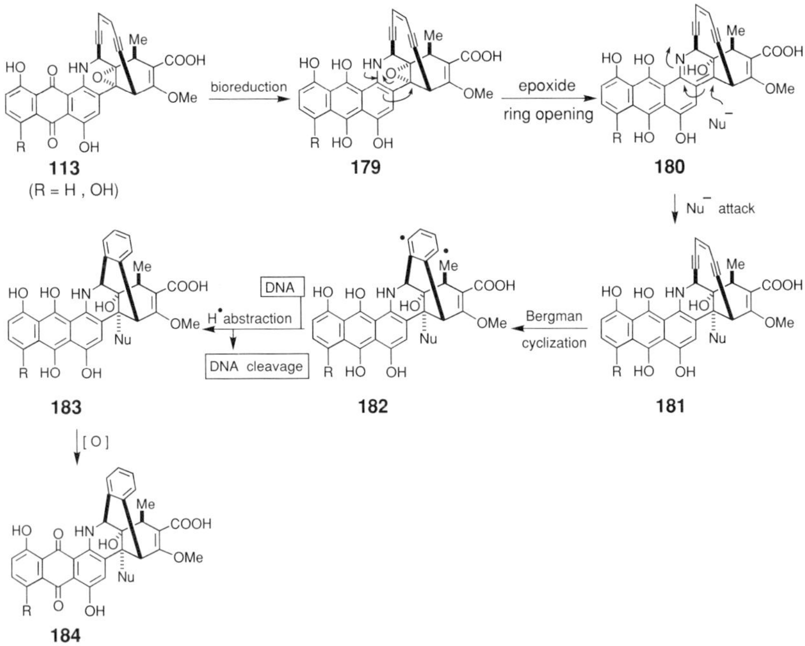

As for the neocarzinostatin chromophore **114**, the mechanism for its action has been proposed by Myers and Proteau[78] to involve an enyne–allene cyclization similar to that discussed earlier in Schemes 19 to 21. In brief, epoxide ring opening triggered by conjugate nucleophilic attack on chromophore **114** generates the enyne–cumulene system **185** which spontaneously cyclizes in a fashion related to the Bergman cyclization, forming intermediate diradical **186**, thus causing damage to the genetic material (Scheme 26).

Scheme 26

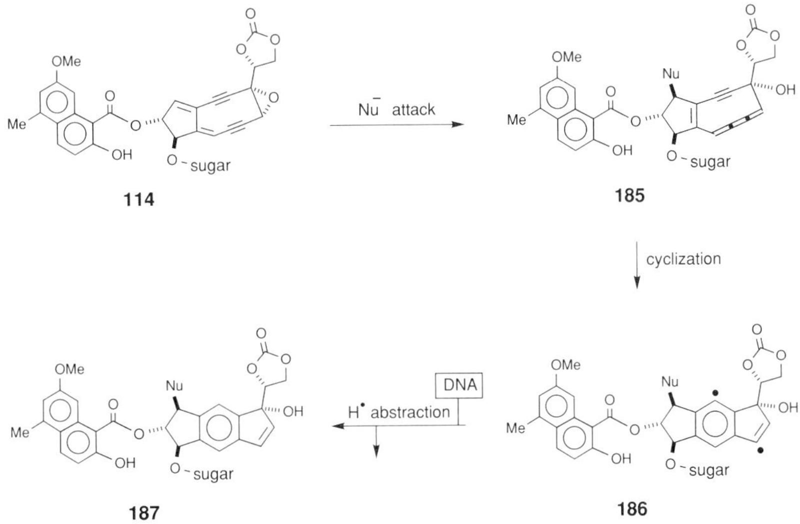

In retrospect, no one would ever have thought that the purely physical organic chemistry studies carried out since the early 1970s by the group of Bergman would ultimately rise to such importance in the scientific community. In devising these molecules (**111** to **114**) for their own defence mechanism it is as though the microorganisms, through natural selection and development, had learned how to apply physical organic chemistry to their needs a very long time ago. Unfortunately it has been humans who have had to wait until the end of the 20th century (towards the end of the 1980s and the beginning of the 1990s) to discover these molecules and study their properties. To be sure, forefront research chemists are now busy with the synthesis of these molecules and their derivatives, trying to attach all kinds of "triggering devices" onto the structures for envisioned pharmaceutical purposes. In the future, if we look back upon the present era, the enediyne anticancer–antibiotics episode will probably be one of the most complete and one of the best stories ever told of natural product chemistry.

IV. FINAL REMARKS

Although the main body of cyclization reactions revolves around cationic, radical, anionic, and metal-complex intermediates, it is recognized that the reaction is by no means restricted to this circle. This is evident in several examples in each of the chapters, in which can be seen radical–cationic, radical–anionic, or anionic–metal-complex intermediates. As a matter of fact, one of the most important modes of ring formation involves the application of pericyclic reactions such as the electrocyclic, sigmatropic, ene-, and, in particular, the cycloaddition reactions. However, since the nature of pericyclic reactions imposes concerted reorganization of orbitals on all the atoms taking part in the reaction, it has become a subject of debate as to whether these reactions can be regarded as "cyclization" in its true classical sense or not. In spite of this, certain examples in this book have included pericyclic reactions, which, however, are not here given a separate, independent treatise.

Because of their usefulness in organic synthesis, pericyclic reactions have enjoyed wide popularity and continuous development. Innumerable reviews have been written, the most recent including that of the [2+2] cycloaddition in β-lactam synthesis[79] and the [4+2] cycloaddition (Diels–Alder) reaction.[80–83] Both these types of addition have featured as key steps in the total synthesis of various natural products,[84–86] including the well-known synthesis of Ginkolides by Corey and Rao.[87] Meanwhile, the present trends of development of the cycloaddition reaction are in the directions of asymmetric addition[88–94] and reaction acceleration through the innovative use of various techniques and additives including Lewis acid[95–97] and transition metal[98,99] catalysis, high-pressure[100–103] aqueous media,[104,105] microwave,[106–109] and even ultrasound.[110] As for the ene-[111–118] and sigmatropic[119–122] reactions, they, too, have seriously invaded the recent scene of total organic synthesis. There can be no mistaking that ring construction reactions have completely infiltrated into the present army of organic chemical transformations, and risen to become an important instrument for the shaping of the future of organic chemistry.

REFERENCES

1. Corey, E. J., and Nicolaou, K. C., An efficient and mild lactonization method for the synthesis of macrolides, *J. Am. Chem. Soc.*, 96, 5614, 1974.
2. Corey, E. J., and Brunelle, D. J., New reagents for the conversion of hydroxy acids to macrolactones by the double activation method, *Tetrahedron Lett.*, 3409, 1976.
3. Nicolaou, K. C., Synthesis of macrolides, *Tetrahedron*, 33, 683, 1977.

4. Masamune, S., Kim, C. U., Wilson, K. E., Spessard, G. O., Georghiou, P. E., and Bates, G. S., Syntheses of macrolide antibiotics. I. Methymycin, *J. Am. Chem. Soc.*, 97, 3512, 1975.
5. Masamune, S., Yamamoto, H., Kamata, S., and Fukuzawa, A., Syntheses of macrolide antibiotics. II. Methymycin, *J. Am. Chem. Soc.*, 97, 3513, 1975.
6. Masamune, S., Kamata, S., and Schilling, W., Syntheses of macrolide antibiotics. III. Direct ester and lactone synthesis from *S-tert*-butyl thioate (thiol ester), *J. Am. Chem. Soc.*, 97, 3515, 1975.
7. Masamune, S., Bates, G. S., and Corcoran, J. W., Macrolides. Recent progress in chemistry and biochemistry, *Angew. Chem. Int. Ed. Engl.*, 16, 585, 1977.
8. Mukaiyama, T., Usui, M., and Saigo, K., The facile synthesis of lactones, *Chem. Lett.*, 49, 1976.
9. Inanaga, J., Hirata, K., Saeki, H., Katsuki, T., and Yamaguchi, M., A rapid esterification by means of mixed anhydride and its application to large-ring lactonization, *Bull. Chem. Soc. Jpn.*, 52, 1989, 1979.
10. Mitsunobu, O., The use of diethyl azodicarboxylate and triphenylphosphine in synthesis and transformation of natural products, *Synthesis*, 1, 1981.
11. Neises, B., and Steglich, W., Simple method for the esterification of carboxylic acids, *Angew. Chem. Int. Ed. Engl.*, 17, 522, 1978.
12. Boden, E. P., and Keck, G. E., Proton-transfer steps in Steglich esterification: a very practical new method for macrolactonization, *J. Org. Chem.*, 50, 2394, 1985.
13. Bartra, M., and Vilarrasa, J., Cyclization of 9-substituted decanoic acid derivatives to 9-decanolide and 9-decanelactam, *J. Org. Chem.*, 56, 5132, 1991.
14. Justus, K., and Steglich, W., First synthesis of a strained 14-membered biaryl ether lactone by macrolactonization, *Tetrahedron Lett.*, 32, 5781, 1991.
15. Nicolaou, K. C., Prasad, C. V. C., and Ogilvie, W. W., Novel stereocontrolled synthesis of the nonacene ring system of brevetoxin A. Conformational–reactivity effects in nine-membered rings, *J. Am. Chem. Soc.*, 112, 4988, 1990.
16. Keck, G. E., and Murry, J. A., Total synthesis of (–)-colletol, *J. Org. Chem.*, 56, 6606, 1991.
17. Kageyama, M., Tamura, T., Nantz, M. H., Roberts, J. C., Somfai, P., Whritenour, D. C., and Masamune, S., Synthesis of bryostatin 7, *J. Am. Chem. Soc.*, 112, 7407, 1990.
18. Evans, D. A., Kaldor, S. W., Jones, T. K., Clardy, J., and Stout, T. J., Total synthesis of the macrolide antibiotic cytovaricin, *J. Am. Chem. Soc.*, 112, 7001, 1990.
19. White, J. D., Vedananda, T. R., Kang, M., and Choudhry, S. C., Total synthesis of (+)-aplasmomycin, *J. Am. Chem. Soc.*, 108, 8105, 1986.
20. Battiste, M. A., Rocca, J. R., Wydra, R. L., Tumlinson, J. H., III, and Chuman, T., Total synthesis and structure proof of (3*E*,8*E*)-suspensolide, *Tetrahedron Lett.*, 29, 6565, 1988.
21. Khan, K. M., and Knight, D. W., [1.3], [3.3] and tandem [1.3]–[3.3] rearrangements of 11- and 12-membered trienolides, *J. Chem. Soc., Chem. Commun.*, 1699, 1991.
22. Kim, B. M., and Sharpless, K. B., A short route to β-lactams: use of cyclic sulfites from *syn*-2,3-dihydroxy esters, *Tetrahedron Lett.*, 31, 4317, 1990.
23. Smith, A. B., III, Leahy, J. W., Noda, I., Remiszewski, S. W., Liverton, N. J., and Zibuck, R., Total synthesis of the latrunculins, *J. Am. Chem. Soc.*, 114, 2995, 1992.
24. White, J. D., and Kawasaki, M., Total synthesis of (+)-latrunculin A, *J. Am. Chem. Soc.*, 112, 4991, 1990.
25. Smith, A. B., III, Noda, I., Remiszewski, S. W., Liverton, N. J., and Zibuck, R., Total synthesis of (+)-latrunculin A, *J. Org. Chem.*, 55, 3977, 1990.
26. Rubin, M. B., The chemistry of vicinal polyketones, *Chem. Rev.*, 75, 177, 1975.
27. Wasserman, H. H., and Vu, C. B., The use of vicinal tricarbonyl derivatives in alkaloid synthesis, *Pure & Appl. Chem.*, 62, 1409, 1990.

28. Wasserman, H. H., and Han, W. T., Vicinal tricarbonyl products from singlet oxygen reactions. Application to the synthesis of carbacephams, *Tetrahedron Lett.*, 25, 3743, 1984.
29. Wasserman, H. H., Fukuyama, J., Murugesan, N., van Duzer, J., Lombardo, L., Rotello, V., and McCarthy, K., The chemistry of vicinal tricarbonyls. A stable vinyl tricarbonyl hydrate as a di- and trielectrophile, *J. Am. Chem. Soc.*, 111, 371, 1989.
30. Wasserman, H. H., and Vu, C. B., Formation of vicinal tricarbonyl compounds by selective oxidation of ylides using potassium peroxymonosulfate, *Tetrahedron Lett.*, 31, 5205, 1990.
31. Wasserman, H. H., and Han, W. T., A synthesis of antibiotic (±)-PS-5, *Tetrahedron Lett.*, 25, 3747, 1984.
32. Wasserman, H. H., and Han, W. T., Penem synthesis through C_3–N ring closure of a β-lactam precursor, *J. Am. Chem. Soc.*, 107, 1444, 1985.
33. Cooke, M. D., Moore, K. W., Ross, B. C., and Turner, S. E., Synthesis of 2-methoxy- and 2-phenoxypenems, *Tetrahedron Lett.*, 24, 3373, 1983.
34. Wasserman, H. H., Amici, R., Frechette, R., and van Duzer, J. H., The chemistry of vicinal tricarbonyl compounds. Applications in the synthesis of isoquinoline alkaloids, *Tetrahedron Lett.*, 30, 869, 1989.
35. Wasserman, H. H., and Kelly, T. A., The chemistry of vicinal tricarbonyl compounds. Short syntheses of eudistomins T, I and M, *Tetrahedron Lett.*, 30, 7117, 1989.
36. Wasserman, H. H., and Kuo, G.-H., The chemistry of vicinal tricarbonyl compounds. Applications in the synthesis of vincamine-related alkaloids, *Tetrahedron Lett.*, 30, 873, 1989.
37. Wasserman, H. H., van Duzer, J. H., and Vu, C. B., The chemistry of vicinal tricarbonyls. Formation of carbazole derivatives, *Tetrahedron Lett.*, 31, 1609, 1990.
38. Wasserman, H. H., Cook, J. D., Fukuyama, J. M., and Rotello, V. M., The chemistry of vicinal tricarbonyls. Use of vinyl tricarbonyl esters in the formation of 3-hydroxypyrrole-2-carboxylates, *Tetrahedron Lett.*, 30, 1721, 1989.
39. Wasserman, H. H., Cook, J. D., and Vu, C. B., Intramolecular alkylation of 3-hydroxypyrrole-2-carboxylates. Formation of 5/5, 5/6, and 5/7 ring systems related to pyrrolidine alkaloids, *Tetrahedron Lett.*, 31, 4945, 1990.
40. Wasserman, H. H., Cook, J. D., and Vu, C. B., The chemistry of vicinal tricarbonyls. New routes to indolizidines, *J. Org. Chem.*, 55, 1701, 1990.
41. Wasserman, H. H., Vu, C. B., and Cook, J. D., Vicinal tricarbonyls in synthesis. New routes to indolizidines, *Tetrahedron*, 48, 2101, 1992.
42. Wasserman, H. H., Henke, S. L., Luce, P., and Nakanishi, E., Addition of amino lactams to vinyl vicinal tricarbonyls. Formation of tricyclic 2-azadethiapenams and 3-azadethiacephams, *J. Org. Chem.*, 55, 5821, 1990.
43. Wasserman, H. H., Henke, S. L., Nakanishi, E., and Schulte, G., Addition of amino amides to vinyl vicinal tricarbonyls. Formation of tricyclic 3-azadethiacephams, *J. Org. Chem.*, 57, 2641, 1992.
44. Wasserman, H. H., and Kuo, G.-H., The chemistry of vicinal tricarbonyls. An efficient synthesis of (±)-vasicine, *Tetrahedron Lett.*, 32, 7131, 1991.
45. Lee, M. D., Ellestad, G. A., and Borders, D. B., Calicheamicins: discovery, structure, chemistry, and interaction with DNA, *Acc. Chem. Res.*, 24, 235, 1991.
46. Lee, M. D., Dunne, T. S., Siegel, M. M., Chang, C. C., Morton, G. O., and Borders, D. B., Calichemicins, a novel family of antitumor antibiotics. I. Chemistry and partial structure of calichemicin γ_1^{I}, *J. Am. Chem. Soc.*, 109, 3464, 1987.
47. Lee, M. D., Dunne, T. S., Chang, C. C., Ellestad, G. A., Siegel, M. M., Morton, G. O., McGahren, W. J., and Borders, D. B., Calichemicins, a novel family of antitumor antibiotics. II. Chemistry and structure of calichemicin γ_1^{I}, *J. Am. Chem. Soc.*, 109, 3466, 1987.
48. Lee, M. D., Manning, J. K., Williams, D. R., Kuck, N. A., Testa, R. T., and Borders, D. B., Calicheamicins, a novel family of antitumor antibiotics. III. Isolation, purification and characterization of calicheamicins β_1^{Br}, γ_1^{Br}, α_2^{I}, α_3^{I}, β_1^{I}, γ_1^{I} and δ_1^{I}, *J. Antibiot.*, 42, 1070, 1989.

49. Golik, J., Clardy, J., Dubay, G., Groenewold, G., Kawaguchi, H., Konishi, M., Krishnan, B., Ohkuma, H., Saitoh, K., and Doyle, T. W., Esperamicins, a novel class of potent antitumor antibiotics. II. Structure of esperamicin X, *J. Am. Chem. Soc.*, 109, 3461, 1987.
50. Golik, J., Dubay, G., Groenewold, G., Kawaguchi, H., Konishi, M., Krishnan, B., Ohkuma, H., Saitoh, K., and Doyle, T. W., Esperamicins, a novel class of potent antitumor antibiotics. III. Structures of esperamicins A_1, A_2 and A_{1b}, *J. Am. Chem. Soc.*, 109, 3462, 1987.
51. Konishi, M., Ohkuma, H., Matsumoto, K., Tsuno, T., Kamei, H., Miyaki, T., Oki, T., Kawaguchi, H., VanDuyne, G. D., and Clardy, J., Dynemicin A, a novel antibiotic with the anthraquinone and 1,5-diyn-3-ene subunit, *J. Antibiot.*, 42, 1449, 1989.
52. Konishi, M., Ohkuma, H., Tsuno, T., Oki, T., VanDuyne, G. D., and Clardy, J., Crystal and molecular structure of dynemicin A : a novel 1,5-diyn-3-ene antitumor antibiotic, *J. Am. Chem. Soc.*, 112, 3715, 1990.
53. Goldberg, I. H., Mechanism of neocarzinostatin action: role of DNA microstructure in determination of chemistry of bistranded oxidative damage, *Acc. Chem. Res.*, 24, 191, 1991.
54. Koide, Y., Ishii, F., Hasuda, K., Koyama, Y., Edo, K., Katamine, S., Kitame, F., and Ishida, N., Isolation of a non-protein component and a protein component from neocarzinostatin (NCS) and their biological activities, *J. Antibiot.*, 33, 342, 1980.
55. Nicolaou, K. C., and Dai, W.-M., Chemistry and biology of the enediyne anticancer antibiotics, *Angew. Chem. Int. Ed. Engl.*, 30, 1387, 1991.
56. Stinson, S., Structure of potent new anticancer agents solved, *Chem. En. News*, June 8, 1987, p. 17.
57. Bergman, R. G., Reactive 1,4-dehydroaromatics, *Acc. Chem. Res.*, 6, 25, 1973.
58. Lockhart, T. P., Comita, P. B., and Bergman, R. G., Kinetic evidence for the formation of discrete 1,4-dehydrobenzene intermediates. Trapping by inter- and intramolecular hydrogen atom transfer and observation of high-temperature CIDNP, *J. Am. Chem. Soc.*, 103, 4082, 1981.
59. Lockhart, T. P., and Bergman, R. G., Evidence for the reactive spin state of 1,4-dehydrobenzenes, *J. Am. Chem. Soc.*, 103, 4091, 1981.
60. Nicolaou, K. C., Zuccarello, G., Ogawa, Y., Schweiger, E. J., and Kumazawa, T., Cyclic conjugated enediynes related to calicheamicins and esperamicins: calculations, synthesis, and properties, *J. Am. Chem. Soc.*, 110, 4866, 1988.
61. Magnus, P., Lewis, R. T., and Huffman, J. C., Synthesis of a remarkably stable bicyclo[7.3.1]diynene esperamicin A_1 / calicheamicin *g* system. Structural requirements for facile formation of a 1,4-diyl, *J. Am. Chem. Soc.*, 110, 6921, 1988.
62. Snyder, J. P., The cyclization of calicheamicin–esperamicin analogues: a predictive biradicaloid transition state, *J. Am. Chem. Soc.*, 111, 7630, 1989.
63. Magnus, P., Fortt, S., Pitterna, T., and Snyder, J. P., Synthetic and mechanistic studies on esperamicin A_1 and calicheamicin γ_1. Molecular strain rather than π-bond proximity determines the cycloaromatization rates of bicyclo[7.3.1] enediynes, *J. Am. Chem. Soc.*, 112, 4986, 1990.
64. Snyder, J. P., and Tipsword, G. E., Proposal for blending classical and biradical mechanisms in antitumor antibiotics: dynemicin A, *J. Am. Chem. Soc.*, 112, 4040, 1990.
65. Snyder, J. P., Monocyclic enediyne collapse to 1,4-diyl biradicals: a pathway under strain control, *J. Am. Chem. Soc.*, 112, 5367, 1990.
66. Semmelhack, M. F., Gallagher, J., and Cohen, D., Bioreductive alkylation as a trigger for toxic effects of dynemicin, *Tetrahedron Lett.*, 31, 1521, 1990.
67. Semmelhack, M. F., Neu, T., and Foubelo, F., Arene 1,4-diradical formation from *o*-dialkynylarenes, *Tetrahedron Lett.*, 33, 3227, 1992.
68. Wong, H. N. C., and Sondheimer, F., 5,12-Dihydro-6,11-didehydronaphthacene, a derivative of 1,4-didehydronaphthalene, *Tetrahedron Lett.*, 21, 217, 1980.
69. Magnus, P., and Carter, P. A., A model for the proposed mechanism of action of the potent antitumor antibiotic esperamicin A_1, *J. Am. Chem. Soc.*, 110, 1626, 1988.

70. Sakai, Y., Nishiwaki, E., Shishido, K., Shibuya, M., and Kido, M., Synthesis of a sulfur-containing 10-membered enediyne model compound related to the esperamicin / calicheamicin / dynemicin aglycones, *Tetrahedron Lett.*, 32, 4363, 1991.
71. Sakai, Y., Bando, Y., Shishido, K., and Shibuya, M., Cycloaromatization of a 10-membered enediyne derivative *via* an allenic sulfone intermediate and its DNA cleaving activity, *Tetrahedron Lett.*, 33, 957, 1992.
72. Myers, A. G., Kuo, E. Y., and Finney, N. S., Thermal generation of α,3-dehydrotoluene from (*Z*)-1,2,4-heptatrien-6-yne, *J. Am. Chem. Soc.*, 111, 8057, 1989.
73. Nagata, R., Yamanaka, H., Okazaki, E., and Saito, I., Biradical formation from acyclic conjugated eneyne–allene system related to neocarzinostatin and esperamicin–calichemicin, *Tetrahedron Lett.*, 30, 4995, 1989.
74. Nagata, R., Yamanaka, H., Murahashi, E., and Saito, I., DNA cleavage by acyclic eneyne–allene systems related to neocarzinostatin and esperamicin–calicheamicin, *Tetrahedron Lett.*, 31, 2907, 1990.
75. Nicolaou, K. C., Maligres, P., Shin, J., de Leon, E., and Rideout, D., DNA cleavage and antitumor activity of designed molecules with conjugated phosphine oxide–allene–ene–yne functionalities, *J. Am. Chem. Soc.*, 112, 7825, 1990.
76. Bharucha, K. N., Marsh, R. M., Minto, R. E., and Bergman, R. G., Double cycloaromatization of (*Z,Z*)-deca-3,7-diene-1,5,9-triyne: evidence for the intermediacy and diradical character of 2,6-didehydronaphthalene, *J. Am. Chem. Soc.*, 114, 3120, 1992.
77. Grissom, J. W., and Calkins, T. L., The tandem Bergman-radical cyclization: a new method for ring annulation, *Tetrahedron Lett.*, 33, 2315, 1992.
78. Myers, A. G., and Proteau, P. J., Evidence for spontaneous, low-temperature biradical formation from a highly reactive neocarzinostatin chromophore–thiol conjugate, *J. Am. Chem. Soc.*, 111, 1146, 1989.
79. van der Steen, F. H., and van Koten, G., Syntheses of 3-amino-2-azetidinones: a literature survey, *Tetrahedron*, 47, 7503, 1991.
80. Oppolzer, W., Camphor derivatives as chiral auxiliaries in asymmetric synthesis, *Tetrahedron*, 43, 1969, 1987.
81. Oppolzer, W., Metal-directed stereoselective functionalizations of alkenes in organic synthesis, *Pure & Appl. Chem.*, 60, 39, 1988.
82. Okamura, W. H., and Curtin, M. L., Pericyclization of vinylallenes in organic synthesis: on the intramolecular Diels–Alder reaction, *Synlett*, 1, 1990.
83. Deslongchamps, P., Transannular Diels–Alder reaction on macrocycles: a general strategy for the synthesis of polycyclic compounds, *Aldrichimica Acta*, 24, 43, 1991.
84. Shishido, Y., and Kibayashi, C., The total synthesis of (-)-indolizidines 205A and 235B, *J. Chem. Soc., Chem. Commun.*, 1237, 1991.
85. Takano, S., Sato, T., Inomata, K., and Ogasawara, K., The enantiocontrolled total synthesis of natural (–)-goniomitine, *J. Chem. Soc., Chem. Commun.*, 462, 1991.
86. Charlton, J. L., and Koh, K., Asymmetric synthesis of (–)-neopodophyllotoxin, *J. Org. Chem.*, 57, 1514, 1992.
87. Corey, E. J., and Rao, K. S., Enantioselective total synthesis of ginkgolide derivatives lacking the *tert*-butyl group, an essential structural subunit for antagonism of platelet activating factor, *Tetrahedron Lett.*, 32, 4623, 1991.
88. Waldmann, H., (*S*)-Proline benzyl ester as chiral auxiliary in Lewis acid catalyzed asymmetric Diels–Alder reactions, *J. Org. Chem.*, 53, 6133, 1988.
89. Linz, G., Weetman, J., Hady, A. F. A., and Helmchen, G., Asymmetric Diels–Alder reactions: EPC-synthesis of a stable sarkomycin precursor (cyclosarkomycin), *Tetrahedron Lett.*, 30, 5599, 1989.
90. Oppolzer, W., Wills, M., Kelly, M. J., Signer, M., and Blagg, J., Chiral toluene-2,α-sultam auxiliaries: asymmetric Diels–Alder reactions of *N*-enoyl derivatives, *Tetrahedron Lett.*, 31, 5015, 1990.

91. Tripathy, R., Carroll, P. J., and Thornton, E. R., Asymmetric Diels–Alder reactions with chiral dienes. Control of facial selectivity through hydrogen bonding, *J. Am. Chem. Soc.*, 112, 6743, 1990.
92. Nouguier, R., Gras, J.-L., Giraud, B., and Virgili, A., Methyl 3,4-*O*-methylene-β-D-arabinoside as a new chiral template for the asymmetric Diels–Alder reaction, *Tetrahedron Lett.*, 32, 5529, 1991.
93. Corey, E. J., Cheng, X.-M., and Cimprich, K. A., 1-Mesityl-2,2,2-trifluoroethanol, an outstanding new chiral controller for catalyzed Diels–Alder reactions, *Tetrahedron Lett.*, 32, 6839, 1991.
94. Gras, J.-L., and Pellissier, H., Synthèse asymétrique assistée par ligandes. I. Modélisation d'auxiliaires chiraux pour la réaction de Diels–Alder enantiosélective, *Tetrahedron Lett.*, 32, 7043, 1991.
95. Grieco, P. A., Nunes, J. J., and Gaul, M. D., Dramatic rate accelerations of Diels–Alder reactions in 5M lithium perchlorate–diethyl ether: the cantharidin problem reexamined, *J. Am. Chem. Soc.*, 112, 4595, 1990.
96. Smith, D. A., and Houk, K. N., $LiBF_4$: a mild Lewis acid for intramolecular Diels–Alder reactions, *Tetrahedron Lett.*, 32, 1549, 1991.
97. Devine, P. N., and Oh, T., Asymmetric Diels–Alder reactions of carboxylic ester dienophiles promoted by chiral Lewis acids, *J. Org. Chem.*, 57, 396, 1992.
98. Wender, P. A., and Jenkins, T. E., Nickel-catalyzed intramolecular [4+2] dienyne cycloadditions: an efficient new method for the synthesis of polycycles containing cyclohexa-1,4-dienes, *J. Am. Chem. Soc.*, 111, 6432, 1989.
99. Jolly, R. S., Luedtke, G., Sheehan, D., and Livinghouse, T., Novel cyclization reactions on transition-metal templates. The catalysis of intramolecular [4+2] cycloadditions by low valent rhodium complexes, *J. Am. Chem. Soc.*, 112, 4965, 1990.
100. Dauben, W. G., and Kozikowski, A. P., Organic reactions at high pressure. Cycloadditions of enamines and dienamines, *J. Am. Chem. Soc.*, 96, 3664, 1974.
101. van Eldik, R., Asano, T., and le Noble, W. J., Activation and reaction volumes in solution. II, *Chem. Rev.*, 89, 549, 1989.
102. Engler, T. A., Sampath, U., Naganathan, S., Velde, D. V., Takusagawa, F., and Yohannes D., A new general synthetic approach to diterpenes: application to syntheses of (±)-taxodione and (±)-royleanone, *J. Org. Chem.*, 54, 5712, 1989.
103. Aben, R. W. M., Minuti, L., Scheeren, H. W., and Taticchi, A., High pressure–Lewis acid catalyzed Diels–Alder reactions of 3-methylcyclohex-2-en-1-one. A straightforward route to *cis* and *trans* angularly methylated octalones, *Tetrahedron Lett.*, 32, 6445, 1991.
104. Rideout, D. C., and Breslow, R., Hydrophobic acceleration of Diels–Alder reactions, *J. Am. Chem. Soc.*, 102, 7816, 1980.
105. Grieco, P. A., Yoshida, K., and Garner, P., Aqueous intermolecular Diels–Alder chemistry: reactions of diene carboxylates with dienophiles in water at ambient temperature, *J. Org. Chem.*, 48, 3137, 1983.
106. Gedye, R., Smith, F., Westaway, K., Ali, H., Baldisera, L., Laberge, L., and Rousell, J., The use of microwave ovens for rapid organic synthesis, *Tetrahedron Lett.*, 27, 279, 1986.
107. Giguere, R. J., Bray, T. L., Duncan, S. M., and Majetich, G., Application of commercial microwave ovens to organic synthesis, *Tetrahedron Lett.*, 27, 4945, 1986.
108. Bose, A. K., Manhas, M. S., Ghosh, M., Shah, M., Raju, V. S., Bari, S. S., Newaz, S. N., Banik, B. K., Chaudhary, A. G., and Barakat, K. J., Microwave-induced organic reaction enhancement chemistry. II. Simplified techniques, *J. Org. Chem.*, 56, 6968, 1991.
109. Abramovitch, R. A., Applications of microwave energy in organic chemistry. A review, *Org. Prep. Proc. Int.*, 23, 685, 1991.
110. Lee, J., and Snyder, J. K., Ultrasound-promoted Diels–Alder reactions: syntheses of tanshinone IIA, nortanshinone, and (±)-tanshindiol B, *J. Am. Chem. Soc.*, 111, 1522, 1989.
111. Oppolzer, W., and Snieckus, V., Intramolecular ene reactions in organic synthesis, *Angew. Chem. Int. Ed. Engl.*, 17, 476, 1978.

112. Snider, B. B., Lewis-acid-catalyzed ene reactions, *Acc. Chem. Res.*, 13, 426, 1980.
113. Maruoka, K., Ooi, T., and Yamamoto, H., Unprecedented stereochemical control in the organoaluminum-promoted intramolecular ene reactions of δ,ε-unsaturated aldehydes, *J. Am. Chem. Soc.*, 112, 9011, 1990.
114. Abouabdellah, A., Bégué, J.-P., Bonnet-Delpon, D., and Lequeux, T., Lewis acid induced ene cyclization of ω-olefinic trifluoromethyl ketones: access to bicyclic compounds bearing a CF_3 group, *J. Org. Chem.*, 56, 5800, 1991.
115. Novák, M., Novák, J., and Salemink, C. A., New type of intramolecular imino–ene cyclization during pyrolysis of (–)-cocaine, *Tetrahedron Lett.*, 32, 4405, 1991.
116. Mikami, K., Terada, M., Sawa, E., and Nakai, T., Asymmetric catalysis by chiral titanium perchlorate for carbonyl–ene cyclization, *Tetrahedron Lett.*, 32, 6571, 1991.
117. Beak, P., Song, Z., and Resek, J. E., Ene reactions of dialkyl dioxosuccinate esters, *J. Org. Chem.*, 57, 944, 1992.
118. Desai, S. R., Gore, V. K., and Bhat, S. V., Unprecedented tandem Michael–ene reaction of 2-formylcyclohexa-2,5-dienone and subsequent unusual autoxidation, *J. Org. Chem.*, 57, 2467, 1992.
119. Harano, K., Ono, K., Nishimoto, M., Eto, M., and Hisano, T., A simple one-pot synthesis of hydroisobenzothiophenes *via* three-step sequential pericyclic reactions of xanthates, *Tetrahedron Lett.*, 32, 2387, 1991.
120. Pirrung, M. C., Brown, W. L., Rege, S., and Laughton, P., Total synthesis of (+)-griseofulvin, *J. Am. Chem. Soc.*, 113, 8561, 1991.
121. Paquette, L. A., and Maleczka, R. E., Jr., Enantioselective total synthesis of (–)-9-*epi*-ambrox, a potent ambergris-type olfactory agent, *J. Org. Chem.*, 56, 912, 1991.
122. Maleczka, R. E., Jr., and Paquette, L. A., Adaptation of oxyanionic sigmatropy to the convergent enantioselective synthesis of ambergris-type odorants, *J. Org. Chem.*, 56, 6538, 1991.

INDEX

A

B

C

D

E

F

G

H

I

J

K

L

M

N

O

P

T

V

W

X

Y

Z